Precalculus

Paul Sisson

Student Solutions Manual

Hawkes Publishing

ISBN: 0-918091-99-3

Table of Contents

Chapter One: Section 1.1
Solutions to Odd-numbered Exercises

1. (a) Natural numbers:
$\{19, 2^5\}$

(b) Whole numbers:
$\left\{\frac{0}{15}, 19, 2^5\right\}$

(c) Integers:
$\left\{-33, \frac{0}{15}, 19, 2^5\right\}$

(d) Rational numbers:
$\left\{-33, -4.3, \frac{0}{15}, 19, 2^5\right\}$

(e) Irrational numbers:
$\{-\sqrt{3}\}$

(f) Real numbers:
$\left\{-33, -4.3, -\sqrt{3}, \frac{0}{15}, 19, 2^5\right\}$

(g) Undefined:
$\frac{15}{0}$

3. Let $x = 2.\overline{3}$. Then $10x = 23.\overline{3}$, so $10x = 21 + 2.\overline{3}$, or $10x = 21 + x$. Thus, $9x = 21$, or $x = \frac{7}{3}$

5. Let $x = 0.\overline{41836}$. Then $100{,}000x = 41{,}836.\overline{41836}$, so $100{,}000x = 41{,}836 + 0.\overline{41836}$, or $100{,}000x = 41{,}836 + x$. Thus, $99{,}999x = 41{,}836$, or $x = \frac{41{,}836}{99{,}999}$

7. Let $x = -1.\overline{01}$. Then $100x = -101.\overline{01}$, so $100x = -100 - 1.\overline{01}$, or $100x = -100 + x$. Thus, $99x = -100$, or $x = -\frac{100}{99}$.

9. Using a scale of 1 unit:

−4.5 −1 0 2.5

11. Using a scale of 5 units:

−24 2 15

13. $>, \geq$, because −3.4 is greater than −3.5.

15. $\leq, \geq$, because 3 = 3.

17. $>, \geq$, because $-\frac{1}{4}$ is greater than $-\frac{1}{3}$.

19. Here is one way:
$\{3n \mid n \text{ is an integer and } -2 \leq n \leq 3\}$

21. Here is one way:
$\{n \mid n \text{ is a prime number}\}$

23. Here is one way:
$\left\{\frac{1}{n} \mid n \text{ is an odd integer}\right\}$

25. $(-5, 2] \cup (2, 4] = (-5, 4]$

27. $[3, 5] \cap [2, 4] = [3, 4]$

29. $\mathbb{Q} \cap \mathbb{Z} = \mathbb{Z}$

31. $(-\infty, \infty) \cap [-\pi, 21) = [-\pi, 21)$

33. $(3, 5] \cup [5, 9] = (3, 9]$

35. $\mathbb{N} \cup \mathbb{Z} \cap \mathbb{Q} = \mathbb{Z}$

37. $[-3, 19)$

39. $(0, \infty)$

41. $[1, 2]$

43. $[5, 14)$

45. $(0, 2)$

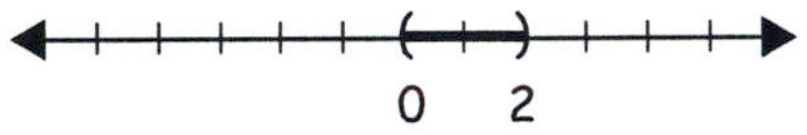

47. $(-\infty, 7]$

49. $-|-11| = -11$

51. $-|4-9| = -|-5| = -5$

53. $\sqrt{|-4|} = \sqrt{4} = 2$

55. $|-\sqrt{2}| = \sqrt{2}$

57. $-3.1 < -2.9$ or $-3.1 \leq -2.9$

59. $-4 < 100$ or $4 \leq 100$

61. $\frac{1}{3} > \frac{1}{4}$ or $\frac{1}{3} \geq \frac{1}{4}$

63. The terms of the expression are $3x^2y^3, -2\sqrt{x+y}$ and $7z$.

65. -2 and $\sqrt{x+y}$

67. The coefficients of the expression are $1, 8.5$ and -14.

69. Substitute the given values for x, y and z:

$$3(-1)^2(2)^3 - 2\sqrt{-1+2} + 7(-2) =$$
$$24 - 2 - 14 = 8$$

71. Substitute the given values for x and y:

$$\frac{|-3|\sqrt{2}}{(-3)^3(2)^2} - \frac{3(2)}{-3} = -\frac{\sqrt{2}}{36} + 2$$

73. Substitute the given values for x and y:

$$\left|-(-3)^2 + 2(-3)(-5) - (-5)^2\right| =$$
$$|-9 + 30 - 25| = 4$$

75. Commutative property of addition.

77. Associative property of addition.

79. Associative property of multiplication.

81. Distributive property.

83. Multiplicative cancellation property. Use $\frac{1}{5}$.

85. Additive cancellation property. Use x.

87. Zero-factor property.

89. Multiplicative cancellation. Use 6.

91. Begin with 3. Add 7 and multiply the result by 3:
$3(3+7)$
Subtract 5, take the square root ...
$\sqrt{3(3+7)-5}$ ...
raise the result to the 3rd power and multiply by $-\frac{1}{5}$:
$-\frac{1}{5}\left(\sqrt{3(3+7)-5}\right)^3$

93. $\dfrac{(x-pq^2)^3}{2q^3}=\dfrac{(-5-(2)(-3)^2)^3}{2(-3)^3}$
≈ 225.31

95. $\sqrt{p^2q-q^3}-|p+q^2|$
$=\sqrt{(-5)^2(2)-(2)^3}-|(-5)+(2)^2|$
$=\sqrt{42}-1$
≈ 5.48

End of Section 1.1

Chapter One: Section 1.2
Solutions to Odd-numbered Exercises

1. $(-2)^4 = 16$

3. $3^2 3^2 = 3^4 = 81$

5. $4 \cdot 4^2 = 4^3 = 64$

7. $\frac{8^2}{4^3} = \frac{64}{64} = 1$

9. $\frac{7^4}{7^5} = \frac{1}{7}$

11. $\frac{x^5}{x^2} = x^3$

13. $\frac{3t^{-2}}{t^3} = \frac{3}{t^5}$

15. $\frac{1}{7x^{-5}} = \frac{x^5}{7}$

17. $\frac{2n^3}{n^{-5}} = 2n^8$

19. $\frac{x^7 y^{-3} z^{12}}{x^{-1} z^9} = \frac{x^8 z^3}{y^3}$

21. $\frac{s^3}{s^{-2}} = s^5$

23. $x^{(y^0)} \cdot x^9 = x^1 \cdot x^9 = x^{10}$

25. $$\frac{-9^0 \left(x^2 y^{-2}\right)^{-3}}{3x^{-4} y} = \frac{-1\left(x^{-6} y^6\right)}{3x^{-4} y} = -\frac{y^5}{3x^2}$$

27. $\frac{\left(3z^{-2} y\right)^0}{3zy^2} = \frac{1}{3zy^2}$

29. $\frac{3^{-1}}{\left(3^2 xy^2\right)^{-2}} = \frac{3^4 x^2 y^4}{3} = 27x^2 y^4$

31. $\left[\left(12x^{-6} y^4 z^3\right)^5\right]^0 = 1$

33. $-1.76 \times 10^{-5} = -0.0000176$

35. $0.00000021 = 2.1 \times 10^{-7}$

37. $5100 = 5.1 \times 10^3$

39. $3.1212 \times 10^2 = 312.12$

41. $\left(2.3 \times 10^{13}\right)\left(2 \times 10^{12}\right) = 4.6 \times 10^{25}$

43. $$\left(2 \times 10^{-13}\right)\left(5.5 \times 10^{10}\right)\left(-1 \times 10^3\right) = -11 \times 10^0 = -11$$

45. $$\left(6 \times 10^{21}\right)\left(5 \times 10^{-19}\right)\left(5 \times 10^4\right) = 150 \times 10^6 = 1.5 \times 10^8$$

47. $\frac{4 \times 10^{-6}}{\left(5 \times 10^4\right)\left(8 \times 10^{-3}\right)} = \frac{10^{-6}}{10(10)} = 1 \times 10^{-8}$

49. $$a^n \cdot a^m = \underbrace{a \cdot \ldots \cdot a}_{n \text{ times}} \cdot \underbrace{a \cdot \ldots \cdot a}_{m \text{ times}} = \underbrace{a \cdot \ldots \cdot a}_{n+m \text{ times}} = a^{n+m}.$$

51. $$(ab)^n = \underbrace{(ab) \cdot \ldots \cdot (ab)}_{n \text{ times}} = \underbrace{a \cdot \ldots \cdot a}_{n \text{ times}} \underbrace{b \cdot \ldots \cdot b}_{n \text{ times}} = a^n b^n$$

53. The volume of a cylinder is the product of the area of the base and the height:
$$\frac{1}{2}(bh)l = \frac{1}{2}bhl$$

55. The volume of the capsule is the sum of volume of the two end hemispheres $\left(\text{together a sphere with volume } \frac{4}{3}(\pi r^3)\right)$ and a cylinder with volume $\pi r^2 h$. Given that the height h is 16 inches and that the radius r is 3 inches, the volume of the capsule is
$$\frac{4}{3}\pi(3)^3 + \pi 3^2 \cdot 16 = 36\pi + 144\pi$$
$$= 180\pi \text{ in}^3.$$

57. The volume of a rectangular solid is the product of length, width and depth:
$6lw$ ft^3.

59. The room has two pairs of opposite walls that have the same area: $7N$ and $7M$. Then, the total area of the four walls is $14N + 14M$, or $14(N+M)$ft^2.

61. $\left[(5xy^0z^{-2})^{-1}(25x^4z)^2\right]^{-2} =$
$$(5xy^0z^{-2})^2(25x^4z)^{-4} =$$
$$\frac{25x^2}{z^4}\cdot\frac{1}{25^4x^{16}z^4} = \frac{1}{25^3x^{14}z^8}$$
$$\text{or } \frac{1}{15{,}625x^{14}z^8}$$

63. $\left((2x^{-2}yz^3)^{-1}\right)^2 = (2x^{-2}yz^3)^{-2}$
$$= \frac{x^4}{4y^2z^6}$$

65. $-\sqrt{9} = -3$

67. $\sqrt{-25}$ is not a real number.

69. $-\sqrt[6]{64} = -\sqrt[6]{2^6} = -2$

71. $\sqrt{\frac{1}{4}} = \sqrt{\left(\frac{1}{2}\right)^2} = \frac{1}{2}$

73. $\sqrt[4]{\sqrt{16} - \sqrt[3]{-27} + \sqrt{81}} = \sqrt[4]{4-(-3)+9}$
$$= \sqrt[4]{16}$$
$$= 2$$

75. $\sqrt{9x^2} = 3|x|$

77. $\sqrt[4]{\frac{x^8z^4}{16}} = |z|\sqrt[4]{\frac{(x^2)^4}{2^4}} = \frac{x^2|z|}{2}$

79. $\sqrt[7]{x^{14}y^{49}z^{21}} = \sqrt[7]{(x^2)^7(y^7)^7(z^3)^7}$
$$= x^2y^7z^3$$

81. $\sqrt{\frac{x^2}{4x^4y^6}} = \sqrt{\frac{1}{4x^2(y^3)^2}}$
$$= \frac{1}{2|xy^3|}$$

83. $\sqrt[3]{\frac{a^3b^{12}}{27c^6}} = \frac{a}{3}\sqrt[3]{\frac{(b^4)^3}{(c^2)^3}}$
$$= \frac{ab^4}{3c^2}$$

85. $\frac{-\sqrt{3a^3}}{\sqrt{6a}} = \frac{-\sqrt{3a^3}}{\sqrt{6a}}\cdot\frac{\sqrt{6a}}{\sqrt{6a}}$
$$= \frac{-\sqrt{3^2\cdot 2(a^2)^2}}{6a}$$
$$= \frac{-3a^2\sqrt{2}}{6a}$$
$$= -\frac{|a|\sqrt{2}}{2}$$

87. $\frac{10}{\sqrt{7}-\sqrt{2}} = \frac{10}{\sqrt{7}-\sqrt{2}}\cdot\frac{\sqrt{7}+\sqrt{2}}{\sqrt{7}+\sqrt{2}}$
$$= \frac{10(\sqrt{7}+\sqrt{2})}{7-2}$$
$$= 2(\sqrt{7}+\sqrt{2})$$
$$= 2\sqrt{7}+2\sqrt{2}$$

89. $\dfrac{x-y}{\sqrt{x}+\sqrt{y}}=\dfrac{x-y}{\sqrt{x}+\sqrt{y}}\cdot\dfrac{\sqrt{x}-\sqrt{y}}{\sqrt{x}-\sqrt{y}}$
$=\dfrac{(x-y)(\sqrt{x}-\sqrt{y})}{x-y}$
$=\sqrt{x}-\sqrt{y}$

91. $\dfrac{1}{2-\sqrt{x}}=\dfrac{1}{2-\sqrt{x}}\cdot\dfrac{2+\sqrt{x}}{2+\sqrt{x}}$
$=\dfrac{2+\sqrt{x}}{4-x}$

93. $\sqrt[3]{-16x^4}+5x\sqrt[3]{2x}=-2x\sqrt[3]{2x}+5x\sqrt[3]{2x}$
$=3x\sqrt[3]{2x}$

95. $\sqrt{7x}-\sqrt[3]{7x}$; cannot be combined

97. $-x^2\sqrt[3]{54x}+3\sqrt[3]{2x^7}=$
$-3x^2\sqrt[3]{2x}+3\sqrt[3]{2x\left(x^2\right)^3}=$
$-3x^2\sqrt[3]{2x}+3x^2\sqrt[3]{2x}=0$

99. $\sqrt[3]{\sqrt[4]{x^{36}}}=\sqrt[3]{x^{\frac{36}{4}}}=\sqrt[3]{|x^9|}$
$=|x^3|$

101. $32^{\frac{-3}{5}}=\dfrac{1}{\left(32^{\frac{1}{5}}\right)^3}=\dfrac{1}{2^3}$
$=\dfrac{1}{8}$

103. $\dfrac{(x-z)^y}{(x-z)^4}$
$=(x-z)^{y-4}$

105. $(-8)^{\frac{2}{3}}=\left((-8)^{\frac{1}{3}}\right)^2=(-2)^2$
$=4$

107. $\dfrac{\sqrt[3]{a^2}}{\sqrt[3]{a^5}}=\sqrt[3]{\dfrac{a^2}{a^5}}=\sqrt[3]{\dfrac{1}{a^3}}$
$=\dfrac{1}{a}$

109. $\sqrt[16]{y^4}=|y|^{\frac{4}{16}}=|y|^{\frac{1}{4}}$
$=\sqrt[4]{|y|}$

111. $\sqrt[3]{x^7}\sqrt[9]{x^6}=x^{\frac{7}{3}}\cdot x^{\frac{6}{9}}$
$=x^{\frac{7}{3}}x^{\frac{2}{3}}$
$=x^{\frac{9}{3}}$
$=x^3$

113. Show $\sqrt[n]{\dfrac{a}{b}}=\dfrac{\sqrt[n]{a}}{\sqrt[n]{b}}$.
$\sqrt[n]{\dfrac{a}{b}}=\left(\dfrac{a}{b}\right)^{\frac{1}{n}}$
$=\dfrac{a^{\frac{1}{n}}}{b^{\frac{1}{n}}}$
$=\dfrac{\sqrt[n]{a}}{\sqrt[n]{b}}$

115. Area of one equilateral triangle:
$\dfrac{s^2\sqrt{3}}{4}\Rightarrow$ the four sides have a total area of $4\left(\dfrac{s^2\sqrt{3}}{4}\right)=s^2\sqrt{3}$.
The area of the base is s^2.
The total surface area is
$s^2\sqrt{3}+s^2$
For $s=43$, the surface area is
$43^2\sqrt{3}+43^2\approx 5052\text{ cm}^2$, or
$\sim 0.5\text{ m}^2$
$\left(10{,}000\text{ cm}^2=1\text{ m}^2\right)$

End of Section 1.2

Chapter One: Section 1.3
Solutions to Odd-numbered Exercises

1. This expression is not a polynomial because it has non-integer exponents.

3. This expression is a 4-term polynomial of degree 11.

5. The expression '8' is a monomial of degree zero.

7. The expression is a binomial of degree 4.

9. The expression is a trinomial of degree 2.

11. The expression is a binomial of degree 5.

13. $\pi z^5 + 8z^2 - 2z + 1$ is of degree 5 with leading coefficient π.

15. $2s^6 - 10s^5 + 4s^3$ is of degree 6 with leading coefficient 2.

17. $9y^6 - 3y^5 + y - 2$ is of degree 6 with leading coefficient 9.

19. $\left(-4x^3y + 2zx - 3y\right) - \left(2xz + 3y + x^2z\right) =$
$-4x^3y + 2zx - 3y - 2xz - 3y - x^2z =$
$-4x^3y - 6y - x^2z$

21. $\left(x^2y - xy - 6y\right) + \left(y^2x + yx + 6x\right) =$
$x^2y + xy^2 + 6x - 6y$

23. $\left(a^2b + 2ab + ab^2\right) - \left(b^2a + 5ba + ba^2\right) =$
$a^2b + 2ab + ab^2 - b^2a - 5ba - ba^2 =$
$-3ab$

25. $\left(3a^2b + 2a - 3b\right)\left(ab^2 + 7ab\right) =$
$3a^3b^3 + 21a^3b^2 + 2a^2b^2$
$+14a^2b - 3ab^3 - 21ab^2$

27. $(3a + 4b)(a - 2b) = 3a^2 - 6ab + 4ab - 8b^2$
$= 3a^2 - 2ab - 8b^2$

29. $3a^2b + 3a^3b - 9a^2b^2 = 3a^2b(1 + a - 3b)$

31. $2x^6 - 14x^3 + 8x = 2x\left(x^5 - 7x^2 + 4\right)$

33. $a^3 + ab - a^2b - b^2 = \left(a^3 + ab\right) - \left(a^2b + b^2\right)$
$= a\left(a^2 + b\right) - b\left(a^2 + b\right)$
$= \left(a^2 + b\right)(a - b)$

35. $z + z^2 + z^3 + z^4 = z\left[(1 + z) + \left(z^2 + z^3\right)\right]$
$= z\left[(1 + z) + z^2(1 + z)\right]$
$= \left(z^3 + z\right)(z + 1)$

37. $nx^2 - 2y - 2x^2 + ny = \left(nx^2 - 2x^2\right) + (ny - 2y)$
$= x^2(n - 2) + y(n - 2)$
$= (n - 2)\left(x^2 + y\right)$

39. $25x^4y^2 - 9 = \left(5x^2y - 3\right)\left(5x^2y + 3\right)$

41. $x^3 - 1000y^3 =$
$(x - 10y)\left(x^2 + 10xy + 100y^2\right)$

43. $x^2 + 2x - 15 = (x + 5)(x - 3)$

45. $x^2 - 2x + 1 = (x - 1)^2$

47. $x^2 - 4x + 4 = (x - 2)^2$

49. $6x^2 + 5x - 6 = 6x^2 + 9x - 4x - 6$
$= 3x(2x + 3) - 2(2x + 3)$
$= (2x + 3)(3x - 2)$

51. $25y^2 + 10y + 1 = 25y^2 + 5y + 5y + 1$
$= 5y(5y + 1) + (5y + 1)$
$= (5y + 1)^2$

53. $6y^2 - 13y - 8 = 6y^2 - 16y + 3y - 8$
$= 2y(3y - 8) + (3y - 8)$
$= (2y + 1)(3y - 8)$

55. $(2x-1)^{\frac{-3}{2}} + (2x-1)^{\frac{-1}{2}} =$
$(2x-1)^{\frac{-3}{2}}\left(1 + (2x-1)^1\right) = 2x(2x-1)^{\frac{-3}{2}}$

57. $7a^{-1} - 2a^{-3}b = a^{-3}\left(7a^2 - 2b\right)$

Use computer algebra system for items 59 - 65.

59. $-25x^5 + 15x^4 - 72x^3 - 35x^2 - 9x - 36$

61. $x^3 - 3x^3y^2z + x^3z - 2x^2y + 6x^2y^3z$
$-2x^2yz + xyz - 3xy^3z^2 + xyz^2$

63. $\left(x^2 - y\right)(3x + y)\left(x - y^2\right)$

65. $(p - 3q)(2p - q)(p + q)(p + 2q)$

End of Section 1.3

Chapter One: Section 1.4
Solutions to Odd-numbered Exercises

1. $\sqrt{-25} = i\sqrt{25} = 5i$

3. $-\sqrt{-27} = -i\sqrt{9 \cdot 3} = -3i\sqrt{3}$

5. $\sqrt{-32x} = i\sqrt{16 \cdot 2x} = 4i\sqrt{2x}$

7. $\sqrt{-29} = i\sqrt{29}$

9. $$\begin{aligned}(4-2i)-(3+i) &= 4-2i-3-i \\ &= 1-3i\end{aligned}$$

11. $$\begin{aligned}(4-i)(2+i) &= 8+4i-2i-i^2 \\ &= 8+2i+1 \\ &= 9+2i\end{aligned}$$

13. $$\begin{aligned}(3-i)^2 &= 9-6i+i^2 \\ &= 8-6i\end{aligned}$$

15. $(3i)^2 = 9i^2 = -9$

17. $$\begin{aligned}(7i-2)+\left(3i^2-i\right) &= 7i-2-3-i \\ &= -5+6i\end{aligned}$$

19. $(3+i)(3-i) = 9-i^2 = 9+1 = 10$

21. $$\begin{aligned}(9-4i)(9+4i) &= 81-16i^2 \\ &= 81+16 \\ &= 97\end{aligned}$$

23. $i^{11}\left(\frac{6}{i^3}\right) = 6i^8 = 6$

25. $$\begin{aligned}\frac{1+2i}{1-2i} &= \frac{1+2i}{1-2i}\cdot\frac{1+2i}{1+2i} \\ &= \frac{1+4i+4i^2}{1+4} \\ &= \frac{1+4i-4}{5} \\ &= \frac{-3+4i}{5} \text{ or, } -\frac{3}{5}+\frac{4i}{5}\end{aligned}$$

27. $$\begin{aligned}\frac{i}{2+i} &= \frac{i}{2+i}\cdot\frac{2-i}{2-i} \\ &= \frac{2i-i^2}{4+1} \\ &= \frac{1+2i}{5}, \text{ or } \frac{1}{5}+\frac{2}{5}i\end{aligned}$$

29. $$\begin{aligned}(2+5i)^{-1} &= \frac{1}{2+5i}\cdot\frac{2-5i}{2-5i} \\ &= \frac{2-5i}{4-25i^2} \\ &= \frac{2-5i}{4+25} \\ &= \frac{2-5i}{29}, \text{ or } \frac{2}{29}-\frac{5}{29}i\end{aligned}$$

31. $\frac{1}{i^{27}} = \frac{1}{i^{27}}\cdot\frac{i}{i} = \frac{i}{i^{28}} = \frac{i}{1} = i$

33. $$\begin{aligned}(2-3i)^{-1} &= \frac{1}{2-3i}\cdot\frac{2+3i}{2+3i} \\ &= \frac{2+3i}{4-9i^2} \\ &= \frac{2+3i}{13}, \text{ or } \frac{2}{13}+\frac{3}{13}i\end{aligned}$$

35. $i^{-4} = \frac{1}{i^4} = 1$

37. $$\begin{aligned}\left(\sqrt{-9}\right)\left(\sqrt{-2}\right) &= 3i \cdot i\sqrt{2} \\ &= -3\sqrt{2}\end{aligned}$$

39. $$\begin{aligned}\frac{\sqrt{18}}{\sqrt{-2}} &= \frac{3\sqrt{2}}{i\sqrt{2}}\cdot\frac{i\sqrt{2}}{i\sqrt{2}} \\ &= -\frac{6i}{2} \\ &= -3i\end{aligned}$$

41. $$\begin{aligned}\left(3+\sqrt{-2}\right)^2 &= 9+6i\sqrt{2}+2i^2 \\ &= 9+6i\sqrt{2}-2 \\ &= 7+6i\sqrt{2}\end{aligned}$$

43. $\frac{3-2i}{1+i}=\frac{3-2i}{1+i}\cdot\frac{1-i}{1-i}$
$=\frac{3-3i-2i+2i^2}{1-i^2}$
$=\frac{1-5i}{2}$, or $\frac{1}{2}-\frac{5}{2}i$

45. $\frac{2500}{(3+i)^4}=\frac{2500}{28+96i}\cdot\frac{28-96i}{28-96i}$
$=7-24i$

47. $(1+i)^5(3-i)^2=(-4-4i)(8-6i)$
$=-56-8i$

End of Section 1.4

Chapter One: Section 1.5
Solutions to Odd-numbered Exercises

1.
$$-3(2t-4)=7(1-t)$$
$$-6t+12=7-7t$$
$$t=-5$$
Solution: $\{-5\}$

3.
$$\frac{y+5}{4}=\frac{1-5y}{6}$$
$$12\left(\frac{y+5}{4}\right)=12\left(\frac{1-5y}{6}\right)$$
$$3(y+5)=2(1-5y)$$
$$3y+15=2-10y$$
$$13y=-13$$
$$y=-1$$
Solution: $\{-1\}$

5.
$$3w+5=2(w+3)-4$$
$$3w+5=2w+6-4$$
$$w=-3$$
Solution: $\{-3\}$

7.
$$\frac{4s-3}{2}+\frac{7}{4}=\frac{8s+1}{4}$$
$$4\left(\frac{4s-3}{2}+\frac{7}{4}\right)=4\left(\frac{8s+1}{4}\right)$$
$$8s-6+7=8s+1$$
$$8s+1=8s+1$$
This equation is an identity and is true for all reals, R.

9.
$$\frac{4z-3}{2}+\frac{3}{8}=\frac{8z+3}{4}$$
$$8\left(\frac{4z-3}{2}+\frac{3}{8}\right)=8\left(\frac{8z+3}{4}\right)$$
$$16z-12+3=16z+6$$
$$16z-9=16z+6$$
This equation is never true.
Its solution is the empty set, $\varnothing$.

11.
$$\frac{6}{7}(m-4)-\frac{11}{7}=1$$
$$7\left(\frac{6}{7}(m-4)-\frac{11}{7}\right)=7(1)$$
$$6m-24-11=7$$
$$6m=42$$
$$m=7$$
Solution: $\{7\}$

13.
$$0.6x+0.08=2.3$$
$$0.6x=2.22$$
$$x=3.7$$
Solution: $\{3.7\}$

15.
$$0.73x+0.42(x-2)=0.35x$$
$$0.73x+0.42x-0.84=0.35x$$
$$0.8x=0.84$$
$$x=1.05$$
Solution: $\{1.05\}$

17.
$$|3x-2|-1=|5-x|$$
[1] $3x-2-1=\pm(5-x)$, or

[2] $-(3x-2)-1=\pm(5-x)$

[1A] $3x-2-1=5-x$
$$3x-3=5-x$$
$$4x=8$$
$x=2$ [This checks]

[1B] $3x-2-1=-5+x$
$$2x=-2$$
$x=-1$ [Does not check]

[2A] $-3x+2-1=5-x$
$$-2x=4$$
$x=-2$ [This checks]

[2B] $-3x+2-1=-5+x$
$$-3x+1=-5+x$$
$$-4x=-6$$
$x=\frac{3}{2}$ [Does not check]

19. $|6x-2|=0$

$6x-2=0$ or $-(6x-2)=0$

[These are equivalent]

$6x=2$

$x=\frac{1}{3}$

21. $|2x-109|=731$

$2x-109=731$ or $-(2x-109)=731$

$2x=840$ or $-2x+109=731$

$x=420$ or $-2x=622$

Both answers check. $x=-311$

23. $|5x-3|=7$

$5x-3=7$ or $-(5x-3)=7$

$5x=10$ or $-5x+3=7$

$x=2$ or $x=-\frac{4}{5}$

Both answers check.

25. $-|6x+1|=11$

This equation has no solution. This type of equation is called a contradiction. Note that its equivalent form is $|6x+1|=-11$. Since $|6x-1|\geq 0$, there can be no solution to the equation.

27. $|x-3|-|x-7|=0$

Geometrically: Find a value of x that is equidistant from 7 and 3.

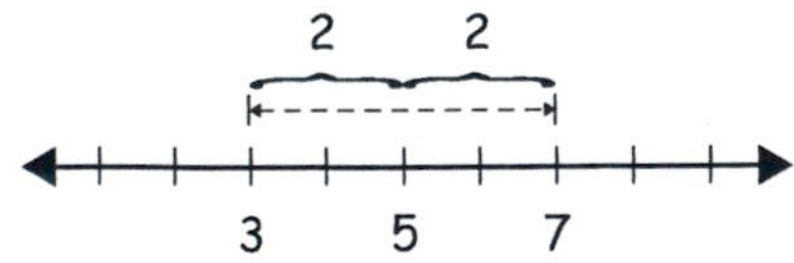

Solution is $\{5\}$.

Algebraically:

$x-3=\pm(x-7)$ or $-(x-3)=\pm(x-7)$

$x-3=x-7$ [Contradiction], or

$-x+3=x-7$

$-2x=-10$

$x=5$ [This checks].

29. $|x|=|x+1|$. Notice that this is the same as $|x-0|=|x-(-1)|$.

Geometrically: Find a value of x that is equidistant between 0 and -1:

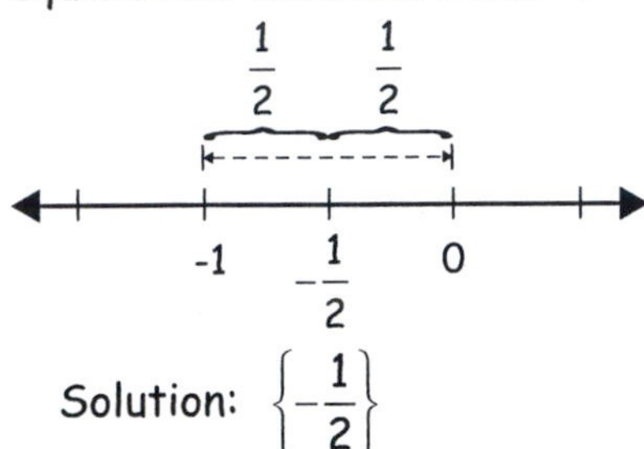

Solution: $\left\{-\frac{1}{2}\right\}$

Algebraically: The equation has two initial possibilities.

$x=\pm(x+1)$ or $-x=\pm(x+1)$.

There are only two distinct equations; that is, two are equivalent to the other two.

$x=x+1$, which is a contradiction, and $-x=x+1$, which yields $x=-\frac{1}{2}$.

31. $\left|x+\frac{1}{4}\right|=\left|x-\frac{3}{4}\right|$. Notice that this is the same as $\left|x-\left(-\frac{1}{4}\right)\right|=\left|x-\frac{3}{4}\right|$.

Geometrically: Find a value of x that is equidistant between $-\frac{1}{4}$ and $\frac{3}{4}$:

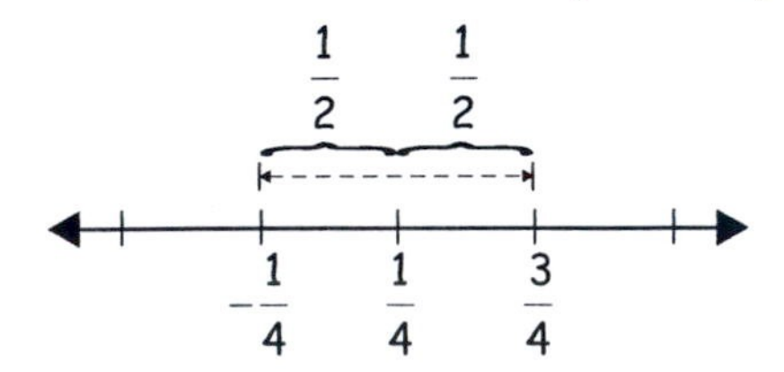

Solution: $\left\{\frac{1}{4}\right\}$

Algebraically: The equation has two initial possibilities.

$x+\frac{1}{4}=\pm\left(x-\frac{3}{4}\right)$ or

$-\left(x+\frac{1}{4}\right)=\pm\left(x-\frac{3}{4}\right)$.

There are only two distinct equations; that is, two are equivalent to the other two:

$x+\frac{1}{4}=x-\frac{3}{4}$, which is a contradiction,

and $-x-\frac{1}{4}=x-\frac{3}{4}$, which yields $x=\frac{1}{4}$.

33. $PV=nRT$, for T:

$$T=\frac{PV}{nR}$$

35. $A=\frac{1}{2}(B+b)h$, for B:

$$2A=(B+b)h$$
$$\frac{2A}{h}=B+b$$
$$B=\frac{2A}{h}-b$$

37. $V=\frac{1}{3}s^2h$, for h:

$$3V=s^2h$$
$$h=\frac{3V}{s^2}$$

39. $d=rt_1+rt_2$, for r:

$$d=r(t_1+t_2)$$
$$r=\frac{d}{t_1+t_2}$$

41. Use $d=rt$: $t=\frac{d}{r}$

$$t=\frac{95}{15}$$

$t=\frac{19}{3}$ hours, or $t=6$ hrs 20 min.

43. Use $d=rt$. The distance of the two trains added together is equal to 630 miles when they pass each other:

$$95t+85t=630$$
$$180t=630$$

$t=3.5$ hours

45. Use $I=Prt$, where $t=1$ year:

$I=12{,}800-10{,}000$, so

$$2{,}800=10{,}000r$$

$r=0.28$, or 28%.

47. First, change the measures to one unit:

Will: 6 ft., 4 in. = 76 in.

Matt: 6 ft., 7 in. = 79 in.

Will's Ht. to Matt's Ht.: $\frac{76}{79}=96.2\%$

Matt's Ht. to Will's Ht.: $\frac{79}{76}=103.9\%$

49. Let $n=$ smallest of the integers

Then, $n+1=$ second integer and

$n+2=$ the third integer:

$$n+n+1+n+2=288$$
$$3n+3=288$$
$$n=95$$
$$n+1=96$$
$$n+2=97$$

51. Let $p=$ original price.

Then, $p-0.30p=15.05$, or

$$0.70p=15.05$$
$$p=\$21.50$$

53. $s=2\pi r^2+2\pi rh$, for h:

$$s-2\pi r^2=2\pi rh$$

$h=\frac{s-2\pi r^2}{2\pi r}$, or

$$h=\frac{s}{2\pi r}-r$$

55. $ab-a(c-3)=ab+c$, for a:

$$-a(c-3)=c$$
$$a=\frac{-c}{c-3}$$

End of Section 1.5

Chapter One: Section 1.6
Solutions to Odd-numbered Exercises

1. $4+3t \le t-2$

$2t \le -6$

$t \le -3$

Solution is $(-\infty, -3]$.

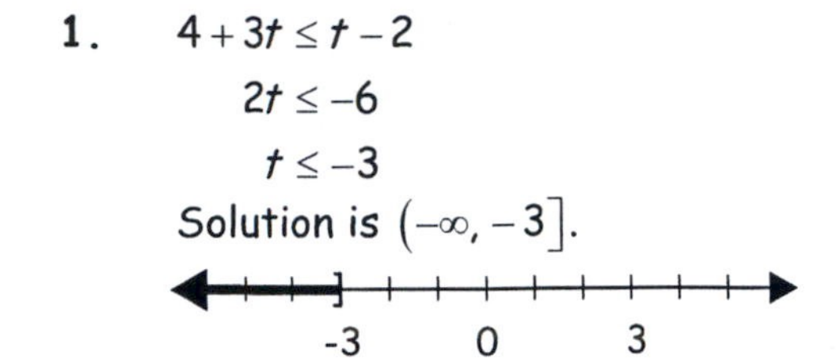

3. $4.2x - 5.6 < 1.6 + x$

$3.2x < 7.2$

$x < 2.25$

Solution is $(-\infty,\ 2.25)$.

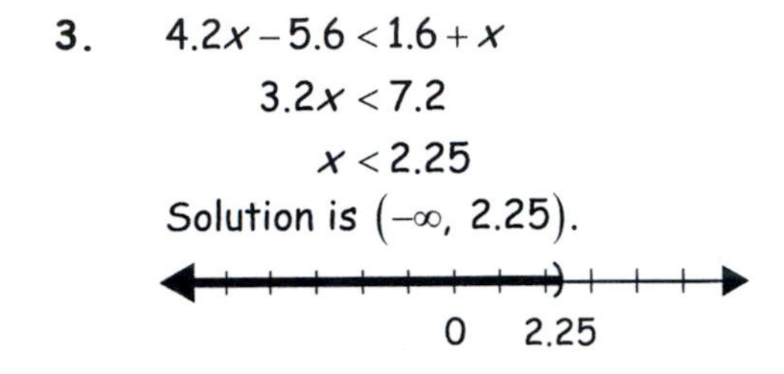

5. $-2(3-x) < -2x$

$-6+2x < -2x$

$4x < 6$

$x < \dfrac{3}{2}$

Solution is $\left(-\infty, \dfrac{3}{2}\right)$.

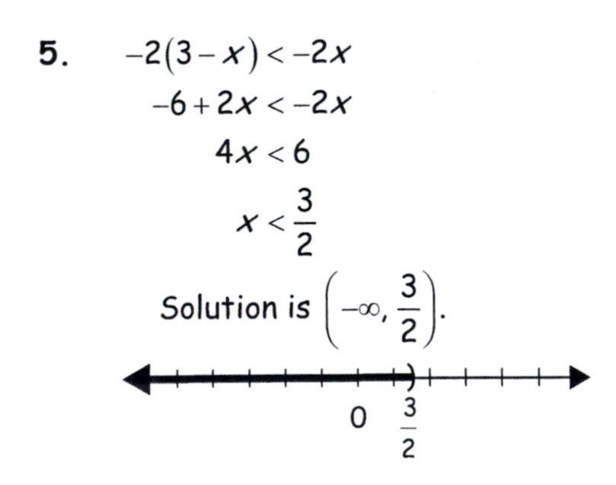

7. $4w+7 \le -7w+4$

$11w \le -3$

$w \le -\dfrac{3}{11}$

Solution is $\left(-\infty, -\dfrac{3}{11}\right]$.

9. $\dfrac{6f-2}{5} < \dfrac{5f-3}{4}$

$20\left(\dfrac{6f-2}{5}\right) < 20\left(\dfrac{5f-3}{4}\right)$

$4(6f-2) < 5(5f-3)$

$24f - 8 < 25f - 15$

$-f < -7$

$f > 7$

Solution is $(7, \infty)$

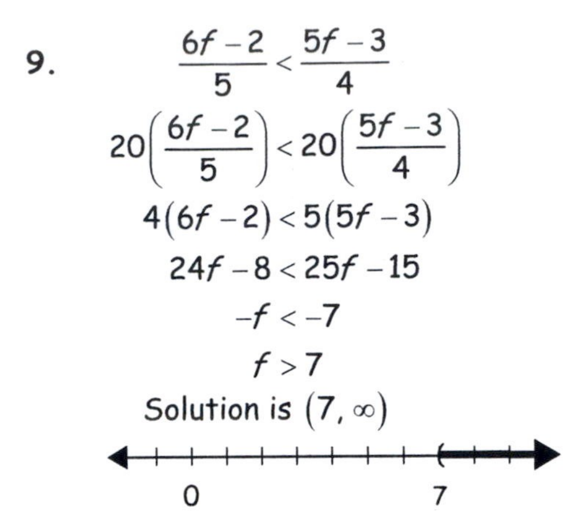

11. $0.04n + 1.7 < 0.13n - 1.45$

$3.15 < 0.09n$

$35 < n$

Solution is $(35, \infty)$.

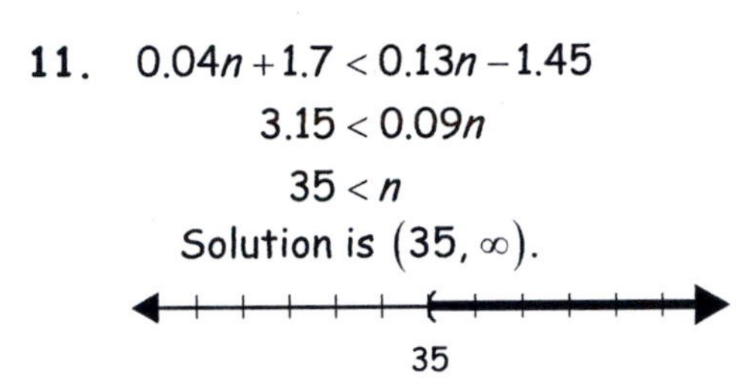

13. $-4 < 3x - 7 \le 8$

$3 < 3x \le 15$

$1 < x \le 5$

Solution is $(1, 5]$.

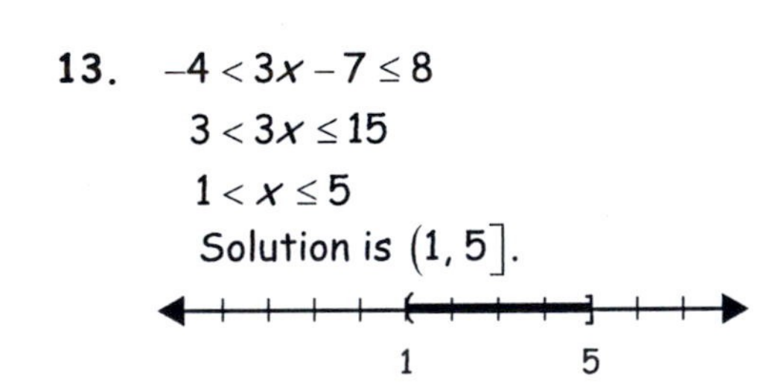

15. $-8 \le \dfrac{z}{2} - 4 < -5$

$-4 \le \dfrac{z}{2} < -1$

$-8 \le z < -2$

Solution is $[-8, -2)$.

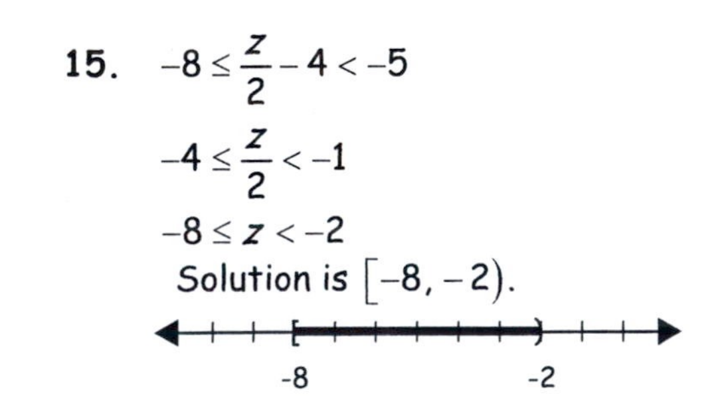

17. $5 \le 2m - 3 \le 13$

$8 \le 2m \le 16$

$4 \le m \le 8$

Solution is $[4, 8]$.

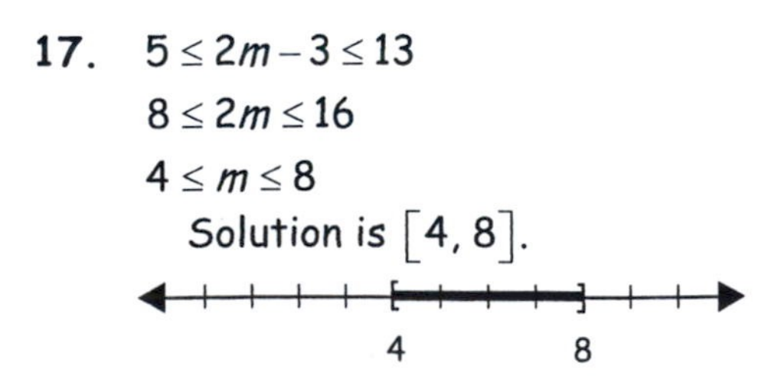

19. $\frac{1}{3} < \frac{7}{6}(l-3) < \frac{2}{3}$

$6\left(\frac{1}{3}\right) < 6\left[\frac{7}{6}(l-3)\right] < 6\left(\frac{2}{3}\right)$

$2 < 7(l-3) < 4$

$2 < 7l - 21 < 4$

$23 < 7l < 25$

$\frac{23}{7} < l < \frac{25}{7}$

Solution is $\left(\frac{23}{7}, \frac{25}{7}\right)$.

Note: Scale of graph is 1/7.

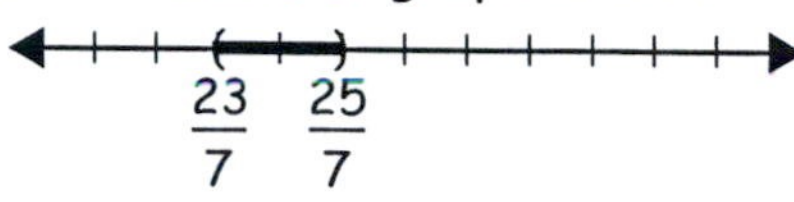

21. $0.08 < 0.03c + 0.13 \le 0.16$

$-0.05 < 0.03c \le 0.03$

$\frac{-0.05}{0.03} < c \le \frac{0.03}{0.03}$

$-\frac{5}{3} < c \le 1$

Solution is $\left(-\frac{5}{3}, 1\right]$.

Note: Scale of graph is 1/3.

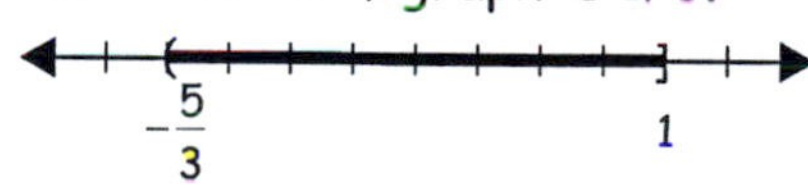

23. $4 + |3-2y| > 6$

$|3-2y| > 2$

$3-2y > 2$ or $-(3-2y) > 2$

$-2y > -1$ or $-3+2y > 2$

$y < \frac{1}{2}$ or $y > \frac{5}{2}$

Solution is $\left(-\infty, \frac{1}{2}\right) \cup \left(\frac{5}{2}, \infty\right)$

Note: Scale of graph is 1/2.

25. $2|z+5| < 12$

$-12 < 2(z+5) < 12$

$-6 < z+5 < 6$

$-11 < z < 1$

Solution is $(-11, 1)$.

Note: Scale of graph is 2.

27. $|4-2x| > 11$

$4-2x > 11$ or $-(4-2x) > 11$

$-2x > 7$ or $-4+2x > 11$

$x < -\frac{7}{2}$ or $x > \frac{15}{2}$

Solution is $\left(-\infty, -\frac{7}{2}\right) \cup \left(\frac{5}{2}, \infty\right)$

29. $6 - 5|x+2| \ge -4$

$-5|x+2| \ge -10$

$|x+2| \le 2$

$-2 \le x+2 \le 2$

$-4 \le x \le 0$

Solution is $[-4, 0]$.

31. $-3|4-t| < -6$

$|4-t| > 2$

$4-t > 2$ or $-(4-t) > 2$

$-t > -2$ or $-4+t > 2$

$t < 2$ or $t > 6$

Solution is $(-\infty, 2) \cup (6, \infty)$.

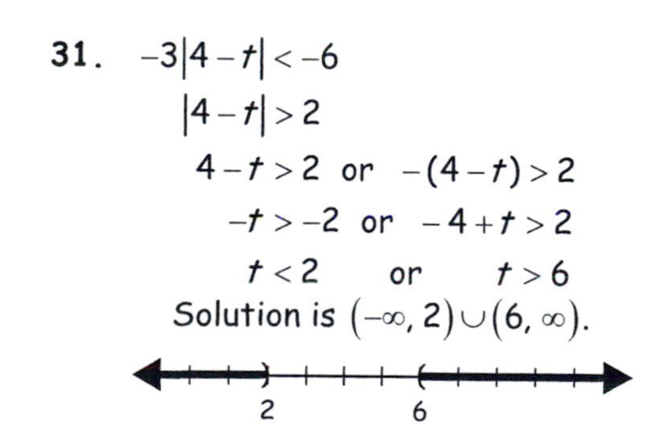

33. $3|4-t| < -6$

$|4-t| < -2$

No solution. Solution set: $\varnothing$

35. $2 < |6w - 2| + 7$

$-5 < |6w - 2|$

Since the absolute value of any number is always non-negative, this expression is true for all values of w.

Solution is $(-\infty, \infty)$.

37. Let $x =$ the number of games Larry must win in a row to have an overall winning percentage greater than 50%.

Then, $\frac{10 + x}{30 + x} > \frac{1}{2}$. Solving:

$2(10 + x) > 30 + x$

$20 + 2x > 30 + x$

$x > 10$

The interval is $[11, \infty)$.

39. Let $x =$ rate for the 4th quarter to provide an annual rate less than 5%.

Then, $\frac{5.2 + 4.3 + 4.7 + x}{4} < 5$. Solving:

$5.2 + 4.3 + 4.7 + x < 20$

$x < 5.8$

Solution is $[0, 5.8)$

End of Section 1.6

Chapter One: Section 1.7
Solutions to Odd-numbered Exercises

1.
$$2x^2 - x = 3$$
$$2x^2 - x - 3 = 0$$
$$(2x-3)(x+1) = 0$$
$$2x - 3 = 0 \text{ or } x + 1 = 0$$
$$x = \frac{3}{2} \text{ or } x = -1$$
$$\left\{-1, \frac{3}{2}\right\}$$

3.
$$9x - 5x^2 = -2$$
$$5x^2 - 9x - 2 = 0$$
$$(5x+1)(x-2) = 0$$
$$5x + 1 = 0 \text{ or } x - 2 = 0$$
$$x = -\frac{1}{5} \text{ or } x = 2$$
$$\left\{-\frac{1}{5}, 2\right\}$$

5.
$$2x^2 - 3x = x^2 + 18$$
$$x^2 - 3x - 18 = 0$$
$$(x-6)(x+3) = 0$$
$$x - 6 = 0 \text{ or } x + 3 = 0$$
$$x = 6 \text{ or } x = -3$$
$$\{-3, 6\}$$

7.
$$15x^2 + x = 2$$
$$15x^2 + x - 2 = 0$$
$$(5x+2)(3x-1) = 0$$
$$5x + 2 = 0 \text{ or } 3x - 1 = 0$$
$$x = -\frac{2}{5} \text{ or } x = \frac{1}{3}$$
$$\left\{-\frac{2}{5}, \frac{1}{3}\right\}$$

9.
$$3x^2 + 33 = 2x^2 + 14x$$
$$x^2 - 14x + 33 = 0$$
$$(x-11)(x-3) = 0$$
$$x - 11 = 0 \text{ or } x - 3 = 0$$
$$x = 11 \text{ or } x = 3$$
$$\{3, 11\}$$

11.
$$3x^2 - 7x = 0$$
$$x(3x-7) = 0$$
$$x = 0 \text{ or } x = \frac{7}{3}$$
$$\left\{0, \frac{7}{3}\right\}$$

13.
$$(x-3)^2 = 9$$
$$x - 3 = \pm 3$$
$$x = 3 \pm 3$$
$$x = 0 \text{ or } x = 6$$
$$\{0, 6\}$$

15.
$$x^2 - 6x + 9 = -16$$
$$(x-3)^2 = -16$$
$$x - 3 = \pm\sqrt{-16}$$
$$x = 3 \pm 4i$$
$$\{3 \pm 4i\}$$

17.
$$(8t-3)^2 = 0$$
$$8t - 3 = 0$$
$$t = \frac{3}{8}$$
$$\left\{\frac{3}{8}\right\}$$

19.
$$(2x-1)^2 = 8$$
$$2x - 1 = \pm\sqrt{8}$$
$$2x = 1 \pm 2\sqrt{2}$$
$$x = \frac{1}{2} \pm \sqrt{2}$$
$$\left\{\frac{1}{2} \pm \sqrt{2}\right\}$$

21. $x^2-4x+4=49$

$$(x-2)^2=49$$
$$x-2=\pm 7$$
$$x=2\pm 7$$
$$x=-5 \text{ or } x=9$$
$$\{-5, 9\}$$

23. $(3x-6)^2=4x^2$

$$3x-6=\pm 2x$$
$$3x\pm 2x=6$$
$$5x=6 \text{ or } 3x-2x=6$$
$$x=\frac{6}{5} \text{ or } x=6$$
$$\left\{\frac{6}{5}, 6\right\}$$

25. Solve by completing the square:

$$y^2+11=12y$$

Rewrite:

$$y^2-12y+11=0$$
$$y^2-12y+\left(\frac{12}{2}\right)^2-\left(\frac{12}{2}\right)^2+11=0$$
$$(y-6)^2=25$$
$$y-6=\pm 5$$
$$y=11, 1$$

Solution Set: $\{1, 11\}$

27. Solve by completing the square:

$$-6z^2+4z=-1$$

Rewrite:

$$-6\left(z^2-\frac{2}{3}z\right)=-1$$
$$z^2-\frac{2}{3}z+\left(\frac{2}{6}\right)^2-\left(\frac{2}{6}\right)^2=\frac{1}{6}$$
$$\left(z-\frac{1}{3}\right)^2=\frac{5}{18}\Rightarrow$$
$$z-\frac{1}{3}=\pm\sqrt{\frac{5}{18}}$$
$$z=\frac{1}{3}\pm\frac{\sqrt{10}}{6}$$
$$z=\frac{2\pm\sqrt{10}}{6}$$

Solution Set: $\left\{\frac{2\pm\sqrt{10}}{6}\right\}$

29. Solve by completing the square:

$$2b^2+10b+5=0$$

Rewrite:

$$2(b^2+5b)=-5$$
$$b^2+5b+\left(\frac{5}{2}\right)^2=\left(\frac{5}{2}\right)^2-\frac{5}{2}$$
$$\left(b+\frac{5}{2}\right)^2=\frac{25-10}{4}$$
$$b+\frac{5}{2}=\pm\frac{\sqrt{15}}{2}$$
$$b=\frac{-5\pm\sqrt{15}}{2}$$

Solution Set: $\left\{\frac{-5\pm\sqrt{15}}{2}\right\}$

31. $x^2+8x+7=-8$ (Complete the square)

$$x^2+8x=-15$$
$$x^2+8x+16=-15+16$$
$$(x+4)^2=1$$
$$x+4=\pm 1$$
$$x=-4\pm 1$$
$$x=-5 \text{ or } x=-3$$
$$\{-5, -3\}$$

33. $2x^2+7x-15=0$ (Complete the square)

$$x^2+\frac{7}{2}x=\frac{15}{2}$$
$$x^2+\frac{7}{2}x+\frac{49}{16}=\frac{15}{2}+\frac{49}{16}$$
$$\left(x+\frac{7}{4}\right)^2=\frac{169}{16}$$
$$x+\frac{7}{4}=\pm\frac{13}{4}$$
$$x=\frac{3}{2} \text{ or } x=-5$$
$$\left\{-5,\frac{3}{2}\right\}$$

35. $u^2+10u+9=0$ (Complete the square)

$$u^2+10u=-9$$
$$u^2+10u+25=-9+25$$
$$(u+5)^2=16$$
$$u+5=\pm4$$
$$u=-1 \text{ or } u=-9$$
$$\{-9,-1\}$$

37. $4x^2-3x=-1$ (By quadratic formula)

$$4x^2-3x+1=0$$
$$x=\frac{-(-3)\pm\sqrt{(-3)^2-4(4)(1)}}{2(4)}$$
$$x=\frac{3\pm\sqrt{-7}}{8}$$
$$\left\{\frac{3\pm i\sqrt{7}}{8}\right\}$$

39. $2.1y^2-3.5y=4$ (By quadratic formula)

Rewrite: $2.1y^2-3.5y-4=0$

$$y=\frac{-(-3.5)\pm\sqrt{(-3.5)^2-4(2.1)(-4)}}{2(2.1)}$$
$$y=\frac{3.5\pm\sqrt{45.85}}{4.2}$$
$$\approx\{-0.78,2.45\}$$

41. $a(a+2)=-1$ (Use quadratic formula)

$$a^2+2a+1=0$$
$$a=\frac{-2\pm\sqrt{2^2-4(1)(1)}}{2(1)}$$
$$a=\frac{-2\pm\sqrt{0}}{2}$$
$$\{-1\}$$

43. $6x^2+5x-4=3x-2$

$$6x^2+2x-2=0$$
$$x=\frac{-2\pm\sqrt{2^2-4(6)(-2)}}{2(6)}$$
$$x=\frac{-2\pm\sqrt{4\cdot13}}{2(6)}$$
$$x=\frac{-1\pm\sqrt{13}}{6}$$
$$\left\{\frac{-1\pm\sqrt{13}}{6}\right\}$$

45. $4x^2-14x-27=3$

$$4x^2-14x-30=0$$
$$x=\frac{-(-14)\pm\sqrt{(-14)^2-4(4)(-30)}}{2(4)}$$
$$x=\frac{14\pm\sqrt{676}}{8}$$
$$x=\frac{14\pm26}{8}$$
$$x=-\frac{3}{2} \text{ or } x=5$$
$$\left\{-\frac{3}{2},5\right\}$$

47. Use the formula : $h = -\frac{1}{2}gt^2 + v_0t + h_0$,
where $g = 32, v_0 = 40$ and $h_0 = 144$.
Find t when $h = 0$:
$-16t^2 + 40t + 144 = 0$
This equation is equivalent to
$2t^2 - 5t - 18 = 0$.
Solve by factoring:
$(2t - 9)(t + 2) = 0$ [Disregard $t = -2$]
$t = \frac{9}{2} = 4.5$ seconds

49. Use the formula : $h = -\frac{1}{2}gt^2 + v_0t + h_0$,
where $g = 9.8, v_0 = 20$ and $h_0 = 24$.
Find t when $h = 7$:
$-4.9t^2 + 20t + 24 = 7$
This equation is equivalent to
$4.9t^2 - 20t - 17 = 0$.
Solve by using the
quadratic formula:

$$t = \frac{20 \pm \sqrt{20^2 - 4(4.9)(-17)}}{2(4.9)}$$

$$t \approx \frac{20 + 27.078}{9.8} \approx 4.8 \text{ seconds}$$

[Disregard the negative t value]

51. Use the quadratic formula:

$$x = \frac{-(-6) \pm \sqrt{(-6)^2 - 4(1)(13)}}{2(1)}$$

$$x = \frac{6 \pm \sqrt{-16}}{2}$$

$x = 3 \pm 2i$
Then, factors of $x^2 - 6x + 13$ are
$(x - 3 - 2i)$ and $(x - 3 + 2i)$.
In other words,
$x^2 - 6x + 13 = (x - 3 - 2i)(x - 3 + 2i)$.

53. Use the quadratic formula:

$$x = \frac{-12 \pm \sqrt{(12)^2 - 4(4)(1)}}{2(4)}$$

$$x = \frac{-12 \pm \sqrt{128}}{8}$$

$$x = \frac{-3 \pm 2\sqrt{2}}{2}$$

Then, factors of $4x^2 + 12x + 1$ are
$(2x + 3 - 2\sqrt{2})$ and $(2x + 3 + 2\sqrt{2})$.
In other words,
$4x^2 + 12x + 1 = (2x + 3 - 2\sqrt{2})(2x + 3 + 2\sqrt{2})$.

55. $(x - (-3))(x - 8) = (x + 3)(x - 8)$
$= x^2 - 5x - 24$
Then, $b = -5$ and $c = -24$.

57. Let $A = x^2 - 1$. Substitute in the
given equation and solve for A. Then,
substitute back to solve for x:
$A^2 + A - 12 = 0$
$(A + 4)(A - 3) = 0 \Rightarrow A = -4;\ A = 3$
Then, back substituting
$x^2 - 1 = -4 \Rightarrow x^2 = -3 \Rightarrow x = \pm i\sqrt{3}$
and
$x^2 - 1 = 3 \Rightarrow x^2 = 2 \Rightarrow x = \pm\sqrt{2}$
Solution Set: $\{\pm\sqrt{2}, \pm i\sqrt{3}\}$

59. Let $A = x^2 - 2x + 1$. Substitute in the given equation and solve for A. Then, substitute back to solve for x:

$$A^2 + A - 12 = 0$$

$$(A+4)(A-3) = 0 \Rightarrow A = -4;\ A = 3$$

Then, back substituting

$$x^2 - 2x + 1 = -4 \Rightarrow x^2 - 2x + 5 = 0 \Rightarrow$$

$$x = \frac{2 \pm \sqrt{4 - 4(5)}}{2} = 1 \pm 2i \text{ and}$$

$$x^2 - 2x + 1 = 3 \Rightarrow x^2 - 2x - 2 = 0 \Rightarrow$$

$$x = \frac{2 \pm \sqrt{4 - 4(-2)}}{2} = 1 \pm \sqrt{3}$$

Solution Set: $\{1 \pm 2i, 1 \pm \sqrt{3}\}$

61. Let $A = x^{\frac{1}{3}}$. Substitute in the given equation and solve for A. Then, substitute back to solve for x:

$$2A^2 - 7A + 3 = 0$$

$$(2A-1)(A-3) = 0 \Rightarrow A = \frac{1}{2};\ A = 3$$

Then, back substituting

$$x^{\frac{1}{3}} = \frac{1}{2} \Rightarrow x = \frac{1}{8} \text{ and}$$

$$x^{\frac{1}{3}} = 3 \Rightarrow x = 27$$

Solution Set: $\left\{\frac{1}{8}, 27\right\}$

63. Let $A = t^2 - t$. Substitute in the given equation and solve for A. Then, substitute back to solve for t:

$$A^2 - 8A + 12 = 0$$

$$(A-6)(A-2) = 0 \Rightarrow A = 6;\ A = 2$$

Then, back substituting

$$t^2 - t = 6 \Rightarrow t^2 - t - 6 = 0 \Rightarrow$$

$$(t-3)(t+2) = 0 \Rightarrow t = 3;\ t = -2$$

and

$$t^2 - t = 2 \Rightarrow t^2 - t - 2 = 0 \Rightarrow$$

$$(t-2)(t+1) = 0 \Rightarrow t = 2;\ t = -1$$

Solution Set: $\{-2, -1, 2, 3\}$

65. Let $A = x^{\frac{1}{3}}$. Substitute in the given equation and solve for A. Then, substitute back to solve for x:

$$3A^2 - A - 2 = 0$$

$$(3A+2)(A-1) = 0 \Rightarrow A = -\frac{2}{3};\ A = 1$$

Then, back substituting

$$x^{\frac{1}{3}} = -\frac{2}{3} \Rightarrow x = -\frac{8}{27} \text{ and}$$

$$x^{\frac{1}{3}} = 1 \Rightarrow x = 1$$

Solution Set: $\left\{-\frac{8}{27}, 1\right\}$

67. Let $A = x^2 - 13$. Substitute in the given equation and solve for A. Then, substitute back to solve for x:

$$A^2 + A - 12 = 0$$

$$(A+4)(A-3) = 0 \Rightarrow A = -4;\ A = 3$$

Then, back substituting

$$x^2 - 13 = -4 \Rightarrow x^2 = 9 \Rightarrow x = \pm 3 \text{ and}$$

$$x^2 - 13 = 3 \Rightarrow x^2 = 16 \Rightarrow x = \pm 4$$

Solution Set: $\{\pm 3, \pm 4\}$

69. Solve by factoring:

$$2x^3 + x^2 + 2x + 1 = 0$$

$$x^2(2x+1) + (2x+1) = 0$$

$$(x^2+1)(2x+1) = 0$$

$$x^2 + 1 = 0 \quad \text{or} \quad 2x + 1 = 0$$

$$x^2 = -1 \qquad 2x = -1$$

$$x = \pm i \qquad x = -\frac{1}{2}$$

Solution Set: $\left\{-\frac{1}{2}, \pm i\right\}$

71. Factor and solve:

$$y^3 + 8 = (y+2)(y^2 - 2y + 4) = 0 \Rightarrow$$

$$y + 2 = 0 \text{ or } y^2 - 2y + 4 = 0$$

$$y = -2 \qquad y = \frac{2 \pm \sqrt{4 - 4(4)}}{2} = 1 \pm \sqrt{3}i$$

Solution Set: $\{-2, 1 \pm \sqrt{3}i\}$

73. Factor and solve:

$16a^4 = 81 \Rightarrow 4a^2 = \pm 9$

$a^2 = \pm\frac{9}{4} \Rightarrow a = \pm\sqrt{\frac{9}{4}};\ a = \pm\sqrt{-\frac{9}{4}}$ or

$a = \pm\frac{3}{2};\ a = \pm\frac{3}{2}i$

Solution Set: $\left\{\pm\frac{3}{2}, \pm\frac{3}{2}i\right\}$

75. Factor and solve:

$6x^3 + 8x^2 = 14x$

Rewrite: $2x\left(3x^2 + 4x - 7\right) = 0 \Rightarrow$

$x = 0$ or $3x^2 + 4x - 7 = 0 \Rightarrow$

$(3x+7)(x-1) = 0 \Rightarrow x = -\frac{7}{3};\ x = 1$

Solution Set: $\left\{-\frac{7}{3}, 0, 1\right\}$

77. Factor and solve: $27x^3 + 64 = 0$

$(3x+4)\left(9x^2 - 12x + 16\right) = 0 \Rightarrow$

$3x + 4 = 0 \Rightarrow x = -\frac{4}{3}$, or

$9x^2 - 12x + 16 = 0$

Use the quadratic formula:

$x = \frac{12 \pm \sqrt{144 - 4(9)(16)}}{2(9)}$

$x = \frac{12 \pm \sqrt{-432}}{18}$

$x = \frac{2 \pm 2\sqrt{3}i}{3}$

Solution Set: $\left\{-\frac{4}{3}, \frac{2 \pm 2\sqrt{3}i}{3}\right\}$

79. Factor and solve:

$x^4 + 5x^2 - 36 = 0$

Let $u = x^2$ and rewrite:

$u^2 + 5u - 36 = 0$

$(u+9)(u-4) = 0$

$u = -9$ or $u = 4$

$x^2 = -9$ or $x^2 = 4$

$x = \pm 3i$ or $x \pm 2$

Solution Set: $\{\pm 2, \pm 3i\}$

81. Factor and solve:

$(x-3)^{-\frac{1}{2}} + 2(x-3)^{\frac{1}{2}} = 0$

$(x-3)^{-\frac{1}{2}}\left(1 + 2(x-3)\right) = 0 \Rightarrow$

$(x-3)^{-\frac{1}{2}} = 0$, or $2x - 5 = 0$

There is no value of x that would make the first equation true. For the second equation, $x = \frac{5}{2}$.

Solution Set: $\left\{\frac{5}{2}\right\}$

83. Factor and solve:

$(2x-5)^{\frac{1}{3}} - 3(2x-5)^{-\frac{2}{3}} = 0$

$(2x-5)^{-\frac{2}{3}}\left((2x-5) - 3\right) = 0 \Rightarrow$

$(2x-5)^{-\frac{2}{3}} = 0$, or $2x - 8 = 0$

There is no value of x that would make the first equation true. For the second equation, $x = 4$.

Solution Set: $\{4\}$

85. Factor and solve:

$x^{\frac{11}{2}} - 6x^{\frac{9}{2}} + 9x^{\frac{7}{2}} = 0$

$x^{\frac{7}{2}}\left(x^2 - 6x + 9\right) = 0 \Rightarrow$

$x^{\frac{7}{2}} = 0$ or $(x-3)^2 = 0 \Rightarrow$

$x = 0$ or $x = 3$

Solution Set: $\{0, 3\}$

87. $(x+2)(x)(x-6)=0$

$(x^2+2x)(x-6)=0$

$x^3-4x^2-12x=0 \Rightarrow$

$b=-4,\ c=-12,\ d=0$

89. $\left(x+\frac{3}{5}\right)\left(x-\frac{2}{3}\right)(x-1)=0$

$\left(x^2-\frac{1}{15}x-\frac{6}{15}\right)(x-1)=0$

$x^3-\frac{16}{15}x^2-\frac{5}{15}x+\frac{6}{15}=0$

Multiply both sides by 15:

$15x^3-16x^2-5x+6=0$

$a=15,\ b=-16,\ c=-5$

Use a computer algebra system for 91 - 95.

91. $\left\{\frac{1}{10}\left(3\pm\sqrt{331}i\right)\right\}$

93. $\{-1.796, 1.067\}$

95. $\left\{\frac{2}{3}\left(2\pm\sqrt{2}i\right)\right\}$

End of Section 1.7

Chapter One: Section 1.8
Solutions to Odd-numbered Exercises

1. $$\frac{2x^2+7x+3}{x^2-2x-15}=\frac{(2x+1)(x+3)}{(x-5)(x+3)}$$
$$=\frac{2x+1}{x-5}$$
$$x\neq -3,5$$

3. $$\frac{x^3+2x^2-3x}{x+3}=\frac{x\left(x^2+2x-3\right)}{x+3}$$
$$=\frac{x(x+3)(x-1)}{x+3}$$
$$=x(x-1)$$
$$x\neq -3$$

5. $$\frac{x^2+5x-6}{x^2+4x-5}=\frac{(x+6)(x-1)}{(x+5)(x-1)}$$
$$=\frac{x+6}{x+5}$$
$$x\neq -5,1$$

7. $$\frac{x+1}{x^3+1}=\frac{x+1}{(x+1)\left(x^2-x+1\right)}$$
$$=\frac{1}{x^2-x+1}$$
$$x\neq -1$$

9. $$\frac{2x^2+11x+5}{x+5}=\frac{(2x+1)(x+5)}{x+5}$$
$$=2x+1$$
$$x\neq -5$$

11. $$\frac{2x^2+11x-21}{x+7}=\frac{(2x-3)(x+7)}{x+7}$$
$$=2x-3$$
$$x\neq -7$$

13. $$\frac{x-3}{x+5}+\frac{x^2+3x+2}{x-3}$$
$$=\frac{(x-3)(x-3)+\left(x^2+3x+2\right)(x+5)}{(x+5)(x-3)}$$
$$=\frac{x^3+9x^2+11x+19}{(x+5)(x-3)}$$

15. $$\frac{x+2}{x-3}-\frac{x-3}{x+5}-\frac{1}{x^2+2x-15}$$
$$=\frac{(x+2)(x+5)}{(x-3)(x+5)}-\frac{(x-3)(x-3)}{(x-3)(x+5)}-\frac{1}{(x-3)(x+5)}$$
$$=\frac{\left(x^2+7x+10\right)-x^2+6x-9-1}{(x-3)(x+5)}$$
$$=\frac{13x}{(x-3)(x+5)}$$

17. $$\frac{x^2+1}{x-3}+\frac{x-5}{x+3}$$
$$=\frac{\left(x^2+1\right)(x+3)+(x-5)(x-3)}{(x-3)(x+3)}$$
$$=\frac{x^3+4x^2-7x+18}{x^2-9}$$

19. $$\frac{y-2}{y+1}\cdot\frac{y^2-1}{y-2}=\frac{y-2}{y+1}\cdot\frac{(y-1)(y+1)}{y-2}$$
$$=y-1$$

21. $$\frac{2x^2-5x-12}{x-3}\cdot\frac{x^2-x-6}{x-4}$$
$$=\frac{(2x+3)(x-4)}{x-3}\cdot\frac{(x-3)(x+2)}{x-4}$$
$$=(2x+3)(x+2)$$

23. $$\frac{3b^2+9b-84}{b^2-5b+4}\div\frac{5b^2+37b+14}{-10b^2+6b+4}$$
$$=\frac{3\left(b^2+3b-28\right)}{(b-4)(b-1)}\cdot\frac{-2\left(5b^2-3b-2\right)}{(5b+2)(b+7)}$$
$$=\frac{3(b+7)(b-4)}{(b-4)(b-1)}\cdot\frac{-2(5b+2)(b-1)}{(5b+2)(b+7)}$$
$$=-6$$

25. $$\frac{\frac{3}{x}+\frac{x}{3}}{2-\frac{1}{x}}=\frac{\frac{3}{x}+\frac{x}{3}}{2-\frac{1}{x}}\cdot\frac{3x}{3x}$$
$$=\frac{x^2+9}{6x-3}$$

27. $$\frac{6x-6}{3-\frac{3}{x^2}}=\frac{6x-6}{3-\frac{3}{x^2}}\cdot\frac{x^2}{x^2}$$
$$=\frac{6x^2(x-1)}{3(x^2-1)}$$
$$=\frac{6x^2\cancel{(x-1)}}{3\cancel{(x-1)}(x+1)}$$
$$=\frac{2x^2}{x+1}$$

29. $$\frac{\frac{1}{r}-\frac{1}{s}}{r+\frac{1}{r}}=\frac{\frac{1}{r}-\frac{1}{s}}{r+\frac{1}{r}}\cdot\frac{rs}{rs}$$
$$=\frac{s-r}{r^2s+s}$$

31. $$\frac{\frac{m}{n}-\frac{n}{m}}{m-n}=\frac{\frac{m}{n}-\frac{n}{m}}{m-n}\cdot\frac{nm}{nm}$$
$$=\frac{m^2-n^2}{nm(m-n)}$$
$$=\frac{m+n}{mn}$$

33. $$\frac{3}{x-2}+\frac{2}{x+1}=1$$
$$(x-2)(x+1)\left(\frac{3}{x-2}+\frac{2}{x+1}\right)=(x-2)(x+1)$$
$$3(x+1)+2(x-2)=x^2-x-2$$
$$3x+3+2x-4=x^2-x-2$$
$$x^2-6x-1=0$$
$$x=\frac{-(-6)\pm\sqrt{(-6)^2-4(1)(-1)}}{2(1)}$$
$$x=\frac{6\pm\sqrt{40}}{2}$$
$$x=3\pm\sqrt{10}$$

35. $$\frac{y}{y-1}+\frac{2}{y-3}=\frac{y^2}{y^2-4y+3}$$
Multiply both sides by the LCM $(y-1)(y-3)$.
$$y(y-3)+2(y-1)=y^2$$
$$y^2-3y+2y-2=y^2$$
$$y=-2$$

37. $$\frac{2}{2b+1}+\frac{2b^2-b+4}{2b^2-7b-4}=\frac{b}{b-4}$$
$$\frac{2}{2b+1}+\frac{2b^2-b+4}{(2b+1)(b-4)}=\frac{b}{b-4}$$
Multiply both side by the LCM $(2b+1)(b-4)$.
$$2(b-4)+2b^2-b+4=b(2b+1)$$
$$2b-8+2b^2-b+4=2b^2+b$$
$$-4\neq 0$$
[Contradiction]
Solution is the empty set $\varnothing$.

39. $$\frac{1}{x-3}+\frac{1}{x+3}=\frac{2x}{x^2-9}$$
Multiply both side by the LCM $(x-3)(x+3)$.
$$x+3+x-3=2x$$
$$0=0$$
Solution set is the set of all real numbers except -3 and 3:
$(-\infty,-3)\cup(-3,3)\cup(3,\infty)$

41. $\frac{2}{n+3}+\frac{3}{n+2}=\frac{6}{n}$

Multiply both sides by the LCM

$(n+3)(n+2)n$

$$\frac{2}{n+3}+\frac{3}{n+2}=\frac{6}{n}$$

$2n(n+2)+3n(n+3)=6(n+2)(n+3)$

$2n^2+4n+3n^2+9n=6n^2+30n+36$

$n^2+17n+36=0$

$x=\frac{-17\pm\sqrt{17^2-4(1)(36)}}{2(1)}$

$x=\frac{-17\pm\sqrt{145}}{2}$

Solution is $\left\{\frac{-17\pm\sqrt{145}}{2}\right\}$

43. Joanne's rate is $\frac{1}{5}$ and

Lisa's rate is $\frac{1}{7}$. Let $x=$ time (hrs)

it takes working together. Then,

$\frac{1}{5}+\frac{1}{7}=\frac{1}{x}$

$7x+5x=35$

$12x=35$

$x=\frac{35}{12}$, or $2\frac{11}{12}$ hours

2 hours 55 minutes

45. Let $x=$ the father's time to plow

Then, $x+2=$ the son's time to plow

$\frac{1}{x}+\frac{1}{x+2}=\frac{1}{5}$

$5(x+2)+5x=x^2+2x$

$5x+10+5x=x^2+2x$

$x^2-8x-10=0$

$x=\frac{8\pm\sqrt{64-4(1)(-10)}}{2}$

$x=4\pm\frac{\sqrt{104}}{2}\approx 9.1,\ -1.1$

Disregard the negative value.

It would take the father approximately 9.1 hours alone.

47. Let $x=$ number of weeks it takes for the lake to be emptied with the "feeder" river replenishing at 1/30.

$\frac{1}{12}-\frac{1}{30}=\frac{1}{x}$

$30x-12x=12(30)$

$18x=360$

$x=20$ weeks

49. Let $x=$ number of hours for Janice's bucket to be filled with Jimmy taking some out.

$\frac{1}{0.5}-\frac{1}{1.5}=\frac{1}{x}$

$1.5x-0.5x=0.5(1.5)$

$x=0.75$ or 45 minutes

Use a computer algebra system for items 51 - 55.

51. $\frac{x+10}{(x-5)(x+5)}$

53. $\frac{1+x}{5x-1}$

55. $\frac{x^2+xy+y^2}{x^2y^2}$

57. $\sqrt{4-x}-x=2$

$\sqrt{4-x}=x+2$

$4-x=x^2+4x+4$

$x^2+5x=0$

$x(x+5)=0$

$x=0$ or ~~$x=-5$~~

-5 is extraneous

Solution is $\{0\}$.

59.
$$\sqrt{x+10}+1=x-1$$
$$\sqrt{x+10}=x-2$$
$$x+10=x^2-4x+4$$
$$x^2-5x-6=0$$
$$(x-6)(x+1)=0$$
$x=6$ or $\cancel{x=-1}$ (extraneous)

Solution is $\{6\}$.

61.
$$\sqrt{x^2-4x+4}+2=3x$$
$$\left(\sqrt{x^2-4x+4}\right)^2=(3x-2)^2$$
$$x^2-4x+4=9x^2-12x+4$$
$$8x^2-8x=0$$
$$8x(x-1)=0$$
$x-1=0$ or $\cancel{x=0}$ (extraneous)

Solution is $\{1\}$.

63.
$$\sqrt[3]{3-2x}-\sqrt[3]{x+1}=0$$
$$\sqrt[3]{3-2x}=\sqrt[3]{x+1}$$
$$3-2x=x+1$$
$$3x=2$$
$x=\frac{2}{3}$ (checks)

Solution is $\left\{\frac{2}{3}\right\}$

65.
$$\sqrt[4]{2x+3}=-1$$
$$2x+3=1$$
$$2x=-2$$
$\cancel{x=-1}$ (extraneous)

Solution set is $\varnothing$.

67.
$$\sqrt{2b-1}+3=\sqrt{10b-6}$$
$$2b-1+6\sqrt{2b-1}+9=10b-6$$
$$6\sqrt{2b-1}=8b-14$$
$$36(2b-1)=64b^2-224b+196$$
$$72b-36=64b^2-224b+196$$
$$64b^2-296b+232=0$$
$$8b^2-37b+29=0$$
$$(8b-29)(b-1)=0$$
$b=\frac{29}{8}$ or $\cancel{b=1}$ (extraneous)

Solution is $\left\{\frac{29}{8}\right\}$.

69.
$$\sqrt{3-3x}-3=\sqrt{3x+2}$$
$$\left(\sqrt{3-3x}-3\right)^2=\left(\sqrt{3x+2}\right)^2$$
$$3-3x-6\sqrt{3-3x}+9=3x+2$$
$$-6\sqrt{3-3x}=6x-10$$
$$\left(-6\sqrt{3-3x}\right)^2=(6x-10)^2$$
$$36(3-3x)=36x^2-120x+100$$
$$108-108x=36x^2-120x+100$$
$$36x^2-12x-8=0$$
$$4(9x^2-3x-2)=0$$
$$9x^2-3x-2=0$$
$$x=\frac{3\pm\sqrt{9-4(9)(-2)}}{2(9)}$$
$$x=\frac{3\pm 9}{18}$$
$\cancel{x=\frac{2}{3}}$ or $\cancel{x=-\frac{1}{3}}$ (both extraneous)

Solution set is $\varnothing$.

71.
$$\sqrt{x^2-10}-1=x+1$$
$$\sqrt{x^2-10}=x+2$$
$$x^2-10=x^2+4x+4$$
$$4x=-14$$
$\cancel{x=-\frac{7}{2}}$ (extraneous)

Solution set is $\varnothing$.

73.
$$\sqrt[5]{7t^2+2t}=\sqrt[5]{5t^2+4}$$
$$7t^2+2t=5t^2+4$$
$$2t^2+2t-4=0$$
$$2(t^2+t-2)=0$$
$$(t+2)(t-1)=0$$
$$t=-2 \text{ or } t=1$$
Solution is $\{-2,1\}$

75.
$$(x+3)^{1/4}+2=0$$
$$x+3=(-2)^4$$
$$x+3=16$$
~~$x=13$~~ (extraneous)
Solution set is $\varnothing$.

77.
$$(2x-1)^{2/3}=x^{1/3}$$
$$(2x-1)^2=x$$
$$4x^2-4x+1=x$$
$$4x^2-5x+1=0$$
$$(4x-1)(x-1)=0$$
$$x=\frac{1}{4} \text{ or } x=1$$
Solution is $\left\{\frac{1}{4},1\right\}$

79.
$$(3x-5)^{1/5}=(x+1)^{1/5}$$
$$3x-5=x+1$$
$$2x=6$$
$$x=3$$
Solution is $\{3\}$.

81.
$$(x^2+21)^{-3/2}=\frac{1}{125}$$
$$\frac{1}{(x^2+21)^{3/2}}=\frac{1}{125}$$
$$(x^2+21)^{3/2}=125$$
$$(x^2+21)^{1/2}=5$$
$$x^2+21=25$$
$$x^2=4$$
$$x=\pm 2$$
Solution is $\{-2,2\}$.

83.
$$z^{4/3}-\frac{16}{81}=0$$
$$z^{4/3}=\frac{16}{81}$$
Take the fourth root of both sides:
$$z^{1/3}=\pm\frac{2}{3}$$
Cube both sides:
$$z=\pm\frac{8}{27}$$
Solution is $\left\{-\frac{8}{27},\frac{8}{27}\right\}$.

85.
$$c=\sqrt{a^2+b^2}, \text{ for } a$$
$$c^2=a^2+b^2$$
$$c^2-b^2=a^2$$
$$a=\sqrt{c^2-b^2} \quad (a>0)$$

87.
$$T^2=\frac{4\pi^2r^3}{GM}, \text{ for } r$$
$$\frac{GMT^2}{4\pi^2}=r^3$$
$$r=\sqrt[3]{\frac{GMT^2}{4\pi^2}}, \text{ or } r=\left(\frac{GMT^2}{4\pi^2}\right)^{1/3}$$

89.
$$Z=\sqrt{R^2+\left(\omega L-\frac{1}{\omega C}\right)^2}, \text{ for } L$$
$$Z^2=R^2+\left(\omega L-\frac{1}{\omega C}\right)^2$$
$$Z^2-R^2=\left(\omega L-\frac{1}{\omega C}\right)^2$$
$$\sqrt{Z^2-R^2}=\omega L-\frac{1}{\omega C}$$
$$\sqrt{Z^2-R^2}+\frac{1}{\omega C}=\omega L$$
$$L=\frac{\sqrt{Z^2-R^2}}{\omega}+\frac{1}{\omega^2 C}$$

End of Section 1.8

Chapter One Test
Solutions to Odd-Numbered Exercises

1. (a) natural numbers

$\{7, 3^3\}$

(b) whole numbers

$\left\{7, \frac{0}{5}, 3^3\right\}$

(c) integers

$\left\{7, \frac{0}{5}, 3^3, -1, -\sqrt{4}\right\}$

(d) rational numbers

$\left\{-15.75, 7, \frac{0}{5}, 3^3, -1, -\sqrt{4}\right\}$

(e) irrational numbers

None

(f) real numbers

$\left\{-15.75, 7, \frac{0}{5}, 3^3, -1, -\sqrt{4}\right\}$

(g) undefined

$\left\{\frac{-8}{0}\right\}$

3. Plot $\{-3, 0, 3, 5\}$

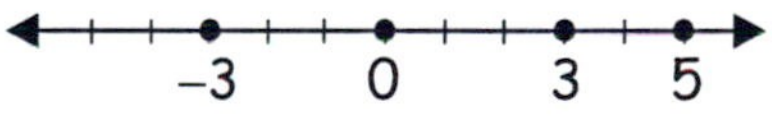

5. $6.1 \underline{\le ,} < 8.3$

7. $\{-2, -1, 0, 1, 2, 3\} = \{x \mid -2 \le x \le 3, x \in \mathbb{Z}\}$

9. $[4, 8] \cup (8, 11] = [4, 11]$

11. $(-2, 3] \cap [0, 3) = [0, 3)$

13. $\{x \mid -7 < x \le 9\} = (-7, 9]$

15. Let $x = 7.\overline{6}$, then $10x = 76.\overline{6} = 69 + 7.\overline{6}$. So, $10x = 69 + x$. Then, $9x = 69$, or $x = \frac{23}{3}$.

17. $-|11-2| = -|9| = -9$

19. $\left|\sqrt{5} - \sqrt{11}\right| = \sqrt{11} - \sqrt{5}$

21. $7y^2 - \frac{1}{3}\pi xy + 8x^3$, for $x = -2$ and $y = 2$:

$$7(2)^2 - \frac{1}{3}\pi(-2)(2) + 8(-2)^3 =$$
$$28 + \frac{4\pi}{3} - 64 = \frac{4\pi}{3} - 36$$

23. Zero-factor property.

25.
$$\left[\left(3y^{-2}z\right)^{-1}\right]^{-3} = \left(3^{-1}y^2z^{-1}\right)^{-3}$$
$$= 3^3y^{-6}z^3$$
$$= \frac{27z^3}{y^6}$$

27. $52{,}240{,}000 = 5.224 \times 10^7$

29.
$$\frac{\sqrt{3a^3}}{\sqrt{12a}} = \sqrt{\frac{3a^3}{12a}}$$
$$= \sqrt{\frac{a^2}{4}}$$
$$= \frac{|a|}{2}$$

31. $\sqrt{16x^2} = 4|x|$

33.
$$\sqrt[4]{\frac{a^9b^{-4}}{81}} = \sqrt[4]{\frac{a \cdot a^8}{81b^4}}$$
$$= \frac{a^2\sqrt[4]{a}}{3|b|}$$

35. $5xyz + 7x^2 + y$ is a third-degree trinomial.

37.
$$\left(5x^2y + 7xy - z\right) - \left(2x^2y + z - 4xz\right) =$$
$$3x^2y + 7xy - 2z + 4xz$$

39. $(5x^2y+2xy-3)(4x+2y)=$
$20x^3y+10x^2y^2+8x^2y+4xy^2-12x-6y$

41. $nx+3mx-2ny-6my=$
$(nx+3mx)-(2ny+6my)=$
$x(n+3m)-2y(n+3m)=$
$(x-2y)(3m+n)$

43. Factor:
$10a^3b-15a^3b^3+5a^2b^4=$
$5a^2b(2a-3ab^2+b^3)$

45. $x^2-x-12=(x-4)(x+3)$

47. $(8i+3i^3)-(4i^4-i^6)=(8i-3i)-(4-(-1))$
$=5i-5$

49. $\frac{3+4i}{3-4i}=\frac{3+4i}{3-4i}\cdot\frac{3+4i}{3+4i}$
$=\frac{9+24i+16i^2}{9-16i^2}$
$=\frac{9+24i-16}{9+16}$
$=\frac{24i-7}{25}$, or $-\frac{7}{25}+\frac{24}{25}i$

51. $\frac{2i}{3-5i}=\frac{2i}{3-5i}\cdot\frac{3+5i}{3+5i}$
$=\frac{6i+10i^2}{9-25i^2}$
$=\frac{6i-10}{34}$
$=\frac{3i-5}{17}$, or $-\frac{5}{17}+\frac{3}{17}i$

53. $\left(2-\sqrt{-4}\right)^2=(2-2i)^2$
$=4-8i+4i^2$
$=4-8i-4$
$=-8i$

55. Solve: $5(3y-2)=(4y+4)+2y$
$15y-10=6y+4$
$9y=14$
$y=\frac{14}{9}$ Conditional

57. Solve: $\frac{x-7}{2}=\frac{5-x}{4}$
$2x-14=5-x$
$3x=19$
$x=\frac{19}{3}$ Conditional

59. $|2x-7|=1$
$2x-7=1$ or $-(2x-7)=1$
$2x=8$ or $-2x=-6$
$x=4$ or $x=3$
Solution is $\{3, 4\}$

61. Solve: $h=-16t^2+v_0t$ for v_0
$v_0t=h+16t^2$
$v_0=\frac{h+16t^2}{t}$

63. Use $d=rt$ and $20=\frac{1}{3}$ hr.
Add the two distances:
$90\left(1\frac{1}{3}\right)+95\left(1\frac{1}{3}\right)=90\left(\frac{4}{3}\right)+95\left(\frac{4}{3}\right)$
≈ 246.7 mi.

65. $-8<3x-5\le 16$
$-3<3x\le 21$
$-1<x\le 7$
$(-1, 7]$

67. $-5|3+t|>-10$
$|3+t|<2$
$-2<3+t<2$
$-5<t<-1$
$(-5, -1)$

69. Let $x =$ maximum monthly car payment

$x < 1800 - (550 + 80 + 420 + 250 + 80)$

$x < 1800 - 1380$

$x < \$420$

71. Solve: $(x-3)^2 = 12$

$x - 3 = \pm 2\sqrt{3}$

$x = 3 \pm 2\sqrt{3}$

Solution: $\{3 \pm 2\sqrt{3}\}$

73. $2x^2 + 7x = x^2 + 2x - 6$

$x^2 + 5x + 6 = 0$

$(x+3)(x+2) = 0$

$x = -3$ or $x = -2$

Solution is $\{-3, -2\}$

75. $(x^2+2)^2 - 7(x^2+2) + 12 = 0$

$A^2 - 7A + 12 = 0$

$(A-3)(A-4) = 0$

$x^2 + 2 - 4 = 0$ or $x^2 + 2 - 3 = 0$

$x^2 = 2$ or $x^2 = 1$

$x = \pm\sqrt{2}$ or $x = \pm 1$

Solution is $\{-\sqrt{2}, -1, 1, \sqrt{2}\}$

77. $x^3 - 4x^2 - 2x + 8 = 0$

$(x^3 - 4x^2) - (2x - 8) = 0$

$x^2(x-4) - 2(x-4) = 0$

$(x^2 - 2)(x-4) = 0$

$x^2 - 2 = 0$ or $x = 4$

$x = \pm\sqrt{2}$

Solution is $\{-\sqrt{2}, \sqrt{2}, 4\}$

79. Solve: $4x^{\frac{18}{7}} - 2x^{\frac{11}{7}} - 3x^{\frac{4}{7}} = 0$

Factor out $x^{\frac{4}{7}}$:

$x^{\frac{4}{7}}(4x^2 - 2x - 3) = 0 \Rightarrow$

$x^{\frac{4}{7}} = 0$ or $4x^2 - 2x - 3 = 0$

$x = 0$ or

$$x = \frac{2 \pm \sqrt{4 - 4(4)(-3)}}{2(4)}$$

$$x = \frac{2 \pm 2\sqrt{13}}{8}$$

$$x = \frac{1 \pm \sqrt{13}}{4}$$

Solution Set: $\left\{0, \frac{1 \pm \sqrt{13}}{4}\right\}$

81. $$\frac{2z^2 - 8z - 42}{z+3} = \frac{2(z^2 - 4z - 21)}{z+3} = \frac{2(z+3)(z-7)}{z+3} = 2z - 14 \quad [z \neq 3]$$

83. $$\frac{3a^3 + 5}{5a+1} + \frac{a-5}{5a+1} = \frac{3a^3 + a}{5a+1}$$

85. $$\frac{y^2 + 8y - 1}{7y+1} \cdot \frac{y-2}{y+2} = \frac{y^3 + 6y^2 - 17y + 2}{7y^2 + 15y + 2}$$

87. $$\frac{\frac{x}{3} - \frac{3}{x}}{-\frac{3}{x} + 1} = \frac{\frac{x}{3} - \frac{3}{x}}{-\frac{3}{x} + 1} \cdot \frac{3x}{3x}$$

$$= \frac{x^2 - 9}{3x - 9}$$

$$= \frac{(x-3)(x+3)}{3(x-3)}$$

$$= \frac{x+3}{3}$$

89. $\frac{y}{y-1}+\frac{1}{y-4}=\frac{y^2}{y^2-5y+4}$

Multiply both sides by LCM $(y-1)(y-4)$.

$$y(y-4)+y-1=y^2$$
$$y^2-4y+y-1=y^2$$
$$-3y=1$$
$$y=-\frac{1}{3}$$

Solution is $\left\{-\frac{1}{3}\right\}$.

91. Let $x=$ hours needed for Mandy and Heather working together. Then,

$\frac{1}{3}+\frac{1}{4}=\frac{1}{x}$ (Multiply by 12 to simplify)

$$4+3=\frac{12}{x}$$
$$7x=12$$

$x=\frac{12}{7}$ hours working together

93. Solve: $\sqrt{5x-1}+\sqrt{x-1}=4$

Rewrite:

$$\sqrt{5x-1}=4-\sqrt{x-1}$$

Square both sides.

$$5x-1=16-8\sqrt{x-1}+x-1$$
$$4x-16=-8\sqrt{x-1}$$
$$x-4=-2\sqrt{x-1}$$

Square both sides (again):

$$x^2-8x+16=4(x-1)$$
$$x^2-12x+20=0$$
$$(x-2)(x-10)=0\Rightarrow$$
$$x-2=0 \text{ or } x-10=0$$

Check these "solutions" in the original equation-the squaring process may have introduced an extraneous solution:
To demonstrate that 10 is not a solution to the original equation, substitute $x=10$ to check:

$$\sqrt{5(10)-1}+\sqrt{10-1}=\sqrt{49}+\sqrt{9}$$
$$=7+3=10(\neq 4)$$

Solution Set: $\{2\}$

95. $(2x-5)^{\frac{1}{6}}=(x-2)^{\frac{1}{6}}$

$$2x-5=x-2$$
$$x=3$$

Solution is $\{3\}$.

97. Solve: [First, raise to the 3rd power]

$$\left(2x^2-18x+67\right)^{\frac{1}{3}}=3$$
$$2x^2-18x+67=27$$
$$2x^2-18x+40=0$$
$$x^2-9x+20=0$$
$$(x-4)(x-5)=0\Rightarrow x=4 \text{ or } x=5$$

Solution Set: $\{4, 5\}$

End of Chapter 1 Test

Chapter Two: Section 2.1
Solutions to Odd-numbered Exercises

1. Plot $\{(-3,2),(5,-1),(0,-2),(3,0)\}$.

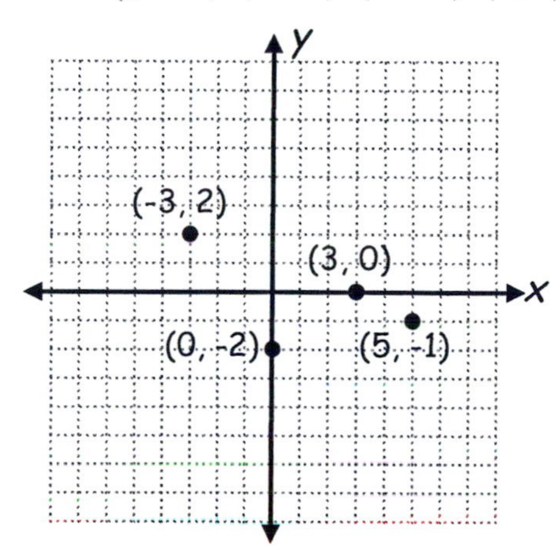

3. Plot $\{(3,4),(-2,-1),(-1,-3),(-3,0)\}$.

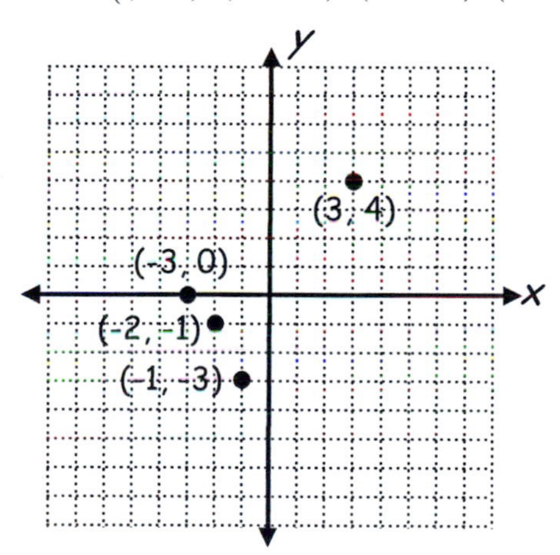

5. Quadrant III

7. Quadrant IV

9. Positive x-axis

11. Quadrant III

13. Quadrant IV

15. Quadrant II

17. Quadrant IV

19. Quadrant I

21. Negative y-axis

23. Substitute the known value into $6x+4y=12$ and solve for the unknown value for each row of the table.

x	y
0	-3
2	0
3	$\frac{3}{2}$
4	3

Plot the results :

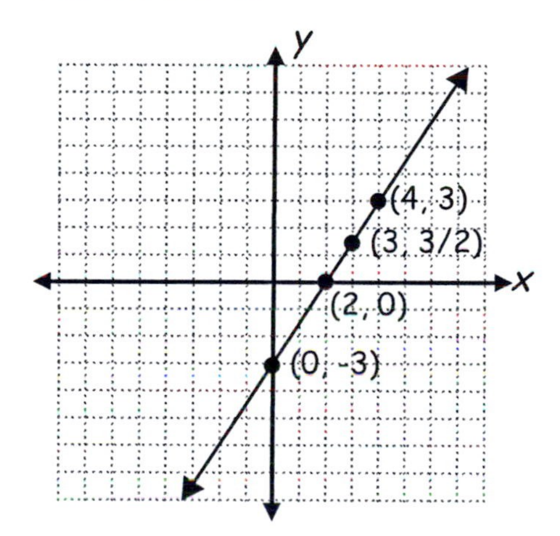

25. Substitute the known value into $x=y^2$ and solve for the unknown value for each row of the table.

x	y
0	0
1	±1
4	±2
9	±3
2	$-\sqrt{2}$

Plot the results :

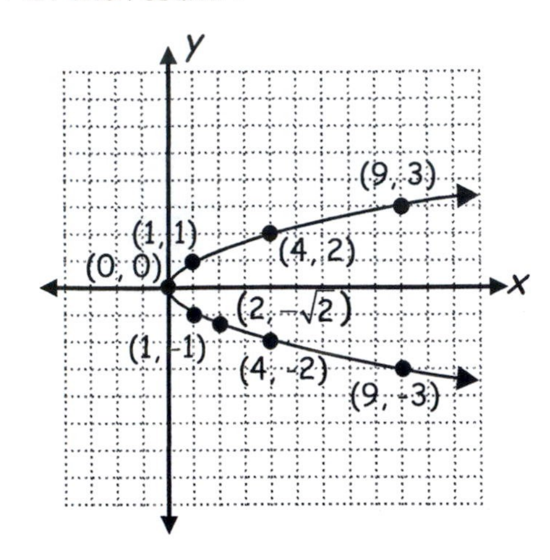

27. Substitute the known value into $x^2 + y^2 = 9$ and solve for the unknown value for each row of the table.

x	y
0	± 3
± 3	0
-1	$\pm 2\sqrt{2}$
1	$\pm 2\sqrt{2}$
$\pm\sqrt{5}$	2

Plot the results :

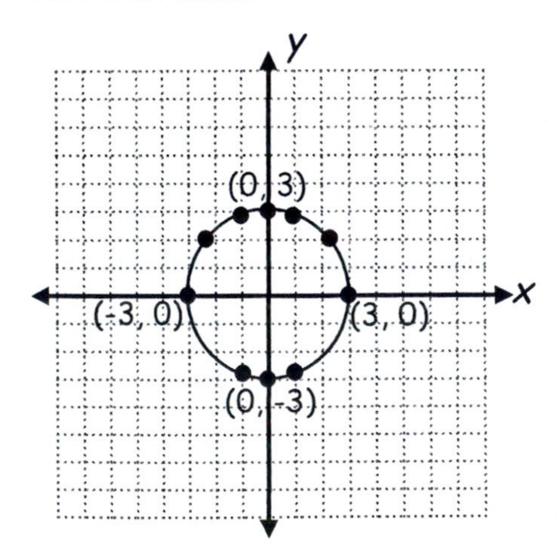

29. **(a)** $d = \sqrt{(-2-(-5))^2 + (3-(-2))^2}$

$d = \sqrt{9+25}$

$d = \sqrt{34}$

(b) Midpoint:

$$\left(\frac{-2+(-5)}{2}, \frac{3+(-2)}{2}\right)$$

$$\left(-\frac{7}{2}, \frac{1}{2}\right)$$

31. **(a)** $d = \sqrt{(0-3)^2 + (7-0)^2}$

$d = \sqrt{9+49}$

$d = \sqrt{58}$

(b) Midpoint:

$$\left(\frac{0+3}{2}, \frac{7+0}{2}\right)$$

$$\left(\frac{3}{2}, \frac{7}{2}\right)$$

33. **(a)** $d = \sqrt{(-2-0)^2 + (0-(-2))^2}$

$d = \sqrt{4+4}$

$d = 2\sqrt{2}$

(b) Midpoint:

$$\left(\frac{-2+0}{2}, \frac{0+(-2)}{2}\right) =$$

$$(-1, -1)$$

35. **(a)** $d = \sqrt{(13-(-7))^2 + (-14-(-2))^2}$

$d = \sqrt{20^2 + (-12)^2}$

$d = \sqrt{400+144}$

$d = 4\sqrt{34}$

(b) Midpoint:

$$\left(\frac{13+(-7)}{2}, \frac{-14+(-2)}{2}\right) =$$

$$(3, -8)$$

37. **(a)** $d = \sqrt{(-3-5)^2 + (-3-(-9))^2}$

$d = \sqrt{64+36}$

$d = 10$

(b) Midpoint:

$\left(\frac{-3+5}{2}, \frac{-3+(-9)}{2}\right) =$

$(1, -6)$

39. Find all three distances between the three points and find the sum:

$d_1 = \sqrt{(-1-2)^2 + (-2-(-2))^2}$
$= 3$

$d_2 = \sqrt{(-1-2)^2 + (-2-2)^2}$
$= 5$

$d_3 = \sqrt{(2-2)^2 + (-2-2)^2}$
$= 4$

$d_1 + d_2 + d_3 = 12$

41. Find all three distances between the three points and find the sum:

$d_1 = \sqrt{(3-(-7))^2 + (-4-0)^2}$
$= \sqrt{116}$
$= 2\sqrt{29}$

$d_2 = \sqrt{(3-(-2))^2 + (-4-(-5))^2}$
$= \sqrt{26}$

$d_3 = \sqrt{(-7-(-2))^2 + (0-(-5))^2}$
$= 5\sqrt{2}$

$d_1 + d_2 + d_3 = 2\sqrt{29} + \sqrt{26} + 5\sqrt{2}$

43. Find all three distances between the three points and find the sum:

$d_1 = \sqrt{(-12-(-7))^2 + (-3-9)^2}$
$= 13$

$d_2 = \sqrt{(-12-9)^2 + (-3-(-3))^2}$
$= 21$

$d_3 = \sqrt{(-7-9)^2 + (9-(-3))^2}$
$= 20$

$d_1 + d_2 + d_3 = 54$

45. Find the three distances between the points to determine which sides are congruent (equal in length). Then, find the midpoint of the third side. Using the vertex opposite the base and the midpoint you find on the base, determine the length of the height. Use the formula $A = \frac{1}{2}bh$:

$d_1 = \sqrt{(2-(-2))^2 + (5-2)^2}$
$= 5$

$d_2 = \sqrt{(2-1)^2 + (5-(-2))^2}$
$= 5\sqrt{2}$

$d_3 = \sqrt{(-2-1)^2 + (2-(-2))^2}$
$= 5$

Midpoint:

$\left(\frac{2+1}{2}, \frac{5+(-2)}{2}\right) = \left(\frac{3}{2}, \frac{3}{2}\right)$

Height:

$h = \sqrt{\left(-2-\frac{3}{2}\right)^2 + \left(2-\frac{3}{2}\right)^2} = \frac{5}{2}\sqrt{2}$

$\text{Area} = \frac{1}{2}(5\sqrt{2})\left(\frac{5}{2}\sqrt{2}\right) = \frac{25}{2}$ sq. units

47. Given that the triangle is a right triangle, we apply the Pythagorean Theorem to find the distance D from the beach to home:

$D = \sqrt{87^2 + 116^2}$

$D = 145$ miles

End of Section 2.1

Chapter Two: Section 2.2
Solutions to Odd-numbered Exercises

For items 1 - 23, simplify and determine whether the equation can be put in the $ax+by=c$ form.

1. $3x+2(x-4y)=2x-y$

$3x+(-7)y=-0$

Yes. Linear

3. $9x^2-(x+1)^2=y-3$

$9x^2-x^2-2x-1=y-3$

$8x^2-2x-y=-2$

Not linear

5. $8-4xy=x-2y$

$x+4xy-2y=8$

Not linear

7. $\frac{6}{x}-\frac{5}{y}=2$

$6y-5x=2xy$

Not linear

9. $2y-(x+y)=y+1$

$(-1)x+0\cdot y=1$

Yes. Linear

11. $x^2-(x-1)^2=y$

$2x+(-1)y=1$

Yes. Linear

13. $x(y+1)=16-y(1-x)$

$xy+x=16-y+xy$

$x+y=16$

Yes. Linear

15. $x-2x^2+3=\frac{x-7}{2}$

$x-4x^2=-13$

Not linear

17. $13x-17y=y(7-2x)$

$13x-24y=-2xy$

Not linear

19. $x-1=\frac{2y}{x}-x$

$x^2-x=2y-x^2$

$2x^2-x-2y=0$

Not linear

21. $x-x(1+x)=y-3x$

$x-x-x^2=y-3x$

$-x^2-y+3x=0$

Not linear

23. $\frac{2y-5}{14}=\frac{x-3}{9}$

$18y-45=14x-42$

$-14x+18y=3$

Yes. Linear

For items 25 - 39, determine the x and y intercepts, if possible, and graph the equation. To determine the x-intercept substitute 0 for y and solve for x. To determine the y-intercept, substitute 0 for x and solve for y.

25. $4x-3y=12$

x-intercept: $(3, 0)$

y-intercept: $(0, -4)$

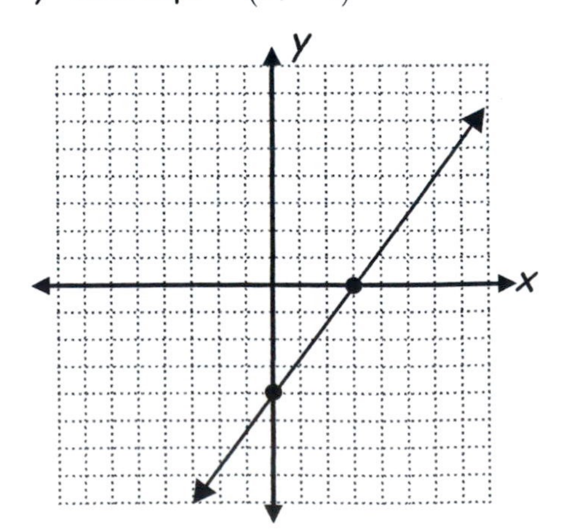

27. $5 - y = 10x$

x-intercept: $\left(\frac{1}{2}, 0\right)$

y-intercept: $(0, 5)$

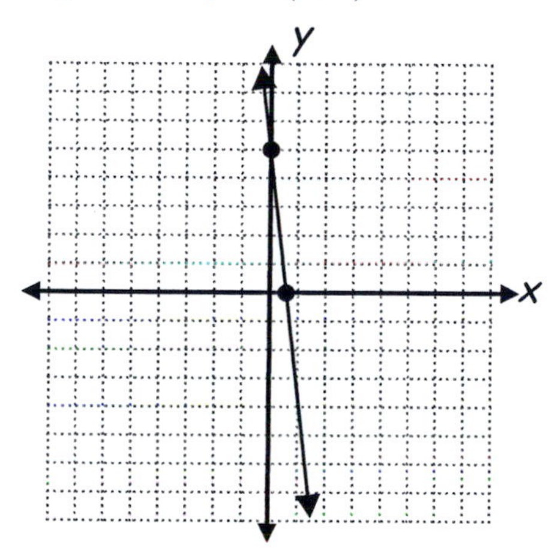

29. $3y = 9$

No x-intercept

y-intercept: $y = 3$

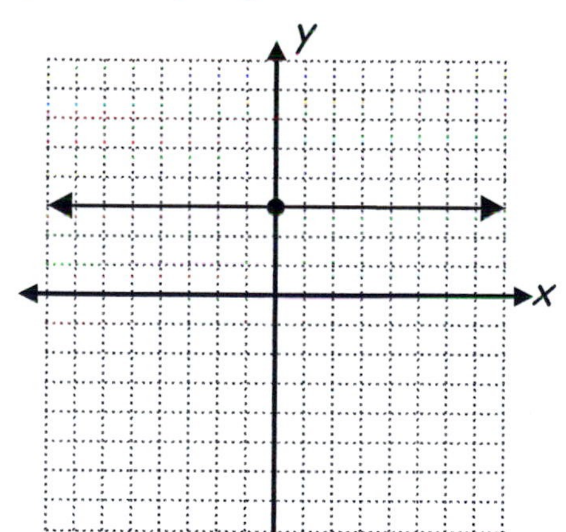

31. $x + 2y = 7$

x-intercept: $(7, 0)$

y-intercept: $\left(0, \frac{7}{2}\right)$

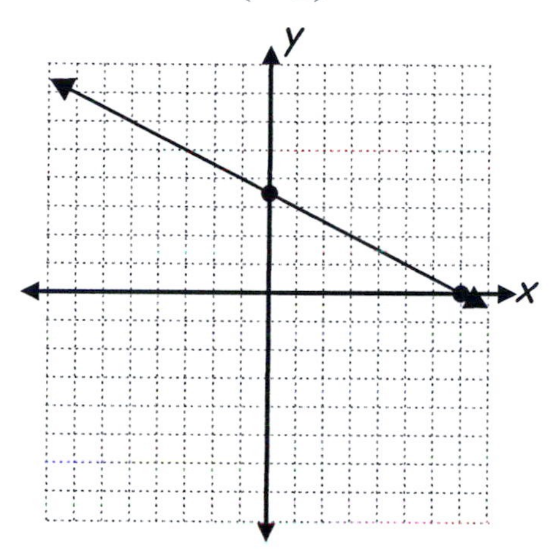

33. $y = -x$

x-intercept: $(0,0)$

y-intercept: $(0,0)$

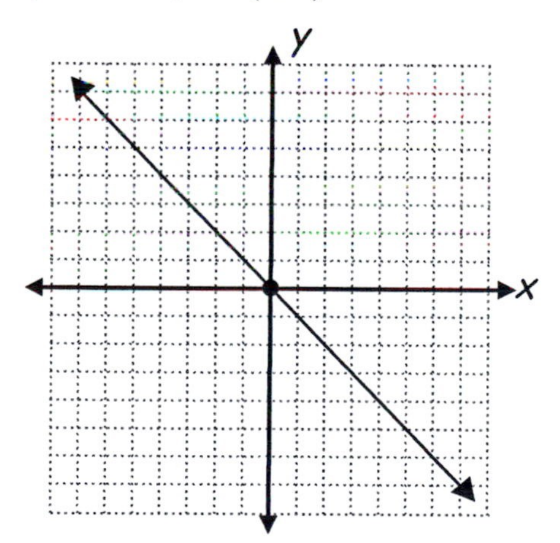

35. $3y+7x=7(3+x)$
$3y=21$
$y=7$
No x-intercept
y-intercept: $y=7$

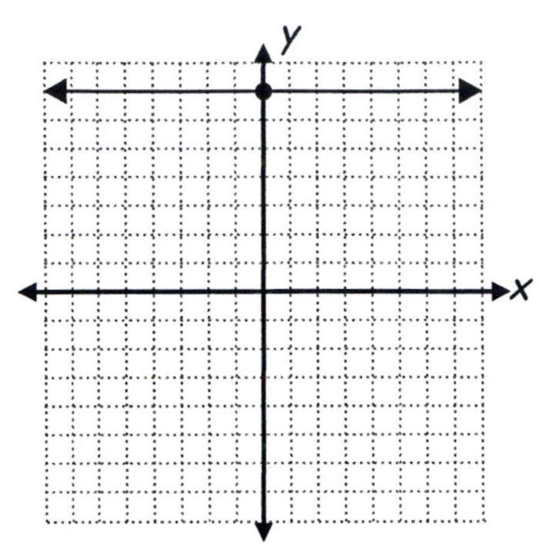

37. $x+y=1+2y$
$x-y=1$
x-intercept: $(1,0)$
y-intercept: $(0,-1)$

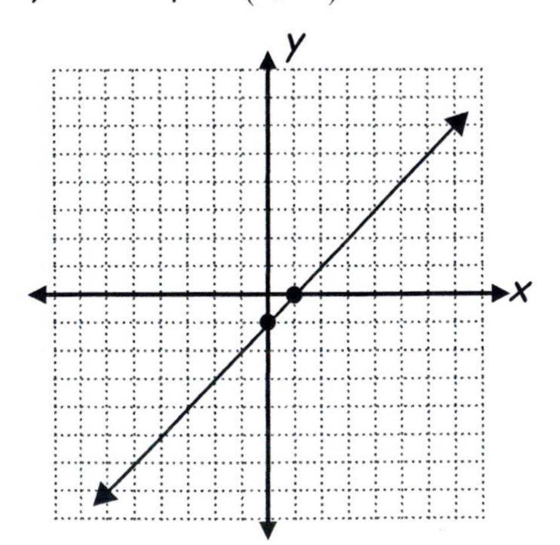

39. $3(x+y)+1=x-5$
$3x+3y=x-6$
$2x+3y=-6$
x-intercept: $(-3,0)$
y-intercept: $(0,-2)$

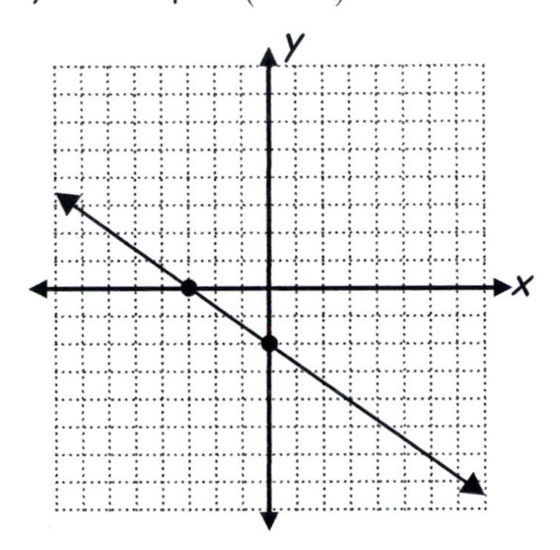

End of Section 2.2

Chapter Two: Section 2.3
Solutions to Odd-numbered Exercises

For items 1 - 11, determine the slope of the line passing through the given points. Using the equation for the slope of a line, $m = \frac{y_2 - y_1}{x_2 - x_1}$.

1. For $(0, -3)$ and $(-2, 5)$

$$m = \frac{5-(-3)}{-2-0} = -4$$

3. For $(4, 5)$ and $(-1, 5)$

$$m = \frac{5-5}{4-(-1)} = 0$$

5. For $(3, -5)$ and $(3, 2)$

$$m = \frac{-5-2}{3-3} = \frac{-7}{0} \text{ Undefined}$$

7. For $(-2, 1)$ and $(-5, -1)$

$$m = \frac{1-(-1)}{-2-(-5)} = \frac{2}{3}$$

9. For $(0, -21)$ and $(-3, 0)$

$$m = \frac{0-(-21)}{-3-0} = -7$$

11. For $(29, -17)$ and $(31, -29)$

$$m = \frac{-17-(-29)}{29-31} = -6$$

For items 13 - 23, determine the slope of the given equation. First, solve for y, if possible. The slope is the resulting coefficient of the x-term.

13. $8x - 2y = 11$

$$y = 4x - \frac{11}{2}$$

$$m = 4$$

15. $4y = 13$

$$y = 0 \cdot x + \frac{13}{4}$$

$$m = 0$$

17. $7x = 2$

The slope is undefined.

19. $3(2y-1) = 5(2-x)$

$$y = -\frac{5}{6}x + \frac{13}{6}$$

$$m = -\frac{5}{6}$$

21. $\frac{x+2}{3} + 2(1-y) = -2x$

$$y = \frac{7}{6}x + \frac{4}{3}$$

$$m = \frac{7}{6}$$

23. $x - 7 = \frac{2y-1}{-5}$

$$y = -\frac{5}{2}x + 18$$

$$m = -\frac{5}{2}$$

25. $6x - 2y = 4$

$$y = 3x - 2$$

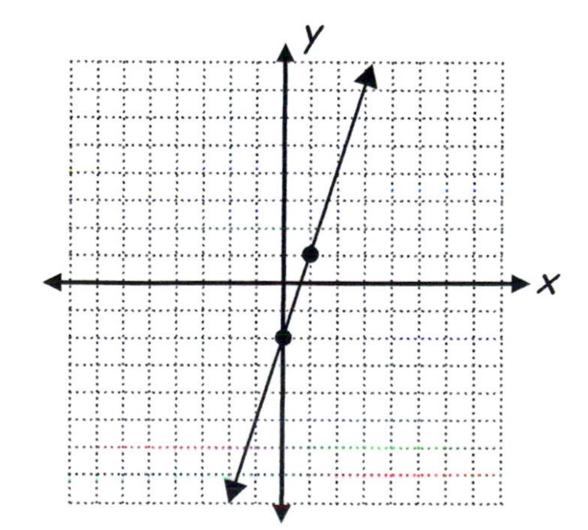

27. $5y-15=0$

$y=3$

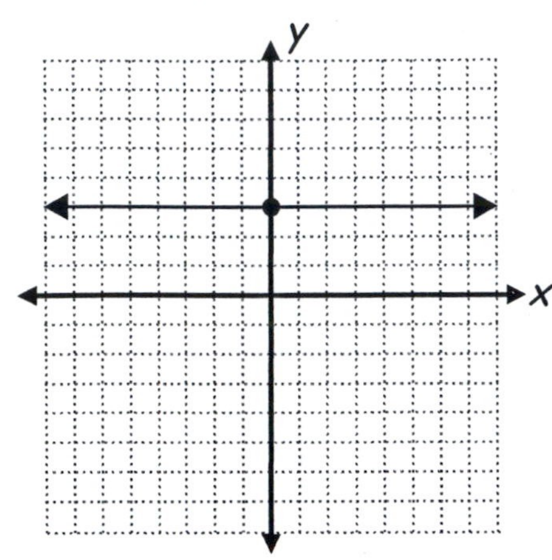

29. $\frac{x-y}{2}=-1$

$y=x+2$

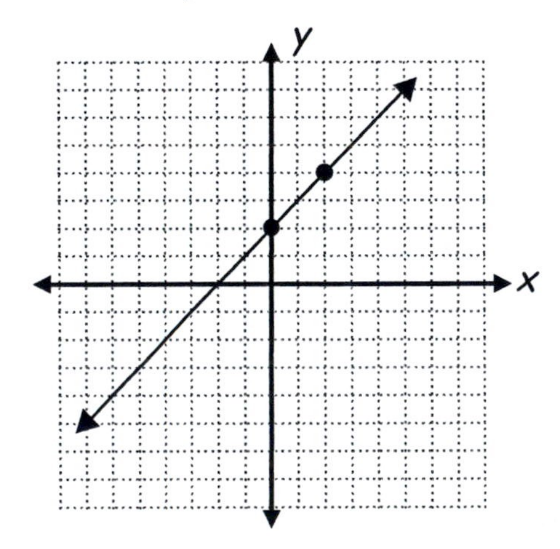

31. $y-(-3)=\frac{3}{4}(x-0)$

$y+3=\frac{3}{4}x$

$y=\frac{3}{4}x-3$

33. $y-(-7)=-\frac{5}{2}(x-0)$

$y+7=-\frac{5}{2}x$

$y=-\frac{5}{2}x-7$

35. $y-(-3)=\frac{3}{2}(x-(-1))$

$y+3=\frac{3}{2}(x+1)$

$2y+6=3x+3$

$3x-2y=3$

37. $y-5=0(x-(-3))$

$y=5$

39. $y-0=\frac{5}{4}(x-6)$

$4y=5x-30$

$5x-4y=30$

For items 41 - 45, find the slope by using the two given points. Choose either one of the two points and use the slope to write the equation.

41. $m=\frac{3-(-1)}{-1-2}=-\frac{4}{3}$

$y-(-1)=-\frac{4}{3}(x-2)$

$3y+3=-4x+8$

$4x+3y=5$

43. $m=\frac{17-(-2)}{2-2}=\frac{21}{0}$ Undefined

$x=2$

45. $m=\frac{-1-(-1)}{8-3}=0$

$y=-1$

47. Given: $C=0.25x+2100$

(a) Cost of $x=500$ bottles

$C=0.25(500)+2100$

$C=\$2,225$

(b) Fixed costs $=\$2,100$

(c) The increase for each bottle is \$0.25.

49. The annual depreciation on the equipment is a fixed amount, say K:
$K=(51500-43200)/3=\$2,766.67$
$V=51,500-2766.67t$ for $t=$ the number of years. To determine the number of years it would take for the value to be 0, set the equation equal to 0 and solve:
$51,500-2766.67t=0 \Rightarrow t\approx 19$ years.

For items 51 - 55 use a computer algebra system.

51. $m=\frac{31}{99}, b=\frac{20}{11}$

53. $m=\frac{-1}{7}, b=\frac{-5}{2}$

55. $m=0, b=0$

End of Section 2.3

Chapter Two: Section 2.4
Solutions to Odd-numbered Exercises

For items 1 - 5, first solve the equation for y to find the slope of the given line. The coefficient of the x-term is the slope. Use the fact that parallel lines have the same slope. Use the point-slope form to write the equation of the requested line.

1. $y-4x=7$

$$y=4x+7;\ m=4$$
$$y-5=4(x-(-1))$$
$$y-5=4x+4$$
$$y=4x+9$$

3. $2-\dfrac{y-3x}{3}=5$

$$6-y+3x=15;$$
$$y=3x-9;\ m=3$$
$$y-(-2)=3(x-0)$$
$$y=3x-2$$

5. $2(y-1)+\dfrac{x+3}{5}=-7$

$$10y-10+x+3=-35;$$
$$y=-\frac{1}{10}x-\frac{14}{5};\ m=-\frac{1}{10}$$
$$y-(0)=-\frac{1}{10}(x-(-5))$$
$$y=-\frac{1}{10}x-\frac{1}{2}$$

For items 7 - 9, first plot the points on a grid to determine which segments to test for being parallel.

7.

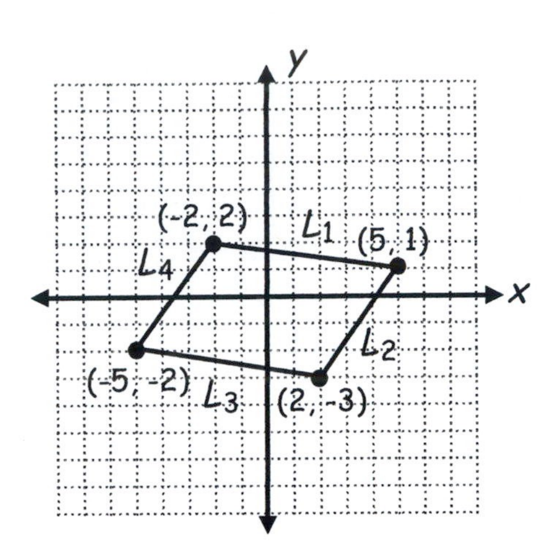

$$m_1=\frac{2-1}{-2-5}=-\frac{1}{7}$$
$$m_3=\frac{-2-(-3)}{-5-2}=-\frac{1}{7}$$

L_1 and L_3 are parallel.

$$m_2=\frac{1-(-3)}{5-2}=\frac{4}{3}$$
$$m_4=\frac{-2-2}{-5-(-2)}=\frac{4}{3}$$

L_2 and L_4 are parallel.

The quadrilateral is a parallelogram.

9.

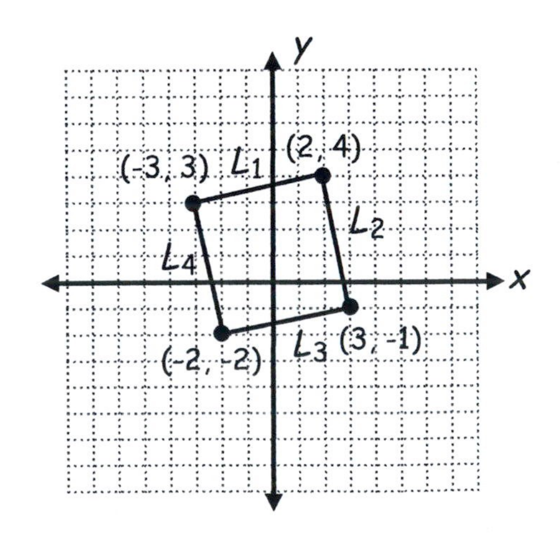

$$m_1 = \frac{4-3}{2-(-3)} = \frac{1}{5}$$

$$m_3 = \frac{-2-(-1)}{-2-3} = \frac{1}{5}$$

L_1 and L_3 are parallel.

$$m_2 = \frac{-1-4}{3-2} = -5$$

$$m_4 = \frac{3-(-2)}{-3-(-2)} = -5$$

L_2 and L_4 are parallel.

The quadrilateral is a parallelogram.

For 11 - 15, solve for y to determine the slope of the lines.

11. $2x-3y=(x-1)-(y-x)$

$2x-3y=2x-1-y$

$y=\frac{1}{2}; \quad m=0$

$-2y-x=9$

$y=-\frac{1}{2}x-\frac{9}{2}; \quad m=-\frac{1}{2}$

Not parallel

13. $x-5y=2$

$y=\frac{1}{5}x-\frac{2}{5}; \quad m=\frac{1}{5}$

$5x-y=2$

$y=5x-2; \quad m=5$

Not parallel

15. $\frac{x-y}{2}=\frac{x+y}{3}$

$3x-3y=2x+2y$

$y=\frac{1}{5}x; \quad m=\frac{1}{5}$

$\frac{2x+3}{5}-4y=1+2y$

$2x+3-20y=5+10y$

$y=\frac{1}{15}x-\frac{1}{15}; \quad m=\frac{1}{15}$

Not parallel

For items 17 - 21, solve the given equation for y; use the negative reciprocal, if possible, as the slope of the requested line with the given point.

17. $3x+2y=3y-7$

$y=3x+7; \quad m=3$

$y-(-2)=-\frac{1}{3}(x-3)$

$y+2=-\frac{1}{3}x+1$

$y=-\frac{1}{3}x-1$

19. $x+y=5$

$y=-x+5; \quad m=-1$

$(y-0)=1(x-0)$

$y=x$

21. $6y+2x=1$

$y=-\frac{1}{3}x+\frac{1}{6}; \quad m=-\frac{1}{3}$

$y-(-12)=3(x-(-4))$

$y+12=3x+12$

$y=3x$

For items 23 - 27, determine the slopes and observe whether one is the negative reciprocal of the other.

23. $x-5y=2$ or $y=\frac{1}{5}x-\frac{2}{5}; \quad m=\frac{1}{5}$

$5x-y=2$ or $y=5x-2; \quad m=5$

Not perpendicular

25. $\frac{3x-y}{3}=x+2$

$3x-y=3x+6$ or $y=-6; \quad m=0$

This line is horizontal.

$x=9$ means the line is vertical and has an undefined slope. The lines are perpendicular.

27. $\frac{x-1}{2}+\frac{3y+2}{3}=-9$

$3x-3+6y+4=-54$

$6y=-3x-55$

$y=-\frac{1}{2}x-\frac{55}{6};\ m=-\frac{1}{2}$

$3y-5x=x+5$

$3y=6x+5$

$y=2x+\frac{5}{3};\ m=2$

Yes, perpendicular

For items 29 - 31, find the slopes of the lines from the pairs of points, in the given respective orders.

29. $m_1=\frac{2-(-2)}{-2-(-5)}=\frac{4}{3}$

$m_2=\frac{-2-(-3)}{-5-2}=-\frac{1}{7}$

These segments are not perpendicular, so the quadrilateral is not a rectangle.

31. $m_1=\frac{-3-2}{3-1}=-\frac{5}{2}$

$m_2=\frac{-1-(-3)}{9-3}=\frac{2}{6}=\frac{1}{3}$

These segments are not perpendicular, so the quadrilateral is not a rectangle.

33. The solution is based on making the slopes of the lines (cables) the same.

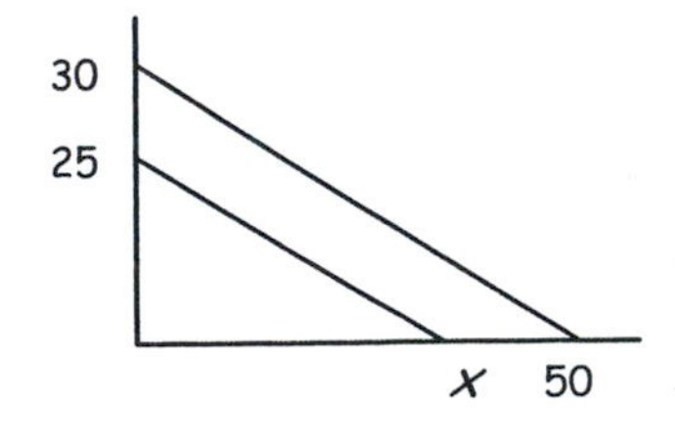

slope $=\frac{30-0}{0-50}=-\frac{3}{5}$

Let x = the distance the second cable should be from the base.

Then, $-\frac{3}{5}=\frac{25-0}{0-x}$. Solve for x.

$3x=125$

$x\approx 41.67$ ft.

For items 35 - 39, use a computer algebra system.

35. $m_1=\frac{2}{3},\ m_2=\frac{-3}{2}$

The lines are perpendicular.

37. $m_1=\frac{17}{13},\ m_2=\frac{-17}{13}$

Neither parallel nor perpendicular.

39. $m_1=\frac{-13}{18},\ m_2=\frac{18}{13}$

The lines are perpendicular.

End of Section 2.4

Chapter Two: Section 2.5
Solutions to Odd-numbered Exercises

1. To graph the solution set of $x - 3y < 6$, first graph the line $x - 3y = 6$. Use a dotted line to indicate the strict inequality. Use a test point, $(0,0)$, to determine on which side of the line the solution set is located.
The test: $0 - 3(0) = 0 < 6$
This is true, so the solution set is the set of all points on the same side of the line as the test point.

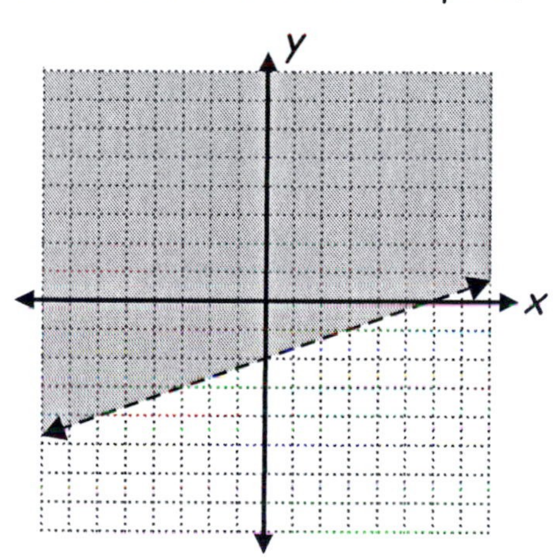

3. To graph the solution set of $3x - y \le 2$, first graph the line $3x - y = 2$. Use a solid line to indicate the equality. Use a test point, $(0,0)$, to determine on which side of the line the solution set is located.
The test: $3(0) - 0 = 0 \le 2$
This is true, so the solution set is the set of all points on the same side of the line as the test point.

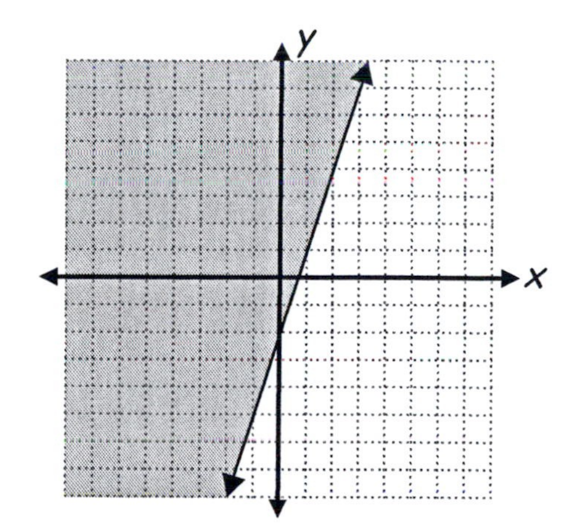

5. To graph the solution set of $y < -2$, first graph the line $y = -2$. Use a dotted line to indicate the strict equality. Use a test point, $(0, 0)$, to determine on which side of the line the solution set is located.
The test: $0 \not< -2$
The test indicates that the solution set is the set of all points on the opposite side of the line as the test point.

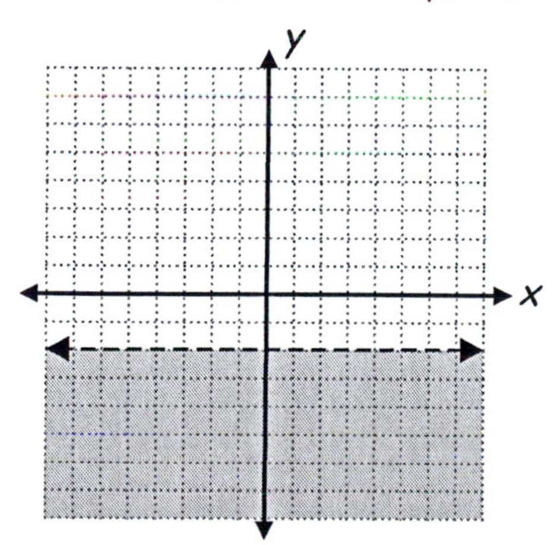

7. To graph the solution set of $x + y < 0$, first graph the line $x + y = 0$. Use a dotted line to indicate the inequality. Use a test point, $(1, 1)$, to determine on which side of the line the solution set is located.

The test: $1+1=2 \nless 0$
The test indicates that the solution set is the set of all points on the opposite side of the line as the test point.

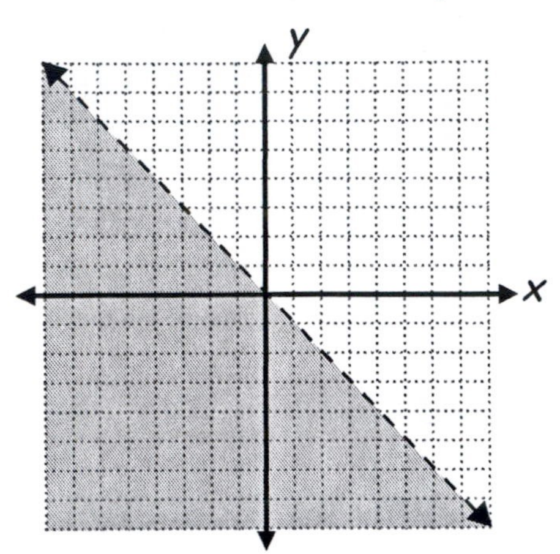

9. To graph the solution set of $y<2x-1$, first graph the line $y=2x-1$. Use a dotted line to indicate the inequality. Use a test point, $(2, 0)$, to determine on which side of the line the solution set is located.
The test: $0<2(2)-1=3$
This is true, so the solution set is the set of all points on the same side of the line as the test point.

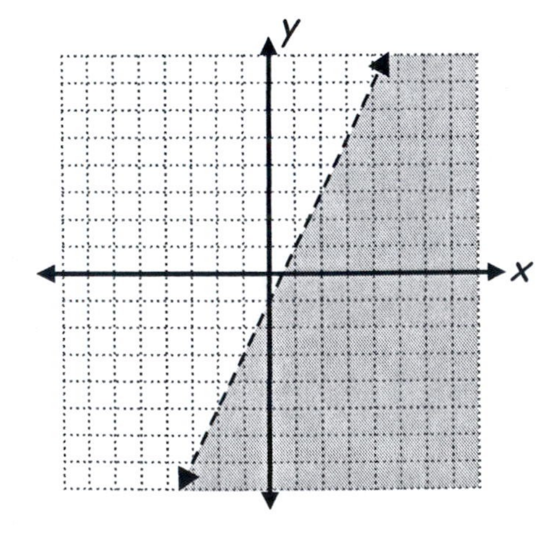

11. To graph the solution set of $-2y \le -x+4$, first graph the line $-2y=-x+4$. Use a solid line to indicate the equality. Use a test point, $(0, 0)$, to determine on which side of the line the solution set is located.
The test: $-2(0)=0 \le -(0)+4=4$
This is true, so the solution set is the set of all points on the same side of the line as the test point.

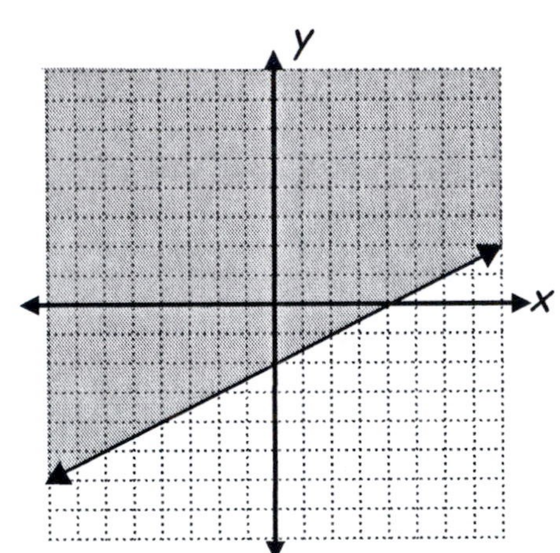

13. To graph the solution set of $5(y+1) \ge -x$, first graph the line $5(y+1)=-x$. Use a solid line to indicate the equality. Use a test point, $(0, 0)$, to determine on which side of the line the solution set is located.
The test: $5(0+1)=5 \ge -(0)=0$
This is true, so the solution set is the set of all points on the same side of the line as the test point.

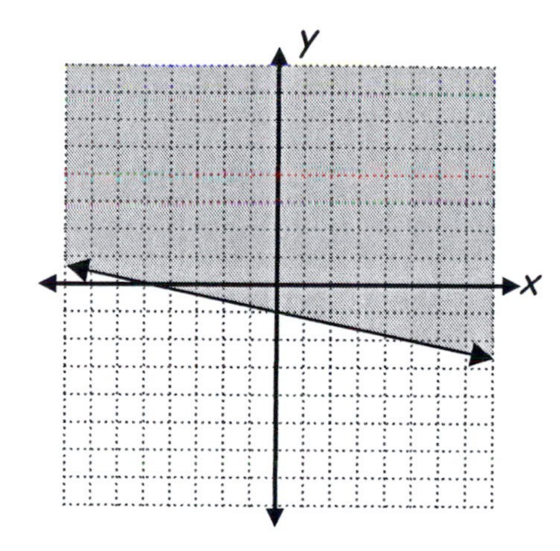

15. To graph the solution set of $x - y < 2y + 3$, first graph the line $x - y = 2y + 3$. Use a dotted line to indicate the inequality. Use a test point, $(0, 0)$, to determine on which side of the line the solution set is located.
The test: $0 - 0 = 0 < 2(0) + 3 = 3$. This is true, so the solution set is the set of all points on the same side of the line as the test point.

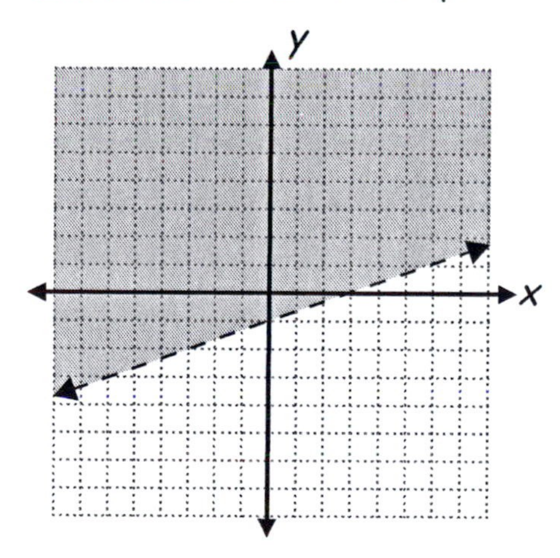

17. To graph the solution set of $x - 3y \geq 6$ or $y > -4$, first, find the solution set of each inequality. The word "or" means that you want the union of the two solution sets.
Find the solution of each line $x - 3y = 6$ and $y = -4$ by graphing each line and choosing a test point, such as $(0,0)$. This point works for both equations, since the point is not contained on either line.
The test point $(0, 0)$ does not satisfy the first inequality, so its solution set is the set of all points on the opposite side of the line. The test point $(0, 0)$ does satisfy the second inequality, so its solution set is the set of all points on the same side of the second line. Shade each of these regions of the grid and observe the union of the two sets:

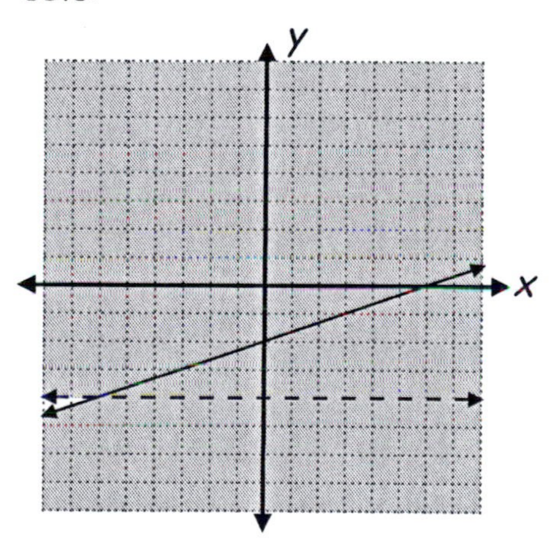

19. To graph the solution set of $x > 1$ and $y > 2$, first, find the solution set of each inequality. The word "and" means that you want the intersection of the two solution sets.
Find the solution of each line $x = 1$ and $y = 2$ by graphing each line and choosing a test point, such as $(0, 0)$. This point works for both equations, since the point is not contained on either line.

The test point $(0, 0)$ does not satisfy the first inequality, so its solution set is the set of all points on the opposite side of the line. The test point $(0, 0)$ does not satisfy the second inequality, so its solution set is the set of all points on the opposite side of the second line. Shade the solution regions for each inequality on the grid and observe the overlap of the two solutions sets. The overlap is the intersection (solution) :

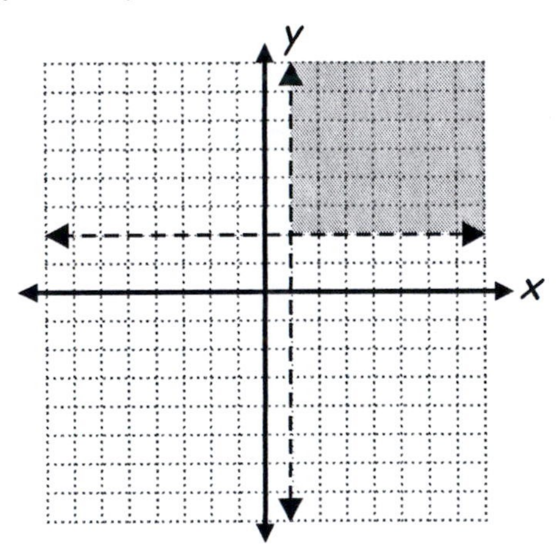

21. To graph the solution set of $x+y>-2$ and $x+y<2$, first, find the solution set of each inequality. The word "and" means that you want the intersection of the two solution sets. Find the solution of each line $x+y=-2$ and $x+y=2$ by graphing each line and choosing a test point, such as $(0, 0)$. This point works for both equations, since the point is not contained on either line.

The test point $(0, 0)$ satisfies the first inequality, so its solution set is the set of all points on the same side of the line. The test point $(0, 0)$ also satisfies the second inequality, so its solution set is the set of all points on the same side of the second line. Shade the solution regions for each inequality on the grid and observe the overlap of the two solutions sets. The overlap is the intersection (solution) :

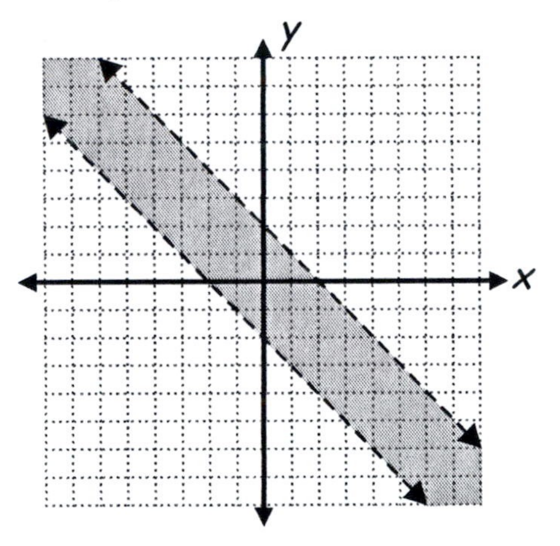

23. To graph the solution set of $y\geq-2$ and $y>1$, first, find the solution set of each inequality. The word "and" means that you want the intersection of the two solution sets. Find the solution of each line $y=-2$ and $y=1$ by graphing each line and choosing a test point, such as $(0, 0)$. This point works for both equations, since the point is not contained on either line.

The test point $(0, 0)$ does satisfy the first inequality, so its solution set are all points on the same side of the line. The test point $(0, 0)$ does not satisfy the second inequality, so its solution set is the set of all points on the opposite side of the second line. Shade the solution regions for each inequality on the grid and observe the overlap of the two solutions sets. The overlap is the intersection (solution) :

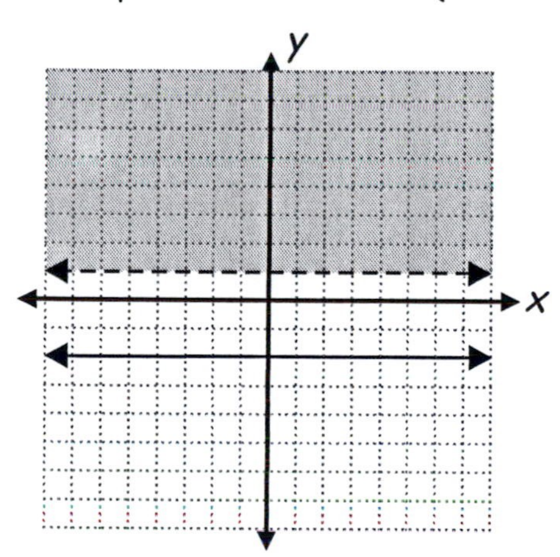

25. To graph the solution set of $y \geq -2x-5$ and $y \leq -6x-9$, first, find the solution set of each inequality. The word "and" means that you want the intersection of the two solution sets. Find the solution of each line $y=-2x-5$ and $y=-6x-9$ by graphing each line and choosing a test point, such as $(0, 0)$. This point works for both equations, since the point is not contained on either line.

The test point $(0, 0)$ does satisfy the first inequality, so its solution set are all points on the same side of the line. The test point $(0, 0)$ does not satisfy the second inequality, so its solution set is the set of all points on the opposite side of the second line. Shade the solution regions for each inequality on the grid and observe the overlap of the two solutions sets. The overlap is the intersection (solution) :

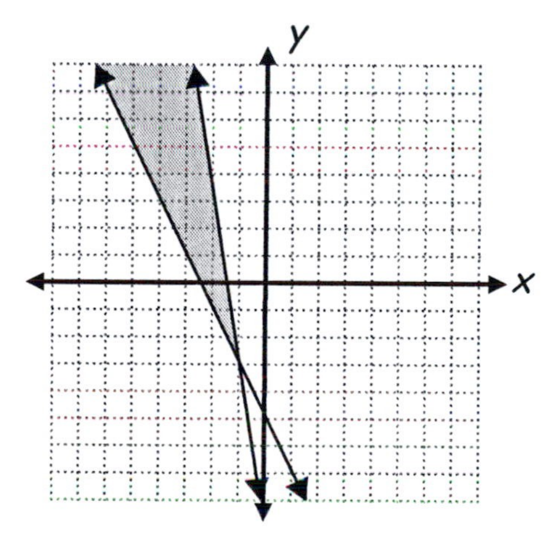

27. To graph the solution set of $y \leq -x$ and $2y+3x > -4$, first, find the solution set of each inequality. The word "and" means that you want the intersection of the two solution sets.

Find the solution of each line $y=-x$ and $2y+3x=-4$ by graphing each line and choosing a test point, such as $(-1, 0)$. This point works for both equations, since the point is not contained on either line.

The test point $(-1, 0)$ satisfies the first inequality, so its solution set is the set of all points on the same side of the line. The test point $(-1, 0)$ also satisfies the second inequality, so its solution set is the set of all points on the same side of the second line. Shade the solution regions for each inequality on the grid and observe the overlap of the two solutions sets. The overlap is the intersection (solution) :

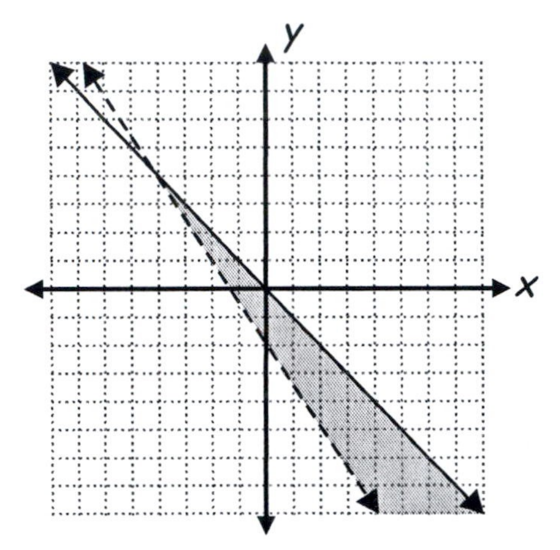

29. Find all ordered pairs for which $x-3>2$ or $x-3<-2$. That is, you need $x>5$ or $x<1$. y-values run the full range of R.

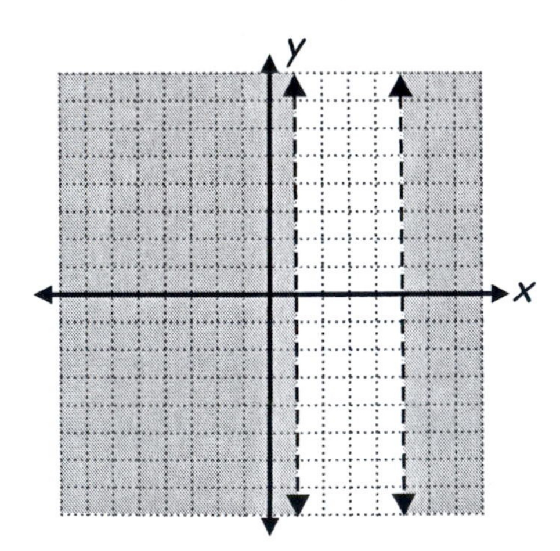

31. Find all ordered pairs for which $2x-4>2$ or $2x-4<-2$. That is, you need $x>3$ or $x<1$. y-values run the full range of R.

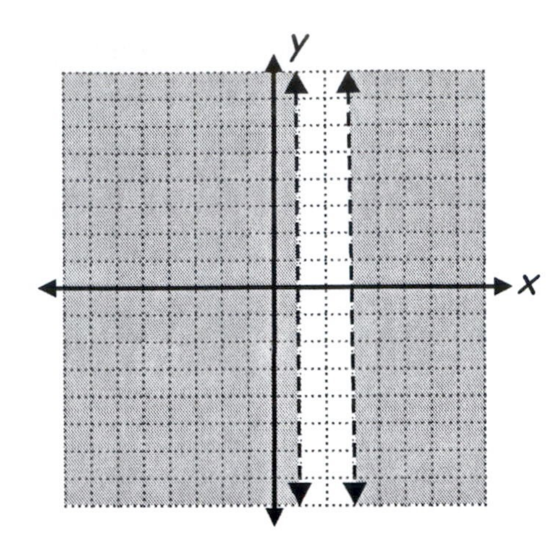

33. To graph the solution set of $|x+1|<2$ and $|y-3|\le 1$, First, find the solution set of each inequality. The word "and" means that you want the intersection of the two solution sets.
You need to identify all ordered pairs (x, y) for which $-2<x+1<2$ and $-1\le y-3\le 1$. That is, you need to have $-3<x<1$ while $2\le y\le 4$.
On the grid, shade the region of the R^2 plane that satisfies the first condition. Then, shade the region of the R^2 plane that satisfies the second condition. The area where the two regions overlap is the solution:

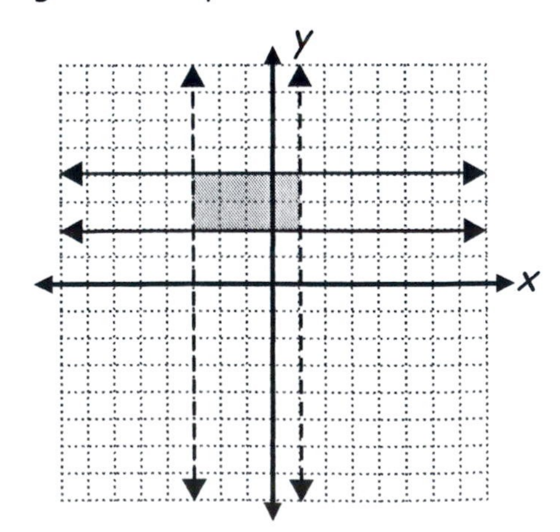

35. The inequality $|x-y|<1$ means that $-1<x-y<1$. This is a disjunctive statement that means $-1<x-y$ and $x-y<1$. First, graph the equation associated with each inequality. Then use a test point for each inequality to determine on which side of the lines their respective solutions are. The point $(0, 0)$ can be used for both inequalities since it is not located on either line. The lines should be dotted to indicate the strict inequality. The solution to the original inequality is the intersection of the two solution sets.

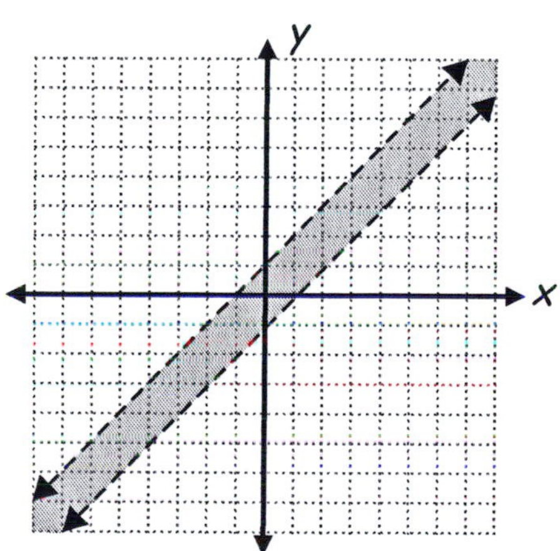

After using the test point $(0, 0)$, you should find that it does satisfy the first inequality--indicating that the solution set is the set of points on the same side of the line $-1=x-y$ as $(0, 0)$. The same is true for the second inequality. The intersection of the two solution sets is their common region shown on the grid.

37. The inequality $|4x-2y-3|\leq 5$ means $-5\leq 4x-2y-3$ and $4x-2y-3\leq 5$. First, graph the "border" lines $-2=4x-2y$ and $4x-2y=8$. Use solid lines. Use a test point such as $(0, 0)$ to determine on which side of the lines, respectively, that the solution sets lie. The test:
$-2\leq 4(0)-2(0)=0$ and
$4(0)-2(0)=0\leq 8$ are both true.
Therefore, the solution sets of each inequality are on the same side of the lines as the point $(0, 0)$. Their intersection is the solution to the original inequality.

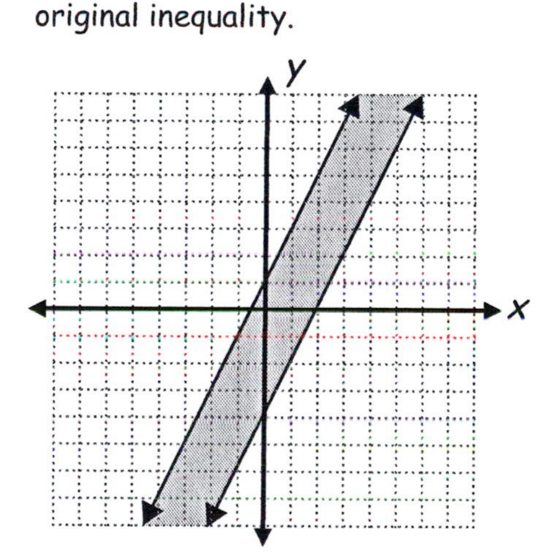

39. The inequalities $|y-3x|\leq 2$ and $|y|<2$ mean $-2\leq y-3x$ and $y-3x\leq 2$, while $-2<y$ and $y<2$. The lines associated with each inequality should be graphed. Use a test point, such as $(0, 0)$ to test each of the four inequalities. The intersection of all four solution sets for each of the inequalities is the solution set for the original condition.

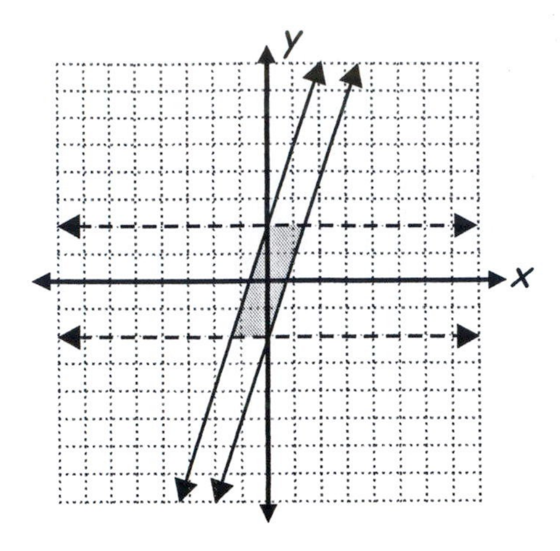

41. Let $x =$ number of lily arrangements.
Let $y =$ number of orchid arrangements.
A linear inequality expressing the cost less than \$150:
$12x + 22y < 150$

43. Let $l =$ the length of the garden.
Let $w =$ the width.
Then, the inequality expressing the limit of 300 feet for the fencing is
$2l + 2w \leq 300$

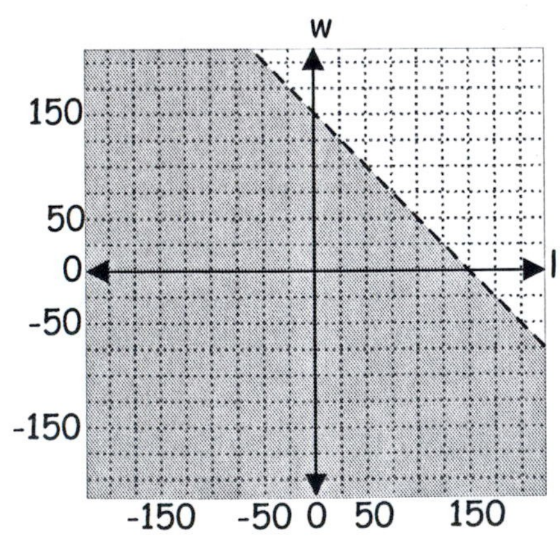

End of Section 2.5

Chapter Two: Section 2.6
Solutions to Odd-numbered Exercises

1. Given: Center at $(-4,-3)$
Radius $=5$
Standard form of the equation:
$(x+4)^2+(y+3)^2=25$

3. Given: Center at $(7,-9)$
Radius $=3$
Standard form of the equation:
$(x-7)^2+(y+9)^2=9$

5. Given: Center at $(0,0)$
Radius $=\sqrt{9}=3$
Standard form of the equation:
$x^2+y^2=9$

7. Given: Center at $(\sqrt{5},\sqrt{3})$
Radius $=4$
Standard form of the equation:
$(x-\sqrt{5})^2+(y-\sqrt{3})^2=16$

9. Given: Center at $(7,2)$
Tangent to the x-axis $\Rightarrow$
Radius $=2$
Standard form of the equation:
$(x-7)^2+(y-2)^2=4$

11. Given: Center at $(-3,8)$
Passes through $(-4,9)$
Use the distance formula:
$r=\sqrt{(9-8)^2+(-4-(-3))^2}$
$r=\sqrt{1+1}=\sqrt{2}$
Standard form of the equation:
$(x+3)^2+(y-8)^2=2$

13. Given: Center at $(4,8)$
Passes through $(1,9)$
Use the distance formula:
$r=\sqrt{(9-8)^2+(1-4)^2}$
$r=\sqrt{1+9}=\sqrt{10}$
Standard form of the equation:
$(x-4)^2+(y-8)^2=10$

15. Given: Center at $(0,0)$
Passes through $(6,-7)$
Use the distance formula:
$r=\sqrt{(-7-0)^2+(6-0)^2}$
$r=\sqrt{49+36}=\sqrt{85}$
Standard form of the equation:
$x^2+y^2=85$

17. Given: Endpoints of a diameter are $(-8,6)$ and $(1,11)$
The radius is at the midpoint of the given diameter and the radius is half the length.
Midpoint (Center):

$$x=\frac{-8+1}{2}=-\frac{7}{2}$$
$$y=\frac{11+6}{2}=\frac{17}{2}$$

$r=\frac{1}{2}(\text{length of the diameter})$
$=\frac{1}{2}\left(\sqrt{(11-6)^2+(1+8)^2}\right)$
$=\frac{1}{2}\sqrt{25+81}$
$=\frac{1}{2}\sqrt{106}\Rightarrow r^2=26.5$

Standard form of the equation:
$\left(x+\frac{7}{2}\right)^2+\left(y-\frac{17}{2}\right)^2=26.5$

19. Given: Endpoints of a diameter are $(-7,-4)$ and $(-5,7)$.
The radius is at the midpoint of the given diameter and the radius is half the length.
Midpoint (Center):

$$x=\frac{-5+(-7)}{2}=-\frac{12}{2}=-6$$

$$y=\frac{7+(-4)}{2}=\frac{3}{2}$$

$$r=\frac{1}{2}(\text{length of the diameter})$$

$$=\frac{1}{2}\left(\sqrt{(7-(-4))^2+(-5-(-7))^2}\right)$$

$$=\frac{1}{2}\sqrt{121+4}$$

$$=\frac{1}{2}\sqrt{125}\Rightarrow r^2=31.25$$

Standard form of the equation:

$$(x+6)^2+\left(y-\frac{3}{2}\right)^2=31.25$$

21. Given: Endpoints of a diameter are $(0,0)$ and $(-13,-14)$.
The radius is at the midpoint of the given diameter and the radius is half the length.
Midpoint (Center):

$$x=\frac{-13}{2}$$

$$y=\frac{-14}{2}=-7$$

$$r=\frac{1}{2}(\text{length of the diameter})$$

$$=\frac{1}{2}\left(\sqrt{(-13)^2+(-14)^2}\right)$$

$$=\frac{1}{2}\sqrt{169+196}$$

$$=\frac{1}{2}\sqrt{365}\Rightarrow r^2=91.25$$

Standard form of the equation:

$$\left(x+\frac{13}{2}\right)^2+(y+7)^2=91.25$$

23. Given: Endpoints of a diameter are $(0,6)$ and $(8,0)$.
The radius is at the midpoint of the given diameter and the radius is half the length.
Midpoint (Center):

$$x=\frac{8+0}{2}=4$$

$$y=\frac{0+6}{2}=3$$

$$r=\frac{1}{2}(\text{length of the diameter})$$

$$=\frac{1}{2}\left(\sqrt{8^2+6^2}\right)$$

$$=\frac{1}{2}\sqrt{64+36}$$

$$=\frac{1}{2}\sqrt{100}\Rightarrow r^2=25$$

Standard form of the equation:

$$(x-4)^2+(y-3)^2=25$$

25. Given (observe from the graph):
Center at $(2,0)$; Radius $=2$
Standard form of the equation:

$$(x-2)^2+y^2=4$$

27. Given (observe from the graph):
Center at $(2,4)$; Radius $=7$
Standard form of the equation:

$$(x-2)^2+(y-4)^2=49$$

29. Given (observe from the graph):
Center at $(-3,-2)$; Radius $=8$
Standard form of the equation:

$$(x+3)^2+(y+2)^2=64$$

31. Center at $(0,0)$; $r=6$

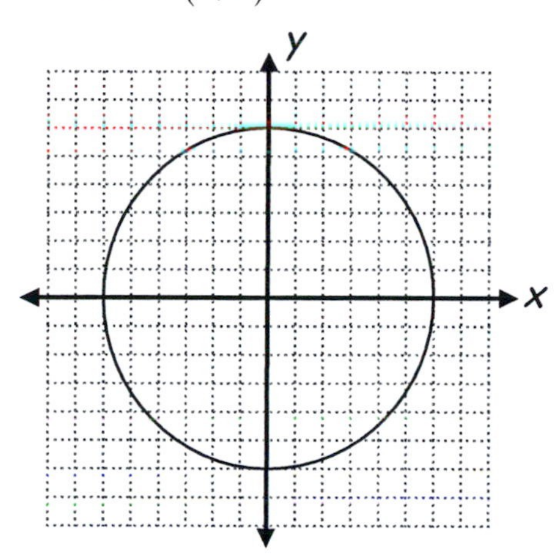

33. Center at $(0,8)$; $r=3$

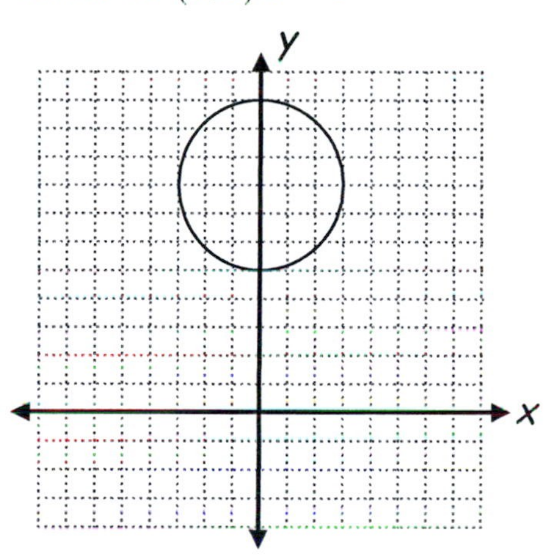

35. Center at $(8,0)$; $r=2\sqrt{2}$

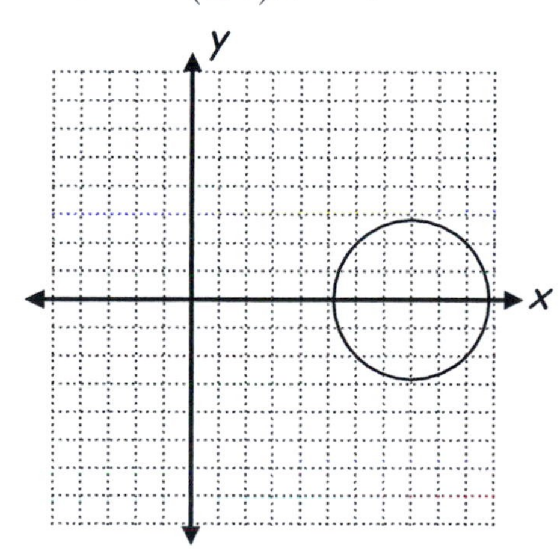

37. Center at $(-5,-4)$; $r=2$

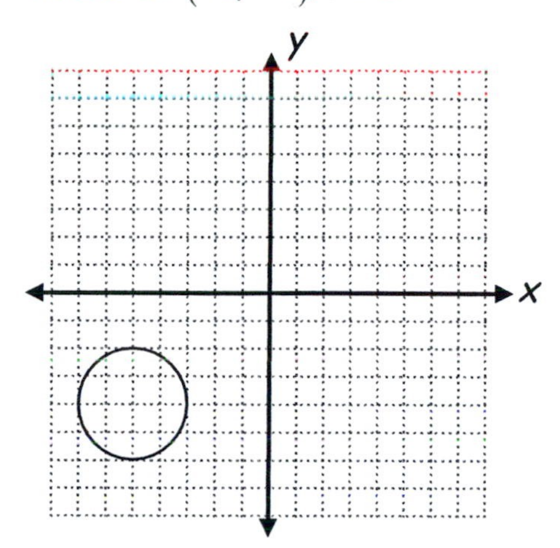

39. Center at $(5,-5)$; $r=\sqrt{5}$

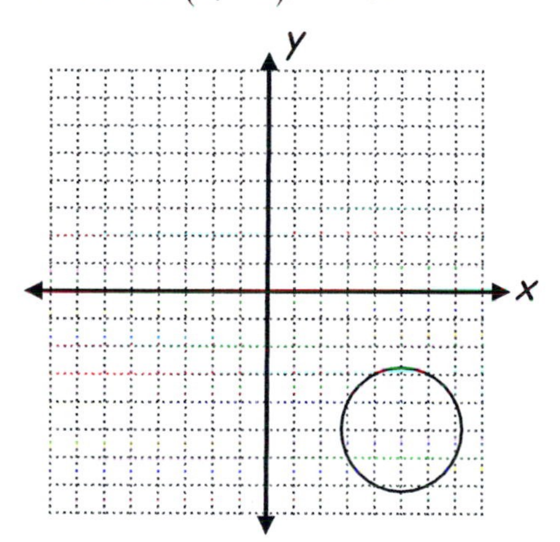

41. Complete the squares in x and y.

$x^2-4x+4+y^2+4y+4=8+4+4$

$(x-2)^2+(y+2)^2=16$

Center at $(2,-2)$; Radius $=4$

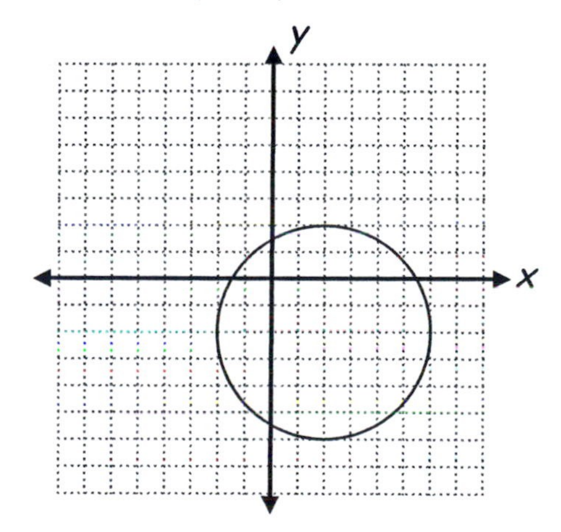

43. Complete the square in y.

$x^2 + y^2 + 10y + 25 = -9 + 25$

$x^2 + (y+5)^2 = 16$

Center at $(0, -5)$; Radius $= 4$

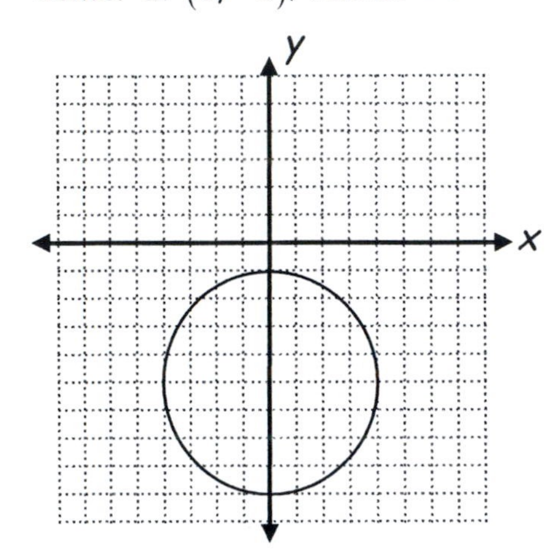

45. Complete the squares in x and y.

$x^2 - 2x + 1 + y^2 + 6y + 9 = -2 + 1 + 9$

$(x-1)^2 + (y+3)^2 = 8$

Center at $(1, -3)$; Radius $= 2\sqrt{2}$

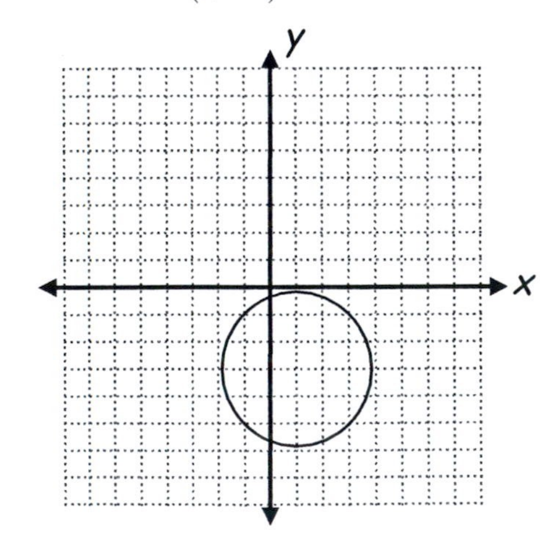

47. First, simplify-divide by 4.

$x^2 + y^2 = 64$

Center at $(0,0)$; Radius $= 8$

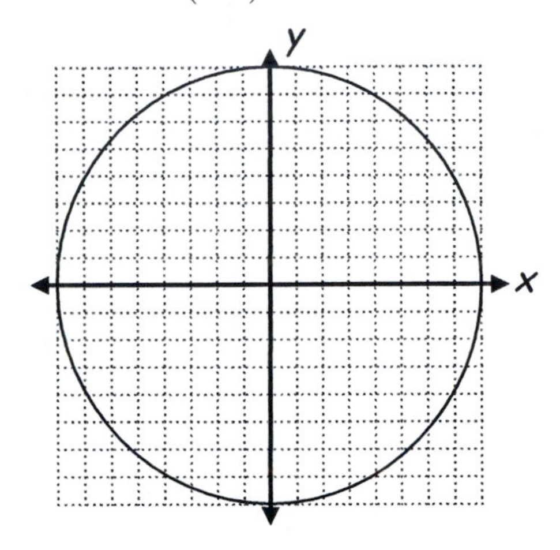

49. Complete the squares in x and y.

$x^2 - 6x + 9 + y^2 + 4y + 4 = 3 + 9 + 4$

$(x-3)^2 + (y+2)^2 = 16$

Center at $(3, -2)$; Radius $= 4$

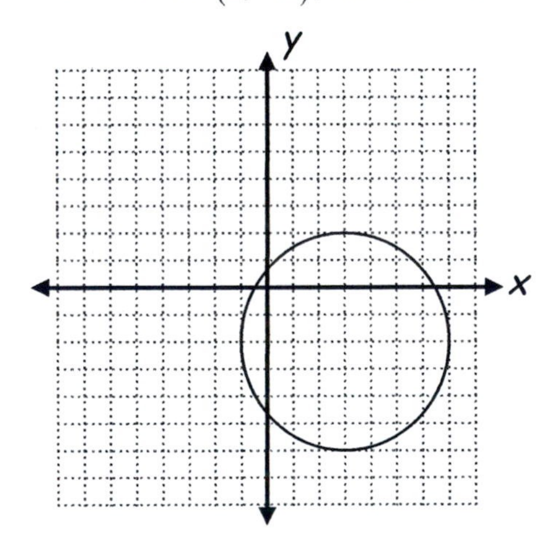

51. Center at $(1, 0)$; $r = 3$

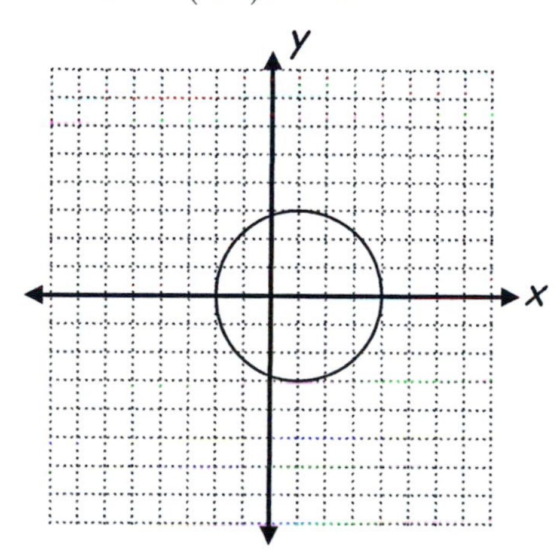

53. Complete the squares in x and y.

$x^2 - 4x + 4 + y^2 + 8y + 16 = 16 + 16 + 4$

$(x-2)^2 + (y+4)^2 = 36$

Center at $(2, -4)$; Radius $= 6$

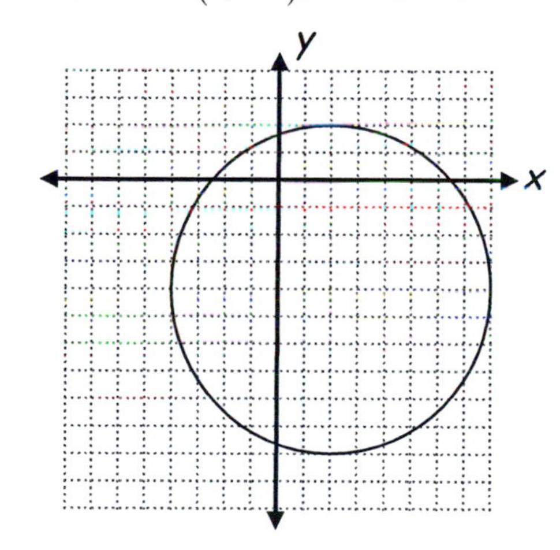

55. First, simplify by dividing by 4.

Then, complete the squares in x and y.

$x^2 - 6x + 9 + y^2 + 6y + 9 = 7 + 9 + 9$

$(x-3)^2 + (y+3)^2 = 25$

Center at $(3, -3)$; Radius $= 5$

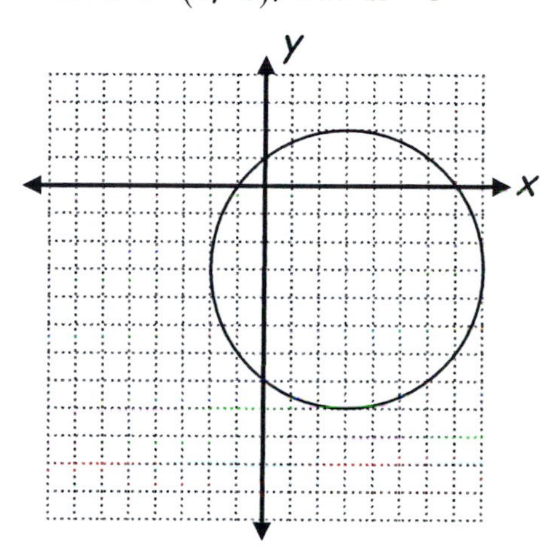

End of Section 2.6

Chapter Two Test
Solutions to Odd-Numbered Exercises

1. $(-5, 6)$ is in quadrant II.

3. $(-2, -4)$ is in quadrant III.

5. Sketch $5x + 2y = 20$

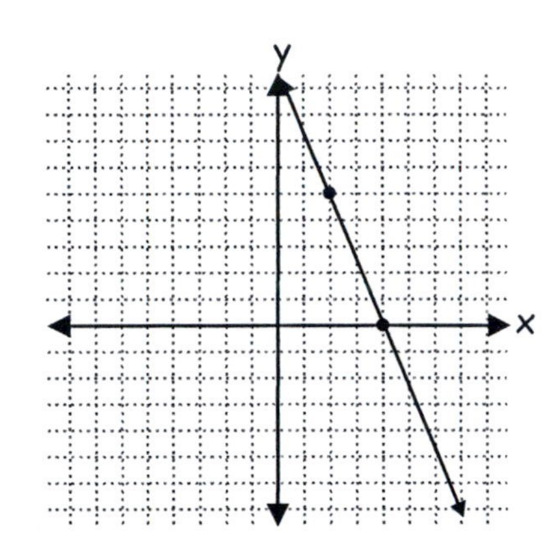

7. Sketch $x^2 + y^2 = 64$

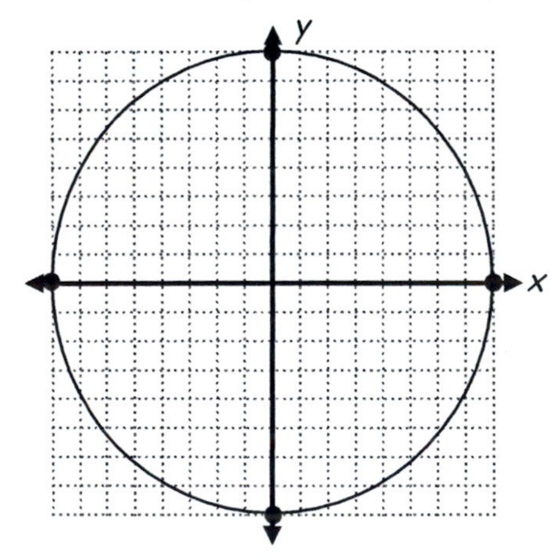

9. **(a)** Distance between $(-4, -3)$ and $(4, -9)$:

$$d = \sqrt{(-4-4)^2 + (-3-(-9))^2} = 10$$

(b) Midpoint:

$$\left(\frac{-4+4}{2}, \frac{-3+(-9)}{2}\right) = (0, -6)$$

11. Find the distances between the given points. The sum is the perimeter.

$$d_1 = \sqrt{(-3-(-3))^2 + (2-0)^2} = 2$$

$$d_2 = \sqrt{(-3-(-6))^2 + (2-(-3))^2} = \sqrt{34}$$

$$d_3 = \sqrt{(-3-(-6))^2 + (0-(-3))^2} = 3\sqrt{2}$$

Perimeter $= 2 + 3\sqrt{2} + \sqrt{34}$

13. $3x + y(4 - 2x) = 8$

$3x + 4y - \underline{2xy} = 8$ Not linear

15. $9x^2 - (3x+1)^2 = y - 3$

$9x^2 - 9x^2 - 6x - 1 = y - 3$

$-6x - y = -2$

Yes. It is linear.

17. $x-$ and $y-$intercepts of $4x + 5y = 2 + 3y$. This is the same as $2x + y = 1$.

$x-$intercept: $\left(\frac{1}{2}, 0\right)$

$y-$intercept: $(0, 1)$

19. Graph: $\frac{3}{4}y = 3$

Rewrite: $y = 4$

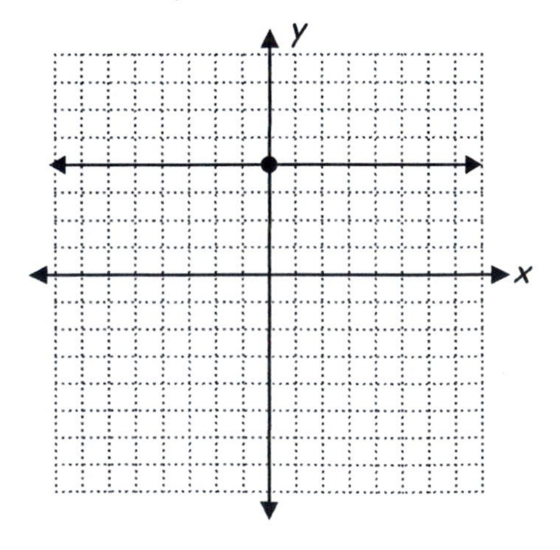

21. Graph: $5y+3(x-2y)-x=2x+10$

Rewrite: $5y+3x-6y-x=2x+10 \Rightarrow$

$$y=-10$$

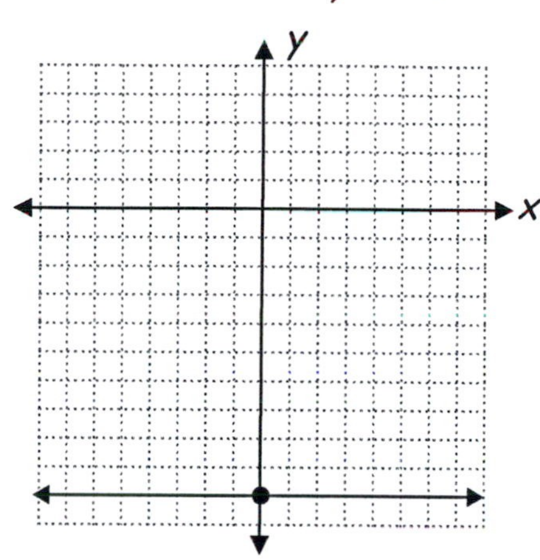

23. $m=\dfrac{6-(-10)}{3-7}=-4$

25. $6x-3y=9$

$$y=2x-3$$

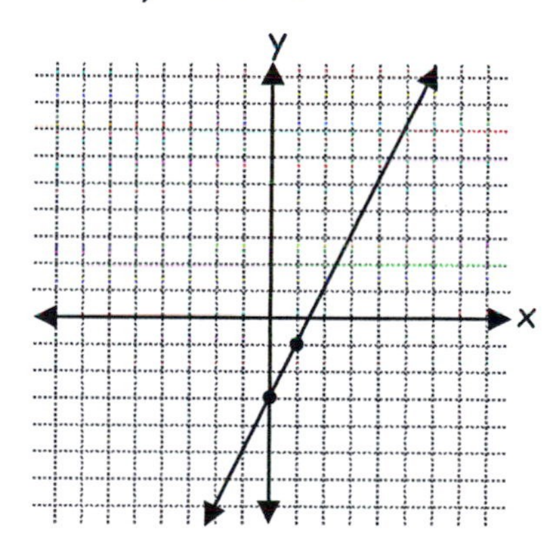

27. $15y-5x=0$

$$y=\frac{1}{3}x$$

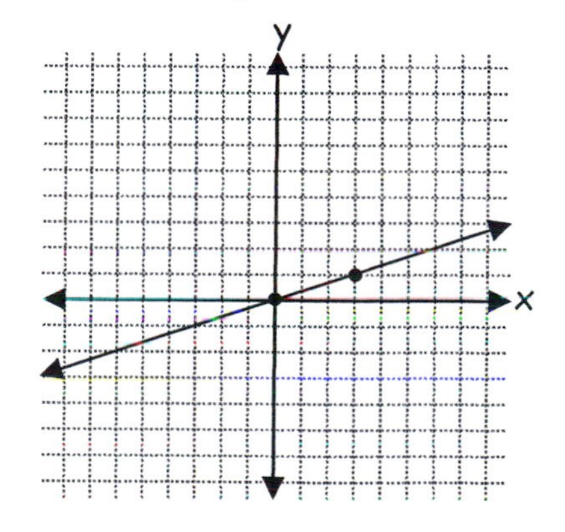

29. Use the point-slope form:

$$y-3=\frac{3}{2}(x-(-2))$$

$$2y-6=3x+6$$

$$3x-2y=-12$$

31. Find the slopes of the two lines by putting them in slope intercept form:

$x-4y=3$, or $y=\dfrac{1}{4}x-\dfrac{3}{4}$

$4x-y=2$, or $y=4x-2$

The lines are neither perpendicular or parallel.

33. Find the slopes of the two lines by putting them in slope intercept form:

$\dfrac{3x-y}{3}=x+2$, or $y=-6$

$\dfrac{y}{3}+x=9$, or $y=-3x+27$

The lines are neither perpendicular or parallel.

35. Find the slopes of the lines containing the first two points and the second and third points. If the figure is a rectangle then the slopes must be negative reciprocals of each other. They are not, so the quadrilateral is not a rectangle.

$m_1=\dfrac{2-(-1)}{-2-(-3)}=3;\ m_2=\dfrac{-1-(-3)}{-3-2}=-\dfrac{2}{5}$

37. Graph the solution sets of each of the inequalities by first graphing the border lines $7x-2y=8$ and $y<5$. Choose a test point such as $(0,0)$ to find the solution set of each. The solution set for the two given conditions is the intersection of the two:

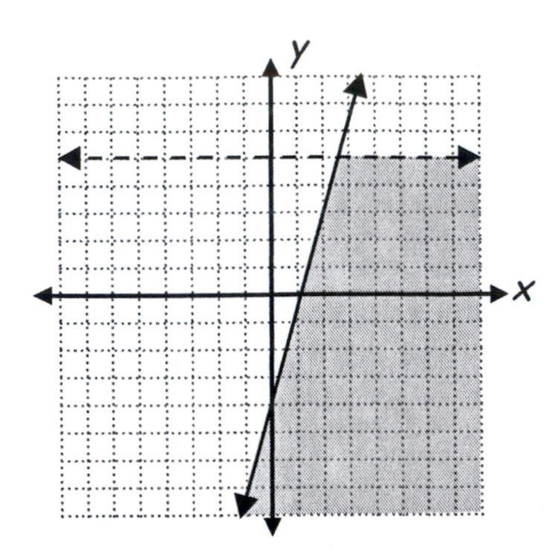

39. $|2x+5|<3$ means $-3<2x+5<3$, or $-4<x<-1$. Graphing this condition on R^2:

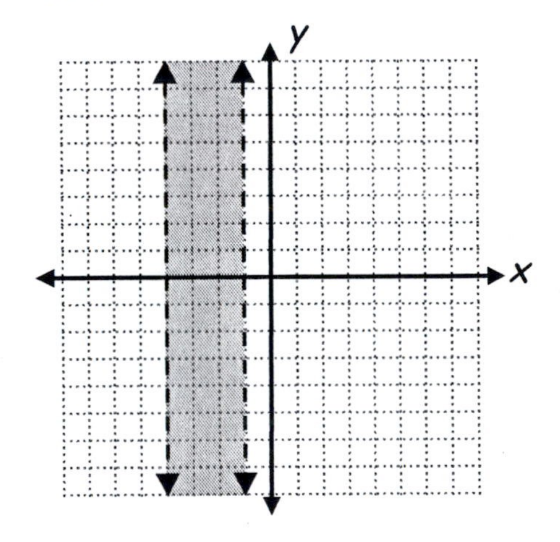

41. Given: A diameter with endpoints $(1,2)$ and $(-5,8)$. The midpoint of the diameter is the Center:

$$\left.\begin{aligned} x &= \frac{-5+1}{2} = -2 \\ y &= \frac{8+2}{2} = 5 \end{aligned}\right\} \Rightarrow \text{Center at } (-2,5)$$

$$\begin{aligned} \text{Radius} &= \frac{1}{2}(\text{Length of the diameter}) \\ &= \frac{1}{2}\sqrt{(-5-1)^2+(8-2)^2} \\ &= \frac{1}{2}\sqrt{36+36} \\ &= \frac{1}{2}\sqrt{72} \Rightarrow r^2 = 18 \end{aligned}$$

The standard form of the equation is

$(x+2)^2+(y-5)^2=18$

43. Given: $x^2+y^2+6x-10y=-5$

Complete the squares and place in standard form:

$$(x^2+6x+9)+(y^2-10y+25)=-5+9+25$$
$$(x+3)^2+(y-5)^2=29$$

Center at $(-3,5)$, Radius $=\sqrt{29}$

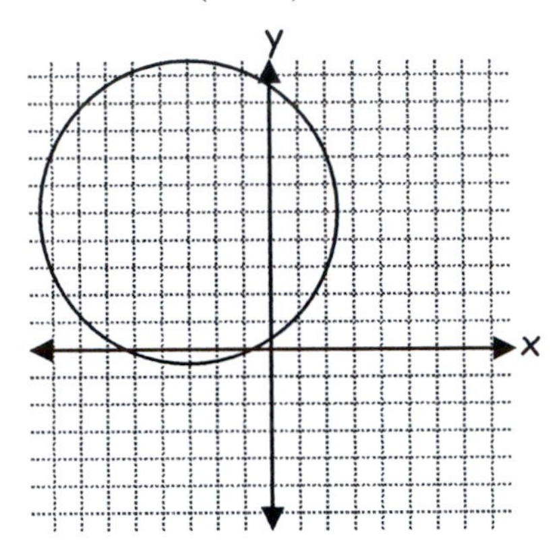

End of Chapter 2 Test

Chapter Three: Section 3.1
Solutions to Odd-numbered Exercises

1. Domain $= \{-2\}$
Range $= \{-9, 0, 3, 5\}$

3. Domain $= \{-2\pi, 1, 3, \pi\}$
Range $= \{0, 2, 4, 7\}$

5. Domain $= \mathbb{Z}$
Range $=$ the even integers

7. Domain $= \mathbb{Z}$
Range $= \{..., -5, -2, 1, 4, 7, ...\}$

9. Domain $= \mathbb{R}$
Range $= \mathbb{R}$

11. Domain $=$ non-negative Reals: $[0, \infty)$
Range $= \mathbb{R}$

13. Domain $= \mathbb{R}$
Range $= \{-1\}$

15. Domain $= \{0\}$
Range $= \mathbb{R}$

17. Domain $= [-3, 1]$
Range $= [0, 4]$

19. Domain $= [0, 3]$
Range $= [1, 5]$

21. Domain $= [-1, 3]$
Range $= [-4, 3]$

23. The Domain is the set of all males who have siblings.
The Range is the set of all people who have brothers.

25. Not a function: $(-2, 5)$ and $(-2, 3)$

27. This is a function.

29. Not a function: $(1, 1)$ and $(1, -3)$

31. This is a function.

33. $y = \dfrac{1}{x}$ describes a function.
For each permissible real number value x, there is exactly one value of y.
Domain $= (-\infty, 0) \cup (0, \infty)$.

35. Rewriting the function as $y^2 = -x$, and solving for $y, y = \pm\sqrt{-x}$. Suppose that $x = -4$. Then, $y = \pm 2$. Having two range values for the same domain value violates the definition of a function.

37. This function is not defined at $x = -2$.
Domain $= (-\infty, -2) \cup (-2, \infty)$.

39. This function is defined for all Reals.
Domain $= \mathbb{R}$.

41. Rewriting the function as $y^2 = 3 + x^2$, and solving for $y, y = \pm\sqrt{3 + x^2}$. Suppose that $x = 1$. Then, $y = \pm 2$. Having two range values for the same domain value violates the definition of a function.

43.
$$6x^2 - x + 3y = x + 2y$$
$$6x^2 - 2x = -y$$
$$y = -6x^2 + 2x$$
$$f(x) = -6x^2 + 2x$$
$$f(-1) = -6(-1)^2 + 2(-1) = -8$$

45. $\dfrac{x+3y}{5}=2$

$x+3y=10$

$3y=-x+10$

$y=-\dfrac{1}{3}x+\dfrac{10}{3}$

$f(x)=\dfrac{-x+10}{3}$

$f(-1)=\dfrac{11}{3}$

47. $y-2x^2=-2\left(x+x^2+5\right)$

$y-2x^2=-2x-2x^2-10$

$y=-2x-10$

$f(x)=-2x-10$

$f(-1)=-2(-1)-10=-8$

49. **(a)** $f(x-1)=(x-1)^2+3(x-1)$

$=x^2-2x+1+3x-3$

$=x^2+x-2$

(b) $f(x+a)-f(x)=$

$(x+a)^2+3(x+a)-x^2-3x=$

$x^2+2ax+a^2+3x+3a-x^2-3x=$

$2ax+a^2+3a$

(c) $f\left(x^2\right)=\left(x^2\right)^2+3x^2$

$=x^4+3x^2$

51. **(a)** $f(x-1)=3(x-1)+2$

$=3x-3+2$

$=3x-1$

(b) $f(x+a)-f(x)=$

$3(x+a)+2-(3x+2)=$

$3x+3a+2-3x-2=3a$

(c) $f\left(x^2\right)=3x^2+2$

53. **(a)** $f(x-1)=2(5-3(x-1))$

$=10-6(x-1)$

$=10-6x+6$

$=16-6x$

(b) $f(x+a)-f(x)=$

$2(5-3(x+a))-(2(5-3x))=$

$10-6(x+a)-10+6x=$

$10-6x-6a-10+6x=-6a$

(c) $f\left(x^2\right)=2\left(5-3x^2\right)$

$=10-6x^2$

55. **(a)** $f(0)=0^2-4(0)+5=5$

(b) $f(-2)=(-2)^2-4(-2)+5=17$

(c) $f(3)=(3)^2-4(3)+5=2$

(d) $f(x+h)=(x+h)^2-4(x+h)+5$

$=x^2+2xh+h^2-4x-4h+5$

(e) $\dfrac{f(x+h)-f(x)}{h}=$

$\dfrac{(x+h)^2-4(x+h)+5-x^2+4x-5}{h}=$

$=\dfrac{2xh+h^2-4h}{h}$

$=2x+h-4$

57. **(a)** $f(0)=\dfrac{3(0)-2}{2}=-1$

(b) $f(-2)=\dfrac{3(-2)-2}{2}=-4$

(c) $f(3)=\dfrac{3(3)-2}{2}=\dfrac{7}{2}$

(d) $f(x+h)=\dfrac{3(x+h)-2}{2}$

$=\dfrac{3x+3h-2}{2}$

(e) $\dfrac{f(x+h)-f(x)}{h}=\dfrac{\frac{3(x+h)-2}{2}-\frac{3x-2}{2}}{h}$

$=\dfrac{\frac{3x+3h-2-3x+2}{2}}{h}$

$=\dfrac{\frac{3h}{2}}{h}$

$=\dfrac{3}{2}$

59. (a) $f(0)=\sqrt{0-8}=\sqrt{-8}=2i\sqrt{2}$
(b) $f(-2)=\sqrt{-2-8}=\sqrt{-10}=i\sqrt{10}$
(c) $f(3)=\sqrt{3-8}=\sqrt{-5}=i\sqrt{5}$
(d) $f(x+h)=\sqrt{x+h-8}$
(e) $\dfrac{f(x+h)-f(x)}{h}=\dfrac{\sqrt{x+h-8}-\sqrt{x-8}}{h}$

61. (a) $f(0)=0^3=0$
(b) $f(-2)=(-2)^3=-8$
(c) $f(3)=(3)^3=27$
(d) $f(x+h)=(x+h)^3$
(e) $\dfrac{f(x+h)-f(x)}{h}=$
$\dfrac{x^3+3x^2h+3xh^2+h^3-x^3}{h}=$
$3x^2+3hx+h^2$

63. (a) $f(0)=6(0)-0^2=0$
(b) $f(-2)=6(-2)-(-2)^2=-16$
(c) $f(3)=6(3)-3^2=9$
(d) $f(x+h)=6(x+h)-(x+h)^2$
$=-x^2+6x-2xh+6h-h^2$
(e) $\dfrac{f(x+h)-f(x)}{h}=$
$\dfrac{6(x+h)-(x+h)^2-6x+x^2}{h}=$
$\dfrac{6h-h^2-2xh}{h}=-2x+6-h$

65. Domain $=\mathbb{Z}$, the integers
Codomain $=\mathbb{Z}$, the integers
Range $=\{\ldots,-6,-3,0,3,6,\ldots\}$

67. Domain $=[0,\infty)$
Codomain $=\mathbb{R}$, the real numbers
Range $=[0,\infty)$

69. Domain $=\mathbb{N}$, the natural numbers
Codomain $=\mathbb{R}$, the real numbers
Range $=\left\{\frac{1}{2},1,\frac{3}{2},2,\frac{5}{2},\ldots\right\}$

71. Domain $=\mathbb{R}$

73. Set $2x+6\geq 0$, and solve:
$x\geq -3$
Domain: $[-3,\infty)$

75. Set $x^2-6x+9=0$, and solve:
$(x-3)(x-3)=0$, so $x=3$.
Then, exclude 3 from the domain.
Domain: $(-\infty,3)\cup(3,\infty)$

For 77-83, use a graphing calculator or computer algebra system.

77. $g(2)=3\sqrt{3}\approx 5.196152$
$g(3)=2\sqrt{7}\approx 5.291503$

79. $g(-19)\approx 7.2901E9$
$g(12)=258{,}474{,}853$

81. $h^5+3h^3+3h+6x+12h^2x+6h^4x+18hx^2+15h^3x^2+12x^3+20h^2x^3+15hx^4+6x^5$

83. $\dfrac{-1}{(x-1)(x+h-1)}$

End of Section 3.1

Chapter Three: Section 3.2
Solutions to Odd-numbered Exercises

For items 1 – 15, graph the linear functions.

1. $f(x) = -5x + 2$
Use the slope-intercept form;
$m = -5$ and $b = 2$.

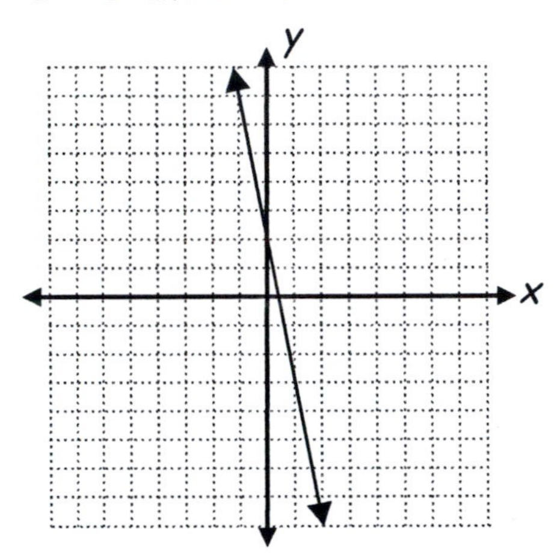

3. $h(x) = -x + 2$
Use the slope-intercept form;
$m = -1$ and $b = 2$.

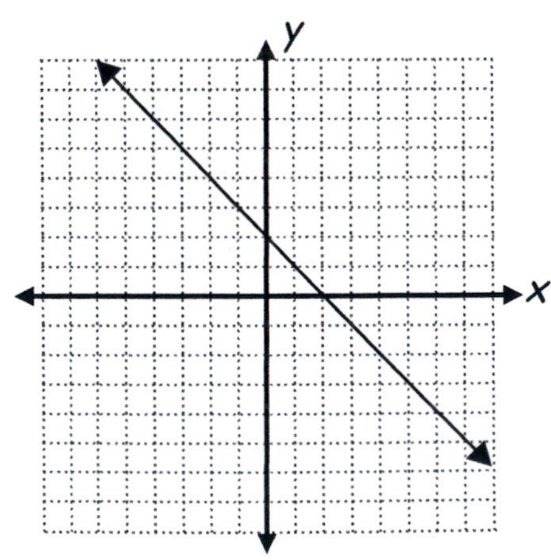

5. $g(x) = -2x + 3$
Use the slope-intercept form;
$m = -2$ and $b = 3$.

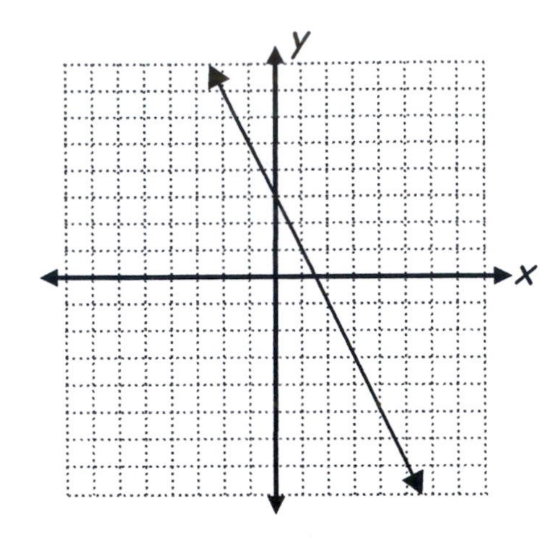

7. $f(x) = 2x - 2$
Use the slope-intercept form;
$m = 2$ and $b = -2$.

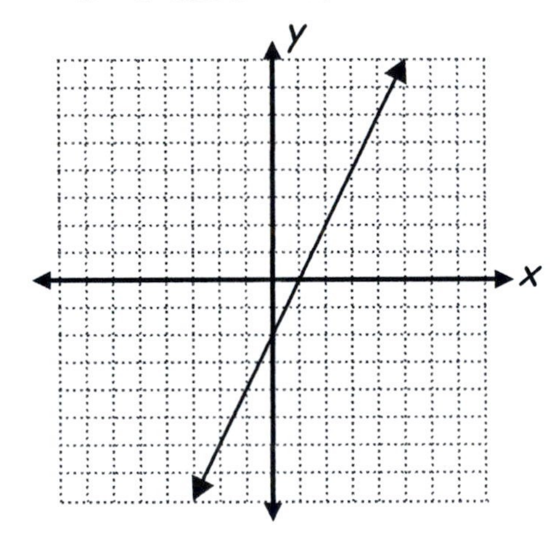

9. $f(x) = -4x + 2$
Use the slope-intercept form;
$m = -4$ and $b = 2$.

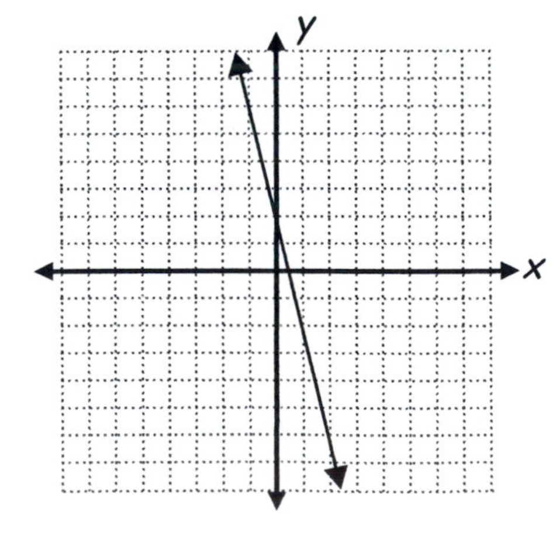

11. $h(x) = 5x - 10$

Use the slope-intercept form; $m = 5$ and $b = -10$.

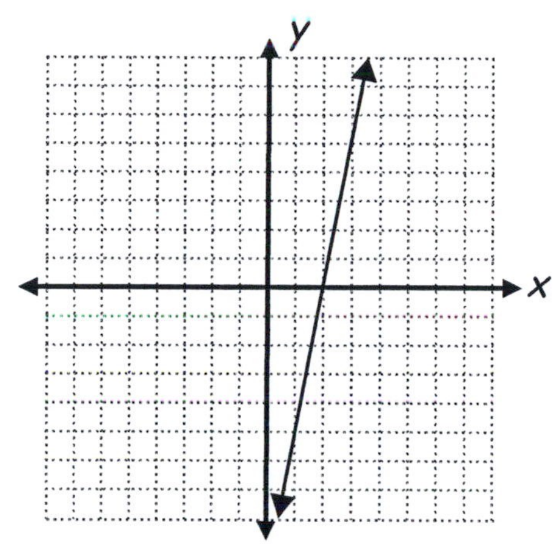

13. $m(x) = -\frac{1}{10}x + \frac{5}{2}$

Use the slope-intercept form; $m = -\frac{1}{10}$ and $b = \frac{5}{2}$.

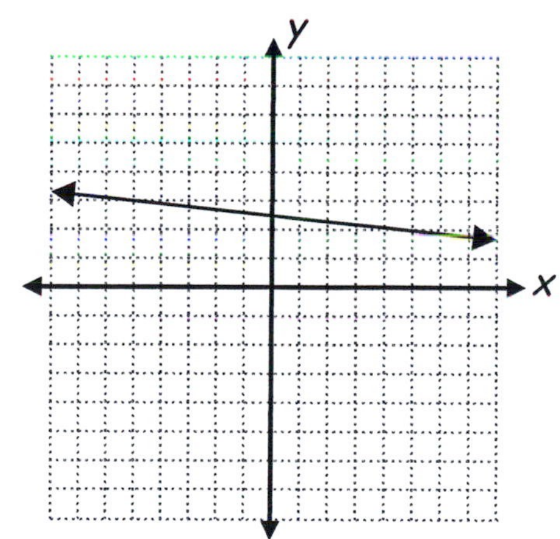

15. $w(x) = -4$

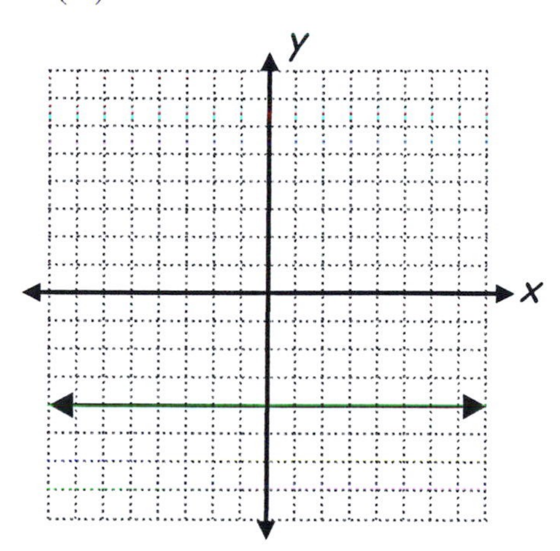

For items 17 - 29, use the form $f(x) = a(x-h)^2 + k$ to find the vertex (h,k). To find the x-intercepts, find the zeros of the quadratic.

17. $g(x) = -(x+2)^2 - 1$

Vertex: $(-2, -1)$

x-intercepts:

$-x^2 - 4x - 5 = 0$

$-(x^2 + 4x + 5) = 0$

$x = \dfrac{-4 \pm \sqrt{16-20}}{2}$

$x = -2 \pm i$ (not real roots)

No x-intercepts

Since there are no x-intercepts it is helpful to find additional points on the graph when graphing. $(-1, -2)$ and $(-3, -2)$ are two such points.

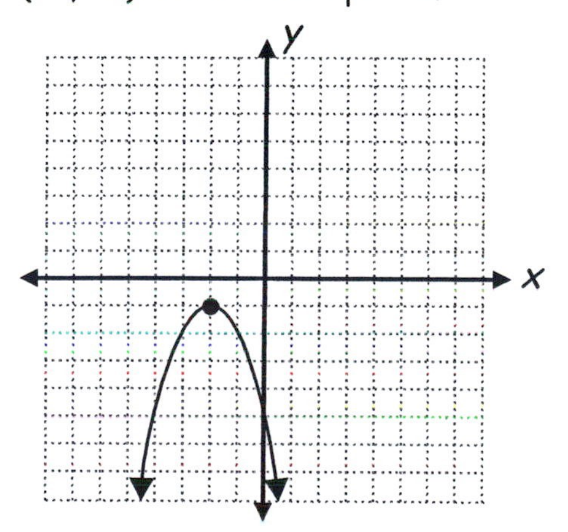

19. $F(x) = 3x^2 + 2$

$= 3(x-0)^2 + 2$

Vertex: $(0, 2)$

x – intercepts:

Set $3x^2 + 2 = 0$. Solve.

$$x^2 = -\frac{2}{3}$$

$$x = \pm\frac{i\sqrt{6}}{3}$$

(not real numbers)

No x-intercepts

Since there are no x-intercepts, it is helpful to find additional points on the graph for graphing. (–1, 5) and (1, 5) are two such points.

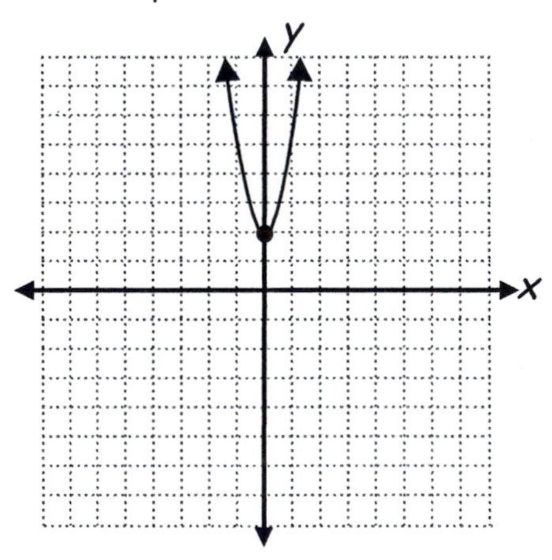

21. $p(x) = -2x^2 + 2x + 12$

$= -2(x^2 - x) + 12$

$= -2\left(x^2 - x + \frac{1}{4}\right) + 12 + \frac{1}{2}$

$= -2\left(x - \frac{1}{2}\right)^2 + \frac{25}{2}$

Vertex: $\left(\frac{1}{2}, \frac{25}{2}\right)$

x-intercepts:

Set $-2x^2 + 2x + 12 = 0$. Solve:

$$x^2 - x - 6 = 0$$

$$(x-3)(x+2) = 0$$

$$x = 3 \text{ and } x = -2$$

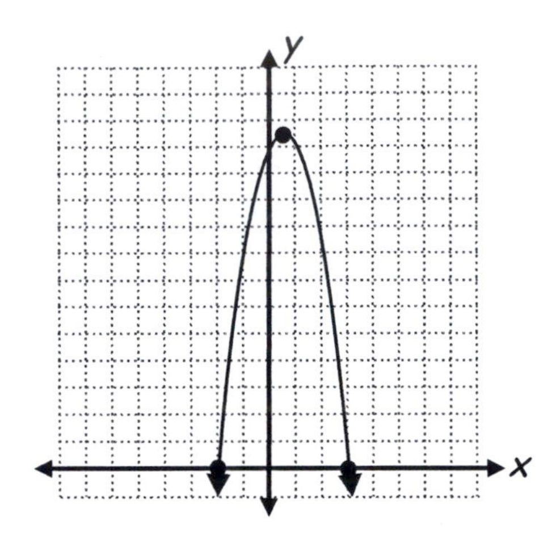

23. $r(x) = -3x^2 - 1$

$= -3(x-0)^2 - 1$

Vertex: $(0, -1)$

x-intercepts:

Set $-3x^2 - 1 = 0$

$$x^2 = -\frac{1}{3}$$

$$x = \pm\frac{i\sqrt{3}}{3} \text{ (not real roots)}$$

No x-intercepts

Since there are no x-intercepts, it is helpful to find additional points on the graph for graphing. $(-1, -4)$ and $(1, -4)$ are two such points.

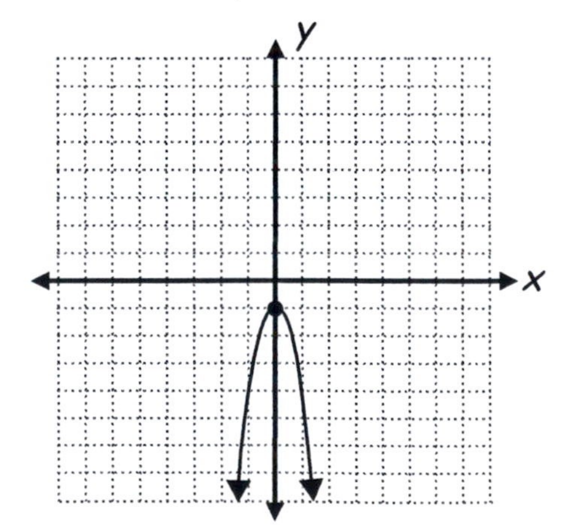

25. $m(x) = x^2 + 2x + 4$

$m(x) = (x^2 + 2x + 1) + 4 - 1$

$m(x) = (x+1)^2 + 3$

Vertex: $(-1, 3)$

x-intercepts:

Set $x^2 + 2x + 4 = 0$. Solve.

$$x = \frac{-2 \pm \sqrt{4 - 4(1)(4)}}{2}$$

$x = -1 \pm i\sqrt{3}$ (not real roots)

No x-intercepts

Since there are no x-intercepts, it is helpful to find additional points on the graph for graphing. $(-2, 4)$ and $(0, 4)$ are two such points.

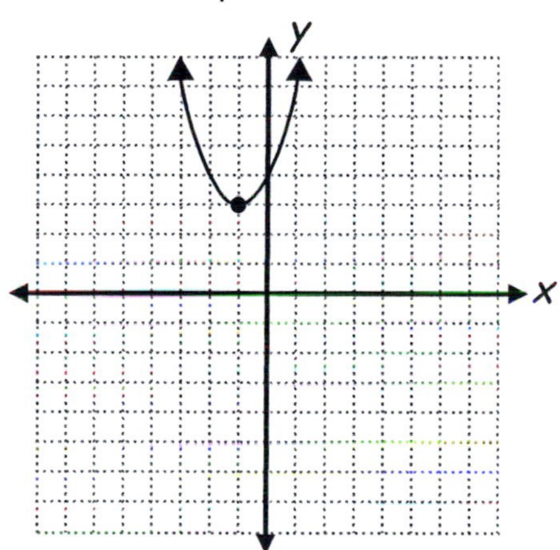

27. $p(x) = -x^2 + 2x - 5$

$= -(x^2 - 2x) - 5$

$= -(x^2 - 2x + 1) - 5 + 1$

$= -(x-1)^2 - 4$

Vertex: $(1, -4)$

x-intercepts:

Set $-x^2 + 2x - 5 = 0$. Solve.

$x^2 - 2x + 5 = 0$

$$x = \frac{2 \pm \sqrt{4 - 4(5)}}{2}$$

$x = 1 \pm 2i$ (not real roots)

No x-intercepts

Since there are no x-intercepts, it is helpful to find additional points on the graph for graphing. $(0, -5)$ and $(2, -5)$ are two such points.

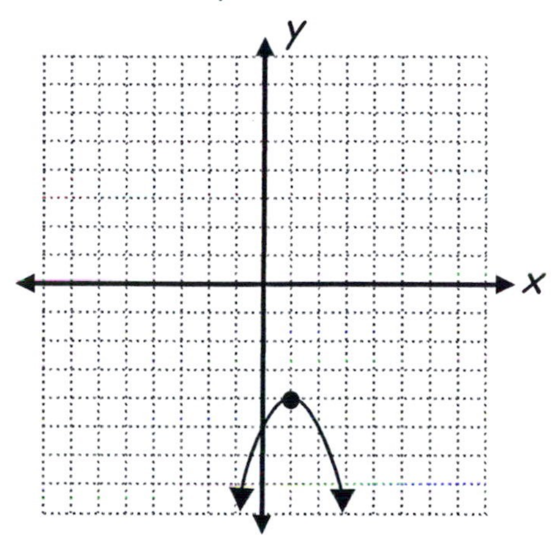

29. $k(x) = 2x^2 - 4x$

$= 2(x^2 - 2x)$

$= 2(x^2 - 2x + 1) - 2$

$= 2(x-1)^2 - 2$

Vertex: $(1, -2)$

x-intercepts:

Set $2x^2 - 4x = 0$

$2x(x-2) = 0$

$x = 0$ and $x = 2$

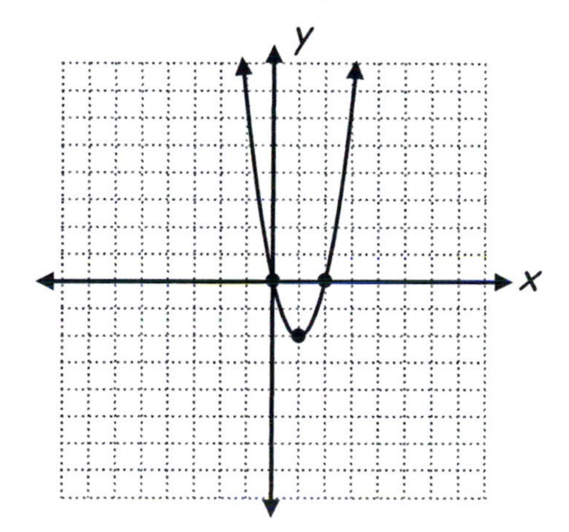

31. Let $x =$ width

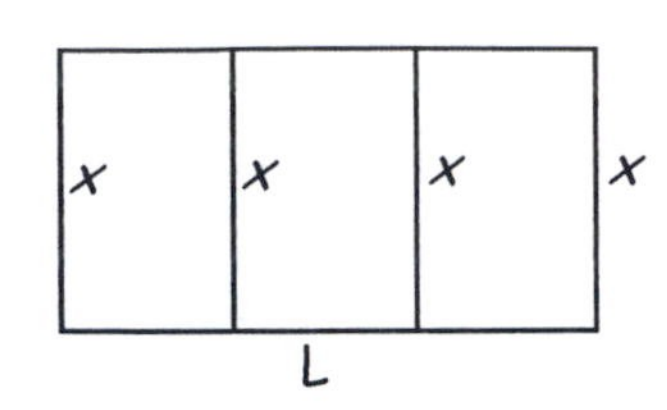

$$\text{Length} = \frac{400-4x}{2} = 200-2x$$

The area:

$A(x) = L \cdot x$

$A(x) = (200-2x)x$

$A(x) = -2x^2 + 200x$

To find the maximum value of $A(x)$, find the vertex by completing the square:

$A(x) = -2(x^2 - 100x)$

$A(x) = -2(x^2 - 100x + 2500) + 5000$

Vertex: $(50, 5000)$

The maximum area is 5000 sq. ft. with a width of 50 ft. and a length of $200 - 2(50) = 100$ ft.

33. Let $x =$ width. Then length $= 10 - x$.

Area $= A(x) = x(10-x)$

$A(x) = 10x - x^2$

To find the maximum area, find the vertex by completing the square:

$$\begin{aligned} A(x) &= -(x^2 - 10x) \\ &= -(x^2 - 10x + 25) + 25 \\ &= -(x-5)^2 + 25 \end{aligned}$$

Vertex: $(5, 25)$

Length: $10 - 5 = 5$ units

The maximum area is 25 sq. units and the dimensions are 5 x 5.

35. Let $y = -2x + 20$

$$\begin{aligned} f(x) &= x^2 + (-2x+20)^2 \\ &= x^2 + 4x^2 - 80x + 400 \\ &= 5(x^2 - 16x + 64) + 400 - 320 \\ &= 5(x-8)^2 + 80 \end{aligned}$$

Vertex: $(8, 80)$

$y = -2(8) + 20 = 4$

$f(x)$ is at a minimum of 80 at $(8, 4)$.

37. $h(t) = -16t^2 + 64t + 48$

To find the maximum height, find the vertex by completing the square:

$$\begin{aligned} h(t) &= -16(t^2 - 4t) + 48 \\ &= -16(t^2 - 4t + 4) + 48 + 64 \\ &= -16(t-2)^2 + 112 \end{aligned}$$

Vertex: $(2, 112)$

Maximum height is 112 ft.

39. $h(t) = -16t^2 + 80t + 64$

To find the maximum height, find the vertex by completing the square:

$$\begin{aligned} h(t) &= -16\left(t^2 - 5t + \frac{25}{4}\right) + 100 + 64 \\ &= -16\left(t - \frac{5}{2}\right)^2 + 164 \end{aligned}$$

Vertex: $\left(\frac{5}{2}, 164\right)$

Maximum height is 164 ft. at $t = \frac{5}{2}$.

41. Let $x=$ the width. Then, $\frac{300-x}{2}$ represents the length:

Area $= A(x) = x\left(\frac{300-x}{2}\right)$

Rewrite and complete the square to find the vertex to expose the maximum area:

$A(x) = -\frac{1}{2}\left(x^2-300x\right)$

$A(x) = -\frac{1}{2}\left(x^2-300x+150^2\right)+\frac{150^2}{2}$

The vertex is $\left(150, \frac{150^2}{2}\right) = (150, 11250)$.

The maximum area is 11,250 ft^2.

43. Using $R(x)=100x-0.1x^2$, find the number of rooms yielding the maximum revenue by finding the vertex:

Complete the square:

$R(x)=-0.1\left(x^2-1000x\right)$

$R(x)=-0.1\left(x^2-1000x+500^2\right)+25{,}000$

$R(x)=-0.1(x-500)^2+25{,}000$

Vertex: $(500, 25000)$

500 rooms produces the maximum revenue of $25,000.

45. Using $C(x)=9000-135x+0.045x^2$, find the number of cars producing the minimum cost by finding the vertex:

Complete the square:

$C(x)=0.045\left(x^2-3000x\right)+9000$

$C(x)=0.045\left(x^2-3000x+1500^2\right)+$
$-101{,}250+9{,}000$

$C(x)=0.045(x-1500)^2-92{,}250$

1500 cars would yield the minimum cost.

47. Let $x=$ the width.

Then, $220-2x$ represents the length.

Area $= A(x) = x(220-2x)$.

$A(x)=-2\left(x^2-110x\right)$

$A(x)=-2\left(x^2-110x+55^2\right)+6050$

$A(x)=-2(x-55)^2+6050$

The maximum area is 6050 ft^2.

49. $f(x)=x^2+1$ opens upward and has a vertex of $(0, 1)$. This matches the graph a.

51. $f(x)=x^2+2x$ can be written as $f(x)=(x+1)^2-1$. So $f(x)$ opens upwards and has a vertex of $(-1, -1)$ which matches the graph b.

53. $f(x)=x^2-2x-3$ can be written as $f(x)=(x-1)^2-4$. So $f(x)$ opens upward and has a vertex of $(1, -4)$, which matches graph f.

55. $f(x)=x^2+2x+2$ can be written as $f(x)=(x+1)^2+1$. So $f(x)$ opens upward and has a vertex of $(-1, 1)$ which matches graph d.

57. Use the "y =" feature and input $y=-x^2-2x+3$. Use the "Calc" feature. Vertex: $(-1, 4)$
To find the x-intercepts, set $y=0$ and solve for x:
$x^2+2x-3=(x+3)(x-1)$
x-intercepts: $x=-3, 1$

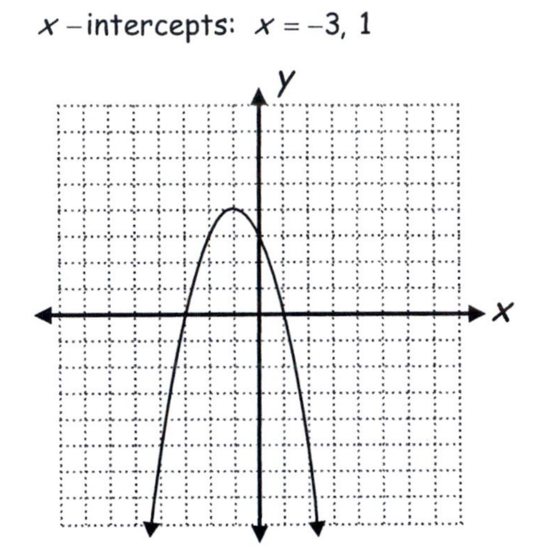

59. Use the "y =" feature and input $y=x^2-4x$. Use the "Calc" feature. Vertex: $(2,-4)$
To find the x-intercepts, set $y=0$ and solve for x:
$x^2-4x=x(x-4)$
x-intercepts: $x=0, 4$

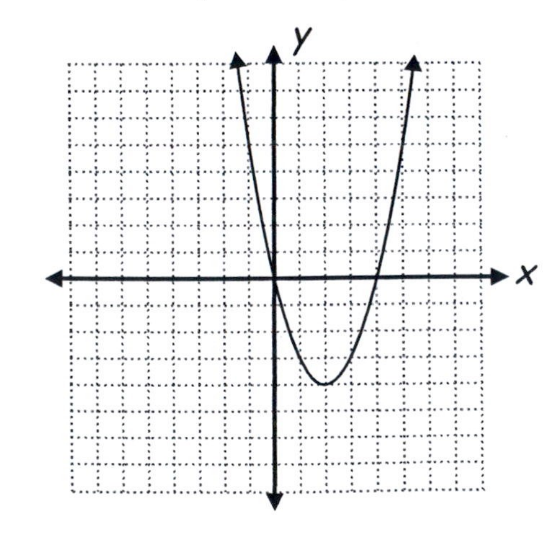

61. Use the "y =" feature and input $y=3x^2+18x$. Use the "Calc" feature. Vertex: $(-3,-27)$
To find the x-intercepts, set $y=0$ and solve for x:
$3x^2+18x=3x(x+6)$
x-intercepts: $x=-6, 0$

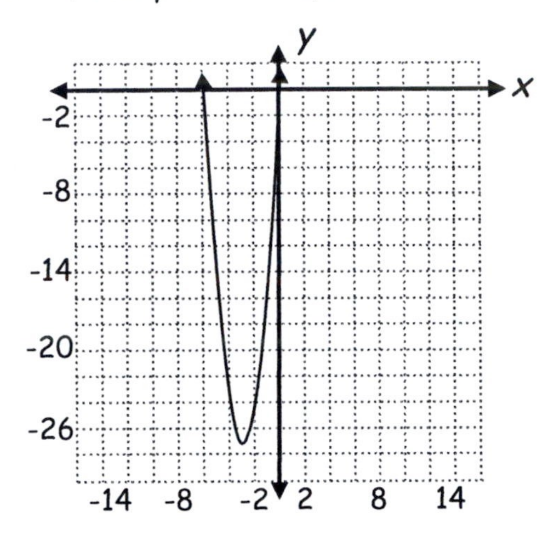

63. Use the "y =" feature and input $y=3x^2-8x+2$. Use the "Calc" feature. Vertex: $(1.\bar{3}, -3.\bar{3})$
To find the x-intercepts, set $y=0$ and solve for x:

$$x=\frac{8\pm\sqrt{64-4(3)(2)}}{2(3)}$$

x-intercepts: $x=\dfrac{4\pm\sqrt{10}}{3}$

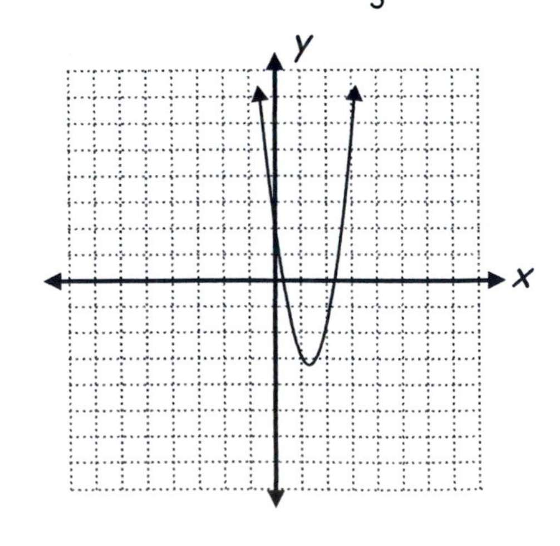

65. Use the "y =" feature and input $y = \frac{1}{2}x^2 + x - 1$. Use the "Calc" feature. Vertex: $(-1, -1.5)$

To find the x-intercepts, set $y = 0$ and solve for x:

$$x = \frac{-1 \pm \sqrt{1+2}}{1}$$

x-intercepts: $x = -1 \pm \sqrt{3}$

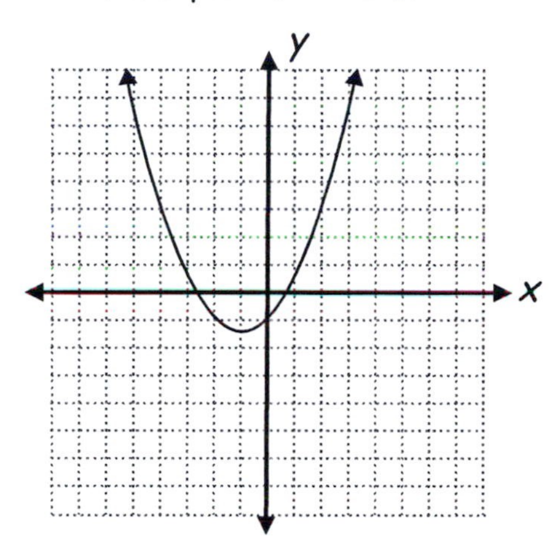

67. Comparison of each of the given quadratics to $y = x^2$:

a. $y = x^2 + 2$: Also opens upward; Shift up 2 units.

b. $y = x^2 - 2$: Also opens upward; Shift down 2 units.

c. $y = x^2 + \frac{9}{2}$: Also opens upward. Shift up $\frac{9}{2}$ units.

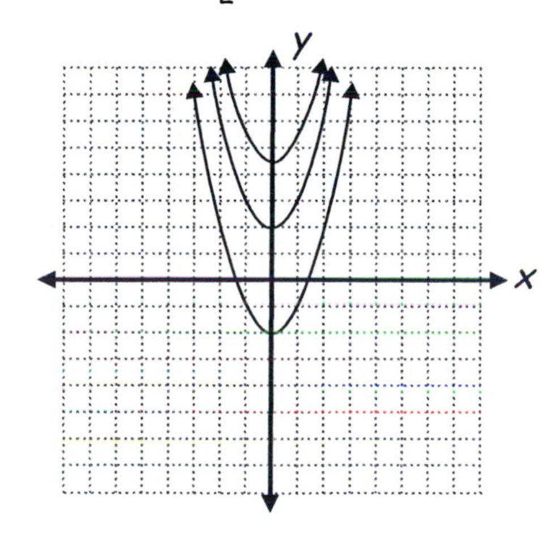

69. Comparison of each of the given quadratics to $y = x^2$:

a. $y = -(x+3)^2 - 3$: Opens downward; Shift left 3 units; shift down 3.

b. $y = (x-2)^2 + 4$: Opens upward; Shift right 2 units; shift up 4.

c. $y = \frac{1}{2}(x+5)^2 + 5$: Opens upward; Shift left 5 units; shift up 5, wider than $y = x^2$.

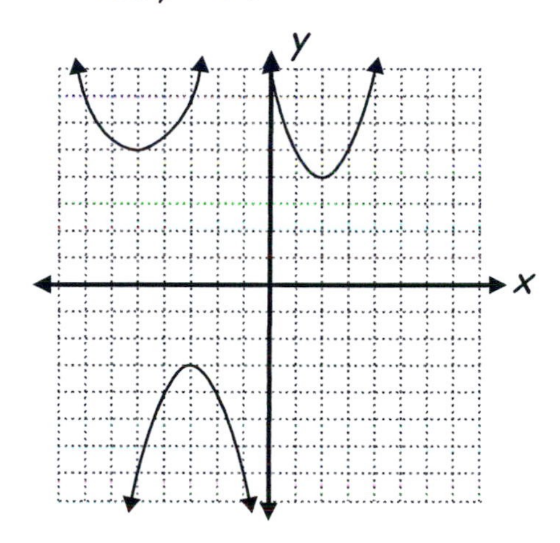

End of Section 3.2

Chapter Three: Section 3.3
Solutions to Odd-numbered Exercises

1. $f(x) = -x^3$

x-intercept: $x = 0$

y-intercept: $y = 0$

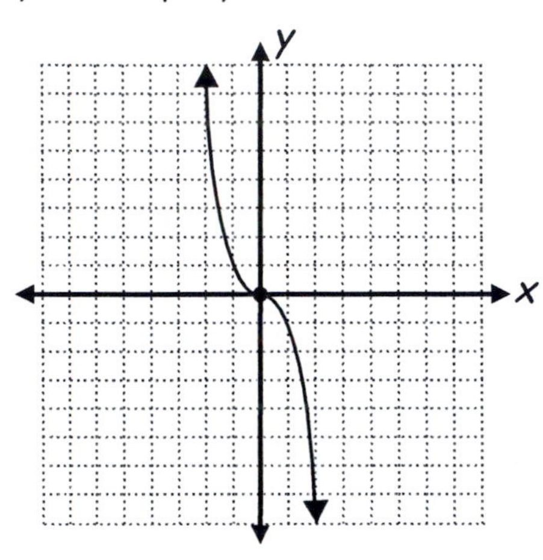

3. $F(x) = \sqrt{x}$

x-intercept: $x = 0$

y-intercept: $y = 0$

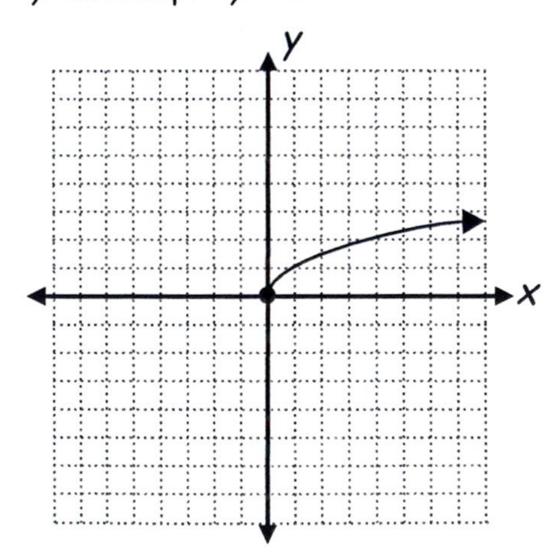

5. $p(x) = -\frac{2}{x}$

x-intercept: none

y-intercept: none

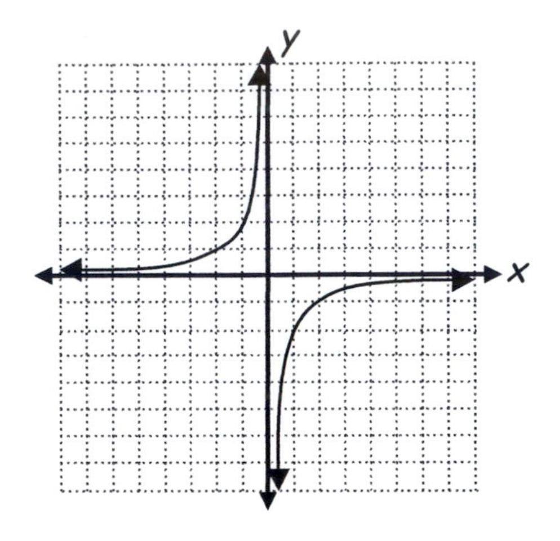

7. $G(x) = -|x|$

x-intercept: $x = 0$

y-intercept: $y = 0$

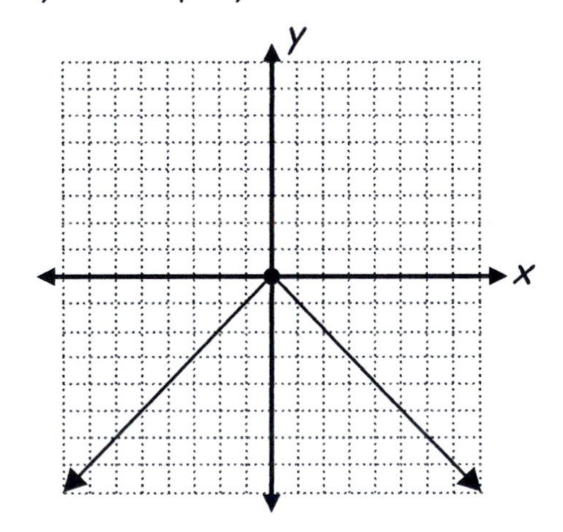

9. $f(x) = 2[\![x]\!]$

x-intercept: $[0, 1)$

y-intercept: $y = 0$

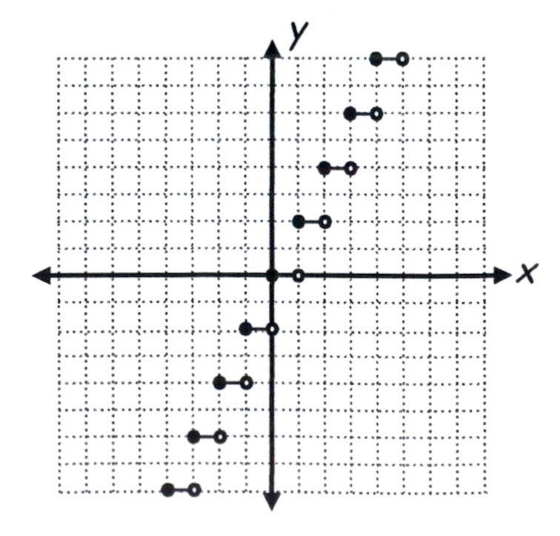

11. $G(x) = \frac{\sqrt{x}}{2}$

x-intercept: $x = 0$

y-intercept: $y = 0$

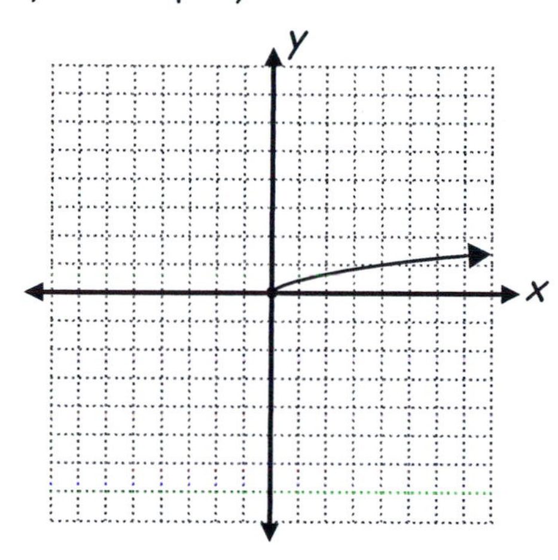

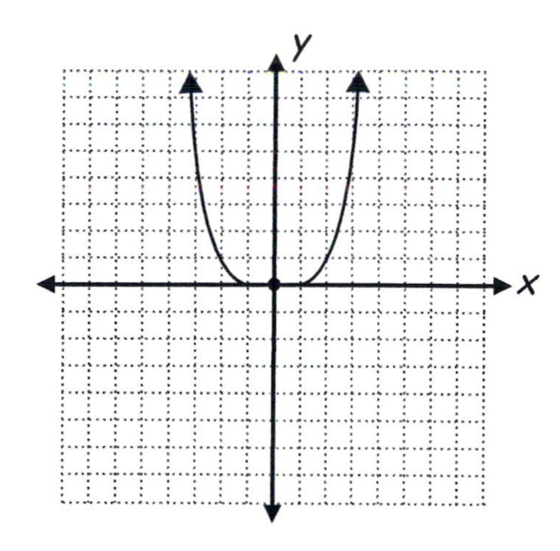

13. $r(x) = 3|x|$

x-intercept: $x = 0$

y-intercept: $y = 0$

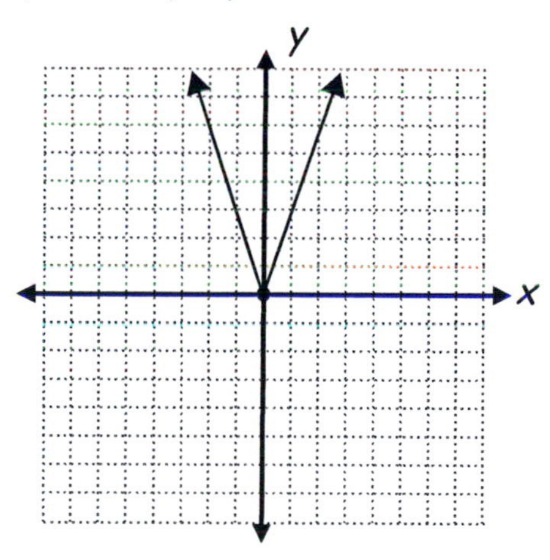

15. $W(x) = \frac{x^4}{16}$

x-intercept: $x = 0$

y-intercept: $y = 0$

17. $h(x) = 2\sqrt[3]{x}$

x-intercept: $x = 0$

y-intercept: $y = 0$

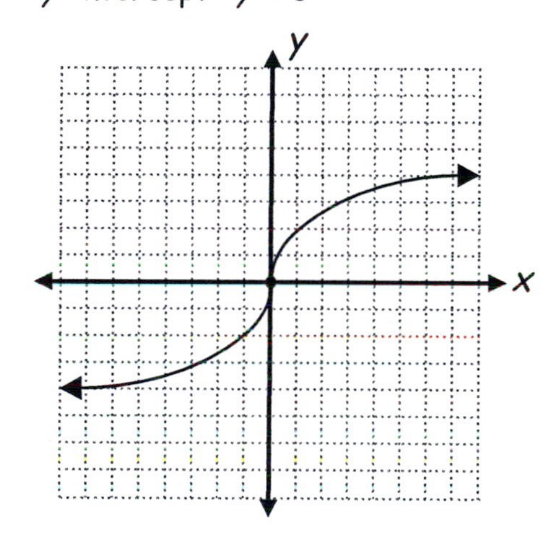

19. $d(x) = 2x^5$

x-intercept: $x = 0$

y-intercept: $y = 0$

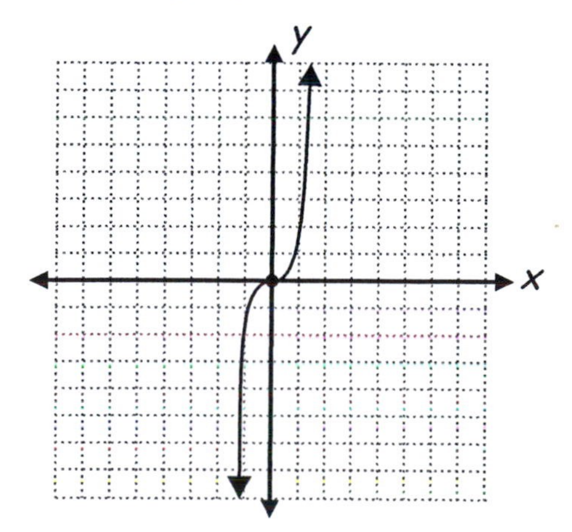

21. $m(x)=\left[\!\left[\dfrac{x}{2}\right]\!\right]$

x-intercept: $[0,2)$

y-intercept: $y=0$

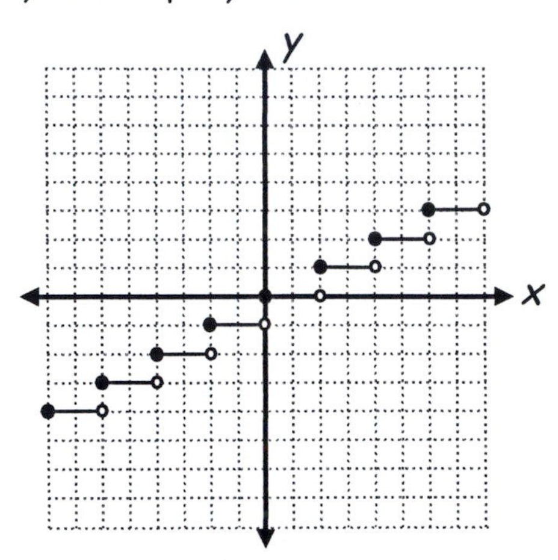

23. $s(x)=-2|x|$

x-intercept: $x=0$

y-intercept: $y=0$

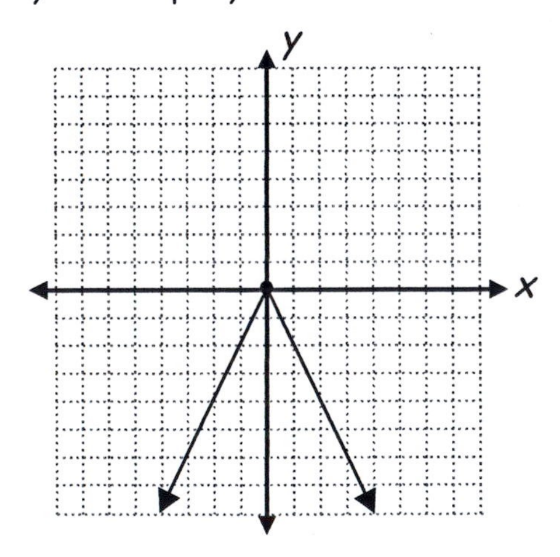

25. $f(x)=\begin{cases}3-x & \text{if } x<-2\\ \sqrt[3]{x} & \text{if } x\geq -2\end{cases}$

x-intercept: $x=0$

y-intercept: $y=0$

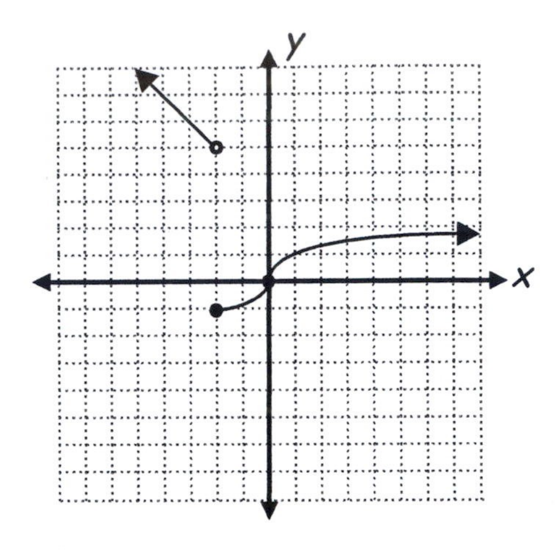

27. $h(x)=\begin{cases}-|x| & \text{if } x<2\\ [\![x]\!] & \text{if } x\geq 2\end{cases}$

x-intercept: $x=0$

y-intercept: $y=0$

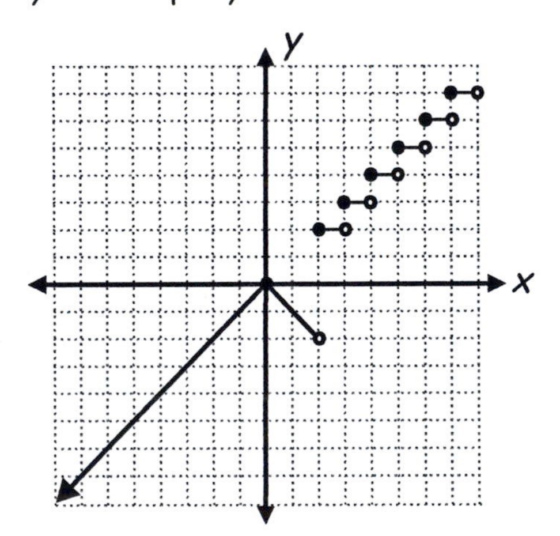

29. $p(x)=\begin{cases}x+1 & \text{if } x<-2\\ x^3 & \text{if } -2\leq x<3\\ -1-x & \text{if } x\geq 3\end{cases}$

x-intercept: $x=0$

y-intercept: $y=0$

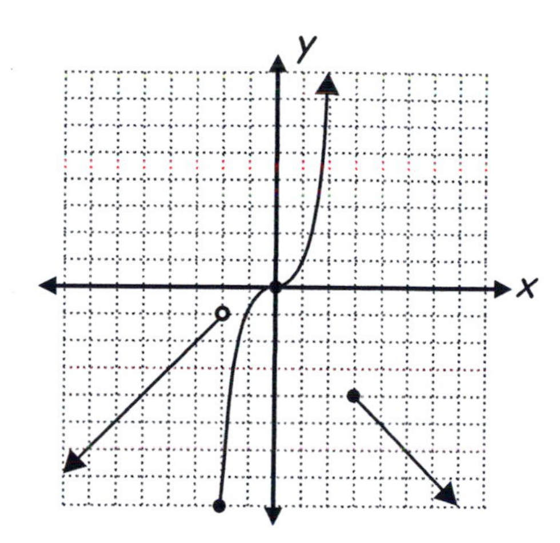

31. $s(x)=\begin{cases} \dfrac{x^2}{3} & \text{if } x<0 \\ -\dfrac{x^2}{3} & \text{if } x \geq 0 \end{cases}$

x-intercept: $x=0$

y-intercept: $y=0$

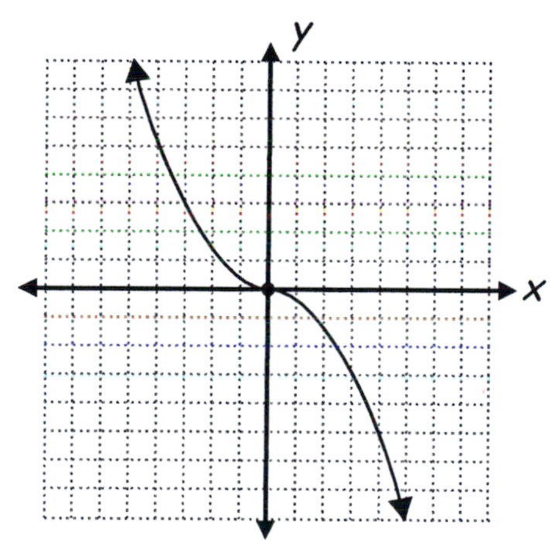

33. $u(x)=\begin{cases} [\![x]\!] & \text{if } x \leq 1 \\ 2x-2 & \text{if } x>1 \end{cases}$

x-intercept: $[0,1)$

y-intercept: $y=0$

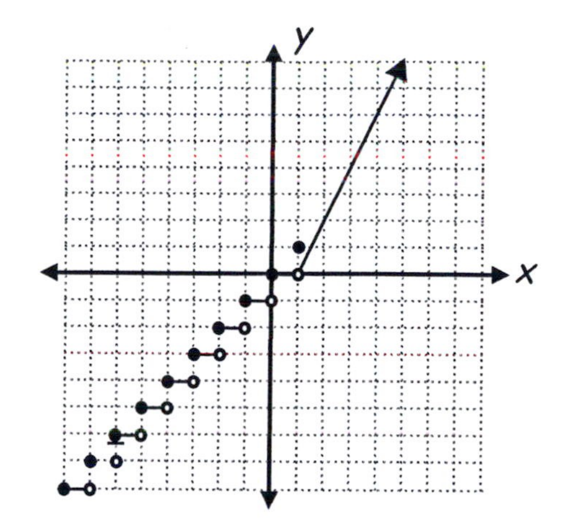

35. $M(x)=\begin{cases} x & \text{if } x \in \mathbb{Z} \\ -x & \text{if } x \notin \mathbb{Z} \end{cases}$

x-intercept: $x=0$

y-intercept: $y=0$

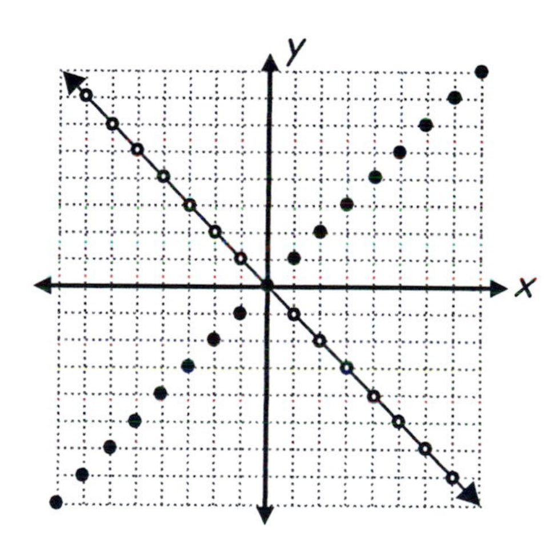

End of Section 3.3

Chapter Three: Section 3.4 Solutions to Odd-numbered Exercises

1. A varies directly as the product of b and h:
$A = kbh$, for some non-zero constant k.

3. W varies inversely as d squared:
$W = k\left(\frac{1}{d^2}\right) = \frac{k}{d^2}$, for some non-zero constant k.

5. r varies inversely as t:
$r = k\left(\frac{1}{t}\right) = \frac{k}{t}$, for some non-zero constant k.

7. x varies jointly as the cube of y and the square of z:
$x = ky^3z^2$, for some non-zero constant k.

9. Given that y varies directly as the square root of x: $y = k\sqrt{x}$, for some non-zero constant k.
Using $y = 36$ when $x = 16$, solve for k:

$$36 = k\sqrt{16}$$
$$k = 9$$

Solve for y when $x = 20$:

$$y = 9\sqrt{20}$$
$$y = 18\sqrt{5}$$

11. Given that y varies directly as the cube root of x: $y = k\sqrt[3]{x}$, for some non-zero constant k.
Using $y = 75$ when $x = 125$, solve for k:

$$75 = k\sqrt[3]{125}$$
$$k = 15$$

Solve for y when $x = 128$:

$$y = 15\sqrt[3]{128}$$
$$y = 60\sqrt[3]{2}$$

13. Given that y varies inversely as the square of x: $y = \frac{k}{x^2}$, for some non-zero constant k.
Using $y = 3$ when $x = 4$, solve for k:

$$3 = \frac{k}{4^2}$$
$$k = 48$$

Solve for y when $x = 8$:

$$y = \frac{48}{8^2}$$
$$y = \frac{3}{4}$$

15. Given that z varies directly as the square of x and inversely as y:
$z = k\frac{x^2}{y}$, for some non-zero constant k.
Using $z = 36$ when $x = 6$ and $y = 7$, solve for k:

$$36 = k\frac{6^2}{7}$$
$$k = 7$$

Solve for z when $x = 12$ and $y = 21$:

$$z = 7\left(\frac{12^2}{21}\right)$$
$$z = 48$$

17. Given that z is jointly proportional to x and y:
$z = kxy$, for some non-zero constant k.
Using $z = 90$ when $x = 1.5$ and $y = 3$, solve for k:

$$90 = k(1.5)(3)$$
$$k = 20$$

Solve for z when $x = 0.8$ and $y = 7$:

$$z = 20(0.8)(7)$$
$$z = 112$$

19. Given: d (distance) $= kt^2$ for some non-zero constant k.

Using $d = 144$ when $t = 3$, find k:

$$144 = k(3)^2$$
$$k = 16$$

Solve for d when $t = 4$:

$$d = 16(4)^2$$
$$d = 256 \text{ ft.}$$

21. Given: $BMI = \frac{kw}{h^2}$ for $w =$ weight (lb) and $h =$ height (in) with some non-zero constant k.

Using $BMI = 24.41$ when $w = 180$ and $h = 72$, find k:

$$24.41 = \frac{k(180)}{72^2}$$
$$k = 703.008$$

Solve for BMI when $w = 120$ and $h = 64$:

$$BMI = 703.008\left(\frac{120}{64^2}\right)$$
$$BMI \approx 20.60$$

23. Given: Load (L) varies directly as the product of width (w) and the square of the height (h) and inversely as the length (l):

I.e., given: $L = \frac{kwh^2}{l}$ for some non-zero constant k.

Using $L = 200$ kg when $l = 10$ m, $w = 10$ cm and $h = 5$ cm, find k:

$$200 = \frac{k(10)(5)^2}{10}$$
$$k = 8$$

Solve for l when $L = 300$ kg:

$$300 = \frac{8(10)(5)^2}{l}$$
$$l \approx 6.7 \text{ meters}$$

25. Given: Time (t) for flow is inversely proportional to the square of the radius (r) of the pipe: $t = \frac{k}{r^2}$, for some non-zero constant k.

Using $t = 25$ sec for $r = 1$ cm, find k:

$$25 = \frac{k}{1}$$
$$k = 25$$

Solve for r when $t = 16$:

$$16 = \frac{25}{r^2}$$
$$r = \frac{5}{4} \text{ cm}$$

27. Given: Circumference (C) varies directly as the diameter (d): $C = kd$, for some non-zero constant k.

Using $d = 13$ when $C = 40.82$, find k:

$$40.82 = k(13)$$
$$k = 3.14$$

Find C when $d = 11$ (radius $= 5.5$)

$$C = 3.14(11)$$
$$C = 34.54 \text{ in.}$$

29. Given: Area (A) varies directly as the sum of the radius (r) times the height (h) and the square of the radius:

I.e., given: $A = k\left(rh + r^2\right)$ for some non-zero constant k.

Using $A = 1099 \text{ in}^2$ when $h = 18$ in., and $r = 7$ in, find k:

$$1099 = k\left(7(18) + 7^2\right)$$
$$k = 6.28$$

Find A when $h = 5$ and $r = 3.2$

$$A = 6.28\left(3.2(5) + 3.2^2\right)$$
$$A = 164.7872 \text{ in}^2$$

31. Given: Voltage (V) is directly proportional to power (p) and inversely proportional to the current (c):

I.e., given: $V = \frac{kp}{c}$ for some non-zero constant k.

Using $V = 18$ volts when $p = 54$ watts and $c = 3$ amps, find k:

$$18 = \frac{k(54)}{3}$$
$$k = 1$$

Solve for p when $V = 18$ and $c = 0.5$:

$$18 = \frac{1(p)}{0.5}$$
$$p = 9 \text{ watts}$$

33. Given: Electric pressue (P) varies directly with the square of the surface charge (σ) and inversely with the permittivity (ε):

$P = \frac{k\sigma^2}{\varepsilon}$, for some non-zero constant k:

Using $P = 6\ N/m^2$ when $\sigma = 6$ coulombs per unit area and $\varepsilon = 3$, find k:

$$6 = \frac{k(6)^2}{3}$$
$$k = \frac{1}{2}$$

Then, $P(\sigma,\varepsilon) = \frac{1}{2}\left(\frac{\sigma^2}{\varepsilon}\right) = \frac{\sigma^2}{2\varepsilon}$

35. Given: a is proportional to $\sqrt{b}$, we have $a = k\sqrt{b}$ for some non-zero constant k.

Using $a = 15$ when $b = 9$, solve for k.

$$15 = k\sqrt{9}$$
$$k = 5$$

So $a = 5\sqrt{b}$. When $b = 12$, we have

$$a = 5\sqrt{12}$$
$$= 10\sqrt{3}$$

37. $a = kb^2$ for some non-zero constant k.

Given: $9 = k(2)^2$. Then, $k = \frac{9}{4}$.

$$a = \frac{9b^2}{4}$$

Solve for a:

$$a = \frac{9}{4}(4)^2$$
$$a = 36$$

39. $a = kbc$ for some non-zero constant k.

Given: $210 = k(14)(5)$. Then, $k = 3$.

$$a = 3bc$$
$$a = 3(6)(6)$$
$$a = 108$$

41. Given: Price (p) varies directly with the number of gallons (G) purchased:

$p = kG$, for some non-zero constant k:

Using $34.40 = k(16)$, solve for k:

$$k = 2.15$$

Then, $p = 2.15G$. Solve for p:

$$p = 2.15(20)$$
$$p = \$43$$

43. Given: Distance (d) of stretch varies directly as the weight (w): $d = kw$, for some non-zero constant k:

Using $d = 9$ cm when $w = 15$ g, solve for k: $9 = k(15)$. So, $k = \frac{3}{5}$

Then, $d = \frac{3}{5}w$.

Solve for d when $w = 20$ g.

$$d = \frac{3}{5}(20)$$
$$d = 12 \text{ cm}$$

45. Given: Volume (V) varies inversely as the pressure (p) on the gas: $V = \frac{k}{p}$, for some non-zero constant k:
Using $V = 100 \text{ cm}^3$ when $p = 8$ g, solve for k: $100 = \frac{k}{8}$. So, $k = 800$.
Solve for V when $p = 4$ g:
$V = \frac{800}{4} = 200 \text{ cm}^3$

47. Given: The resistance (R) of a wire varies directly as its length (L) and inversely as the square of the diameter (d): $R = \frac{kL}{d^2}$, for some non-zero constant k:
Using $20 = \frac{k(500)}{(0.015)^2}$, solve for k:
$k = 9 x 10^{-6}$, or 0.000009
$R = \frac{0.000009(L)}{d^2}$
Solve for R when $L = 1200$ ft. and $d = 0.025$ in.:
$R = \frac{0.000009(1200)}{(0.025)^2} = 17.28$ ohms

End of Section 3.4

Chapter Three: Section 3.5
Solutions to Odd-numbered Exercises

1. Given $f(x)=(x+2)^3$: The more basic function is $F(x)=x^3$.

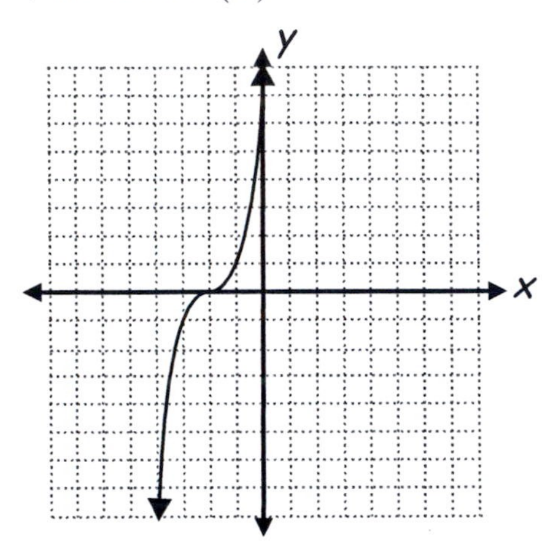

A shift of 2 to the left.
Domain = Range =

3. Given $p(x)=-(x+1)^2+2$: The more basic function is $P(x)=-x^2$.

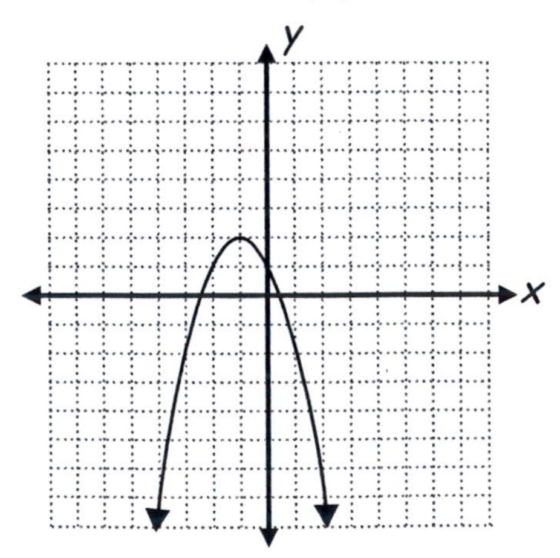

A shift of 1 to the left and up 2.
Domain = $\mathbb{R}$; Range = $(-\infty, 2]$

5. Given $q(x)=(1-x)^2$: The more basic function is $Q(x)=(-x)^2=x^2$.

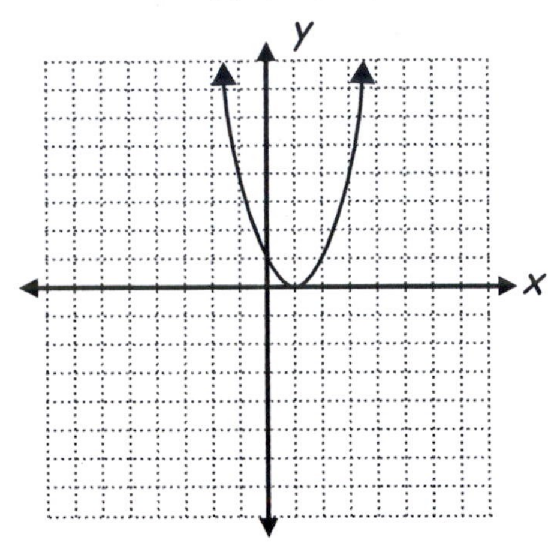

A shift of 1 to the right .
Domain = $\mathbb{R}$; Range = $[0, \infty)$

7. Given $s(x)=\sqrt{2-x}$: The more basic function is $S(x)=\sqrt{x}$.

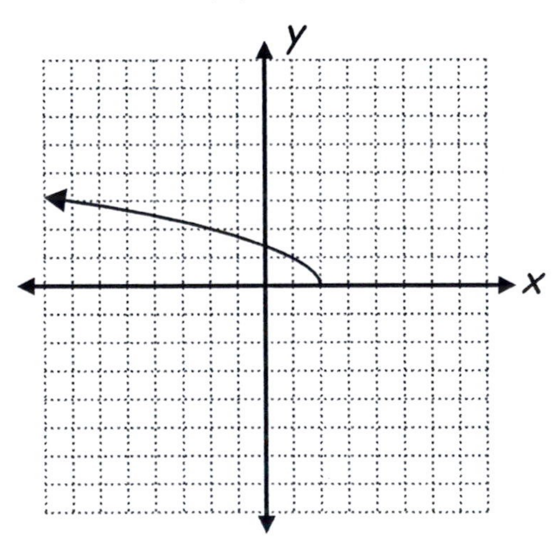

A reflection through the y-axis and shift of 2 to the right .
Domain = $(-\infty, 2]$; Range = $[0, \infty)$

9. Given $w(x)=\frac{1}{(x-3)^2}$: The more basic function is $W(x)=\frac{1}{x^2}$.

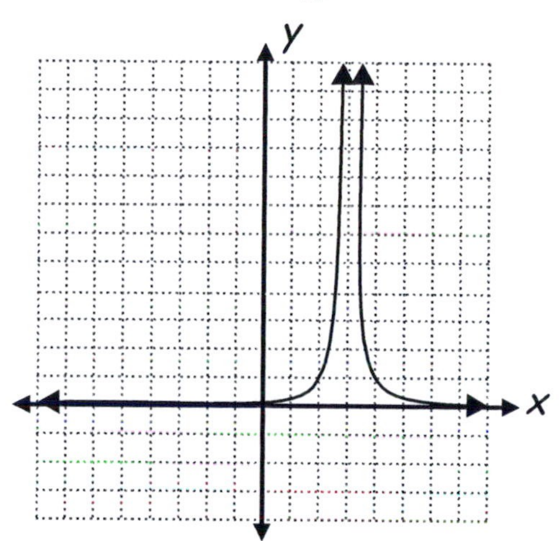

A shift of 3 to the right.
Domain = $(-\infty,3)\cup(3,\infty)$
Range = $(0,\infty)$

11. Given $f(x)=\frac{1}{2-x}$ or $-\frac{1}{x-2}$:
The more basic function is $F(x)=\frac{1}{x}$.

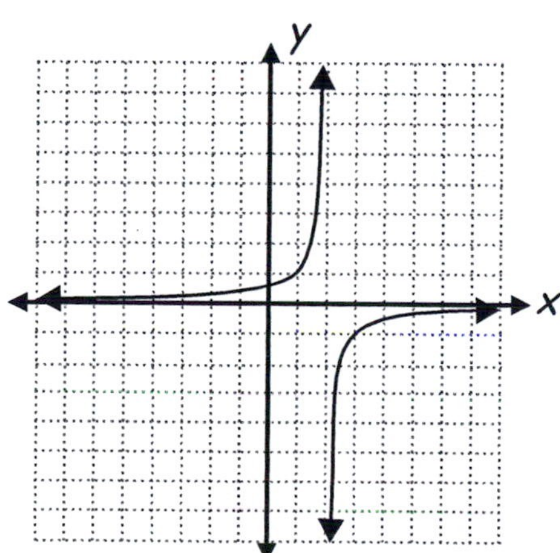

A reflection through the x-axis, and a shift of 2 to the right.
Domain = $(-\infty,2)\cup(2,\infty)$
Range = $(-\infty,0)\cup(0,\infty)$

13. Given $b(x)=[\![x-4]\!]+4$: The more basic function is $B(x)=[\![x]\!]$.

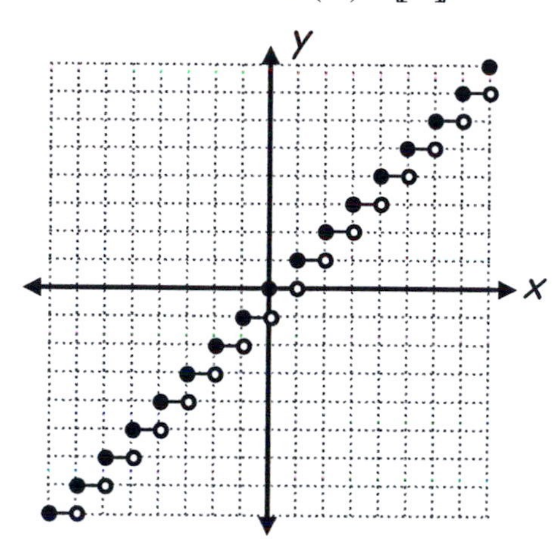

The shift of 4 to the right and the shift of 4 up do not "show" because the shifts correspond identically to the graph of the basic function.
Domain = $\mathbb{R}$; Range = $\mathbb{Z}$

15. Given $S(x)=(3-x)^3$: The more basic function is $s(x)=x^3$.

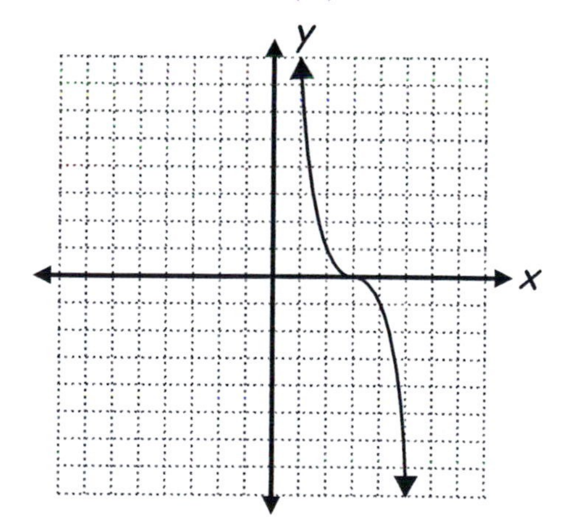

The transformation is a reflection through the y-axis and a shift of 3 to the right.
Domain = $\mathbb{R}$; Range = $\mathbb{R}$

17. Given $h(x) = \frac{x^2}{2} - 3$: The more basic function is $H(x) = x^2$.

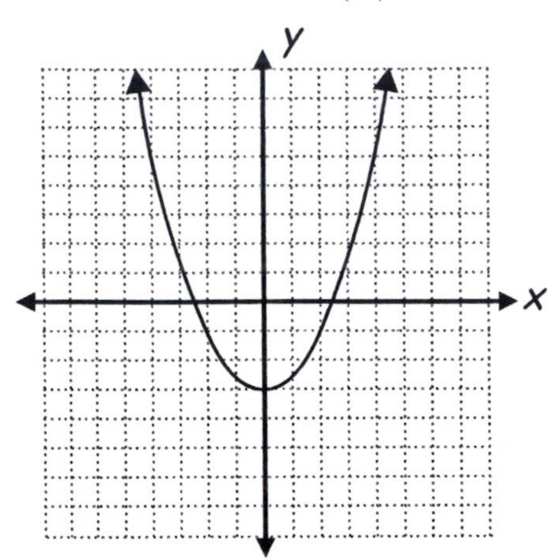

There is a compression of half a unit and a shift of 3 units down.
Domain = $\mathbb{R}$; Range $= [-3, \infty)$

19. Re-write as $g(x) = (x-3)^2$
The more basic function is $G(x) = x^2$.

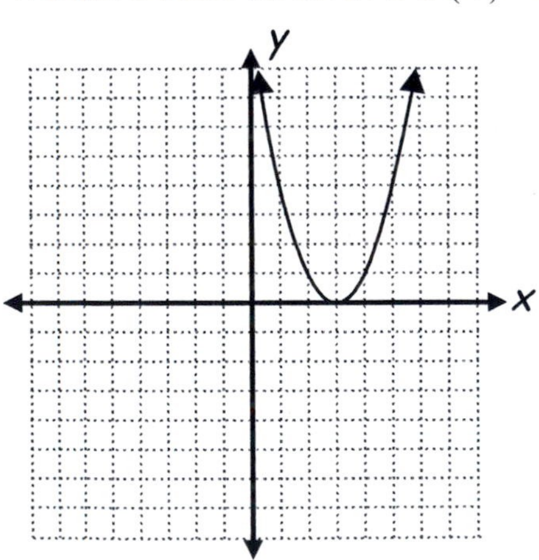

A shift of 3 to the right.
Domain = $\mathbb{R}$; Range $= [-0, \infty)$

21. Given $W(x) = \frac{x-1}{|x-1|}$. The more basic function is $w(x) = \frac{x}{|x|}$.

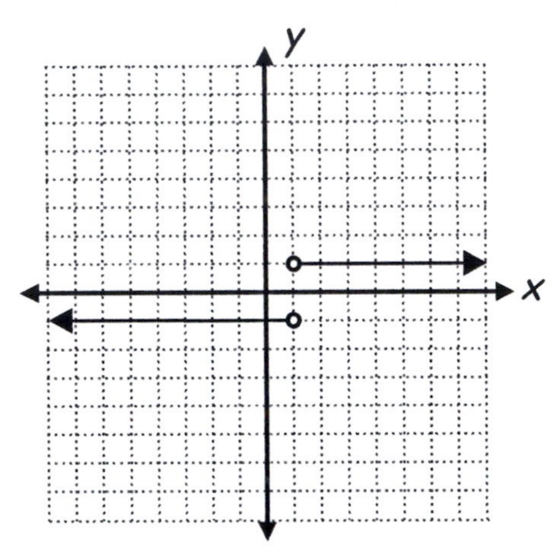

A shift of 1 to the right.
Notice that W is not defined at $x = 1$.
Domain = $(-\infty, 1) \cup (1, \infty)$; Range $= \{-1, 1\}$

23. Given $V(x) = -3\sqrt{x-1} + 2$, the more basic function is $v(x) = \sqrt{x}$.

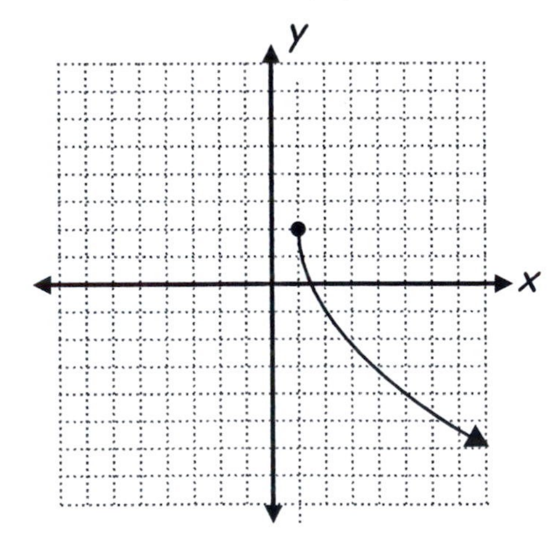

There are four transformations on the basic function:
Reflection about the x-axis;
A shift up of 2;
A shift to the right of 1;
A stretch vertically by a factor of 3.

25. If the function g is the function f, $f(x)=x^2$, moved 4 units to the right and 2 units up, then $g(x)=(x-4)^2+2$.

27. If the function g is the function f, $f(x)=x^2$, moved 2 units to the right and reflected across the y-axis, then $g(x)=(-x-2)^2$.

29. If the function g is the function f, $f(x)=x^3$, moved 10 units to the right and 4 units up, then $g(x)=(x-10)^3+4$.

31. If the function g is the function f, $f(x)=\sqrt{x}$, moved 3 units down and reflected across the y-axis, then $g(x)=\sqrt{-x}-3$.

33. If the function g is the function f, $f(x)=|x|$, moved 8 units to the right, 2 units up and reflected across the x-axis, then $g(x)=-|x-8|+2$.

35. Given $g(x)=x^3$. This function is odd (It is a polynomial with only odd exponents--is the simplest test.).

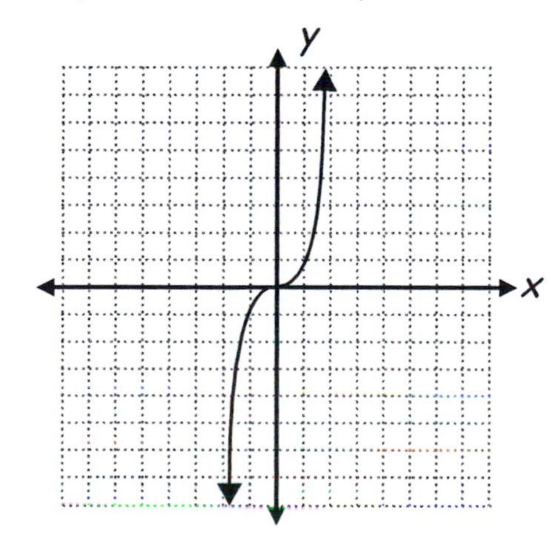

37. Given $w(x)=\sqrt[3]{x}$. This function is odd. To verify, test to determine if it meets the definition:

$$w(-x)=\sqrt[3]{-x}$$
$$=\sqrt[3]{-1(x)}$$
$$=-1\sqrt[3]{x}=-w(x)$$

w satisfies the definition.

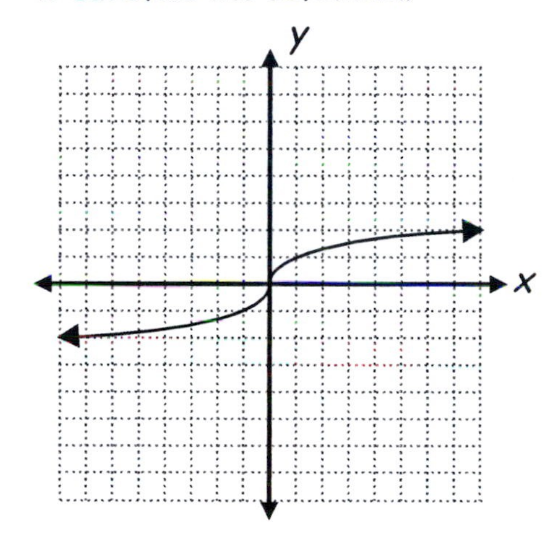

39. Given $3y-2x=1$, you can re-write this as: $y=\frac{2}{3}x+\frac{1}{3}$.

You can test whether the equation has y-axis, x-axis or origin symmetry:

y-axis: substituting $-x$ does not result in an equivalent equation:

$$y=-\frac{2}{3}x+\frac{1}{3}.$$

x-axis: substituting $-y$ does not result in an equivalent equation:

$$-y=\frac{2}{3}x+\frac{1}{3} \text{ or.}$$
$$y=-\frac{2}{3}x-\frac{1}{3}.$$

origin: substituting $-x$ and $-y$ does not result in an equivalent equation:

$$-y=-\frac{2}{3}x+\frac{1}{3} \text{ or}$$
$$y=\frac{2}{3}x-\frac{1}{3}.$$

Thus the equation does not posess y-axis, x-axis or origin symmetry.

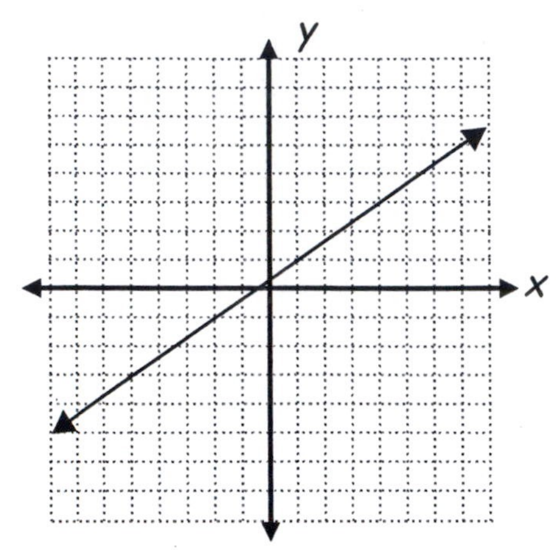

41. Given $F(x)=(x-1)^2$. You can also express this function as $F(x)=x^2-2x+1$, and determine explicitly that it is a polynomial having both odd and even exponents, so it is neither.
You can also test whether the function meets the definitions of odd or even:
It is not odd because

$$f(-x)=(-x-1)^2=x^2+2x+1$$
$$\neq -f(x).$$

It is not even because $f(-x)\neq f(x)$.
Therefore, it is neither

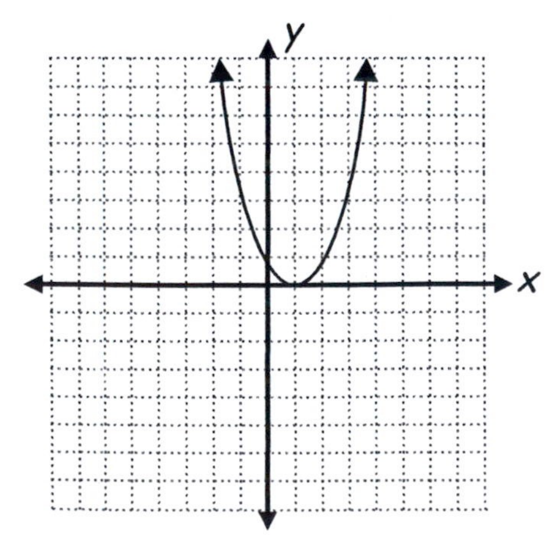

43. For the equation $x=|2y|$, if you replace y with $-y$, you get $x=|2(-y)|$ which is equivalent to $x=|2y|$. Then, the equation is symmetric with respect to the x-axis.

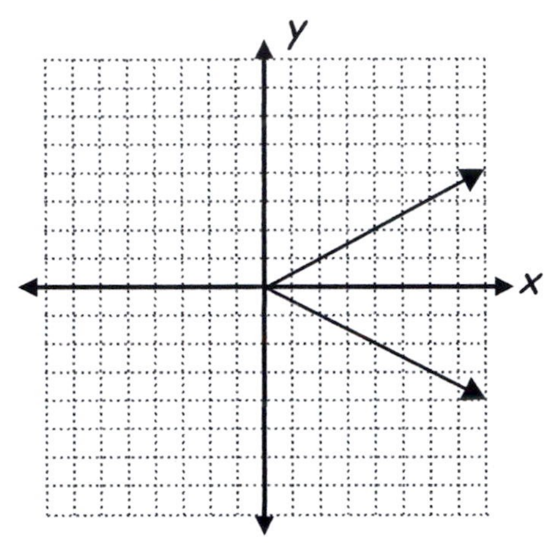

45. The function $s(x)=\left[\!\left[x+\frac{1}{2}\right]\!\right]$ can be tested directly to show that it is neither odd nor even:

If s were odd, then $s(-x)=-s(x)$, for all x in the domain. But, $s\left(-\frac{1}{2}\right)=0$, and $-s\left(\frac{1}{2}\right)=-1$, so s is not odd.
If s were even, then $s(-x)=s(x)$, for all x in the domain. But, $s(-1)=-1$, and $s(1)=1$, s is not odd.
Therefore, the function is neither.

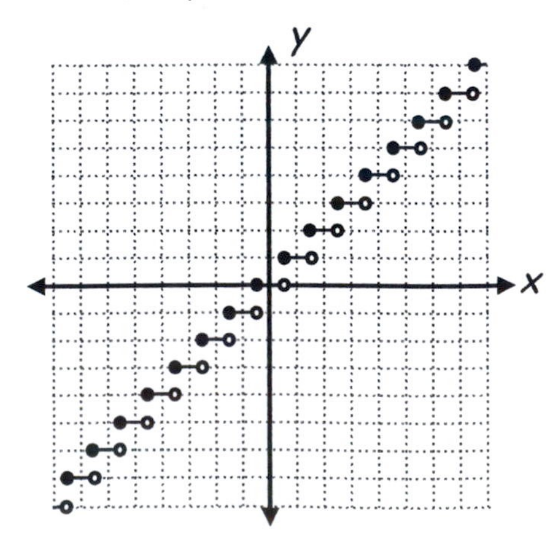

47. Given $xy = 2$, the equation is symmetric with respect to the origin because, replacing x with $-x$ and y with $-y$ gives $(-x)(-y) = xy = 2$, which is equivalent to the original equation.

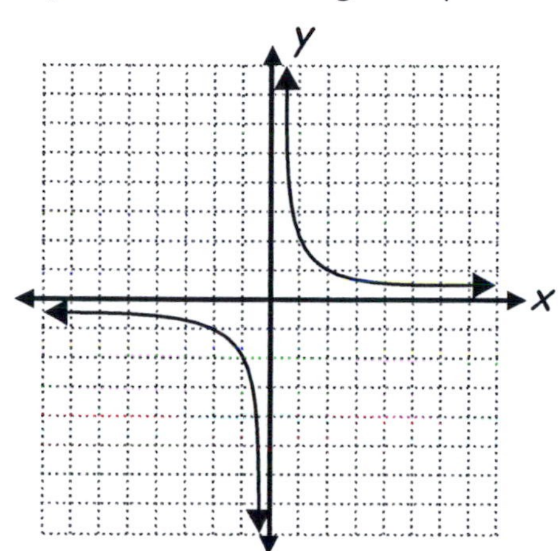

49. The function $f(x) = (x+3)^2$ is a parabola with vertex $(-3, 0)$ and the vertex is a minimum point. This means that the function is decreasing on the interval $(-\infty, -3)$ and increasing on the interval $(-3, \infty)$.

51. The function $h(x) = \dfrac{1}{x-1}$ is not defined at $x = 1$. You should examine the intervals $(-\infty, 1)$ and $(1, \infty)$ by graphing the function:

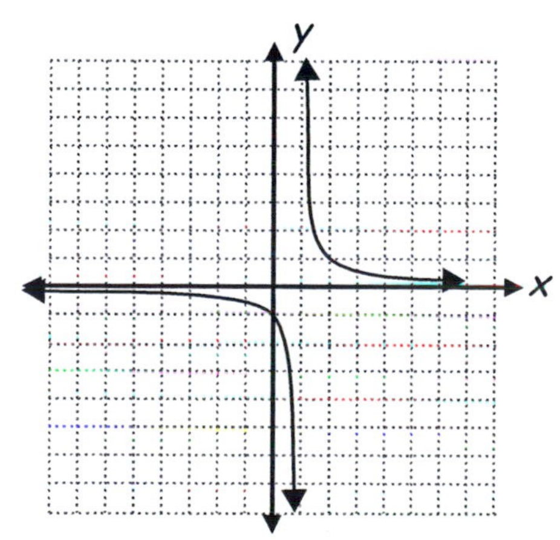

The graph of h indicates that it is decreasing over the intervals $(-\infty, 1)$ and $(1, \infty)$.

53. As a real-valued function, $G(x) = \sqrt{x+1}$ is defined over the interval $[-1, \infty)$, and it is increasing throughout the interval.

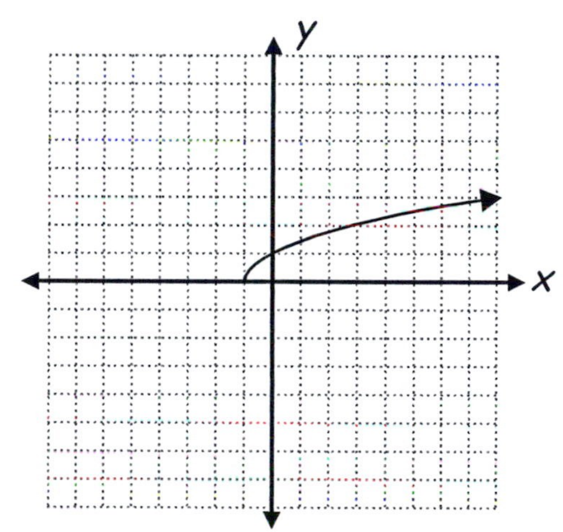

55. $p(x) = -30|x-1|$ is defined over the interval $(-\infty, \infty)$, and it is increasing over $(-\infty, 1)$ and decreasing over $(1, \infty)$.

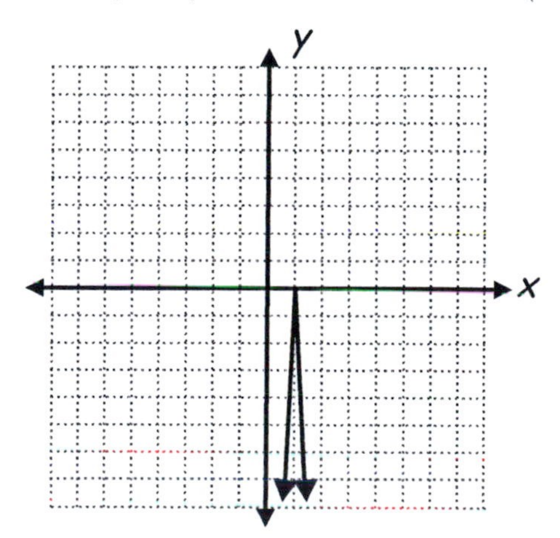

57. The function $r(x)=\dfrac{(x-7)^4}{-2}+4$ is defined over the interval $(-\infty,\infty)$. You should examine the intervals $(-\infty,7)$ and $(7,\infty)$ and realize that the more basic function is $R(x)=x^4$. The given function r is a shift of 7 to the right, a shift up of 4, a compression and a reflection. The function r is increasing on $(-\infty,7)$ and decreasing on $(7,\infty)$.

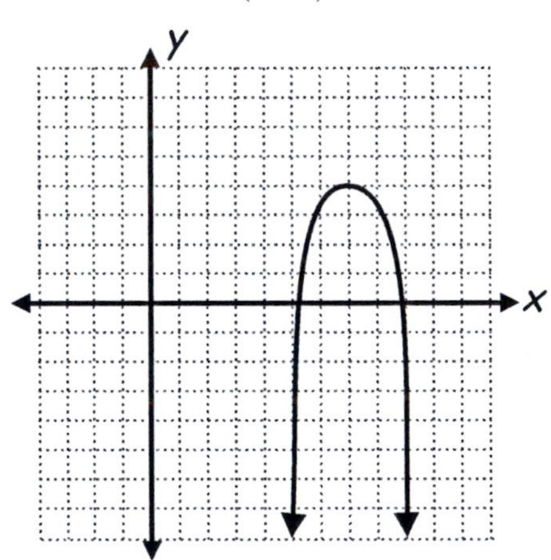

59. The function $Q(x)=\begin{cases}|x-1| \text{ if } x\le 3\\ 5-x \text{ if } x>3\end{cases}$ is defined over the interval $(-\infty,\infty)$. Intervals of special interest for close examination are $(-\infty,1)$, $(1,3)$ and $(3,\infty)$. Graph Q over each of these intervals:

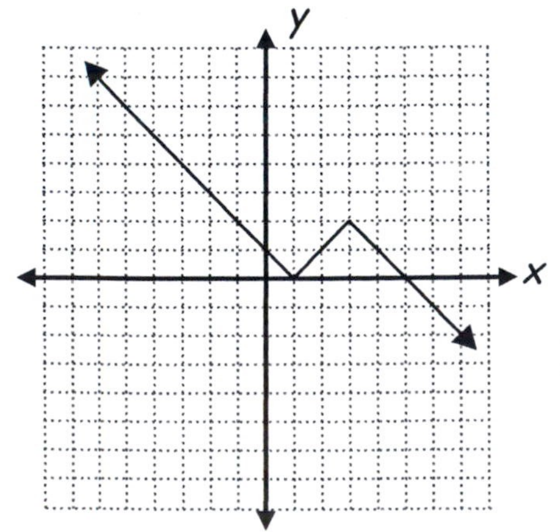

The function Q is decreasing on $(-\infty,1)$, increasing on $(1,3)$ and decreasing on $(3,\infty)$.

61. The cost function is $C(t)=2-0.23[\![-t+1]\!]$, where t is the number of minutes the call lasted. A table of values for the first 9 minutes shows the values are monotonic over the indicated intervals:

$(n,n+1]$	$C(t)$
$(0,1]$	2
$(1,2]$	2.23
$(2,3]$	2.46
$(3,4]$	2.69
$(4,5]$	2.92
$(5,6]$	3.15
$(6,7]$	3.38
$(7,8]$	3.61
$(8,9]$	3.84

For an 8 minute 35 second call, the table indicates that the call would cost \$3.84. Symbolically, $C(8\text{min. }35\text{sec.})=\$3.84$

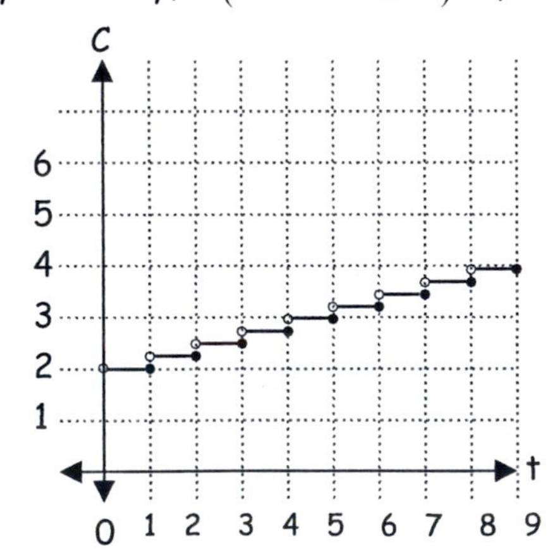

63. Graph of $P(x)=\begin{cases}-2\sqrt{x}+8, & 0\le x<3\\ -x+8, & x\ge 3\end{cases}$:

The graph is decreasing on $(0,3)$ and $(3,\infty)$.

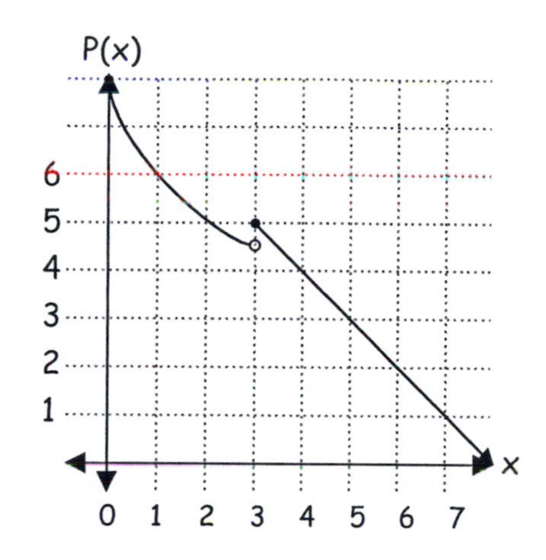

65. Graph of $f(x)=\frac{3}{x+5}-1$:
The basic shape is the graph of $g(x)=\frac{1}{x}$.

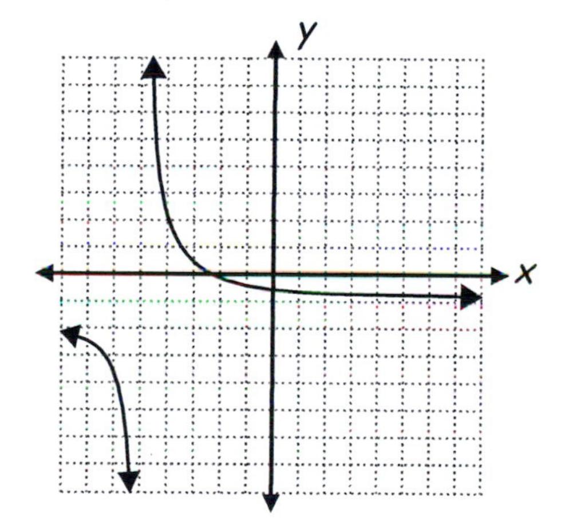

67. Graph of $f(x)=-3|x+2|-4$:
The basic shape is the graph of $g(x)=|x|$.

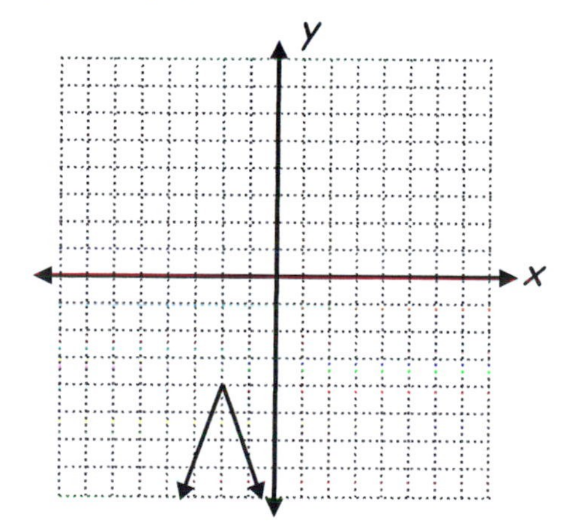

69. Graph of $f(x)=\sqrt[3]{2+x}-1$:
The basic shape is the graph of $g(x)=\sqrt[3]{x}$.

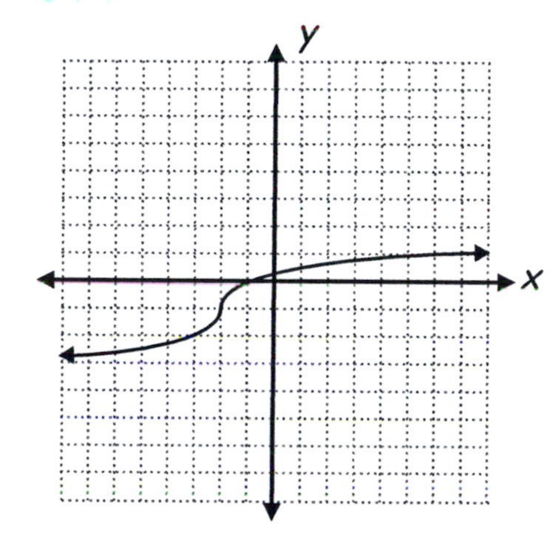

71. The graph shown is a shift of 4 units to the left and down 1 unit of the basic function $f(x)=|x|$. An equation that fits this description is $g(x)=|x+4|-1$.

73. The graph shown is a shift of 6 units to the right and up 2 units and a reflection with respect to the y-axis of the basic function $f(x)=-\sqrt{x}$. An equation that fits this description is $g(x)=-\sqrt{6-x}+2$.

75. The graph shown is a shift of 3 units to the right and up 6 units of the basic function $f(x)=-x^2$. An equation that fits this description is $g(x)=-(x-3)^2+6$.

End of Section 3.5

Chapter Three: Section 3.6
Solutions to Odd-numbered Exercises

1. (a) $(f+g)(-1)=f(-1)+g(-1)$
$=-3+5=2$

(b) $(f-g)(-1)=f(-1)-g(-1)$
$=-3-5=-8$

(c) $(f\cdot g)(-1)=f(-1)\cdot g(-1)$
$=-3\cdot 5=-15$

(d) $\left(\frac{f}{g}\right)(-1)=\frac{f(-1)}{g(-1)}$
$=\frac{-3}{5}=-\frac{3}{5}$

3. (a) $(f+g)(-1)=f(-1)+g(-1)$
$=(-1)^2-3+(-1)=-3$

(b) $(f-g)(-1)=f(-1)-g(-1)$
$=(-1)^2-3-(-1)=-1$

(c) $(f\cdot g)(-1)=f(-1)\cdot g(-1)$
$=\left[(-1)^2-3\right]\cdot(-1)=2$

(d) $\left(\frac{f}{g}\right)(-1)=\frac{f(-1)}{g(-1)}$
$=\frac{(-1)^2-3}{-1}=2$

5. (a) $(f+g)(-1)=f(-1)+g(-1)$
$=15+(-3)=12$

(b) $(f-g)(-1)=f(-1)-g(-1)$
$=15-(-3)=18$

(c) $(f\cdot g)(-1)=f(-1)\cdot g(-1)$
$=15\cdot(-3)=-45$

(d) $\left(\frac{f}{g}\right)(-1)=\frac{f(-1)}{g(-1)}$
$=\frac{15}{-3}=-5$

7. (a) $(f+g)(-1)=f(-1)+g(-1)$
$=(-1)^4+1+(-1)^{11}+2$
$=1+1-1+2=3$

(b) $(f-g)(-1)=f(-1)-g(-1)$
$=1+1-(-1+2)=1$

(c) $(f\cdot g)(-1)=f(-1)\cdot g(-1)$
$=\left[(-1)^4+1\right]\cdot\left[(-1)^{11}+2\right]$
$=2\cdot 1=2$

(d) $\left(\frac{f}{g}\right)(-1)=\frac{f(-1)}{g(-1)}$
$=\frac{(-1)^4+1}{(-1)^{11}+2}=\frac{2}{1}=2$

9. (a) $(f+g)(-1)=f(-1)+g(-1)$
$=3+3=6$

(b) $(f-g)(-1)=f(-1)-g(-1)$
$=3-3=0$

(c) $(f\cdot g)(-1)=f(-1)\cdot g(-1)$
$=3\cdot 3=9$

(d) $\left(\frac{f}{g}\right)(-1)=\frac{f(-1)}{g(-1)}$
$=\frac{3}{3}=1$

11. (a) $(f+g)(-1)=f(-1)+g(-1)$
$=2+3=5$

(b) $(f-g)(-1)=f(-1)-g(-1)$
$=2-3=-1$

(c) $(f\cdot g)(-1)=f(-1)\cdot g(-1)$
$=2\cdot 3=6$

(d) $\left(\frac{f}{g}\right)(-1)=\frac{f(-1)}{g(-1)}$
$=\frac{2}{3}$

13. (a) $(f+g)(x)=f(x)+g(x)$
$=|x|+\sqrt{x}$
Domain: $[0,\infty)$

(b) $\left(\frac{f}{g}\right)(x)=\frac{f(x)}{g(x)}$
$=\frac{|x|}{\sqrt{x}}$
Domain: $(0,\infty)$

15. (a) $(f+g)(x)=f(x)+g(x)$
$=x-1+x^2-1$
$=x^2+x-2$
Domain: $(-\infty,\infty)=\mathbb{R}$

(b) $\left(\frac{f}{g}\right)(x)=\frac{f(x)}{g(x)}$
$=\frac{x-1}{x^2-1}$
$=\frac{1}{x+1}$
Domain: $(-\infty,-1)\cup(-1,1)\cup(1,\infty)$

17. (a) $(f+g)(x)=f(x)+g(x)$
$=3x+x^3-8$
$=x^3+3x-8$
Domain: $(-\infty,\infty)=$

(b) $\left(\frac{f}{g}\right)(x)=\frac{f(x)}{g(x)}$
$=\frac{3x}{x^3-8}$
Domain: $(-\infty,2)\cup(2,\infty)$

19. (a) $(f+g)(x)=f(x)+g(x)$
$=-2x^2+[\![x+4]\!]$
Domain: $(-\infty,\infty)=\mathbb{R}$

(b) $\left(\frac{f}{g}\right)(x)=\frac{f(x)}{g(x)}$
$=\frac{-2x^2}{[\![x+4]\!]}$
Domain: $(-\infty,-4)\cup[-3,\infty)$.
Notice: $[\![x+4]\!]=0$ for
$x\in[-4,-3)$.

For items 21 - 29, $f(x)=\frac{1}{x^2}$ and $g(x)=2x+3$.

21. $(f+g)(-7)=f(-7)+g(-7)$
$=\frac{1}{(-7)^2}+2(-7)+3$
$=\frac{1}{49}-14+3$
$=\frac{1}{49}-\frac{686}{49}+\frac{147}{49}=-\frac{538}{49}$

23. $(f-g)(-5)=f(-5)-g(-5)$

$$=\frac{1}{(-5)^2}-2(-5)-3$$
$$=\frac{1}{25}+10-3$$
$$=\frac{1+250-75}{25}=\frac{176}{25}$$

25. $(fg)(4)=f(4)\cdot g(4)$

$$=\left(\frac{1}{4^2}\right)(2(4)+3)$$
$$=\frac{8+3}{16}=\frac{11}{16}$$

27. $\left(\frac{f}{g}\right)(-2)=\frac{f(-2)}{g(-2)}$

$$=\frac{\frac{1}{(-2)^2}}{2(-2)+3}$$
$$=\frac{\frac{1}{4}}{-1}=-\frac{1}{4}$$

29. $\left(\frac{g}{f}\right)(1)=\frac{g(1)}{f(1)}$

$$=\frac{2(1)+3}{\frac{1}{1^2}}$$
$$=5$$

31. $(f\circ g)(3)=f(g(3))$

$$=f(-5)=2$$

33. $(f\circ g)(3)=f(g(3))$

$$=f\left(\sqrt{3}\right)$$
$$=\left(\sqrt{3}\right)^2-3=3-3=0$$

35. $(f\circ g)(3)=f(g(3))$

$$=f\left(3^3+3^2\right)$$
$$=f(36)$$
$$=2+\sqrt{36}=8$$

37. $(f\circ g)(3)=f(g(3))$

$$=f\left(\sqrt{12-3}\right)$$
$$=f(3)$$
$$=\sqrt{3+6}=3$$

39. $(f\circ g)(3)=f(g(3))$

$$=f(4)=1$$

41. (a) $(f\circ g)(x)=f\left(x^2\right)$

$$=\sqrt{x^2-1}$$

Domain: $(-\infty,-1]\cup[1,\infty)$

(b) $(g\circ f)(x)=g\left(\sqrt{x-1}\right)$

$$=x-1$$

Domain: $[1,\infty)$

Notice: the domain of f is restricted to $x\geq 1$, and this restricts the domain of the composition.

43. (a) $(f\circ g)(x)=f(g(x))$

$$=f\left(\frac{1}{x}\right)$$
$$=\frac{\frac{4}{x}-2}{3}$$
$$=\frac{4-2x}{3x}$$

Domain: $(-\infty,0)\cup(0,\infty)$

(b) $(g\circ f)(x)=g(f(x))$

$$=g\left(\frac{4x-2}{3}\right)$$
$$=\frac{1}{\frac{4x-2}{3}}=\frac{3}{4x-2}$$

Domain: $\left(-\infty,\frac{1}{2}\right)\cup\left(\frac{1}{2},\infty\right)$

45. (a) $$\begin{aligned}(f \circ g)(x) &= f(g(x)) \\ &= f\left(x^3+1\right) \\ &= [\![x^3+1-3]\!] \\ &= [\![x^3-2]\!]\end{aligned}$$

Domain: $\mathbb{R}$

(b) $$\begin{aligned}(g \circ f)(x) &= g(f(x)) \\ &= g([\![x-3]\!]) \\ &= [\![x-3]\!]^3+1\end{aligned}$$

Domain: $\mathbb{R}$

47. (a) $$\begin{aligned}(f \circ g)(x) &= f(g(x)) \\ &= [g(x)]^2+1 \\ &= \left[3x^2+5\right]^2+1 \\ &= 9x^4+30x^2+25+1 \\ &= 9x^4+30x^2+26\end{aligned}$$

Domain: $\mathbb{R}$

(b) $$\begin{aligned}(g \circ f)(x) &= g(f(x)) \\ &= 3[f(x)]^2+5 \\ &= 3\left(x^2+1\right)^2+5 \\ &= 3\left(x^4+2x^2+1\right)+5 \\ &= 3x^4+6x^2+8\end{aligned}$$

Domain: $\mathbb{R}$

49. (a) $$\begin{aligned}(f \circ g)(x) &= f(g(x)) \\ &= f\left(\frac{2}{x}\right) \quad [\Rightarrow x \neq 0] \\ &= \frac{1}{\frac{2}{x}+7} \\ &= \frac{1}{\frac{2+7x}{x}} \\ &= \frac{x}{2+7x} \quad \left(\Rightarrow x \neq -\frac{2}{7}\right)\end{aligned}$$

Domain: $\left(-\infty, -\frac{2}{7}\right) \cup \left(-\frac{2}{7}, 0\right) \cup (0, \infty)$

(b) $$\begin{aligned}(g \circ f)(x) &= g(f(x)) \\ &= g\left(\frac{1}{x+7}\right) \quad [\Rightarrow x \neq -7] \\ &= \frac{2}{\frac{1}{x+7}} \\ &= 2(x+7) = 2x+14\end{aligned}$$

Domain: $(-\infty, -7) \cup (-7, \infty)$

51. (a) $$\begin{aligned}(f \circ g)(x) &= f(g(x)) \\ &= f(3x+1) \\ &= (3x+1)^2 \\ &= 9x^2+6x+1\end{aligned}$$

Domain: $\mathbb{R}$

(b) $$\begin{aligned}(g \circ f)(x) &= g(f(x)) \\ &= f\left(x^2\right) \\ &= 3x^2+1\end{aligned}$$

Domain: $\mathbb{R}$

53. (a) $$\begin{aligned}(f \circ g)(x) &= f(g(x)) \\ &= f\left(x^2+2\right) \\ &= \sqrt{x^2+2-4} \\ &= \sqrt{x^2-2}\end{aligned}$$

Domain: $\left(-\infty, -\sqrt{2}\right] \cup \left[\sqrt{2}, \infty\right)$

(b) $$\begin{aligned}(g \circ f)(x) &= g(f(x)) \\ &= g\left(\sqrt{x-4}\right), \quad [\Rightarrow x \geq 4] \\ &= \left(\sqrt{x-4}\right)^2+2 \\ &= x-4+2 \\ &= x-2\end{aligned}$$

Domain: $[4, \infty)$

55. Given $f(x)=\sqrt[3]{3x^2-1}$

Answers will vary. One composite:
Let $h(x)=\sqrt[3]{x}$ and $g(x)=3x^2-1$,
then $f(x)=(h\circ g)(x)$.

57. Given $f(x)=|x-2|+3$.

Answers will vary. One composite:
Let $h(x)=x-2$ and $g(x)=|x|+3$,
then $f(x)=(g\circ h)(x)$.

59. Given $f(x)=\left|x^3-5x\right|+7$.

Answers will vary. One composite:
Let $h(x)=x^3-5x$ and $g(x)=|x|+7$,
then $f(x)=(g\circ h)(x)$.

61. Given: $V=\pi r^2h$ and $h=3r$.

Substituting for h (in functional form):

$V(r)=\pi r^2(3r)$

$V(r)=3\pi r^3$

63. Given: $V=\frac{1}{3}\pi r^2h$ and $h=\frac{1}{4}t^2,\ t\geq 0$.

Substituting for h (in functional form):

$V(t)=\frac{1}{3}\pi r^2\left(\frac{1}{4}t^2\right)$

$V(t)=\frac{\pi r^2t^2}{12}$

65. $L(x)=M(x)\cdot I(x)+C(x)\cdot I(x)$

$=I(x)(M(x)+C(x))$

67. Given: $f(x)=2x^4-x^2$ and $g(x)=\frac{1}{x^2}$.

$(f\cdot g)(x)=f(x)\cdot g(x)$

$=\left(2x^4-x^2\right)\left(\frac{1}{x^2}\right)$

$=2x^2-1$

Then, find $(f\cdot g)(-x)$:

$(f\cdot g)(-x)=f(-x)\cdot g(-x)$

$=\left[2(-x)^4-(-x)^2\right]\cdot\left[\frac{1}{(-x)^2}\right]$

$=\left(2x^4-x^2\right)\left(\frac{1}{x^2}\right)$

$=2x^2-1$

$=(f\cdot g)(x)$

Therefore, $f\cdot g$ is an even function.

For items 69 - 77, the following process should be used to test whether the given value of c is in the Mandelbrot Set:

Use the composition of functions concept and examine several iterations of the function $f(z)=z^2+c$ for the given value c. After each iteration of f find the distance that the result is from (0, 0); in other words, calculate $\sqrt{a^2+b^2}$ where a is the real part of the result and b is the imaginary coefficient of the result. If at any time, up to 50 iterations, we find that $\sqrt{a^2+b^2}>2$, then we would conclude that the given value c is not in the Mandelbrot Set.

69. Given $c=1$; $f(0)=0^2+1=1=1+0i$

Test magnitude: $\sqrt{1^2+0^2}=1<2$

Continue to the next iteration:

$f^2(0)=f(f(0))=f(1)=1^2+1=2$

Test magnitude: $\sqrt{2^2+0^2}=2\leq 2$

Continue to the next iteration:

$f^3(0)=f\left(f^2(0)\right)=f(2)=2^2+1=5$

Test magnitude: $\sqrt{5^2+0^2}=5>2$

Therefore, c is not in the Mandelbrot set.

71. Given $c=-1$; $f(0)=0^2-1=-1+0i$

Test magnitude: $\sqrt{(-1)^2+0^2}=1<2$

Continue to the next iteration:

$f^2(0)=f(f(0))=f(-1)=(-1)^2-1=0$

Test magnitude: $\sqrt{0^2+0^2}=0<2$

With another iteration, we will be able to foresee a pattern and draw a conclusion:

$f^3(0)=f(f^2(0))=f(0)=-1$

The iteration cycles through the same values, with each result having a magnitude ≤ 2.

Therefore, $c=-1$ is in the Mandelbrot set.

73. Given $c=-2$; $f(0)=0^2+(-2)=-2$

Test magnitude: $\sqrt{(-2)^2+(0)^2}=\sqrt{4}=2$

Continue to the next iteration:

$$f^2(0)=f(f(0))=f(-2)$$
$$=(-2)^2+(-2)=2$$

Test magnitude: $\sqrt{(2)^2+0^2}=\sqrt{4}=2$

Continue to the next iteration:

$$f^3(0)=f(f^2(0))=f(2)$$
$$=2^2+(-2)=2$$

Test magnitude: $\sqrt{(2)^2+0^2}=\sqrt{4}=2$

The iteration result has become constant and for each iteration, the magnitude of the resulting complex number ≤ 2.

Therefore, $c=-2$ is in the Mandelbrot set.

75. Given $c=-1-i$; $f(0)=0^2+(-1-i)=-1-i$

Test magnitude: $\sqrt{(-1)^2+(-1)^2}=\sqrt{2}<2$

Continue to the next iteration:

$$f^2(0)=f(f(0))=f(-1-i)$$
$$=(-1-i)^2+(-1-i)$$
$$=1+2i+i^2-1-i=-1+i$$

Test magnitude: $\sqrt{(-1)^2+1^2}=\sqrt{2}\leq 2$

Continue to the next iteration:

$$f^3(0)=f(f^2(0))=f(-1+i)$$
$$=(-1+i)^2+(-1-i)$$
$$=1-2i+i^2-1-i=-1-3i$$

Test magnitude: $\sqrt{(-1)^2+(-3)^2}=\sqrt{10}$

Since $\sqrt{10}>2$, the complex number $c=-1-i$ is not in the Mandelbrot set.

End of Section 3.6

Chapter Three: Section 3.7
Solutions to Odd-numbered Exercises

1. Given $R = \{(-4, 2), (3, 2), (0, -1), (3, -2)\}$, interchange x and y to obtain R^{-1}:
$R^{-1} = \{(2, -4), (2, 3), (-1, 0), (-2, 3)\}$

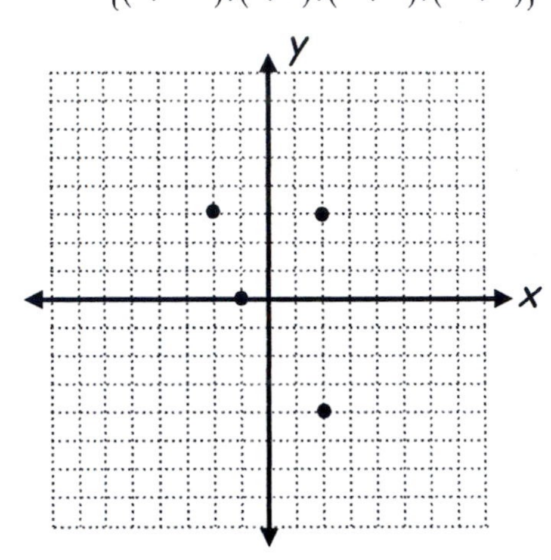

Domain $= \{-2, -1, 2\}$; Range $= \{-4, 0, 3\}$

3. Given $y = x^3$; the inverse satisfies the relation $x = y^3$ (interchanging x and y).
Domain $= \mathbb{R}$; Range $= \mathbb{R}$

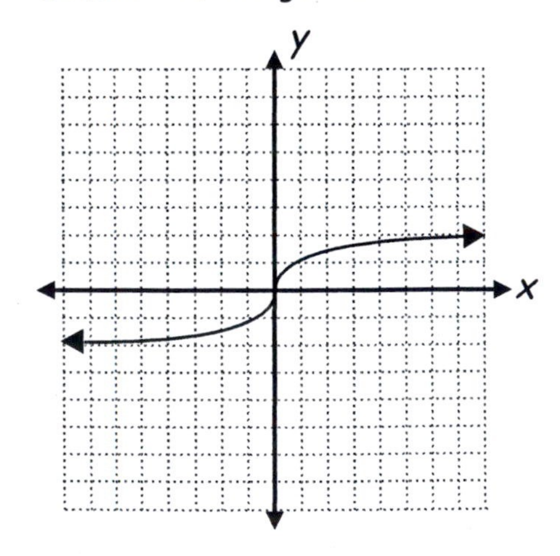

5. Given $x = |y|$; the inverse satisfies the relation $y = |x|$ (interchanging x and y).
Domain $= \mathbb{R}$; Range $= [0, \infty)$

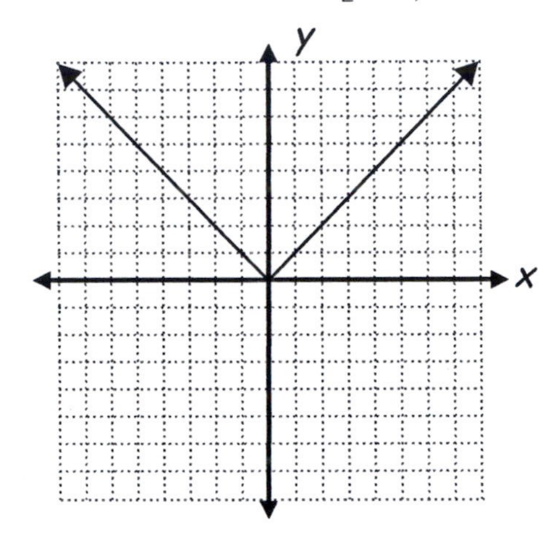

7. Given $y = \frac{1}{2}x - 3$; the inverse satisfies the relation $\left\{(x, y) \mid x = \frac{1}{2}y - 3\right\}$ (interchanging x and y).
Domain $= \mathbb{R}$; Range $= \mathbb{R}$

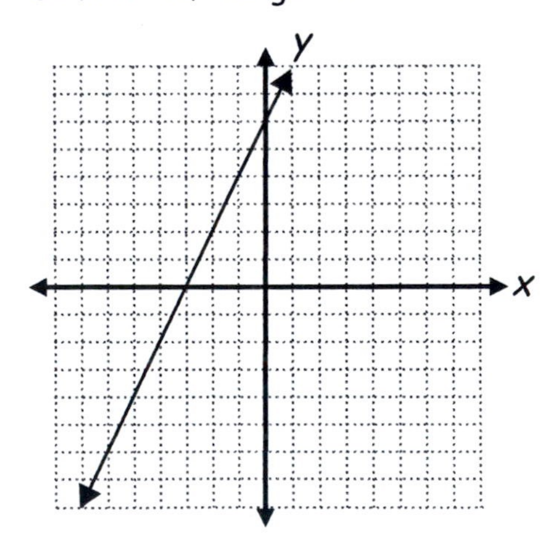

9. Given $y = [\![x]\!]$; the inverse satisfies the relation $\{(x, y) \mid x = [\![y]\!]\}$ (interchanging x and y).
Domain $= \mathbb{Z}$; Range $= \mathbb{R}$

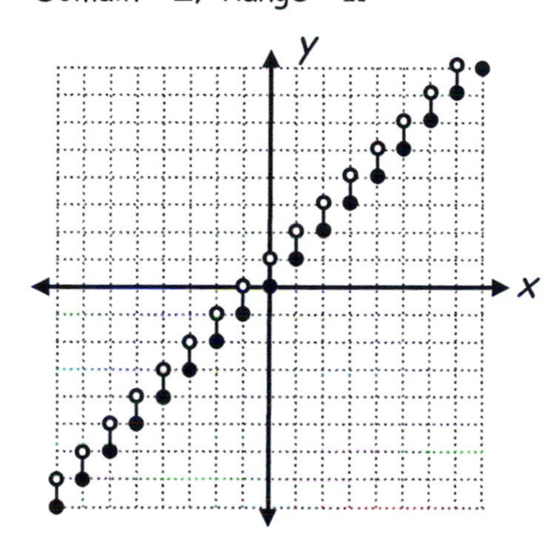

11. Given $x = y^2 - 2$; the inverse satisfies the relation $\{(x,y) \mid y = x^2 - 2\}$ (interchanging x and y).
Domain $= \mathbb{R}$; Range $= [-2, \infty)$

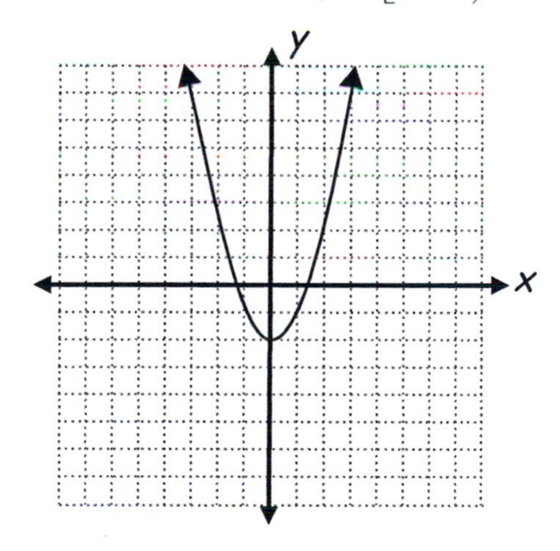

13. Given $f(x) = x^2 + 1$, without a restriction of its domain, the inverse is not a function. Examination of its graph should suggest how to restrict the domain so that its inverse is a function:

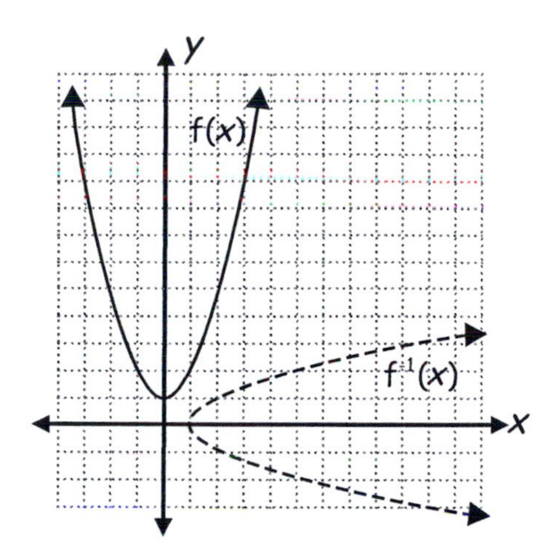

Restrict the domain to $[0, \infty)$ or to $(-\infty, 0]$ and the inverse is a function.

15. Given $h(x) = \sqrt{x+3}$, its inverse $h^{-1}(x) = \{(x,y) \mid x = \sqrt{y+3}\}$ is a function.

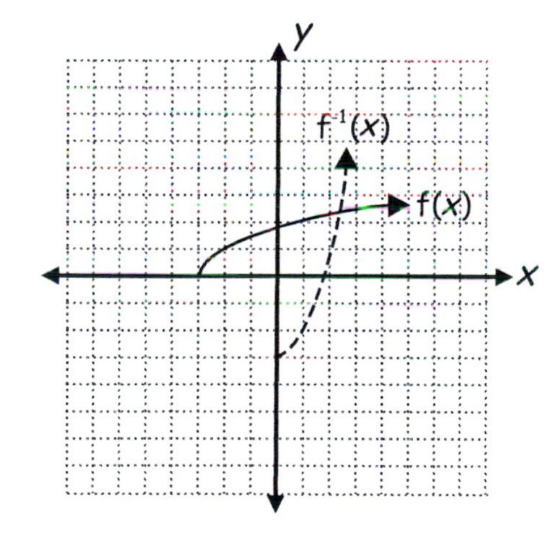

17. Given $G(x) = 3x - 5$, the inverse is also a linear function, G^{-1} described by $\{(x,y) \mid x = 3y - 5\}$. Explicitly, $G^{-1}(x) = \frac{1}{3}x + \frac{5}{3}$.

19. Given $r(x) = -\sqrt{x^3}$, the inverse is also a function, r^{-1} described by $\{(x,y) \mid x = -\sqrt{y^3}\}$. Explicitly, $r^{-1}(x) = \sqrt[3]{x^2}$.

21. Given $f(x) = x^2 - 4x$, without a restriction of its domain, the inverse is not a function. The given function is a parabola, so finding the vertex by completing the square identifies the values of x which can be used to restrict the domain so that its inverse is also a function. Restricting the domain to $[2, \infty)$ or to $(-\infty, 2]$ are two ways to ensure that the inverse of f is also a function.

23. Given $H(x) = |x - 12|$, you can restrict the domain to $[12, \infty)$ and its inverse is a function satisfying

$$\{(x,y) \mid x = |y - 12|\}.$$

25. Given $f(x) = x^{\frac{1}{3}} - 2$, set $y = x^{\frac{1}{3}} - 2$.

Exchange x and y: $x = y^{\frac{1}{3}} - 2$

Solve for y:

$$y^{\frac{1}{3}} = x + 2$$

$$y = (x+2)^3$$

$$f^{-1}(x) = (x+2)^3$$

27. Given $r(x) = \frac{x-1}{3x+2}$, set $y = \frac{x-1}{3x+2}$.

Exchange x and y: $x = \frac{y-1}{3y+2}$

Solve for y:

$$x(3y+2) = y - 1$$

$$3xy + 2x = y - 1$$

$$3xy - y = -2x - 1$$

$$y(3x-1) = -2x - 1$$

$$y = \frac{-2x-1}{3x-1}$$

$$r^{-1}(x) = \frac{-2x-1}{3x-1}$$

29. Given $F(x) = (x-5)^3 + 2$,

set $y = (x-5)^3 + 2$.

Exchange x and y: $x = (y-5)^3 + 2$

Solve for y:

$$(y-5)^3 = x - 2$$

$$y - 5 = \sqrt[3]{x-2}$$

$$y = \sqrt[3]{x-2} + 5$$

$$F^{-1}(x) = (x-2)^{\frac{1}{3}} + 5$$

31. Given $V(x) = \frac{x+5}{2}$, set $y = \frac{x+5}{2}$.

Exchange x and y: $x = \frac{y+5}{2}$

Solve for y:

$$y + 5 = 2x$$

$$y = 2x - 5$$

$$V^{-1}(x) = 2x - 5$$

33. Given $h(x) = x^{\frac{3}{5}} - 2$, set $y = x^{\frac{3}{5}} - 2$.

Exchange x and y: $x = y^{\frac{3}{5}} - 2$

Solve for y:

$$y^{\frac{3}{5}} = x + 2$$

$$y = (x+2)^{\frac{5}{3}}$$

$$h^{-1}(x) = (x+2)^{\frac{5}{3}}$$

35. Given $J(x) = \frac{2}{1-3x}$, set $y = \frac{2}{1-3x}$.

Exchange x and y: $x = \frac{2}{1-3y}$

Solve for y:

$$1 - 3y = \frac{2}{x}$$

$$3y = 1 - \frac{2}{x}$$

$$y = \frac{x-2}{3x}$$

$$J^{-1}(x) = \frac{x-2}{3x}$$

37. Given $h(x)=x^7+6$, set $y=x^7+6$.
Exchange x and y: $x=y^7+6$
Solve for y:

$$y^7=x-6$$
$$y=(x-6)^{1/7}$$
$$h^{-1}(x)=(x-6)^{1/7}$$

39. Given $r(x)=\sqrt[5]{2x}$, set $y=\sqrt[5]{2x}$.
Exchange x and y: $x=\sqrt[5]{2y}$
Solve for y:

$$x^5=2y$$
$$y=\frac{x^5}{2}$$
$$r^{-1}(x)=\frac{x^5}{2}$$

41. Given $f(x)=3(2x)^{\frac{1}{3}}$, set $y=3(2x)^{\frac{1}{3}}$. Exchange x and y:
$x=3(2y)^{\frac{1}{3}}$
Solve for y:

$$x^3=27(2y)$$
$$y=\frac{x^3}{54}$$
$$f^{-1}(x)=\frac{x^3}{54}$$

43. $$f\left(f^{-1}(x)\right)=f\left(\frac{5x+1}{3}\right)=\frac{3\left(\frac{5x+1}{3}\right)-1}{5}=\frac{5x+1-1}{5}=x$$

$$f^{-1}\left(f(x)\right)=f^{-1}\left(\frac{3x-1}{5}\right)=\frac{5\left(\frac{3x-1}{5}\right)+1}{3}=\frac{3x-1+1}{3}=x$$

45. $$f\left(f^{-1}(x)\right)=f\left(\frac{x+7}{x-2}\right)=\frac{2\left(\frac{x+7}{x-2}\right)+7}{\left(\frac{x+7}{x-2}\right)-1}=\frac{\frac{2x+14+7x-14}{x-2}}{\frac{x+7-x+2}{x-2}}=\frac{9x}{9}=x$$

$$f^{-1}\left(f(x)\right)=f^{-1}\left(\frac{2x+7}{x-1}\right)=\frac{\frac{2x+7}{x-1}+7}{\frac{2x+7}{x-1}-2}=\frac{2x+7+7x-7}{2x+7-2x+2}=\frac{9x}{9}=x$$

47. $$f\left(f^{-1}(x)\right)=f\left(\frac{x+3}{2}\right)=2\left(\frac{x+3}{2}\right)-3=x+3-3=x$$

$$f^{-1}\left(f(x)\right)=f^{-1}(2x-3)=\frac{2x-3+3}{2}=\frac{2x}{2}=x$$

49. $$f\left(f^{-1}(x)\right)=f\left(\frac{1}{x}\right)=\frac{1}{1/x}=x$$

$$f^{-1}\left(f(x)\right)=f^{-1}\left(\frac{1}{x}\right)=\frac{1}{1/x}=x$$

51. $f(f^{-1}(x)) = f(\sqrt{x}+2)$

$= (\sqrt{x}+2-2)^2$

$= (\sqrt{x})^2$

$= x$

$f^{-1}(f(x)) = f^{-1}((x-2)^2)$

$= \sqrt{(x-2)^2}+2$

$= x-2+2$

$= x$

For items 53 - 57, draw the line $y = x$. Look for the "mirror image" of the given graph through the line. A starting point is to use a point of intersection of the given graph with either one of the x-axis or y-axis.

53. Graph b is the inverse of the graph of $f(x) = x^3$.

55. Graph e is the inverse of the graph of $f(x) = \sqrt{x-4}$.

57. Graph a is the inverse of the graph of $f(x) = \dfrac{x}{4}$.

59. Use $S = 19$, $A = 1$, $N = 14$, $D = 4$, $Y = 25$, $H = 8$, $O = 15$, $E = 5$:

Letter Formula Code

$S \to f(19) = 4(19) - 3 = 73$

$A \to f(1) = 4(1) - 3 = 1$

$N \to f(14) = 4(14) - 3 = 53$

$D \to f(4) = 4(4) - 3 = 13$

$Y \to f(25) = 4(25) - 3 = 97$

$S \to 73$

$H \to f(8) = 4(8) - 3 = 29$

$O \to f(15) = 4(15) - 3 = 57$

$E \to f(5) = 4(5) - 3 = 17$

$S \to 73$

61. Given: $f(x) = 8x - 7$

Substitute y for $f(x)$ and switch x and y.

$y = 8x - 7$

$x = 8y - 7$. Now, solve for y:

$y = \dfrac{x+7}{8}$. Replace y with f^{-1}:

$f^{-1}(x) = \dfrac{x+7}{8}$

To decode, use the code numbers in turn in f^{-1}, and simplify. Use the normal order of letters in the alphabet to find the corresponding letter:

$f^{-1}(41) = \dfrac{41+7}{8} = 6 \to F$

$f^{-1}(137) = \dfrac{137+7}{8} = 18 \to R$

$f^{-1}(65) = \dfrac{65+7}{8} = 9 \to I$

$f^{-1}(145) = \dfrac{145+7}{8} = 19 \to S$

$f^{-1}(9) = \dfrac{9+7}{8} = 2 \to B$

$f^{-1}(33) = \dfrac{33+7}{8} = 5 \to E$

$f^{-1}(169) = \dfrac{169+7}{8} = 22 \to V$

$f^{-1}(113) = \dfrac{113+7}{8} = 15 \to O$

$f^{-1}(89) = \dfrac{89+7}{8} = 12 \to L$

$f^{-1}(193) = \dfrac{193+7}{8} = 25 \to Y$

$f^{-1}(1) = \dfrac{1+7}{8} = 1 \to A$

$f^{-1}(105) = \dfrac{105+7}{8} = 14 \to N$

$f^{-1}(25) = \dfrac{25+7}{8} = 4 \to D$

$f^{-1}(57) = \dfrac{57+7}{8} = 8 \to H$

Using these results, the decoded message is FRISBEE VOLLEYBALL AND HORSE SHOES.

63. Given: $f(x)=x^3$
Substitute y for $f(x)$ and switch x and y.
$y=x^3$
$x=y^3$. Now, solve for y:
$y=\sqrt[3]{x}$. Replace y with $f^{-1}(x)$:
$f^{-1}(x)=\sqrt[3]{x}$
$f^{-1}(27)=3\to C$
$f^{-1}(1)=1\to A$
$f^{-1}(8000)=20\to T$
$f^{-1}(512)=8\to H$
$f^{-1}(12167)=23\to W$
$f^{-1}(10648)=22\to V$
$f^{-1}(125)=5\to E$
Using these results, the decoded message is CATCH A WAVE.

For items 65 - 69, use a graphing calculator:

65. Given: $f(x)=\sqrt{x+5}$ and $f^{-1}(x)=x^2-5$.

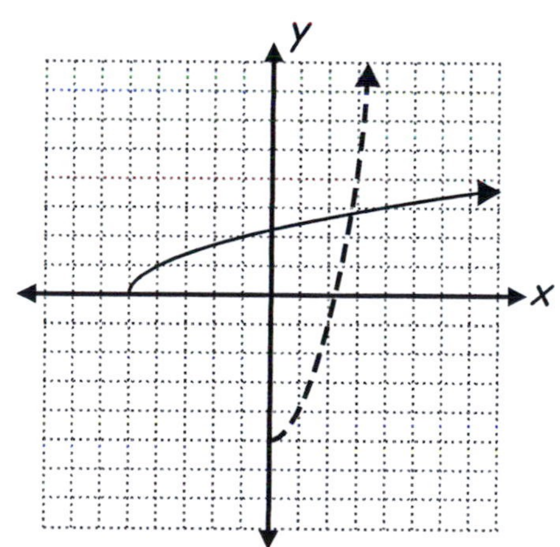

Domain: $[0,\infty)$; Range: $[-5,\infty)$

67. Given: $f(x)=x^2+3$ and $f^{-1}(x)=\sqrt{x-3}$.

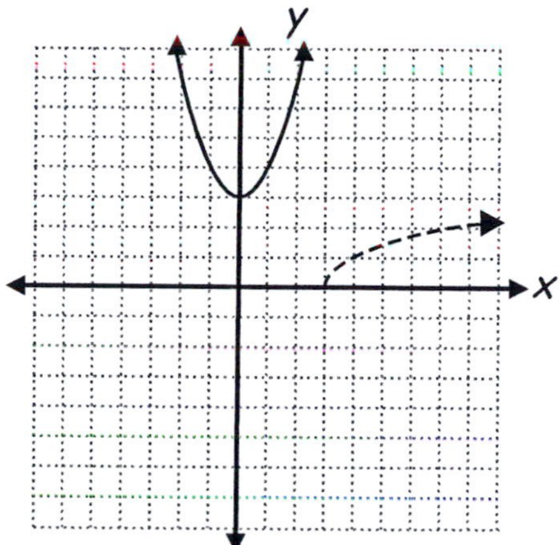

Domain: $[3,\infty)$; Range: $[0,\infty)$

69. Given: $f(x)=\dfrac{2x+1}{x-1}$ and $f^{-1}(x)=\dfrac{x+1}{x-2}$.

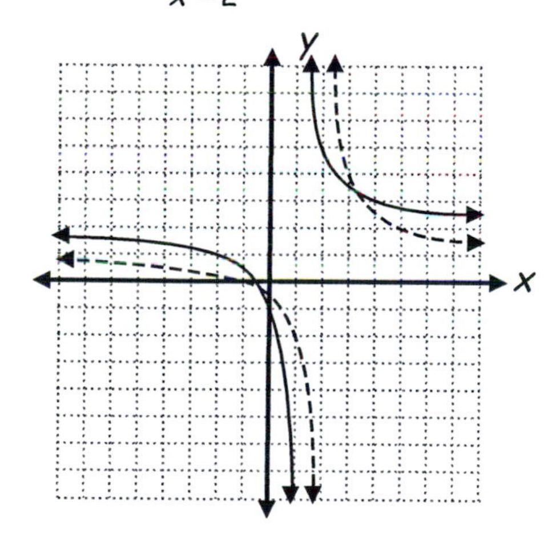

Domain: $(-\infty,2)\cup(2,\infty)$;
Range: $(-\infty,1)\cup(1,\infty)$

End of Section 3.7

Chapter Three Test
Solutions to Odd-Numbered Exercises

1. Domain of $R = \{-2, -3\}$
Range of $R = \{-9, -3, 2, 9\}$
R is not a function because of $(-2, 9)$ and $(-2, 2)$.

3. $x = y^2 - 6$ does not describe a function. Notice that for $y = \pm 1$, $x = -5$. This means that $(-5, 1)$ and $(-5, -1)$ are both contained in the relation.
Domain $= [-6, \infty)$; Range $= \mathbb{R}$

5. Given: $f(x) = (x+5)(2x)$

(a) $$\begin{aligned} f(x-1) &= (x-1+5)(2(x-1)) \\ &= (x+4)(2x-2) \\ &= 2x^2 + 6x - 8 \end{aligned}$$

(b) $$\begin{aligned} &f(x+a) - f(x) = \\ &(x+a+5)(2(x+a)) - 2x^2 - 10x = \\ &(x+a+5)(2x+2a) - 2x^2 - 10x = \\ &2x^2 + 2ax + 10x + 2ax + \\ &2a^2 + 10a - 2x^2 - 10x = \\ &4ax + 2a^2 + 10a \end{aligned}$$

(c) $$\begin{aligned} f(x^2) &= (x^2+5)(2x^2) \\ &= 2x^4 + 10x^2 \end{aligned}$$

7. Given: $g : \mathbb{N} \to \mathbb{R}$ by $g(x) = \dfrac{3x}{4}$.
Dom $= \mathbb{N}$; Codomain $= \mathbb{R}$
Ran $= \left\{\dfrac{3}{4}, \dfrac{3}{2}, \dfrac{9}{4}, \ldots\right\}$

9. To graph the function $f(x) = 7x - 2$, plot at least two points (x, y) that satisfy $y = 7x - 2$.

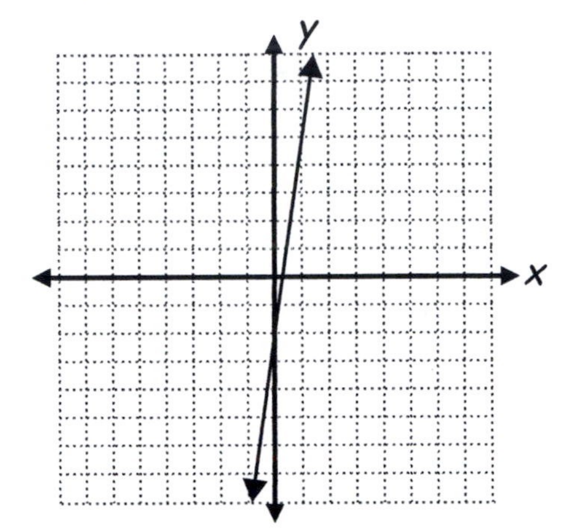

11. The quadratic function $f(x) = (x-1)^2 - 1$ has vertex at $(1, -1)$ and its graph opens upward. To find the x-intercepts, set $f(x) = 0$ and solve for x:

$$\begin{aligned} x^2 - 2x + 1 - 1 &= 0 \\ x^2 - 2x &= 0 \\ x(x-2) &= 0 \end{aligned}$$

The x-intercepts are at $x = 0$ and $x = 2$.

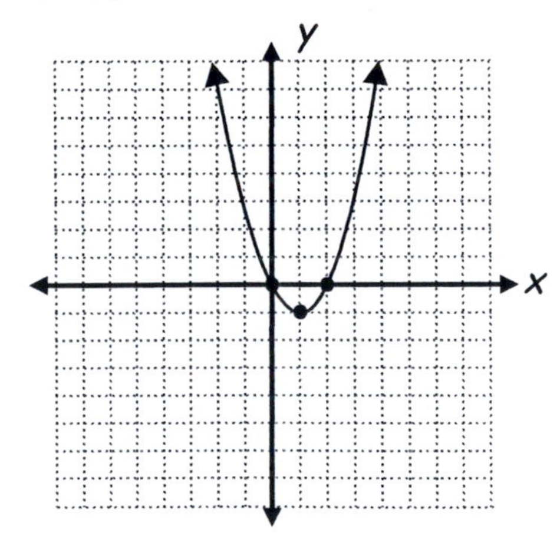

13. Let x = one of the numbers. Then, $15 - x =$ the other number.
Let $y = x(15 - x)$, or $y = 15x - x^2$.
The graph of this quadratic has a maximum value at its vertex.
Complete the square to find the vertex:

$$y = -\left(x^2 - 15x + \left(\frac{15}{2}\right)^2\right) + \left(\frac{15}{2}\right)^2$$

$$y = -\left(x - \frac{15}{2}\right)^2 + \frac{225}{4}$$

The vertex $(h, k) = \left(\frac{15}{2}, \frac{225}{4}\right)$

Therefore, the maximum value occurs at $x = \frac{15}{2}$. The other number is $15 - \frac{15}{2} = \frac{15}{2}$.

15. Graph: $f(x) = \frac{-2}{x^2}$

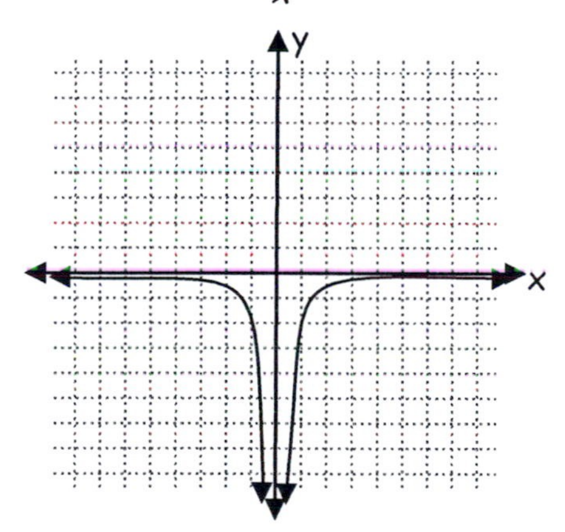

17. Graph: $f(x) = 5|-x|$

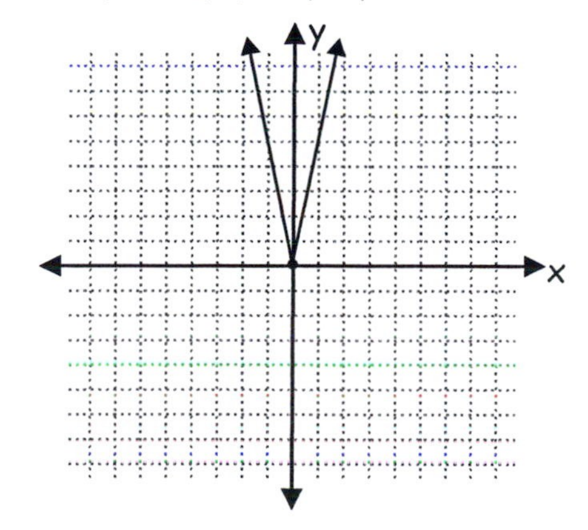

19. Graph of $f(x) = \begin{cases} x^2 \text{ if } x < 1 \\ \frac{1}{x} \text{ if } x \geq 1 \end{cases}$

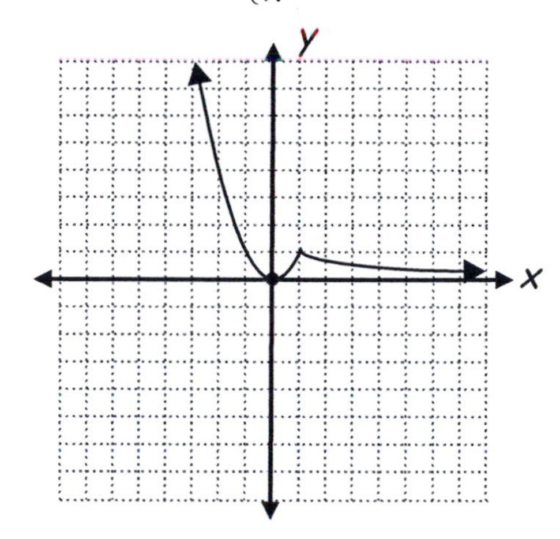

21. y varies jointly with the cube of x and the square root of z: Then,
$y = kx^3\sqrt{z}$, for some non-zero constant k. Solve for k when $y = 270$, $x = 3$ and $z = 25$:

$270 = k \cdot 3^3\sqrt{25}$

$k = 2$

Substitute $k = 2$, $x = 2$ and $z = 9$:

$y = 2(2)^3\sqrt{9} = 48$

23. Let $n(p) =$ the functional relationship between the number of videos rented per month and price p.

$n(p) = k\left(\frac{1}{p}\right)$ for some constant k.

Given $n(3.49) = 1050$, you have that

$1050 = k\left(\frac{1}{3.49}\right)$ for some constant k.

Solving for k, $k = 3664.5$.

$n(2.99) = 3664.5\left(\frac{1}{2.99}\right)$

≈ 1226 videos per month

25. For $f(x)=(x-1)^3+2$, the more basic function is $F(x)=x^3$. The shift is upward 2 and to the right 1:

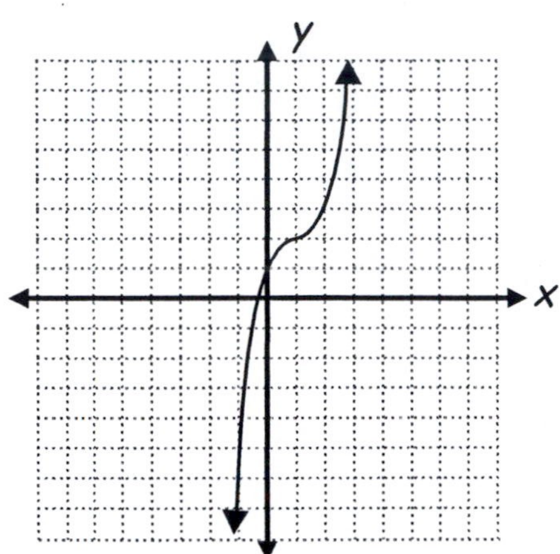

Domain $=\mathbb{R}$; Range $=\mathbb{R}$

27. Test $f: f(x)=\frac{1}{3}x^3$

$$f(-x)=\frac{1}{3}(-x)^3$$
$$=-\frac{1}{3}(x)^3$$
$$=-f(x)$$

Therefore, f is an odd function.

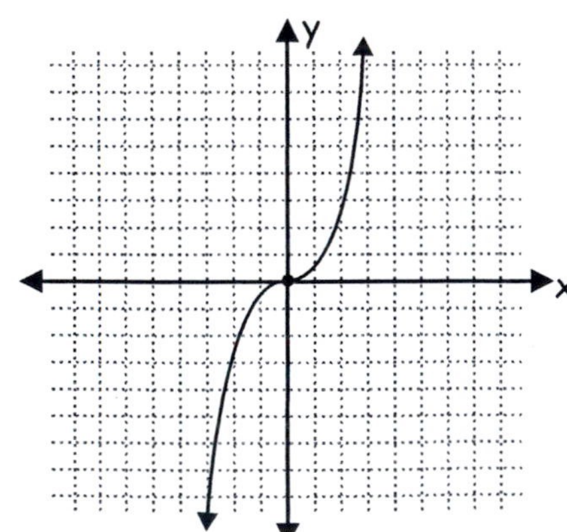

29. $y=|5x|$ has y-axis symmetry.

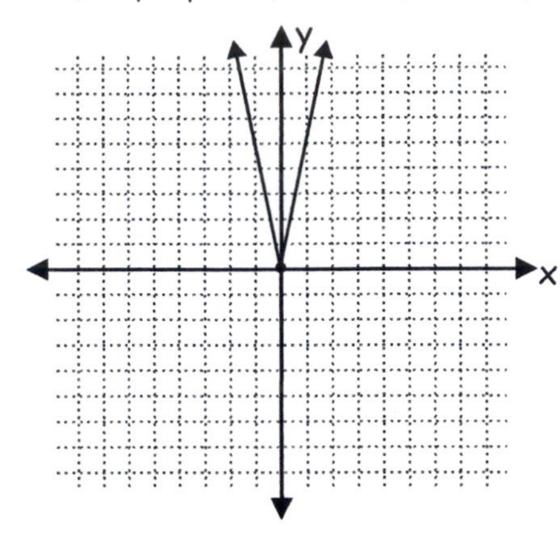

31. Graph $f(x)=(x-2)^4-6$. Notice that the more basic function is $F(x)=x^4$.

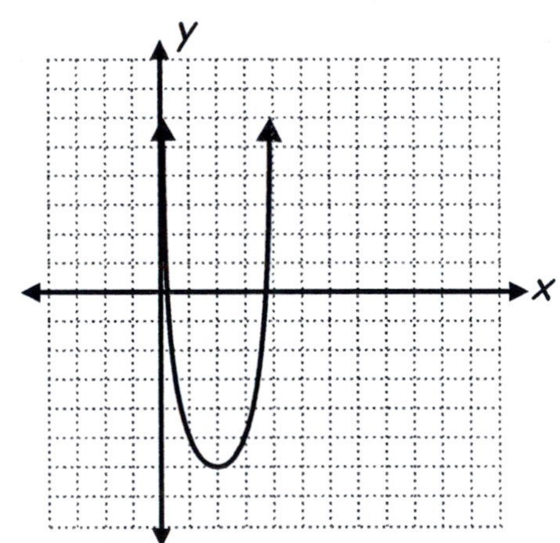

The function f is decreasing on $(-\infty, 2)$ and increasing on $(2, \infty)$.

33. For $f(x)=x^2$ and $g(x)=\sqrt{x}$:

(a) $(f+g)(x)=f(x)+g(x)$
$$=x^2+\sqrt{x}$$
Domain $=[0, \infty)$

(b) $\left(\frac{f}{g}\right)(x)=\frac{f(x)}{g(x)}$
$$=\frac{x^2}{\sqrt{x}}=x^{2-\frac{1}{2}}=x^{\frac{3}{2}}$$
Domain $=(0, \infty)$

35. **(a)** $(f \circ g)(x) = f(g(x))$
$= f(-x-1)$
$= -(-x-1)+1$
$= x+2$

(b) $(g \circ f)(x) = g(f(x))$
$= g(-x+1)$
$= -(-x+1)-1$
$= x-2$

(c) $(f \circ g)(3) = f(g(3))$
$= f(-4)$
$= -(-4)+1 = 5$

37. $f(x) = \dfrac{\sqrt{x+3}+2}{x^2+6x+9}$

To decompose into a composite of two functions: Let $g(W) = \dfrac{\sqrt{W}+2}{W^2}$ and $W = h(x) = x+3$.

Then, $f(x) = g(h(x))$ or $(g \circ h)(x)$

39. Given $c = -i$; $f(0) = 0^2 - i = -i$

Test magnitude: $\sqrt{0^2 + (-1)^2} = 1 < 2$

Continue to the next iteration:

$f^2(0) = f(f(0)) = f(-i) = (-i)^2 - i$
$= -1-i$

Test magnitude: $\sqrt{(-1)^2 + (-1)^2} = \sqrt{2} \le 2$

Continue to the next iteration:

$f^3(0) = f(f(-i)) = f(-1-i)$
$= (-1-i)^2 - i = 1 + 2i + i^2 - i = i$

Test magnitude: $\sqrt{(0)^2 + (1)^2} = 1 < 2$

Continue to the next iteration:

$f^4(0) = f(f(-1-i)) = f(i)$
$= i^2 - i = -1-i$

But, this is the same as $f^2(0)$, so the iteration cycles through the same values. The resulting magnitudes are less than 2, so $c = -i$ is in the Mandelbrot set.

41. The inverse of the given relation R:

$R^{-1} = \{(5,-3), (1,2), (-5,0), (-2,-1)\}$

Dom $= \{-5, -2, 1, 5\}$

Ran $= \{-3, -1, 0, 2\}$

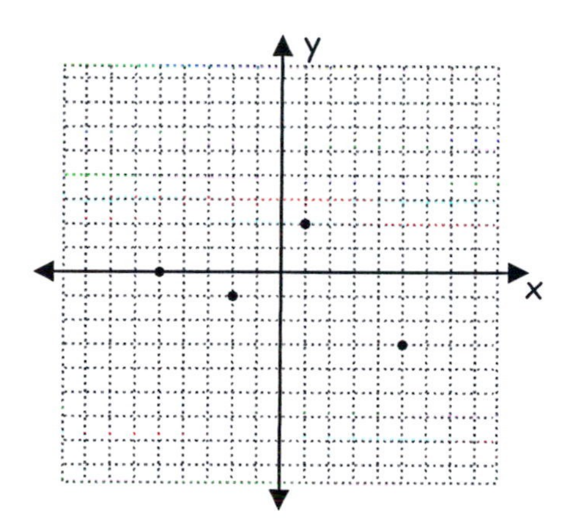

43. Given $r(x) = \dfrac{2}{7x-1}$:

Let $y = \dfrac{2}{7x-1}$.

Interchange x and y:

$x = \dfrac{2}{7y-1}$

Solve for y:

$7y - 1 = \dfrac{2}{x}$

$y = \dfrac{2}{7x} + \dfrac{1}{7}$

$r^{-1}(x) = \dfrac{x+2}{7x}$

45. Given $f(x) = x^{\frac{1}{5}} - 6$:

Let $y = x^{\frac{1}{5}} - 6$.
Interchange x and y:
$x = y^{\frac{1}{5}} - 6$
Solve for y:

$$y^{\frac{1}{5}} = x + 6$$
$$y = (x+6)^5$$
$$f^{-1}(x) = (x+6)^5$$

End of Chapter 3 Test

Chapter Four: Section 4.1
Solutions to Odd-numbered Exercises

1. Substitute $x=-1$ in the given equation:

$$9(-1)^2-4(-1)\stackrel{?}{=}2(-1)^3+15$$
$$9+4\stackrel{?}{=}-2+15$$
$$13=13$$

Therefore, $x=-1$ solves the equation.

3. Substitute $x=2+3i$ in the given equation:

$$(2+3i)^2+13\stackrel{?}{=}4(2+3i)$$
$$4+12i-9+13\stackrel{?}{=}8+12i$$
$$8+12i=8+12i$$

Therefore, $x=2+3i$ solves the equation.

5. Substitute $x=3$ in the given equation:

$$9(3)^2-4(3)\stackrel{?}{=}2(3)^3+15$$
$$81-12\stackrel{?}{=}54+15$$
$$69=69$$

Therefore, $x=3$ solves the equation.

7. Substitute $x=-2$ in the given equation:

$$3(-2)^3+(5-3i)(-2)^2\stackrel{?}{=}(2+5i)(-2)-2i$$
$$-24+20-12i\stackrel{?}{=}-4-10i-2i$$
$$-4-12i=-4-12i$$

Therefore, $x=-2$ solves the equation.

9. Substitute $x=3$ in the given equation:

$$4(3)^5-8(3)^4-12(3)^3\stackrel{?}{=}16(3)^2-25(3)-69$$
$$972-648-324\stackrel{?}{=}144-75-69$$
$$0=0$$

Therefore, $x=3$ solves the equation.

11. Substitute $x=2$ in the given equation:

$$(2)^5-10(2)^4-80(2)^2\stackrel{?}{=}32-80(2)-40(2)^3$$
$$32-160-320=32-160-320$$

Therefore, $x=2$ solves the equation.

13. Substitute $x=2i$ in the given equation:

$$4(2i)^2+32(2i)+(8+i)(2i)^3\stackrel{?}{=}-8$$
$$-16+64i+(8+i)(-8i)\stackrel{?}{=}-8$$
$$-16+64i-64i+8\stackrel{?}{=}-8$$
$$-8=-8$$

Therefore, $x=2i$ solves the equation.

15. Substitute $x=2$ in the given equation:

$$(5-3i)(2)-3(2)\stackrel{?}{=}4-6i$$
$$10-6i-6\stackrel{?}{=}4-6i$$
$$4-6i=4-6i$$

Therefore, $x=2$ solves the equation.

17. Test $x=-5$ by substituting:

$$16(-5)\stackrel{?}{=}(-5)^3+(-5)^2+20$$
$$-80\stackrel{?}{=}-125+25+20$$
$$-80=-80$$

Therefore, $x=-5$ solves the equation.

19. Test $x=2$ by substituting:

$$(2)^4-3(2)^3-10(2)^2\stackrel{?}{=}0$$
$$16-24-40\stackrel{?}{=}0$$
$$-48\neq 0$$

Therefore, $x=2$ is not a zero of the given equation.

21. Test $x=-i$ by substituting:

$$(-i)^3-8i(-i)+30\stackrel{?}{=}15(-i)+2(-i)^2+16i$$
$$i-8+30\stackrel{?}{=}-15i-2+16i$$
$$i+22\neq i-2$$

Therefore, $x=-i$ is not a zero of the given equation.

23. Factor $x^3 - x^2 - 6x$ and set equal to zero:

$x(x^2 - x - 6) = 0$

$x(x-3)(x+2) = 0$

$x = 0;\ x = 3;\ x = -2$

Solution Set: $\{-2, 0, 3\}$

25. Factor $x^4 + x^2 - 2$ and set equal to zero:

$(x^2 - 1)(x^2 + 2) = 0$

$x = \pm 1;\ x = \pm i\sqrt{2}$

Solution Set: $\{-i\sqrt{2}, -1, 1, i\sqrt{2}\}$

27. Factor $9x^2 - 6x + 1$ and set equal to zero:

$(3x-1)(3x-1) = 0$

$x = \frac{1}{3}$

Solution Set: $\left\{\frac{1}{3}\right\}$

29. Factor $x^3 - x^2 - 72x$ and set equal to zero:

$x(x-9)(x+8) = 0$

$x = 0;\ x = 9;\ x = -8$

Solution Set: $\{-8, 0, 9\}$

31. Factor $2x^2 - 11x + 5$ and set equal to zero:

$(2x-1)(x-5) = 0$

$x = \frac{1}{2};\ x = 5$

Solution Set: $\left\{\frac{1}{2}, 5\right\}$

33. Factor $x^4 - 13x^2 + 36$ and set equal to zero:

$(x^2 - 9)(x^2 - 4) = 0$

$(x-3)(x+3)(x-2)(x+2) = 0$

$x = 3;\ x = -3;\ x = 2;\ x = -2$

Solution Set: $\{-3, -2, 2, 3\}$

35. The lead coefficient is 2. The degree is 4. The even degree indicates that $p(x) \to \infty$ as $x \to -\infty$ and as $x \to \infty$.

37. The lead coefficient and degree can be determined by the product $3x(x)(2x)(4x) = 24x^4$. The lead coefficient is 24 and the degree is 4. The even degree indicates that $r(x) \to \infty$ as $x \to -\infty$ and as $x \to \infty$.

39. The lead coefficient and degree can be determined by the product of the lead terms of the binomial factors: $x^3(2x)(-x) = -2x^5$. The lead coefficient is -2 and the degree is 5. The odd degree indicates that $g(x) \to \infty$ as $x \to -\infty$ and $g(x) \to -\infty$ as $x \to \infty$.

41. Given $f(x) = (x-3)(x+2)(x+4)$.

The zeros of f: $x = 3, x = -2, x = -4$

Using $x = 0$, the y-intercept can be determined: $f(0) = -3(2)(4) = -24$.

The y-intercept is $(0, -24)$.

Notice that f is a cubic function and its more basic function is $y = x^3$ so $f(x) \to -\infty$ as $x \to -\infty$ and $f(x) \to \infty$ as $x \to \infty$:

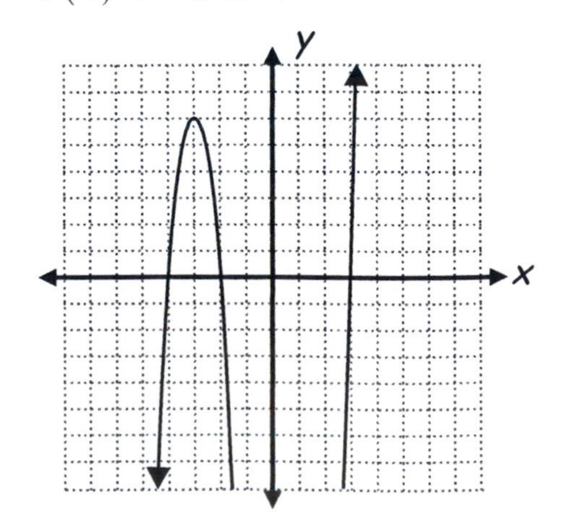

43. Given $f(x)=(x-2)^2(x+5)$.
The zeros of f : $x=2$, $x=-5$.
Using $x=0$, the y-intercept can be determined: $f(0)=(-2)^2(5)=20$.
The y-intercept is $(0, 20)$.
Notice that f is a cubic function and its more basic function is $y=x^3$
So $f(x)\to-\infty$ as $x\to-\infty$
and $f(x)\to\infty$ as $x\to\infty$:

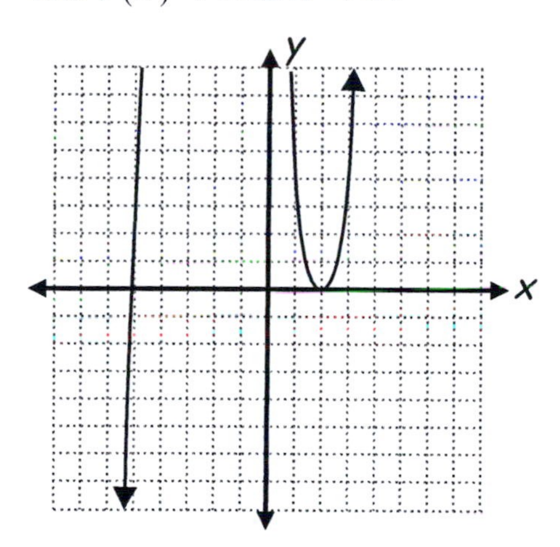

45. Given $r(x)=x^2-2x-3$
Factoring: $r(x)=(x-3)(x+1)$.
The zeros of r : $x=3$, $x=-1$.
Notice that r is a quadratic function.
You can find the vertex by completing the square:
$r(x)=(x^2-2x+1)-1-3$
$r(x)=(x-1)^2-4$
The vertex is $(1,-4)$.
Using $x=0$, the y-intercept can be determined: $r(0)=-3$.
The y-intercept is $(0,-3)$.
Since the degree, 2, is even, and the leading coefficient, 1, is positive, $f(x)\to\infty$ as $x\to-\infty$ and ∞.

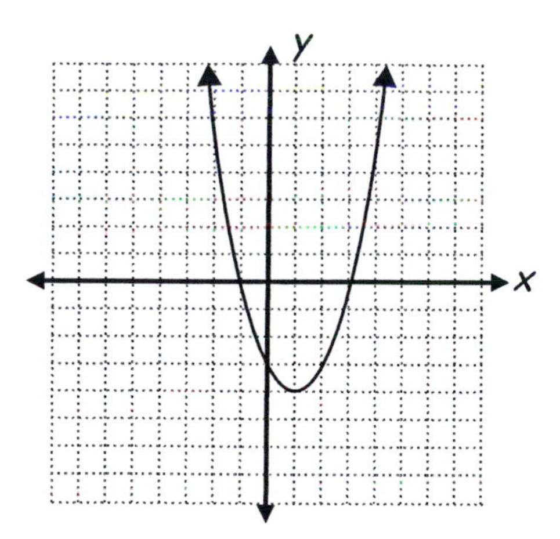

47. Given $f(x)=-(x-2)(x+1)^2(x+3)$
The factored form reveals the zeros:
$x=2$, $x=-1$, $x=-3$.
Using $x=0$, the y-intercept can be determined: $f(0)=-(-2)(1)^2(3)=6$.
The y-intercept is $(0, 6)$.
Since the degree, 4, is even, and the leading coefficient, -1, is negative, $f(x)\to-\infty$ as $x\to-\infty$ and $f(x)\to-\infty$ as $x\to\infty$.

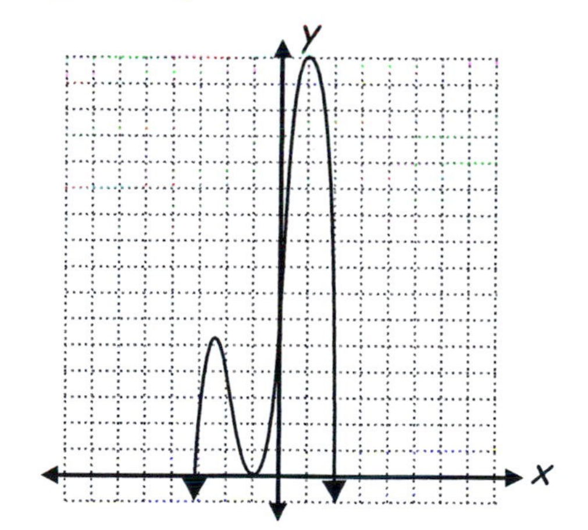

49. $f(x) = x\left[5x^3 + 6x^2 - 25x - 30\right]$

$= x\left[\left(5x^3 + 6x^2\right) - (25x + 30)\right]$

$= x\left[x^2(5x + 6) - 5(5x + 6)\right]$

$= x\left(x^2 - 5\right)(5x + 6)$

Set each factor equal to zero and solve for x:

$x = 0,\ x = \pm\sqrt{5},\ x = -\dfrac{6}{5}$

These values are the x-axis intercepts.

$f(0) = 0$, so the y-intercept is $y = 0$.

$f(x) \to \infty$ as $x \to \infty$ and as $x \to -\infty$.

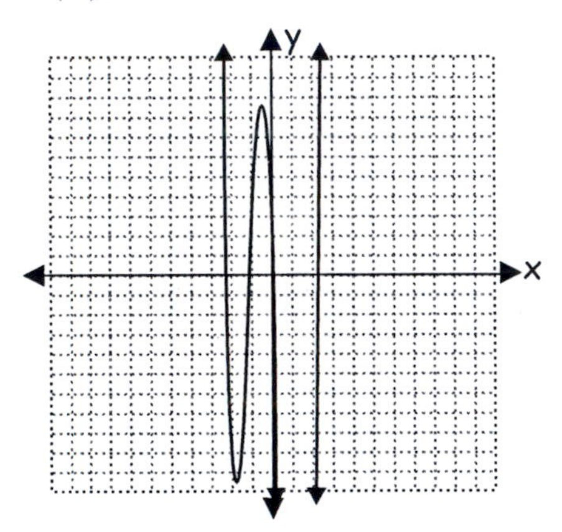

51. $g(x)$ is given in factored form:

Set each factor equal to zero and solve for x:

$2x - 3 = 0 \Rightarrow x = \dfrac{3}{2}$,

$x - 5 \Rightarrow x = 5$,

$1 - x \Rightarrow x = 1$

These values are the x-axis intercepts.

Find $g(0)$ to determine the y-intercept

$g(0) = (-3)(-5)(1)^2 = 15$.

Therefore, the y-intercept is 15.

$g(x) \to \infty$ as $x \to -\infty$ and as $x \to \infty$.

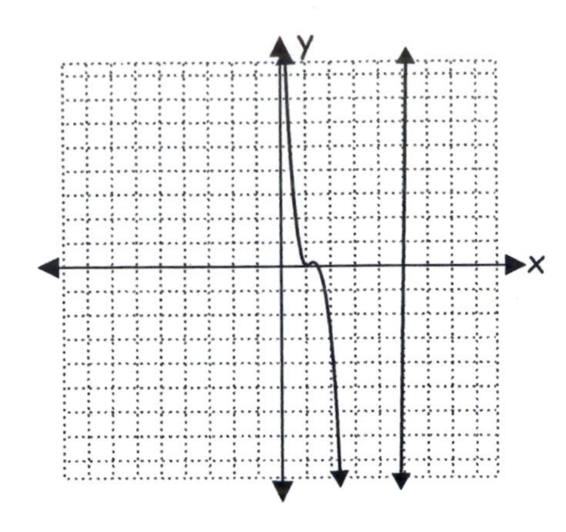

53. Inspection of the given factors of $g(x)$ reveal that $x = -1$ and $x = 3$ are zeros (x-intercepts of the graph of d). In addition, $g(0) = 9$, and $g(x) \to \infty$ as $x \to -\infty$ and as $x \to \infty$.

The graph (d) meets these conditions.

55. Given $f(x) = (x-1)(x+2)(3-x)$.

f is a cubic function (i.e., of degree 3) with zeros at $x = 1$; $x = -2$, and $x = 3$.

Only graph (b) has three distinct zeros and is the characteristic cubic graph.

57. Given $s(x) = (x-1)^3 - 2$.

s is a cubic function (i.e., of degree 3).

The basic form is $f(x) = x^3$, and the given function is a shift right 1 unit and down 2 units.

Only graph (c) fits this description.

59. Factor the left side of the inequality: $(x-3)(x+2) \le 0$. From this form you can determine the graph intersects the x-axis at $x = 3$ and $x = -2$.

The polynomial is a quadratic whose graph opens upward. The graph is below, or on, the x-axis on the interval $[-2, 3]$.

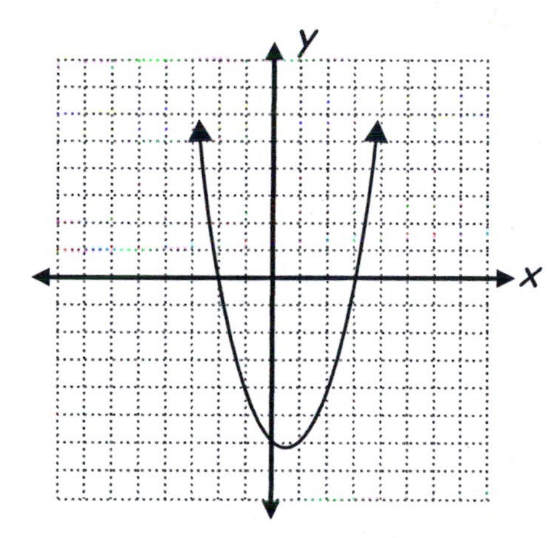

61. The polynomial is already in factored form, so the intercepts are readily determined: $x=-2$ and $x=1$. You might notice that the polynomial is non-negative for all values of x, so to find the intervals over which the polynomial is strictly greater that 0, you exclude the x-intercepts. Thus, $(x+2)^2(x-1)^2>0$ for all x in the set $(-\infty,-2)\cup(-2,1)\cup(1,\infty)$.

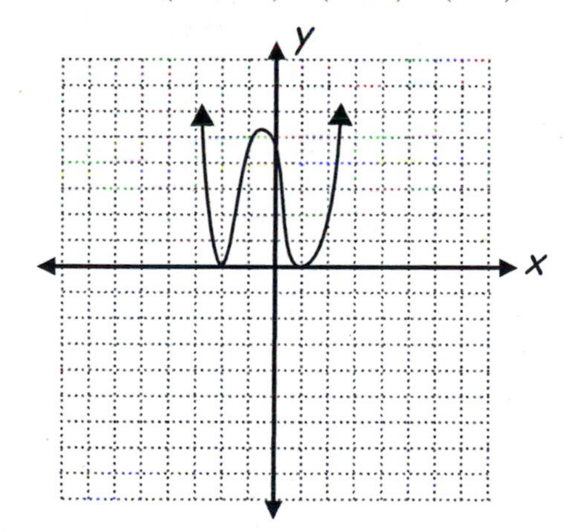

63. The polynomial is already in factored form, so the intercepts are readily determined: $x=2$ and $x=-1$ and $x=-3$.
The y-intercept is $(0,-6)$, which you can find by substituting $x=0$ in the polynomial. It helps to recognize the more basic function, $y=x^3$, and that the graph is that of a "typical" cubic polynomial. From the graph, you can readily determine the intervals that satisfy $(x-2)(x+1)(x+3)\geq 0$:
All $x\in[-3,-1]\cup[2,\infty)$.

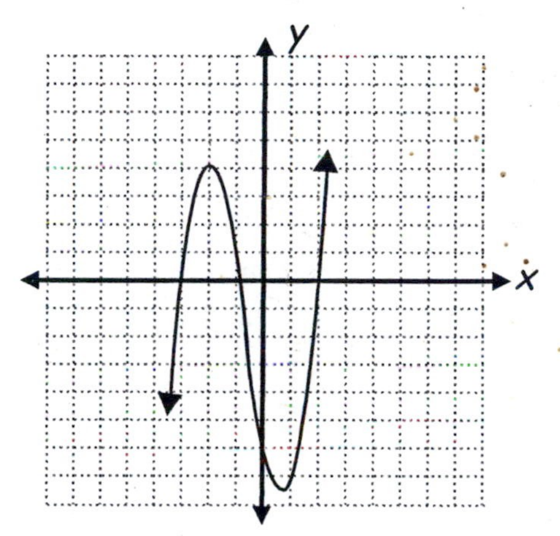

65. Factor the polynomial to find the x-intercepts: $-x(x^2+x-30)>0$, or more completely, $-x(x+6)(x-5)>0$. The x-intercepts are $x=0$, $x=-6$ and $x=5$. The more basic polynomial is $y=-x^3$. The y-intercept is $(0,0)$, which you can find by substituting $x=0$ in the polynomial expression. From the graph you can readily see the intervals over which the value of the polynomial is strictly greater than zero: $x\in(-\infty,-6)\cup(0,5)$

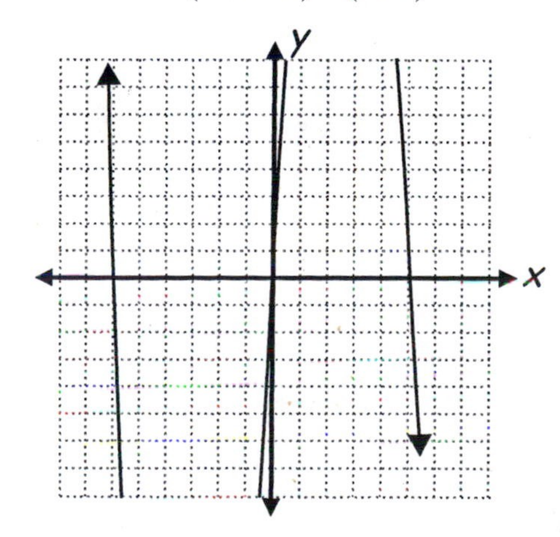

67. Factor the left side: $x^2(x^2+1)>0$. Notice that for all real values of x, $x^2 \ge 0$ and $x^2+1 \ge 0$. This means that $x^2(x^2+1) \ge 0$, for all values of x. However, the problem requires that $x^2(x^2+1)>0$. Notice that this is true except for $x=0$. Then, the solution for the inequality is the set of real numbers, except 0, or $(-\infty, 0)\cup(0, \infty)$. Yet another approach is to determine the zeros of the polynomial, which are $x=0$ and $x=\pm i$. The more basic function is $y=x^4$, the graph of which opens upward. The y-intercept of the given polynomial is $(0, 0)$:

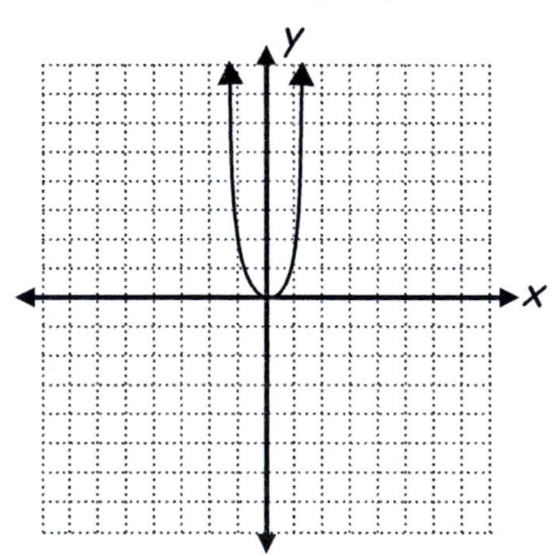

69. Given $L(m)=110m^2-0.35m^4+750$. Suggestion: Use a graphing calculator: Input the expression $-0.35m^4+110m^2+750$ with the use of the ‹y = › key [on TI-83 or TI-84]. Use the CALC feature and select "zero." Enter the requested left and right boundary values of, say 0 and 30, respectively. The calculator will return the solution of ≈ 17.9 months.

71. Given $M(w)=200w^2-0.01w^4+1200$. Suggestion: Use a graphing calculator: Input the expression $-0.01w^4+200w^2+1200$ with the use of the ‹y = › key [on TI-83 or TI-84]. Use the CALC feature and select "zero." Enter the requested left and right boundary values of, say 100 and 200, respectively. The calculator will return the solution of ≈ 141.4 weeks.

End of Section 4.1

Chapter Four: Section 4.2
Solutions to Odd-numbered Exercises

1.

$$3x^2 - x + 1 + \frac{5x-1}{2x^2+2}$$

$$2x^2+2 \overline{\big)\, 6x^4 - 2x^3 + 8x^2 + 3x + 1}$$

$$-\left(6x^4 + 6x^2\right)$$

$$2x^3 + 2x^2 + 3x + 1$$

$$-\left(-2x^3 - 2x\right)$$

$$2x^2 + 5x + 1$$

$$-\left(2x^2 + 2\right)$$

$$5x - 1$$

3.

$$x - 2 - \frac{2}{x^2-4x+4}$$

$$x^2-4x+4 \overline{\big)\, x^3 - 6x^2 + 12x - 10}$$

$$-\left(x^3 - 4x^2 + 4x\right)$$

$$-2x^2 + 8x - 10$$

$$-\left(-2x^2 + 8x - 8\right)$$

$$-2$$

5.

$$4x^2 - 14x + 29 - \frac{65}{x+2}$$

$$x+2 \overline{\big)\, 4x^3 - 6x^2 + x - 7}$$

$$-\left(4x^3 + 8x^2\right)$$

$$-14x^2 + x - 7$$

$$-\left(-14x^2 - 28x\right)$$

$$29x - 7$$

$$-(29x + 58)$$

$$-65$$

7.

$$x^3 + 6x^2 - 2x + 5 + \frac{2x+5}{3x^2-1}$$

$$3x^2-1 \overline{\big)\, 3x^5 + 18x^4 - 7x^3 + 9x^2 + 4x}$$

$$-\left(3x^5 - x^3\right)$$

$$18x^4 - 6x^3 + 9x^2 + 4x$$

$$-\left(18x^4 - 6x^2\right)$$

$$-6x^3 + 15x^2 + 4x$$

$$-\left(6x^3 + 2x\right)$$

$$15x^2 + 2x$$

$$-\left(15x^2 - 5\right)$$

$$2x + 5$$

9.

$$x^2 - ix + 6 + \frac{1+i}{2x-i}$$

$$2x-i \overline{\big)\, 2x^3 - 3ix^2 + 11x + (1-5i)}$$

$$-\left(2x^3 - ix^2\right)$$

$$-2ix^2 + 11x + (1 - 5i)$$

$$-\left(-2ix^2 - x\right)$$

$$12x + (1-5i)$$

$$-(12x - 6i)$$

$$1 + i$$

11.

$$x^2 + 3$$

$$3x+i \overline{\big)\, 3x^3 + ix^2 + 9x + 3i}$$

$$-\left(3x^3 + ix^2\right)$$

$$9x + 3i$$

$$-(9x + 3i)$$

$$0$$

13.

$$\begin{array}{r} 2x^3-3x^2+2x-5 \\ x^2-x+1\overline{)2x^5-5x^4+7x^3-10x^2+7x-5} \\ -\left(2x^5-2x^4+2x^3\right) \qquad\qquad\qquad\quad \\ \hline -3x^4+5x^3-10x^2+7x-5 \\ -\left(-3x^4+3x^3-3x^2\right) \qquad\quad\;\; \\ \hline 2x^3-7x^2+7x-5 \\ -\left(2x^3-2x^2+2x\right) \quad\;\; \\ \hline -5x^2+5x-5 \\ -\left(-5x^2+5x-5\right) \\ \hline 0 \end{array}$$

15.

$$\begin{array}{r} x^3+3x^2+10x+10+\dfrac{22}{x-3} \\ x-3\overline{)\quad x^4+0x^3+x^2-20x-8} \\ -\left(x^4-3x^3\right) \qquad\qquad\qquad \\ \hline 3x^3+x^2-20x-8 \\ -\left(3x^3-9x^2\right) \qquad\qquad \\ \hline 10x^2-20x-8 \\ -\left(10x^2-30x\right) \quad\;\; \\ \hline 10x-8 \\ -(10x-30) \\ \hline 22 \end{array}$$

17.

$$\begin{array}{r} 3x^2+5x+9+\dfrac{45}{3x-5} \\ 3x-5\overline{)\quad 9x^3+0x^2+2x} \\ -\left(9x^3-15x^2\right) \qquad\; \\ \hline 15x^2+2x \\ -\left(15x^2-25x\right) \\ \hline 27x \\ -(27x-45) \\ \hline 45 \end{array}$$

19.

$$\begin{array}{r|rrrrrr} 1 & 32 & -80 & 80 & -40 & 10 & 2 \\ & & 32 & -48 & 32 & -8 & 2 \\ \hline & 32 & -48 & 32 & -8 & 2 & 4 \end{array}$$

$k=1$ is not a zero.

$p(1)=4.$

21.

$$\begin{array}{r|rrrrr} 2 & 12 & -7 & -32 & -7 & 6 \\ & & 24 & 34 & 4 & -6 \\ \hline & 12 & 17 & 2 & -3 & 0 \end{array}$$

$k=2$ is a zero.

$p(2)=0.$

23.

$$\begin{array}{r|rrrrr} \frac{1}{3} & 12 & -7 & -32 & -7 & 6 \\ & & 4 & -1 & -11 & -6 \\ \hline & 12 & -3 & -33 & -18 & 0 \end{array}$$

$k=\dfrac{1}{3}$ is a zero.

$p\left(\dfrac{1}{3}\right)=0.$

25.

$$\begin{array}{r|rrrrr} 1 & 8 & 0 & 0 & -2 & 6 \\ & & 8 & 8 & 8 & 6 \\ \hline & 8 & 8 & 8 & 6 & 12 \end{array}$$

$k=1$ is not a zero.

$p(1)=12.$

27.

$$\begin{array}{r|rrrrrr} -2 & 1 & 0 & 0 & 0 & 0 & 32 \\ & & -2 & 4 & -8 & 16 & -32 \\ \hline & 1 & -2 & 4 & -8 & 16 & 0 \end{array}$$

$k=-2$ is a zero.

$p(-2)=0.$

29.

$$\begin{array}{r|rrr} -3i & 2 & -(3-5i) & +(3-9i) \\ & & -6i & -3+9i \\ \hline & 2 & -3-i & 0 \end{array}$$

$k=-3i$ is a zero.

$p(-3i)=0.$

31.
$$\begin{array}{r|rrr} 3-2i & 1 & -6 & 13 \\ & & 3-2i & -13 \\ \hline & 1 & -3-2i & 0 \end{array}$$

$k = 3-2i$ is a zero.
$p(3-2i) = 0.$

33.
$$\begin{array}{r|rrrr} 6 & 3 & -13 & -28 & -12 \\ & & 18 & 30 & 12 \\ \hline & 3 & 5 & 2 & 0 \end{array}$$

$k = 6$ is a zero.
$p(6) = 0.$

35.
$$\begin{array}{r|rrrr} 5 & 2 & -8 & -23 & 63 \\ & & 10 & 10 & -65 \\ \hline & 2 & 2 & -13 & -2 \end{array}$$

$k = 5$ is not a zero.
$p(5) = -2.$

37.
$$\begin{array}{r|rrrrr} -2 & 1 & -3 & -3 & 11 & -6 \\ & & -2 & 10 & -14 & 6 \\ \hline & 1 & -5 & 7 & -3 & 0 \end{array}$$

$k = -2$ is a zero.
$p(-2) = 0.$

39.
$$\begin{array}{r|rrrr} -5 & 1 & 1 & -18 & 9 \\ & & -5 & 20 & -10 \\ \hline & 1 & -4 & 2 & -1 \end{array}$$

$x^2 - 4x + 2 - \dfrac{1}{x+5}$

41.
$$\begin{array}{r|rrrrrrrrr} -1 & 1 & 1 & 0 & 0 & 0 & -3 & -3 & 0 & 3 \\ & & -1 & 0 & 0 & 0 & 0 & 3 & 0 & 0 \\ \hline & 1 & 0 & 0 & 0 & 0 & -3 & 0 & 0 & 3 \end{array}$$

$x^7 - 3x^2 + \dfrac{3}{x+1}$

43.
$$\begin{array}{r|rrrr} 3+i & 4 & -(16+4i) & (14+4i) & (-6-2i) \\ & & 12+4i & -12-4i & 6+2i \\ \hline & 4 & -4 & 2 & 0 \end{array}$$

$4x^2 - 4x + 2$

45.
$$\begin{array}{r|rrrrrr} 2 & 1 & -3 & 1 & -5 & 0 & 18 \\ & & 2 & -2 & -2 & -14 & -28 \\ \hline & 1 & -1 & -1 & -7 & -14 & -10 \end{array}$$

$x^4 - x^3 - x^2 - 7x - 14 - \dfrac{10}{x-2}$

47.
$$\begin{array}{r|rrrr} -i & 1 & (i-1) & (1-i) & i \\ & & -i & i & -i \\ \hline & 1 & -1 & 1 & 0 \end{array}$$

$x^3 - x^2 + x$

49. $(x+4)$ and $(x-3)$ are factors.
The function is a quadratic and it opens downward. This means the lead coefficient must be negative.
$f(x) = -(x+4)(x-3) = -x^2 - x + 12$
is one function with these properties.

51. $f(x) = a\left(x-(2-3i)\right)\left(x-(2+3i)\right)$
$f(x) = a(x^2 - (2-3i)x - (2+3i)x +$
$\quad (2-3i)(2+3i))$
$f(x) = a\left(x^2 - (2-3i+2+3i)x + 4 + 9\right)$
$f(x) = a\left(x^2 - 4x + 13\right)$, for some real a.
To find a, set $f(0) = -13$ to satisfy the given condiction of the y-intercept.
$f(0) = a(13) = -13.$ So, $a = -1.$
$f(x) = -x^2 + 4x - 13$

53. $f(x) = (x-3)^4$

$= (x-3)^2(x-3)^2$

$= (x^2 - 6x + 9)^2$

$f(x) = x^4 - 12x^3 + 54x^2 - 108x + 81$

55. $f(x) = a(x+3)(x+2)(x-1)^2$, for some constant a.

Set $f(0) = a(3)(2)(-1)^2 = 18$ and solve for a. This satisfies the y-intercept condition:

$6a = 18$

$a = 3$

$f(x) = 3(x+3)(x+2)(x-1)^2$

$f(x) = 3x^4 + 9x^3 - 9x^2 - 21x + 18$

End of Section 4.2

Chapter Four: Section 4.3
Solutions to Odd-numbered Exercises

1. $f(x) = 3x^3 + 5x^2 - 26x + 8$

Factors of a_3: $\pm\{1, 3\}$

Factors of a_0 : $\pm\{1, 2, 4, 8\}$

Potential zeros: $\pm\left\{1, \frac{1}{3}, 2, \frac{2}{3}, 4, \frac{4}{3}, 8, \frac{8}{3}\right\}$

Use synthetic division to test each potential zero. The first one shown below illustrates the failure of one of the potential zeros to be an actual zero. The second division tests 1/3, and the result indicates that it is a zero:

1|	3	5	−26	8
		3	8	−18
	3	8	−18	−10

$p(1) = -10 \neq 0$, so $k = 1$ is not a zero.

Similiar tests for -4, $\frac{1}{3}$ and 2 indicate that these values are rational zeros of $f(x)$. The test is shown for $k = \frac{1}{3}$.

1/3|	3	5	−26	8
		1	2	−8
	3	6	−24	0

$p\left(\frac{1}{3}\right) = 0$, so $k = \frac{1}{3}$ is a zero.

3. $p(x) = x^4 - 5x^3 + 10x^2 - 20x + 24$

Factors of a_4: $\pm\{1\}$

Factors of a_0 : $\pm\{1, 2, 3, 4, 6, 8, 12, 24\}$

There are 16 potential rational zeros.

Potential zeros: $\pm\{1, 2, 3, 4, 6, 8, 12, 24\}$

Use synthetic division to test each potential zero. The first one shown below illustrates the failure of one of the potential zeros to be an actual zero. The second and third divisions test $k = 2$ and $k = 3$, and the results indicate that they are both zeros:

1|	1	−5	10	−20	24
		1	−4	6	−14
	1	−4	6	−14	10

$p(1) = 10 \neq 0$, so $k = 1$ is not a zero.

Similiar tests for 2 and 3 indicate that these values are rational zeros of $p(x)$. The tests are shown with the result of the second one using the result of the first.

2|	1	−5	10	−20	24
		2	−6	8	−24
3|	1	−3	4	−12	0
		3	0	12	
	1	0	4	0	

$p(2) = 0$, so $k = 2$ is a zero;

$p(3) = 0$, so $k = 3$ is a zero.

The last row in the matrix translates as a quadratic whose zeros are readily determined: $x^2 + 4 = 0$

The zeros of the quadratic are $x = \pm 2i$.

The zero set is $\{2, 3, -2i, 2i\}$.

5. $q(x) = x^3 - 10x^2 + 23x - 14$

Factors of a_3: $\pm\{1\}$

Factors of a_0 : $\pm\{1, 2, 7, 14\}$

Potential zeros: $\pm\{1, 2, 7, 14\}$

Use synthetic division to test each potential zero. The first one shown below illustrates the failure of one of the potential zeros to be an actual zero. The second division tests 7, and the result indicates that it is a zero:

−1|	1	−10	23	−14
		−1	11	−34
	1	−11	34	−48

$p(-1) = -48 \neq 0$, so $k = -1$ is not a zero.

Similiar tests for 1, 2 and 7 indicate that these values are rational zeros of $f(x)$. The test is shown for $k=7$.

7	1	−10	23	−14
		7	−21	14
	1	−3	2	0

$q(7)=0$, so $k=7$ is a zero.

7. $s(x)=2x^3-9x^2+4x+15$

Factors of a_3: $\pm\{1, 2\}$

Factors of a_0 : $\pm\{1, 3, 5, 15\}$

There are 16 potential zeros:

$\pm\left\{1, \frac{1}{2}, 3, \frac{3}{2}, 5, \frac{5}{2}, 15, \frac{15}{2}\right\}$

Use synthetic division to test each potential zero. The first one shown below illustrates the failure of one of the potential zeros to be an actual zero. The second division tests $\frac{5}{2}$, and the result indicates that it is a zero:

$\frac{1}{2}$	2	−9	4	15
		1	−4	0
	2	−8	0	15

$s\left(\frac{1}{2}\right)=15\neq 0$, so $k=\frac{1}{2}$ is not a zero.

Similiar tests for -1, $\frac{5}{2}$ and 3 indicate that these values are rational zeros of $s(x)$. The test is shown for $k=\frac{5}{2}$.

$\frac{5}{2}$	2	−9	4	15
		5	−10	−15
	2	−4	−6	0

$s\left(\frac{5}{2}\right)=0$, so $k=\frac{5}{2}$ is a zero.

9. $j(x)=3x^4-3$

First, factor out the 3 to focus on the polynomial factor of j : $j(x)=3\left(x^4-1\right)$

Factors of a_4: $\pm\{1\}$

Factors of a_0 : $\pm\{1\}$

Potential zeros: $\pm\{1\}$

You could factor x^4-1 as the difference of two squares and find its zeros, or use synthetic division as shown here:

You can test the second potential zero on the results of the test of the first potential zero:

−1	1	0	0	0	−1
		−1	1	−1	1
1	1	−1	1	−1	0
		1	0	1	
	1	0	1	0	

Therefore both $k=-1$ and $k=1$ are zeros. The last row indicates that the quadratic x^2+1 is also a factor. It has complex zeros of $\pm i$.

The zero set is $\{-1, 1, -i, i\}$.

11. Solve: $x^4+x-2=-2x^4+x+1$

Re-write the equation with all terms on the left side:

$3x^4-3=0$

To solve, use the zeros in Exercise 9:

The solution set is $\{-1, 1, -i, i\}$.

13. Solve: $x^3 - 3x^2 + 9x + 13 = 0$

Factors of a_3 : ± 1

Factors of a_0 : $\pm\{1, 13\}$

Potential rational zeros: $\pm\{1, 13\}$

Testing $k = 1$:

$$\begin{array}{r|rrrr} 1 & 1 & -3 & 9 & 13 \\ & & 1 & -2 & 7 \\ \hline & 1 & -2 & 7 & 20 \end{array}$$

$k = 1$ is not a zero and is not a solution to the given equation.

Testing $k = -1$:

$$\begin{array}{r|rrrr} -1 & 1 & -3 & 9 & 13 \\ & & -1 & 4 & -13 \\ \hline & 1 & -4 & 13 & 0 \end{array}$$

$k = -1$ is a solution to the given equation. Along with the zero of -1, the last row in the matrix indicates that the left side of the equation factors as $(x+1)(x^2 - 4x + 13) = 0$. Setting the quadratic factor equal to zero, you can solve it using the quadratic formula:

$$x = \frac{4 \pm \sqrt{16 - 4(1)(13)}}{2}$$

$x = 2 \pm 3i$

The solution set is $\{-1,\ 2 \pm 3i\}$

15. Solve: $x^4 + 10x^2 - 20x = 5x^3 - 24$.

Re-write as

$x^4 - 5x^3 + 10x^2 - 20x + 24 = 0$.

The zeros of the polynomial are as shown in Exercise 3. The solution set is $\{2, 3, -2i, 2i\}$.

17. Solve: $2x^3 - 12x^2 + 26x = 40$.

Re-write as

$2x^3 - 12x^2 + 26x - 40 = 0$.

Factors of a_3 : $\pm\{1, 2\}$

Factors of a_0 : $\pm\{1, 2, 4, 5, 8, 10, 20, 40\}$

There are 20 potential rational zeros of the polynomial. Each can be tested by using synthetic division.

Testing $k = 4$:

$$\begin{array}{r|rrrr} 4 & 2 & -12 & 26 & -40 \\ & & 8 & -16 & 40 \\ \hline & 2 & -4 & 10 & 0 \end{array}$$

This result shows that $k = 4$ is a zero, and is, therefore, a solution to the equation. It also shows that $x - 4$ is a factor of the polynomial, and that the polynomial factors as $(x-4)(2x^2 - 4x + 10)$ which factors further as $2(x-4)(x^2 - 2x + 5)$. The latter factor is a quadratic. Set the factor equal to zero and solve by using the quadratic formula.

Solve $x^2 - 2x + 5 = 0$:

$$x = \frac{2 \pm \sqrt{4 - 4(1)(5)}}{2}$$

$x = 1 \pm 2i$

The solution set is $\{4, 1+2i, 1-2i\}$

19. Solve: $x^4 + x^3 + 23x^2 = 50 - 25x$.

Re-write as

$x^4 + x^3 + 23x^2 + 25x - 50 = 0$.

Factors of a_4 : ± 1

Factors of a_0 : $\pm\{1, 2, 5, 10, 25, 50\}$

There are 12 potential rational zeros of the polynomial:

$\pm\{1, 2, 5, 10, 25, 50\}$

Each can be tested by using synthetic division. The work below finds that $k=-2$ is a zero (solution) and uses the resulting coefficients to test $k=1$:
Testing $k=-2$, and then testing $k=1$:

-2	1	1	23	25	-50
		-2	2	-50	50
1	1	-1	25	-25	0
		1	0	25	
	1	0	25	0	

The partially factored form of the polynomial is $(x+2)(x-1)(x^2+25)=0$.
The results above indicated that $k=-2$ and 1 are zeros of the polynomial and therefore, solutions to the equation. The third factor comes from the last row of the synthetic division.
Set $x^2+25=0$ and solve directly:
$x^2=-25$, or $x=\pm 5i$.
The solution set is $\{-2, 1, 5i, -5i\}$.

21. The given polynomial is given in descending order:
$f(x)=x^3+8x^2+17x+10$
There are zero sign changes in $f(x)$.
Therefore there are zero positive real zeros.
Examine the variations of sign change in $f(-x)$: $f(-x)=-x^3+8x^2-17x+10$
There are three sign changes in $f(-x)$.
Therefore the number of negative real zeros is 3 or 1.

23. For $f(x)=x^3-6x^2+3x+10$, there are two variations in sign which indicates 2 or 0 positive real zeros.
For $f(-x)=-x^3-6x^2-3x+10$ there is one variation in sign, which indicates one negative real zero.

25. For $f(x)=x^4-5x^3-2x^2+40x-48$, there are three variations in sign which indicates 3 or 1 positive real zeros.
For $f(-x)=x^4+5x^3-2x^2-40x-48$ there is one variation in sign, which indicates one negative real zero.

27. For $f(x)=x^4-25$, there is one variation in sign which indicates 1 positive real zero.
For $f(-x)=x^4-25$, there is one variation in sign, which indicates one negative real zero.

29. For $f(x)=5x^5-x^4+2x^3+x-9$, there are three variations in sign which indicates 3 or 1 positive real zeros.
For $f(-x)=-5x^5-x^4-2x^3-x-9$, There are no variations in sign, so there are no real negative zeros.

31. Given $f(x)=x^3+4x^2-x-4$:
Answers will vary. Here is one way:
Choose an integer to test as an upper or lower bound. Let's say 2:

2	1	4	-1	4
		2	12	22
	1	6	11	26

Therefore, 2 is an upper bound of the zeros of f.(There are no negative coefficients in the third row of our synthetic division.)
Choose another integer, less than 2, to test as an upper or lower bound.
Let's say, -5:

-5	1	4	-1	4
		-5	5	-20
	1	-1	4	-16

−5 is a lower bound of the zeros of f (The signs of the coefficients in the third row of our synthetic division alternate.)
Therefore the interval $[-5, 2]$ indicates the lower and upper bound of the real zeros of the given polynomial. (Note: 1 is also an upper bound, so the interval $[-5, 1]$ can also be used.)

33. Given $f(x) = x^3 - 6x^2 + 3x + 10$:
Answers will vary. Here is one way:
Choose an integer to test as an upper or lower bound. Let's say 7:

$$\begin{array}{r|rrrr} 7 & 1 & -6 & 3 & 10 \\ & & 7 & 7 & 70 \\ \hline & 1 & 1 & 10 & 80 \end{array}$$

7 is an upper bound of the zeros of f (There are no negative coefficients in the third row of our synthetic division.)
Choose another integer, less than 7, to test as an upper or lower bound. Let's say, -3:

$$\begin{array}{r|rrrr} -3 & 1 & -6 & 3 & 10 \\ & & -3 & 27 & -90 \\ \hline & 1 & -9 & 30 & -80 \end{array}$$

−3 is a lower bound of the zeros of f (The signs of the coefficients in the third row of our synthetic division alternate.)
Therefore, the interval $[-3, 7]$ indicates a lower and upper bound of the real zeros of the given polynomial. (Note: similarly we could show -1 is a lower bound and 6 is an upper bound, so the interval $[-1, 6]$ also applies.)

35. Given $f(x) = x^4 - 5x^3 - 2x^2 + 40x - 48$.
Answers will vary in finding upper and lower bounds of real zeros. Here is one solution:
Choose an integer to test as an upper or lower bound. Let's say 5:

$$\begin{array}{r|rrrrr} 5 & 1 & -5 & -2 & 40 & -48 \\ & & 5 & 0 & -10 & 150 \\ \hline & 1 & 0 & -2 & 30 & 102 \end{array}$$

Therefore, the conditions are not met, and we conclude that 5 is not an upper or lower bound for the real zeros.
Choose another integer to test as an upper or lower bound. Let's say 6:

$$\begin{array}{r|rrrrr} 6 & 1 & -5 & -2 & 40 & -48 \\ & & 6 & 6 & 24 & 384 \\ \hline & 1 & 1 & 4 & 64 & 336 \end{array}$$

Therefore, 6 is an upper bound for the real zeros (All coefficients are positive).
Test -3:

$$\begin{array}{r|rrrrr} -3 & 1 & -5 & -2 & 40 & -48 \\ & & -3 & 24 & -66 & 78 \\ \hline & 1 & -8 & 22 & -26 & 30 \end{array}$$

Therefore, -3 is a lower bound for the real zeros of f (the coefficients alternate in sign).

37. Given $f(x) = x^4 - 25$.
Answers will vary in finding upper and lower bounds of real zeros. Here is one solution:
Choose an integer to test as an upper or lower bound. Let's say 4:

$$\begin{array}{r|rrrrr} 4 & 1 & 0 & 0 & 0 & -25 \\ & & 4 & 16 & 64 & 256 \\ \hline & 1 & 4 & 16 & 64 & 231 \end{array}$$

Therefore, 4 is an upper bound for the real zeros of f.

Test -3:

$$\begin{array}{r|rrrrr} -3 & 1 & 0 & 0 & 0 & -25 \\ & & -3 & 9 & -27 & 81 \\ \hline & 1 & -3 & 9 & -27 & 56 \end{array}$$

Therefore, -3 is a lower bound for the real zeros of f (the coefficients alternate in sign).

The interval $[-3, 4]$ is a boundary interval for the real zeros of f.

39. Given $f(x) = 2x^3 - 7x^2 - 28x - 12$.

Answers will vary in finding upper and lower bounds of real zeros. Here is one solution:

Choose an integer to test as an upper or lower bound. Let's say 7:

$$\begin{array}{r|rrrr} 7 & 2 & -7 & -28 & -12 \\ & & 14 & 49 & 147 \\ \hline & 2 & 7 & 21 & 135 \end{array}$$

Therefore, 7 is an upper bound for the real zeros of f.

Choose another integer to test as an upper or lower bound. Let's say -2:

$$\begin{array}{r|rrrr} -2 & 2 & -7 & -28 & -12 \\ & & -4 & 22 & 12 \\ \hline & 2 & -11 & -6 & 0 \end{array}$$

Notice that -2 is a zero!

Let's try an integer less than -2:

Test -3:

$$\begin{array}{r|rrrr} -3 & 2 & -7 & -28 & -12 \\ & & -6 & 39 & -33 \\ \hline & 2 & -13 & 11 & -45 \end{array}$$

Therefore, -3 is a lower bound for the real zeros of f (the coefficients alternate in sign).

The interval $[-3, 7]$ is a boundary interval for the real zeros of f.

41. From Exercise 31, we know that any real zeros of $f(x) = x^3 + 4x^2 - x - 4$ are bounded by the interval $[-5, 2]$.

Potential rational zeros are $\pm\{1, 2, 4\}$, according to the Rational Zero Theorem.

In this case, it is also possible to factor $f(x)$ by grouping. The solutions can then be realized straightforward:

$$\begin{aligned} f(x) &= x^2(x+4) - (x+4) \\ &= (x^2 - 1)(x+4) \\ &= (x-1)(x+1)(x-4) \end{aligned}$$

Set each factor equal to zero and solve for x:

$x = -1, x = 1, x = -4$

Therefore, the solution set is $\{-4, -1, 1\}$.

43. From Exercise 33, we know that any real zeros of $f(x) = x^3 - 6x^2 + 3x + 10$ are bounded by the interval $[-1, 6]$. Then potential rational zeros are $\{-1, 1, 2, 5\}$, according to the Rational Zero Theorem.

Test $k = -1$:

$$\begin{array}{r|rrrr} -1 & 1 & -6 & 3 & 10 \\ & & -1 & 7 & -10 \\ \hline & 1 & -7 & 10 & 0 \end{array}$$

Therefore, $k = -1$ is a zero.

Test $k = 2$, by using the last row.

$$\begin{array}{r|rrr} 2 & 1 & -7 & 10 \\ & & 2 & -10 \\ \hline & 1 & -5 & 0 \end{array}$$

Therefore, $k = 2$ is a zero.

The last row indicates that $x - 5$ is a factor of $f(x)$, and that $k = 5$ is also a zero.

The solution set is $\{-1, 2, 5\}$.

45. From Exercise 35, we know that any real zeros of

$f(x) = x^4 - 5x^3 - 2x^2 + 40x - 48$

are bounded by the interval $[-3, 6]$.

Potential rational zeros are then $\{-3, -2, -1, 1, 2, 3, 4, 6\}$, according to the Rational Zero Theorem.

Test $k = 2$:

$$\begin{array}{r|rrrrr} 2 & 1 & -5 & -2 & 40 & -48 \\ & & 2 & -6 & -16 & 48 \\ \hline & 1 & -3 & -8 & 24 & 0 \end{array}$$

Therefore, $k = 2$ is a zero.

Test $k = 3$, by using the last row.

$$\begin{array}{r|rrrr} 3 & 1 & -3 & -8 & 24 \\ & & 3 & 0 & -24 \\ \hline & 1 & 0 & -8 & 0 \end{array}$$

Therefore, $k = 3$ is a zero.

The last row indicates that $x^2 - 8$ is a factor of $f(x)$. Solving this quadratic when set equal to zero yields $x = \pm 2\sqrt{2}$.

The solution set is $\{2, 3, 2\sqrt{2}, -2\sqrt{2}\}$

47. From Exercise 37, we know that any real zeros of $f(x) = x^4 - 25$ are bound by the interval $[-3, 4]$.

$x^4 - 25$ factors as $(x^2 - 5)(x^2 + 5)$.

Set each factor equal to zero and solve: $x = \pm\sqrt{5}$ and $x = \pm i\sqrt{5}$, respectively.

The solution set is $\{-\sqrt{5}, \sqrt{5}, i\sqrt{5}, -i\sqrt{5}\}$.

49. From Exercise 39, we know that any real zeros of

$f(x) = 2x^3 - 7x^2 - 28x - 12$

are bounded by the interval $[-3, 7]$.

Potential rational zeros are then

$\left\{-3, -2, -\frac{3}{2}, -1, -\frac{1}{2}, \frac{1}{2}, 1, \frac{3}{2}, 2, 3, 4, 6\right\}$,

according to the Rational Zero Theorem.

Test $k = -2$:

$$\begin{array}{r|rrrr} -2 & 2 & -7 & -28 & -12 \\ & & -4 & 22 & 12 \\ \hline & 2 & -11 & -6 & 0 \end{array}$$

Therefore, $k = -2$ is a zero.

Test $k = -\frac{1}{2}$, by using the last row.

$$\begin{array}{r|rrr} -\frac{1}{2} & 2 & -11 & -6 \\ & & -1 & 6 \\ \hline & 2 & -12 & 0 \end{array}$$

Therefore, $k = -\frac{1}{2}$ is a zero.

The last row indicates that $2x - 12$ is a factor of $f(x)$, and that $k = 6$ is also a zero. The solution set is $\left\{-2, -\frac{1}{2}, 6\right\}$.

51. According to the Intermediate Value Theorem, f has a zero between $x = -2$ and $x = -1$ if $f(-2)$ and $f(-1)$ differ in sign:

$$\begin{aligned} f(-2) &= 5(-2)^3 - 4(-2)^2 - 31(-2) - 6 \\ &= -40 - 16 + 62 - 6 \\ &= 0 \end{aligned}$$

Since $f(-2) = 0$, there is at least one zero in the interval $[-2, -1]$.

53. According to the Intermediate Value Theorem, f has a zero between $x = 2$ and $x = 3$ if $f(2)$ and $f(3)$ differ in sign:

$$\begin{aligned} f(2) &= 2^4 + 2(2^3) - 10(2)^2 - 14(2) + 21 \\ &= 16 + 16 - 40 - 28 + 21 \\ &= -15 \end{aligned}$$

$$\begin{aligned} f(3) &= 3^4 + 2(3)^3 - 10(3)^2 - 14(3) + 21 \\ &= 81 + 54 - 90 - 42 + 21 \\ &= 24 \end{aligned}$$

Therefore, the conditions of the Intermediate Value Theorem are met. There is some value $x = c$ such that $c \in (2, 3)$ and $f(c) = 0$.

55. Re-write the equation as

$f(x) = x^4 + 2x^3 - 10x^2 - 14x + 21 = 0$

The polynomial is exactly equivalent to that of Exercise 53. The Intermediate Value Theorem is applicable.

57. Rational Zero Theorem Proof:

The theorem states that if

$f(x) = a_n x^n + a_{n-1}x^{n-1} + ... + a_1 x + a_0$

is a polynomial with integer coefficients, then any rational zero of f must be of the form $\frac{p}{q}$, where p is a factor of a_0 and q is a factor of a_n.

We start with the assumption that $\frac{p}{q}$ is in reduced form (i.e., that p and q have no common factors other than 1).

Suppose that $\frac{p}{q}$ is a zero of f, where p and q are integers and $q \neq 0$.

Then,

$$f\left(\frac{p}{q}\right) = a_n\left(\frac{p}{q}\right)^n + a_{n-1}\left(\frac{p}{q}\right)^{n-1} + ... + a_1\left(\frac{p}{q}\right) + a_0$$
$$= 0$$

Multiply both sides of the equation by q^n to obtain

$a_n p^n + a_{n-1}p^{n-1}q + ... + a_1 pq^{n-1} + a_o q^n = 0$

Or,

$a_n p^n + a_{n-1}p^{n-1}q + ... + a_1 pq^{n-1} = -a_o q^n$

Since p divides each term on the left side (Therefore, it is a factor of the left side.), p must also be a factor of the right side, $a_0 q^n$. But, since p and q have no common factors, p must divide a_0, and is therefore a factor of a_0.

In similar fashion, the last equation can be written as

$a_{n-1}p^{n-1}q + ... + a_1 pq^{n-1} + a_o q^n = -a_n p^n$

and the argument can be made that q divides each term of the left side and is, therefore, a factor of the left side. Hence, q is a factor of the right side, and since q and p have no common factors other than 1, q must be a factor of a_n.

The argument is complete.

59. Find all real zeros of

$f(x) = -4x^3 - 19x^2 + 29x - 6.$

From the factors of the lead coefficient and the constant term, we can identify the potential rational zeros:

$\pm\left\{\frac{1}{4}, \frac{1}{2}, \frac{3}{4}, 1, \frac{3}{2}, 2, 3, 6\right\}$

Using Descartes' Rule of Signs, we know that there can be two or no positive real zeros, and that there can be no more than one negative real zero. Begin by selecting from among the potential rational zeros, say $x = -6$:

Use synthetic division:

$$\begin{array}{r|rrrr} -6 & -4 & -19 & 29 & -6 \\ & & 24 & -30 & 6 \\ \hline & -4 & 5 & -1 & 0 \end{array}$$

Therefore, -6 is a zero.

Then, $f(x) = (x+6)(-4x^2 + 5x - 1)$
$= -(x+6)(4x^2 - 5x + 1)$
$= -(x+6)(4x-1)(x-1)$

Set each of the factors equal to zero and solve:

$x = \frac{1}{4}, x = 1, x = -6$

Therefore, the zeros are $\left\{-6, \frac{1}{4}, 1\right\}$.

61. Find all real zeros of

$f(x) = 2x^4 + 5x^3 - 9x^2 - 15x + 9.$

From the factors of the lead coefficient and the constant term, we can identify the potential rational zeros:

$\pm\left\{\frac{1}{2}, 1, \frac{3}{2}, 3, \frac{9}{2}, 9\right\}$

Using Descartes' Rule of Signs, we know that there can be two or no positive real zeros, and that there can be two or no negative real zeros.

Note: $f(-x) = 2x^4 - 5x^3 - 9x^2 + 15x + 9$

[Two changes in sign]

Begin by selecting from among the potential rational zeros, say $x = 3$:

Begin by testing 3:

3	2	5	−9	−15	9
		6	33	72	171
	2	11	24	57	180

The non-zero result indicates that, 3 is not a zero.

Test -3:

−3	2	5	−9	−15	9
		−6	3	18	−9
	2	−1	−6	3	0

The zero result indicates that -3 is a zero, and $x + 3$ is a factor of $f(x)$.

$f(x) = (x+3)(2x^3 - x^2 - 6x + 3)$

The potential rational factors of $(2x^3 - x^2 - 6x + 3)$ are $\pm\left\{\frac{1}{2}, 1, \frac{3}{2}, 3\right\}$.

But, we know that 3 is not a factor.

Test $\frac{1}{2}$:

1/2	2	−1	−6	3
		1	0	−3
	2	0	−6	0

Therefore, $\frac{1}{2}$ is a zero.

We also know three factors of $f(x)$, thus far: $f(x) = (x+3)\left(x - \frac{1}{2}\right)(2x^2 - 6)$.

Set $2x^2 - 6 = 0$ and solve for x:

$x = \pm\sqrt{3}$

The solution set is $\left\{-3, \frac{1}{2}, \pm\sqrt{3}\right\}$.

63. Find all real zeros of

$f(x) = 2x^4 + 13x^3 - 23x^2 - 32x + 20.$

From the factors of the lead coefficient and the constant term, we can identify the potential rational zeros:

$\pm\left\{\frac{1}{2}, 1, 2, \frac{5}{2}, 4, 5, 10, 20\right\}$

Using Descartes' Rule of Signs, we know that there can be two or no positive real zeros, and that there can be two or no negative real zeros.

Note:

$f(-x) = 2x^4 - 13x^3 - 23x^2 + 32x + 20$

[Two changes in sign]

Test 2:

2	2	13	−23	−32	20
		4	34	22	−20
	2	17	11	−10	0

The zero result indicates that 2 is a zero, and $x - 2$ is a factor of $f(x)$.

$f(x) = (x-2)(2x^3 + 17x^2 + 11x - 10)$

Test $\frac{1}{2}$:

1/2	2	17	11	−10
		1	9	10
	2	18	20	0

Therefore, $\frac{1}{2}$ is a zero.

We also know three factors of $f(x)$, thus far:

$$f(x)=(x-2)\left(x-\frac{1}{2}\right)\left(2x^2+18x+20\right).$$

Rewrite:

$$f(x)=2(x-2)\left(x-\frac{1}{2}\right)\left(x^2+9x+10\right)$$

Set $\left(x^2+9x+10\right)=0$ and solve for x:

Using the quadratic formula:

$$x=\frac{-9\pm\sqrt{81-4(10)}}{2}=\frac{-9\pm\sqrt{41}}{2}$$

The solution set is $\left\{2,\frac{1}{2},\frac{-9\pm\sqrt{41}}{2}\right\}$.

65. Find all real zeros of

$f(x)=x^5+7x^4+5x^3-43x^2-42x+72.$

From the factors of the lead coefficientand the constant term, we can identify the potential rational zeros:

$\pm\{1, 2, 3, 4, 6, 8, 9, 12, 18, 24, 36, 72\}$

Using Descartes' Rule of Signs, we know that there can be two or no positive real zeros, and that there can be three or one negative real zeros.

Note:

$f(-x)=-x^5+7x^4-5x^3-43x^2+42x+72$

[Three changes in sign]

Test 1:

$$\begin{array}{r|rrrrrr} 1 & 1 & 7 & 5 & -43 & -42 & 72 \\ & & 1 & 8 & 13 & -30 & -72 \\ \hline & 1 & 8 & 13 & -30 & -72 & 0 \end{array}$$

The zero result indicates that 1 is a zero, and $x-1$ is a factor of $f(x)$.

Test 2:

$$\begin{array}{r|rrrrr} 2 & 1 & 8 & 13 & -30 & -72 \\ & & 2 & 20 & 66 & 72 \\ \hline & 1 & 10 & 33 & 36 & 0 \end{array}$$

The zero result indicates that 2 is a zero, and $x-2$ is a factor of $f(x)$.

Test -3:

$$\begin{array}{r|rrrr} -3 & 1 & 10 & 33 & 36 \\ & & -3 & -21 & -36 \\ \hline & 1 & 7 & 12 & 0 \end{array}$$

Therefore, -3 is a zero, and $x+3$ is a factor of $f(x)$.

The last result also means that $x^2+7x+12$ is a factor of $f(x)$. Factor this quadratic, set each factor equal to zero and solve for x:

$(x+3)(x+4)=0$

$x=-3$ and $x=-4$

The zeros of $f(x)$ are $\{-4, -3, 1, 2\}$.

67. Find all real zeros of

$f(x)=x^6-125x^4+4804x^2-57{,}600.$

From the prime factors of the lead coefficient and the constant term, we can generate all the potential rational zeros. $57600=2^8\cdot 3^2\cdot 5^2$

[There are many! We will begin testing some of the smallest.]

Test 5:

$$\begin{array}{r|rrrrrrr} 5 & 1 & 0 & -125 & 0 & 4804 & 0 & -57600 \\ & & 5 & 25 & -500 & -2500 & 11520 & 57600 \\ \hline & 1 & 5 & -100 & -500 & 2304 & 11520 & 0 \end{array}$$

The zero result indicates that 5 is a zero, and $x-5$ is a factor of $f(x)$.

Test -5:

$$\begin{array}{r|rrrrrr} -5 & 1 & 5 & -100 & -500 & 2304 & 11{,}520 \\ & & -5 & 0 & 500 & 0 & -11{,}520 \\ \hline & 1 & 0 & -100 & 0 & 2304 & 0 \end{array}$$

The zero result indicates that -5 is a zero, and $x+5$ is a factor of $f(x)$.

Test 8:

$$\begin{array}{r|rrrrr} 8 & 1 & 0 & -100 & 0 & 2304 \\ & & 8 & 64 & -288 & -2304 \\ \hline & 1 & 8 & -36 & -288 & 0 \end{array}$$

The zero result indicates that 8 is a zero, and $x-8$ is a factor of $f(x)$.

Test -8:

$$\begin{array}{r|rrrr} -8 & 1 & 8 & -36 & -288 \\ & & -8 & 0 & 288 \\ \hline & 1 & 0 & -36 & 0 \end{array}$$

Therefore, -8 is a zero, and $x+8$ is a factor of $f(x)$. This last result also means that x^2-36 is a factor. Set it equal to zero and solve for x: $x=\pm 6$

The zeros of $f(x)$ are $\{-8, -6, -5, 5, 6, 8\}$.

69. Re-write and find all real zeros of $f(x)=x^3-6x^2-7x+60$.

From the factors of the lead coefficient and the constant term, we can identify the potential rational zeros:

$\pm\{1, 2, 3, 4, 5, 6, 10, 12, 15, 20, 30, 60\}$

There are many, so we will begin with some of the smaller possible zeros and test them:

Test 2:

$$\begin{array}{r|rrrr} 2 & 1 & -6 & -7 & 60 \\ & & 2 & -8 & -30 \\ \hline & 1 & -4 & -15 & 30 \end{array}$$

Therefore, 2 is a <u>not</u> solution (i.e., not a zero).

Test -3:

$$\begin{array}{r|rrrr} -3 & 1 & -6 & -7 & 60 \\ & & -3 & 27 & -60 \\ \hline & 1 & -9 & 20 & 0 \end{array}$$

Therefore -3 is a solution.

The last result leaves a quadratic factor. Set it equal to zero and solve:

$$x^2-9x+20=0$$
$$(x-4)(x-5)=0$$
$$x=4, 5$$

The solution set is $\{-3, 4, 5\}$.

71. Re-write and find all real zeros of $f(x)=6x^3-41x^2-9x+14$.

From the factors of the lead coefficient and the constant term, we can identify the potential rational zeros:

$$\pm\left\{\frac{1}{6}, \frac{1}{3}, \frac{1}{2}, \frac{2}{3}, 1, \frac{7}{6}, 2, \frac{7}{3}, \frac{7}{2}, \frac{14}{3}, 7, 14\right\}$$

Test 1:

$$\begin{array}{r|rrrr} 1 & 6 & -41 & -9 & 14 \\ & & 6 & -35 & -44 \\ \hline & 6 & -35 & -44 & -30 \end{array}$$

Therefore 1 is a <u>not</u> solution (i.e., not a zero).

Test 7:

$$\begin{array}{r|rrrr} 7 & 6 & -41 & -9 & 14 \\ & & 42 & 7 & -14 \\ \hline & 6 & 1 & -2 & 0 \end{array}$$

Therefore 7 is a solution.

This result leaves $6x^2+x-2$ as a factor of $f(x)$, or in factored form: $(3x+2)(2x-1)$

Set each factor equal to zero and solve:

$$x=-\frac{2}{3}, \frac{1}{2}$$

Therefore the solution set is $\left\{-\frac{2}{3}, \frac{1}{2}, 7\right\}$.

73. Re-write and find all real zeros of $f(x) = 3x^3 + 15x^2 - 6x - 72$.

From the factors of the lead coefficient and the constant term, (after first factoring at the GCF, 3) we can identify the potential rational zeros:

$\pm\{1, 2, 3, 4, 6, 8, 12, 24\}$

There are many, so we will begin with some of the smaller possible zeros and test them:

Test 2:

2⌋	1	5	−2	−24
		2	14	24
	1	7	12	0

Therefore, 2 is a solution, and $x^2 + 7x + 12$ is a factor (along with $x - 2$).

Factor the quadratic, set the factors equal to zero and solve for x:

$x^2 + 7x + 12 = (x + 3)(x + 4)$

$x + 3 = 0 \Rightarrow x = -3$

$x + 4 = 0 \Rightarrow x = -4$

The solution set is $\{-4, -3, 2\}$

75. Re-write and find all real zeros of $f(x) = x^4 - 3x^3 + 7x^2 - 21x$.

Factor: $x(x^3 - 3x^2 + 7x - 21)$

From the factors of the lead coefficient and the constant term, 21, we can identify the potential rational zeros:

A relatively short list: $\pm\{1, 3, 7, 21\}$

Test 1:

1⌋	1	−3	7	−21
		1	−2	5
	1	−2	5	−16

Therefore, 1 is a <u>not</u> solution (i.e., not a zero).

Test 3:

3⌋	1	−3	7	−21
		3	0	21
	1	0	7	0

Therefore 3 is a solution (i.e., a zero of $f(x)$).

The last result also indicates that $x^2 + 7$ is a factor. Set $x^2 + 7 = 0$ and solve:

$x = \pm i\sqrt{7}$

The solution set is $\{\pm i\sqrt{7}, 0, 3\}$

77. Re-write and find all real zeros of $f(x) = 4x^5 - 5x^4 - 6x^3 + 20x^2 - 25x - 30$.

From the factors of the lead coefficient and the constant term, we can identify the potential rational zeros:

$$\pm\left\{\begin{array}{l}\frac{1}{4}, \frac{1}{2}, \frac{3}{4}, 1, \frac{5}{4}, \frac{3}{2}, 2, \frac{5}{2}, 3, \\ \frac{15}{4}, 5, 6, \frac{15}{2}, 10, 15, 30\end{array}\right\}$$

Using Descartes' Rule of Signs, we know that there can be three or one positive real zeros, and that there can be two or none negative real zeros.

Note:

$f(-x) = -4x^5 - 5x^4 + 6x^3 + 20x^2 + 25x - 30$

[Two changes in sign]

Test 2:

2⌋	4	−5	−6	20	−25	−30
		8	6	0	40	30
	4	3	0	20	15	0

The zero result indicates that 2 is a zero of $f(x)$.

Test $-\frac{3}{4}$:

$$\begin{array}{r|rrrrr} -\frac{3}{4} & 4 & 3 & 0 & 20 & 15 \\ & & -3 & 0 & 0 & -15 \\ \hline & 4 & 0 & 0 & 20 & 0 \end{array}$$

The zero result indicates that $-\frac{3}{4}$ is a zero, and $4x^3 + 20$ is a factor of $f(x)$. Set this factor equal to zero and solve:

$$4x^3 + 20 = 0$$
$$x^3 = -5$$
$$x = \sqrt[3]{-5}$$

The solution set is $\left\{\sqrt[3]{-5}, -\frac{3}{4}, 2\right\}$

End of Section 4.3

Chapter Four: Section 4.4
Solutions to Odd-numbered Exercises

1. $f(x)=(x+1)^4(x-2)^3(x-1)$

f has zeros $x=-1$ (multiplicity 4), $x=2$ (multiplicity 3) and $x=1$.

For $x=0$, $f(0)=1^4(-2)^3(-1)=8$.

Then, the graph has a y-intercept at $(0, 8)$.

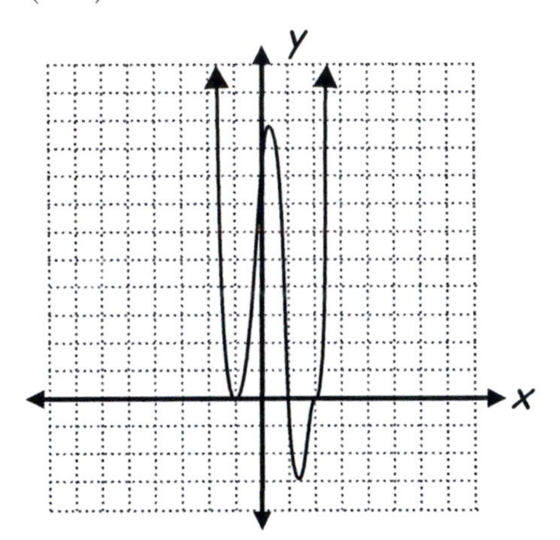

3. $f(x)=-x(x+2)(x-1)^2$

f has zeros $x=0$, $x=-2$ and $x=1$ (multiplicity 2).

For $x=0$, $f(0)=0(0+2)(-1)^2=0$.

Then, the graph has a y-intercept at $(0, 0)$.

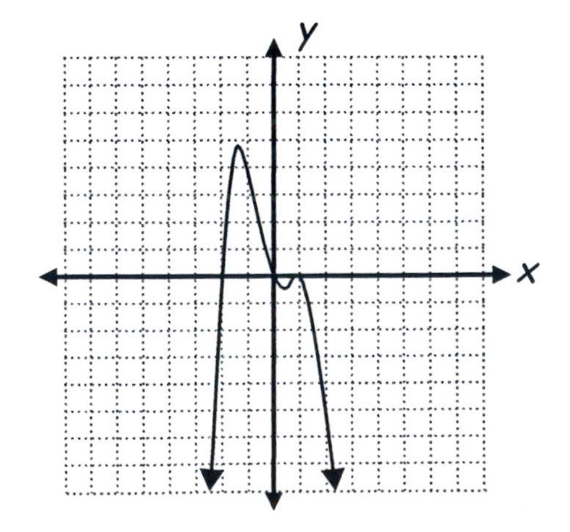

5. $f(x)=(x-1)^4(x-2)(x-3)$

f has zeros $x=1$ (multiplicity 4), $x=2$ and $x=3$.

For $x=0$, $f(0)=(-1)^4(-2)(-3)=6$.

Then, the graph has a y-intercept at $(0, 6)$.

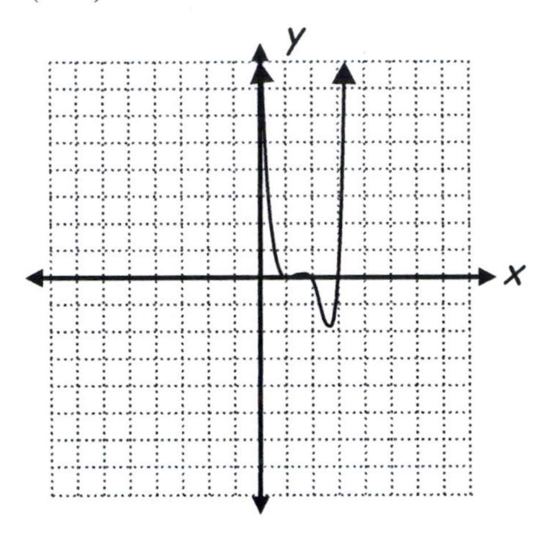

7. $f(x)=x^5+4x^4+x^3-10x^2-4x+8$

Factors of a_5: ± 1

Factors of a_0: $\pm\{1, 2, 4, 8\}$

Potential rational zeros: $\pm\{1, 2, 4, 8\}$

Use synthetic division to test for zeros:

Test $k=1$:

1	1	4	1	–10	–4	8
		1	5	6	–4	–8
	1	5	6	–4	–8	0

Therefore, $x-1$ is a factor.

Test $k=-2$:

–2	1	5	6	–4	–8
		–2	–6	0	8
	1	3	0	–4	0

Therefore, $x+2$ is a factor.

Test $k=-2$ (again):

–2	1	3	0	–4
		–2	–2	4
	1	1	–2	0

Therefore, $x+2$ occurs as a factor twice.

The result of the last row translates to the quadratic factor $x^2 + x - 2$, which factors as $(x+2)(x-1)$.
Therefore, $f(x) = (x-1)^2(x+2)^3$
For the y-intercept $f(0) = 8$.

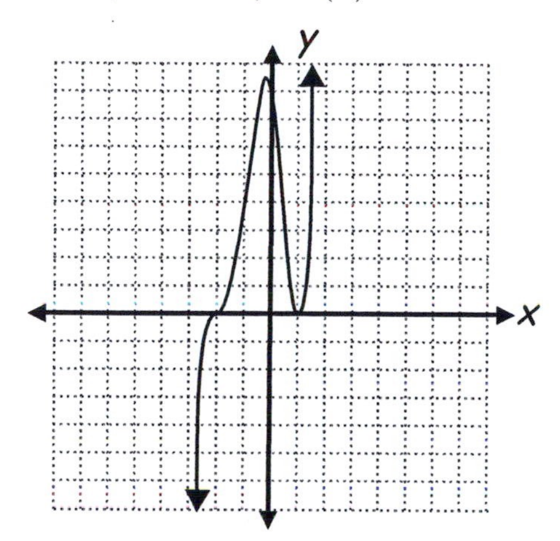

9. $s(x) = -x^4 + 2x^3 + 8x^2 - 10x - 15$
Factors of a_4 : ± 1
Factors of a_0 : $\pm\{1, 3, 5\}$
Potential rational zeros: $\pm\{1, 3, 5\}$
Use synthetic division to test for zeros:
Test $k = -1$:

-1	-1	2	8	-10	-15
		1	-3	-5	15
	-1	3	5	-15	0

Therefore, $x+1$ is a factor.
Test $k = 1$:

1	-1	3	5	-15
		-1	2	7
	-1	2	7	-8

Therefore, $x-1$ is not a factor.
Test $k = 3$:

3	-1	3	5	-15
		-3	0	15
	-1	0	5	0

Therefore, $x-3$ is a factor.

The last row translates to the quadratic $-x^2 + 5$. Set it equal to 0 and solve: $x = \pm\sqrt{5}$. This means that $x - \sqrt{5}$ and $x + \sqrt{5}$ are factors.
$s(x) = -(x-3)(x+1)(x-\sqrt{5})(x+\sqrt{5})$.
$f(0) = -(-3)(1)(-5) = -15$.
Then, $(0, -15)$ is the y-intercept.

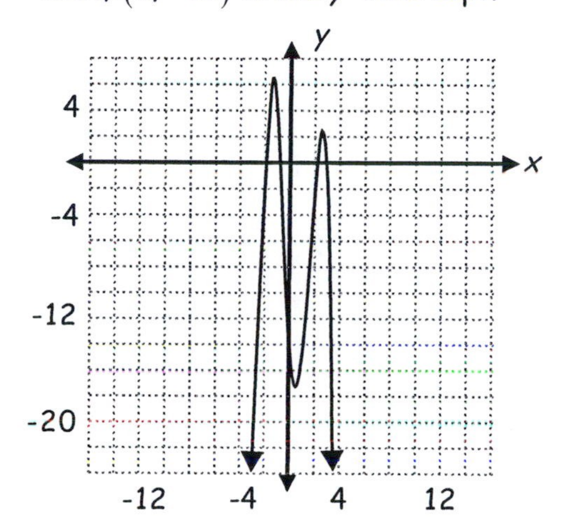

11. $H(x) = x^4 - x^3 - 5x^2 + 3x + 6$
Factors of a_4 : ± 1
Factors of a_0 : $\pm\{1, 2, 3, 6\}$
Potential rational zeros: $\pm\{1, 2, 3, 6\}$
Use synthetic division to test for zeros:
Test $k = -1$:

-1	1	-1	-5	3	6
		-1	2	3	-6
	1	-2	-3	6	0

Therefore, $x+1$ is a factor.
Test $k = 2$:

2	1	-2	-3	6
		2	0	-6
	1	0	-3	0

Therefore, $x-2$ is a factor.

The last row translates to the quadratic $x^2 - 3$. Set it equal to 0 and solve: $x = \pm\sqrt{3}$. This means that $x-\sqrt{3}$ and $x+\sqrt{3}$ are factors.
$H(x) = (x-2)(x+1)(x-\sqrt{3})(x+\sqrt{3})$.
$f(0) = (-2)(1)(-3) = 6$.
Then, $(0, 6)$ is the y-intercept.

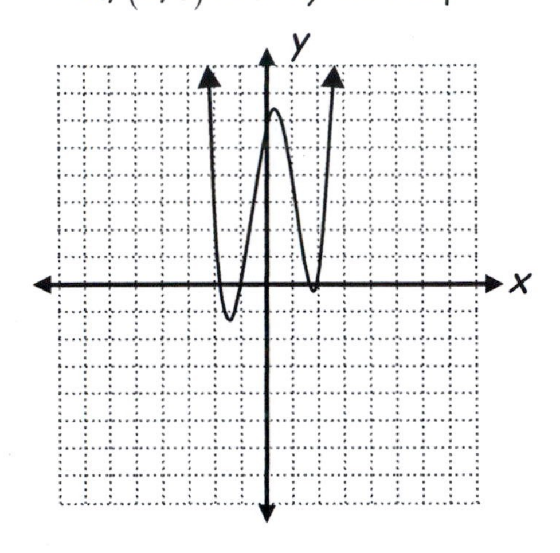

13. Solve $x^5 + 4x^4 + x^3 = 10x^2 + 4x - 8$.
Re-write as
$x^5 + 4x^4 + x^3 - 10x^2 - 4x + 8 = 0$
Factors of a_5 : ± 1
Factors of a_0 : $\pm\{1, 2, 4, 8\}$
Potential rational solutions: $\pm\{1, 2, 4, 8\}$
Test $k = 1$:

1|	1	4	1	−10	−4	8
		1	5	6	−4	−8
	1	5	6	−4	−8	0

Therefore, $x = 1$ is a solution.

1|	1	5	6	−4	−8
		1	6	12	8
	1	6	12	8	0

Therefore, $x = 1$ is a mutiple solution.
Test $k = -2$

−2|	1	6	12	8
		−2	−8	−8
	1	4	4	0

Therefore, $x = -2$ is a solution.

Test $k = -2$ (again):

−2|	1	4	4
		−2	−4
	1	2	0

Therefore, $x = -2$ is a multiple solution. The last row also translates to $x = -2$ (again). The solution set is $\{-2, 1\}$.
1 has multiplicity 2 and -2 has multiplicity 3.

15. Solve $x^4 + x^3 + 3x^2 + 5x - 10 = 0$
Factors of a_4 : ± 1
Factors of a_0 : $\pm\{1, 2, 5, 10\}$
Potential rational solutions: $\pm\{1, 2, 5, 10\}$
Test $k = -2$:

−2|	1	1	3	5	−10
		−2	2	−10	10
	1	−1	5	−5	0

Therefore, $x = -2$ is a solution.
Test $k = 1$

1|	1	−1	5	−5
		1	0	5
	1	0	5	0

Therefore, $x = 1$ is a solution.
The last row in the matrix translates to the quadratic $x^2 + 5$. Set it equal to 0 and solve: $x = \pm i\sqrt{5}$.
The solution set is $\{-2, 1, -i\sqrt{5}, i\sqrt{5}\}$.

17. Solve $x^5 + x^4 - x^3 + 7x^2 - 20x + 12 = 0$.
Factors of a_5 : ± 1
Factors of a_0 : $\pm\{1, 2, 3, 4, 6, 12\}$
Potential rational solutionss:
$\pm\{1, 2, 3, 4, 6, 12\}$

Test $k=-3$:

$$\begin{array}{r|rrrrrr} -3 & 1 & 1 & -1 & 7 & -20 & 12 \\ & & -3 & 6 & -15 & 24 & -12 \\ \hline & 1 & -2 & 5 & -8 & 4 & 0 \end{array}$$

Therefore, $x=-3$ is a solution.

$$\begin{array}{r|rrrrr} 1 & 1 & -2 & 5 & -8 & 4 \\ & & 1 & -1 & 4 & -4 \\ \hline & 1 & -1 & 4 & -4 & 0 \end{array}$$

Therefore, $x=1$ is a solution.

Test $k=1$ (again)

$$\begin{array}{r|rrrr} 1 & 1 & -1 & 4 & -4 \\ & & 1 & 0 & 4 \\ \hline & 1 & 0 & 4 & 0 \end{array}$$

Therefore, $x=1$ occurs twice as a solution.

The last row of the matrix translates to the quadratic x^2+4. Set it equal 0 and solve: $x=\pm 2i$.

The solution set is $\{-3, 1, -2i, 2i\}$.

19. Factor

$f(x)=x^4-9x^3+27x^2-15x-52$.

Given: $3-2i$ is a zero.

The given information means that $3+2i$ is also a zero.

Therefore, $x-3+2i$ and $x-3-2i$ are factors. The potential rational zeros are $\pm\{1, 2, 4, 13, 52\}$.

Test $k=4$:

$$\begin{array}{r|rrrrr} 4 & 1 & -9 & 27 & -15 & -52 \\ & & 4 & -20 & 28 & 52 \\ \hline & 1 & -5 & 7 & 13 & 0 \end{array}$$

Therefore, $x-4$ is a factor.

Test $k=-1$:

$$\begin{array}{r|rrrr} -1 & 1 & -5 & 7 & 13 \\ & & -1 & 6 & -13 \\ \hline & 1 & -6 & 13 & 0 \end{array}$$

Therefore, $x+1$ is a factor.

The complete factorization is

$f(x)=(x-4)(x+1)(x-3-2i)(x-3+2i)$

21. Factor

$f(x)=x^3-(2+3i)x^2-(1-3i)x+(2+6i)$.

Given: 2 is a zero.

The given information means that $x-2$ is a factor.

$$\begin{array}{r|rrrr} 2 & 1 & -(2+3i) & -(1-3i) & 2+6i \\ & & 2 & -6i & -2-6i \\ \hline & 1 & -3i & -1-3i & 0 \end{array}$$

Use the last row and test $k=-1$:

$$\begin{array}{r|rrr} -1 & 1 & -3i & -1-3i \\ & & -1 & 1+3i \\ \hline & 1 & -1-3i & 0 \end{array}$$

Therefore, $x+1$ is a factor.

Furthermore, the last row of the matrix translates to $x-1-3i$, which is (therefore) a factor.

The complete factorization is

$f(x)=(x-2)(x+1)(x-1-3i)$.

23. Factor

$f(x)=x^4-3x^3+5x^2-x-10$.

The potential rational zeros are $\pm\{1, 2, 5, 10\}$.

Test $k=2$ by using synthetic division:

$$\begin{array}{r|rrrrr} 2 & 1 & -3 & 5 & -1 & -10 \\ & & 2 & -2 & 6 & 10 \\ \hline & 1 & -1 & 3 & 5 & 0 \end{array}$$

Therefore, $x-2$ is a factor

Test $k=-1$:

$$\begin{array}{r|rrrr} -1 & 1 & -1 & 3 & 5 \\ & & -1 & 2 & -5 \\ \hline & 1 & -2 & 5 & 0 \end{array}$$

Therefore, $x+1$ is a factor.

The last row translates to the quadratic x^2-2x+5. Set it equal to 0 and solve:

$$x=\frac{2\pm\sqrt{4-4(1)(5)}}{2}$$

$x=1\pm 2i$.

Then, $x-1-2i$ and $x-1+2i$ are factors.

The complete factorization is

$f(x)=(x-2)(x+1)(x-1-2i)(x-1+2i)$

25. Factor

$n(x)=x^4-4x^3+6x^2+28x-91$.

Given: $2+3i$ is a zero.

The given information means that $x-2-3i$ is a factor and that $x-2+3i$ is also a factor.

Using synthetic division in successive steps will expose a quadratic in the final row of the division. The zeros of the quadratic are readily determined. Here is the work:

$2+3i$	1	-4	6	28	-91
		$2+3i$	-13	$-14-21i$	91
$2-3i$	1	$-2+3i$	-7	$14-21i$	0
		$2-3i$	0	$-14+21i$	
	1	0	-7	0	

The last row translates to the quadratic x^2-7. Set it equal to zero and solve: $x=\pm\sqrt{7}$.

Therefore $x-\sqrt{7}$ and $x+\sqrt{7}$ are factors.

The complete factorization is $n(x)=$ $(x-2-3i)(x-2+3i)\left(x-\sqrt{7}\right)\left(x+\sqrt{7}\right)$.

27. Factor $r(x)=x^4+7x^3-41x^2+33x$.

Notice that x is a factor:

$r(x)=x\left(x^3+7x^2-41x+33\right)$

Potential rational factors: $\{1, 3, 11, 33\}$

Test $k=1$:

1	1	7	-41	33
		1	8	-33
	1	8	-33	0

Therefore, $x-1$ is a factor

Test $k=3$:

3	1	8	-33
		3	33
	1	11	0

Therefore, $x-3$ is a factor.

The last row translates to $x+11$.

Therefore $x+11$ is a factor.

$r(x)=(x-1)(x-3)(x+11)$

29. Factor $P(x)=x^3-6x^2+28x-40$.

Potential rational factors:

$\pm\{1, 2, 4, 5, 8, 10, 20, 40\}$

Test $k=2$:

2	1	-6	28	-40
		2	-8	40
	1	-4	20	0

Therefore, $x-2$ is a factor.

The last row translates to $x^2-4x+20$.

Using the quadratic formula, we have

$$x=\frac{4\pm\sqrt{16-4(1)(20)}}{2}=2\pm 4i.$$

Therefore, $(x-2+4i)$ and $(x-2-4i)$ are also factors.

Thus, the complete factorization is

$(x-2)(x-2+4i)(x-2-4i)$.

31. The 3rd degree polynomial would also have the conjugate of $5+i$ as a zero. Thus,

$f(x)=a(x+1)(x-5-i)(x-5+i)$

Given $f(0)=-52$, solve for a:

$$a(1)(-5-i)(-5+i)=-52$$
$$a(1)(25+1)=-52$$
$$26a=-52$$
$$a=-2$$

$f(x)=-2(x+1)(x-5-i)(x-5+i)$

$f(x)=(-2x-2)(x-(5+i))(x-(5-i))$

$f(x)=(-2x-2)(x^2-10x+26)$

$f(x)=-2x^3+18x^2-32x-52$

33. If 1 is a zero of multiplicity 3, then $(x-1)^3$ is a factor of the given polynomial. If -2 is the only other zero, and the polynomial is of degree 5, then $(x+2)^2$ is the other major factor. If 2 is the lead coefficient,

$f(x)=2(x+2)^2(x-1)^3$

$f(x)=2(x^2+4x+4)(x^3-3x^2+3x-1)$

$f(x)=2(x^5+x^4-5x^3-x^2+8x-4)$

$f(x)=2x^5+2x^4-10x^3-2x^2+16x-8$

35. $f(x)=3x(x-6)(x+2i)(x-2i)$.

$2i$ is the conjugate of $-2i$ and it is also a zero of the polynomial.

$f(x)=3x(x-6)(x^2+4)$

$f(x)=(3x^2-18x)(x^2+4)$

$f(x)=3x^4-18x^3+12x^2-72x$

37. **a.** $V(x)=$ Depth x Width x Length

$V(x)=x(10-2x)(18-2x)$

$V(x)=4x(5-x)(9-x)$

b. $V(x)=0$ for $x=0, 5$ and 9.

c. $x=0$ and $x=5$ are possible.

$x=9$ is not possible since $2x>10$.

End of Section 4.4

Chapter Four: Section 4.5
Solutions to Odd-numbered Exercises

1. $f(x)=\dfrac{5}{x-1}$

Vertical asymptote where $x-1=0$:

$x=1$

3. $f(x)=\dfrac{x^2-4}{x+2}$

Factor numerator and simplify:

$f(x)=\dfrac{(x-2)(x+2)}{x+2}=x-2$

No vertical asymptote.

5. $f(x)=\dfrac{x+2}{x^2-9}=\dfrac{x+2}{(x-3)(x+3)}$

Vertical asymptote where $x-3=0$ and $x+3=0$:

$x=3$ and $x=-3$

7. $f(x)=\dfrac{2x^2+2x-4}{x^2+2x+1}=\dfrac{2(x+2)(x-1)}{(x+1)(x+1)}$

Vertical asymptote where $x+1=0$:

$x=-1$

9. $f(x)=\dfrac{3x^2+1}{x-2}$

Vertical asymptote where $x-2=0$:

$x=2$

11. $f(x)=\dfrac{x^2+5}{x^3-27}=\dfrac{x^2+5}{(x-3)\left(x^2+3x+9\right)}$

Vertical asymptote where $x-3=0$:

$x=3$

Note: The zeros of x^2+3x+9 are not real numbers.

13. $f(x)=\dfrac{x^2-1}{x^2-8x+7}=\dfrac{(x+1)(x-1)}{(x-7)(x-1)}$

$=\dfrac{(x+1)}{(x-7)}$

Vertical asymptote where $x-7=0$:

$x=7$

15. $f(x)=\dfrac{x^3-6x^2+11x-6}{(x+2)\left(x^2-2x+4\right)}$

To factor the numerator, use synthetic division:

$$\begin{array}{r|rrrr} 3 & 1 & -6 & 11 & -6 \\ & & 3 & -9 & 6 \\ \hline & 1 & -3 & 2 & 0 \end{array}$$

Then, $x-3$ is a factor of $f(x)$ and the last row indicates that x^2-3x+2 is a factor. This quadratic factors as $(x-2)(x-1)$. In factored form,

$f(x)=\dfrac{(x-2)(x-1)(x-3)}{(x+2)\left(x^2-2x+4\right)}$.

Vertical asymptote where $x+2=0$:

$x=-2$

17. $f(x)=\dfrac{x^2-16}{x^2-4}=\dfrac{(x+4)(x-4)}{(x-2)(x+2)}$

Vertical asymptote where $x-2=0$ and $x+2=0$:

$x=2$ and $x=-2$

19. $f(x)=\dfrac{5}{x-1}$

deg. of num. $<$ deg. of den.

$y=0$ is the horizontal asymptote.

21. $f(x)=\dfrac{x^2-4}{x+2}$

deg. of num. = deg. of den. + 1

Divide the numerator by the denominator to find the oblique asymptote:

$f(x)=\dfrac{(x-2)(x+2)}{x+2}=x-2$

Then, $y=x-2$ is the oblique asymptote.

23. $f(x)=\dfrac{x+2}{x^2-9}$

deg. of num. < deg. of den.

$y=0$ is the horizontal asymptote.

25. $f(x)=\dfrac{2x^2+2x-4}{x^2+2x+1}$

deg. of num. = deg. of den.

$y=\dfrac{2}{1}=2$ is the horizontal asymptote.

27. $f(x)=\dfrac{3x^2+1}{x-2}$

deg. of num. = deg. of den. + 1, so there is an oblique asymptote.

Using synthetic division:

$$\begin{array}{r|rrr} 2 & 3 & 0 & 1 \\ & & 6 & 12 \\ \hline & 3 & 6 & 13 \end{array}$$

(Note: The remainder 13 is irrelevant.)

$y=3x+6$ is the oblique asymptote.

29. $f(x)=\dfrac{x^2+5}{x^3-27}$

deg. of num. < deg. of den.

$y=0$ is the horizontal asymptote.

31. $f(x)=\dfrac{x^2-81}{x^3+7x-12}$

deg. of num. < deg. of den.

$y=0$ is the horizontal asymptote.

33. $f(x)=\dfrac{x^2-9x+4}{x+2}$

deg. of num. = deg. of den. + 1, so there is an oblique asymptote.

Using synthetic division:

$$\begin{array}{r|rrr} -2 & 1 & -9 & 4 \\ & & -2 & 22 \\ \hline & 1 & -11 & 26 \end{array}$$

$y=x-11$ is the oblique asymptote.

35. $f(x)=\dfrac{5x^2-x+12}{x-1}$

deg. of num. = deg. of den. + 1, so there is an oblique asymptote.

Using synthetic division:

$$\begin{array}{r|rrr} 1 & 5 & -1 & 12 \\ & & 5 & 4 \\ \hline & 5 & 4 & 16 \end{array}$$

$y=5x+4$ is the oblique asymptote.

37. Graph $f(x)=\dfrac{5}{x-1}$:

See Exercise 1, and Exercise 19.

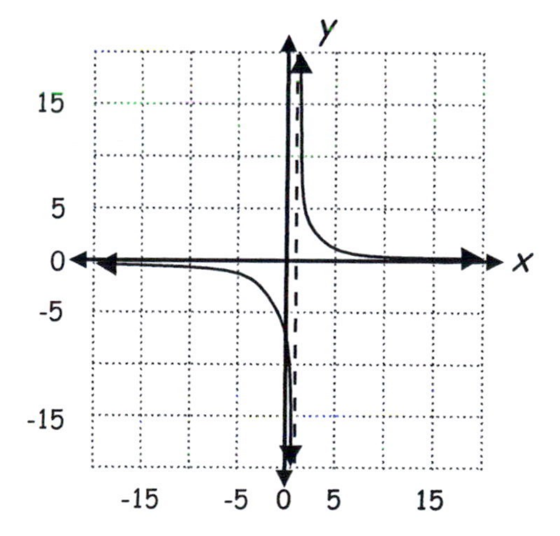

39. Graph $f(x) = \dfrac{x^2 - 4}{x + 2}$:

See Exercise 3, and Exercise 21.

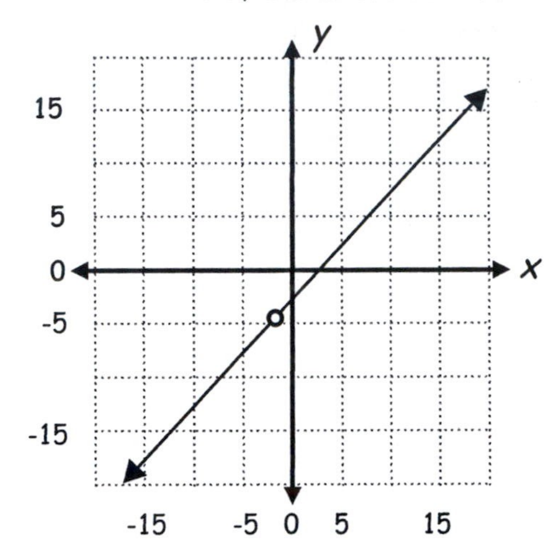

41. Graph $f(x) = \dfrac{x + 2}{x^2 - 9}$:

See Exercise 5 and Exercise 23.

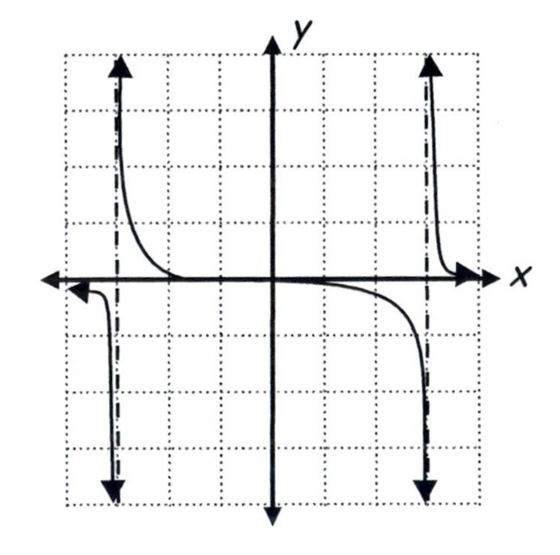

43. Graph $f(x) = \dfrac{2x^2 + 2x - 4}{x^2 + 2x + 1}$

See Exercises 7 and 25.

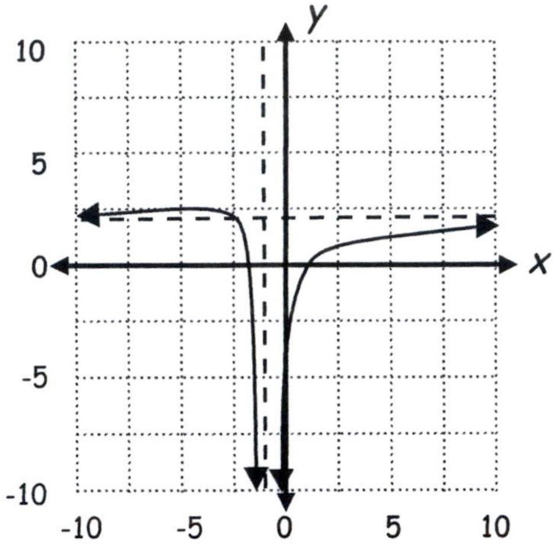

45. Graph $f(x) = \dfrac{3x^2 + 1}{x - 2}$

See Exercises 9 and 27.

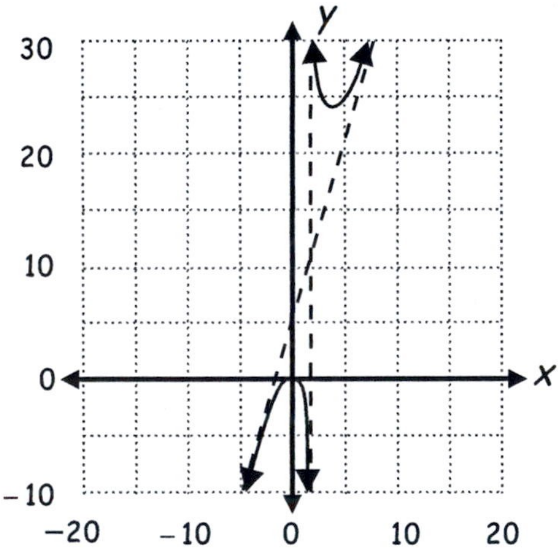

47. Graph $f(x)=\dfrac{x^2+5}{x^3-27}$

See Exercises 11 and 29.

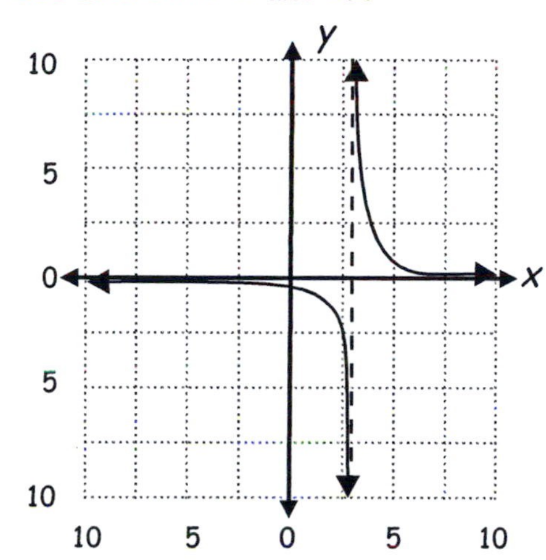

49. Solve $2x<\dfrac{4}{x+1}$.

Re-write in standard form:

$$2x-\frac{4}{x+1}<0$$

$$\frac{2x(x+1)-4}{x+1}<0$$

$$\frac{2(x^2+x-2)}{x+1}<0$$

$$\frac{x^2+x-2}{x+1}<0$$

$$\frac{(x+2)(x-1)}{x+1}<0$$

The denominator tells us that $x=-1$ is a vertical asymptote. But, there's more. Since deg. of num. = deg. of den. $+1$, there is an oblique asymptote. Using synthetic division we can find the equation of the asymptote:

-1	1	1	-2
		-1	0
	1	0	-2

From the last row the equation can be determined: $y=x$

(Remember, the remainder -2 is not relevant to this use of synthetic division.) So looking at the graph, the solution is $(-\infty,-2)\cup(-1,1)$.

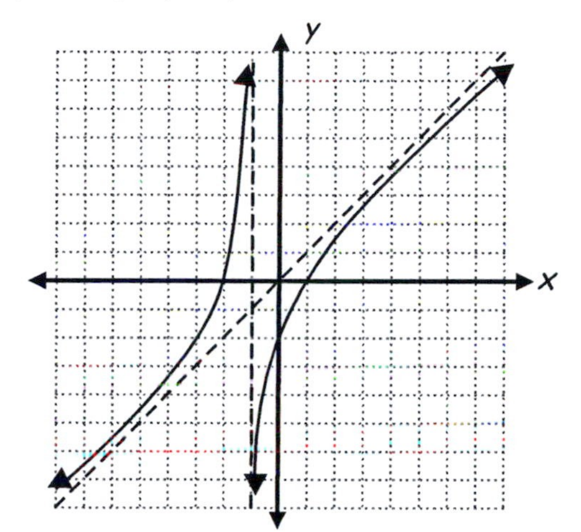

51. Solve $\dfrac{5}{x-2}>\dfrac{3}{x+2}$.

Re-write in standard form:

$$\frac{5}{x-2}-\frac{3}{x+2}>0$$

$$\frac{5(x+2)-3(x-2)}{(x-2)(x+2)}>0$$

$$\frac{2x+16}{(x-2)(x+2)}>0$$

Vertical asymptotes: $x=-2$ and $x=2$

y-intercept of $(0,-4)$. Also, from $x=-8$, the value of the rational function remains negative as $x\to-\infty$.

The solution is $(-8,-2)\cup(2,\infty)$

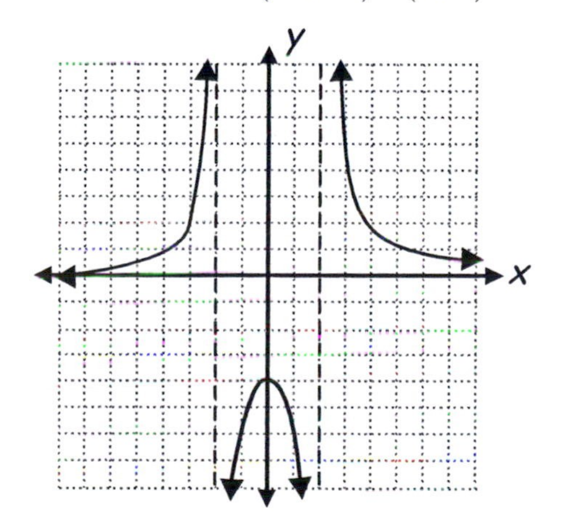

53. Solve $\frac{x}{x^2-x-6} \le \frac{-2}{x^2-x-6}$.

Re-write in standard form:

$$\frac{x+2}{x^2-x-6} \le 0$$

$$\frac{x+2}{(x-3)(x+2)} \le 0$$

$$\frac{1}{x-3} \le 0$$

Note that the rational function is not defined in its original form at $x=-2$, and that it has a vertical asymptote at $x=3$.

The solution is $(-\infty,-2)\cup(-2,3)$.

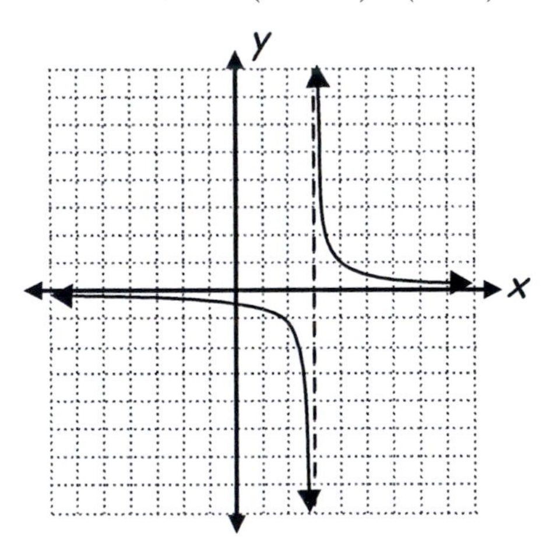

55. Solve $\frac{4}{x-3} \le \frac{4}{x}$.

Re-write in standard form:

$$\frac{4}{x-3}-\frac{4}{x} \le 0$$

$$\frac{4x-4(x-3)}{x(x-3)} \le 0$$

$$\frac{3}{x(x-3)} \le 0$$

Vertical asymptote at $x=0$ and $x=3$.

From the graph, the solution is $(0,3)$:

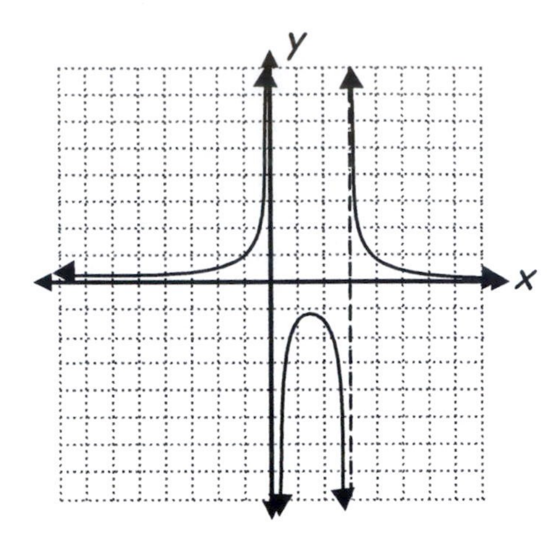

57. Solve $\frac{x}{x^2+3x+2} > \frac{1}{x^2+3x+2}$.

Re-write in standard form:

$$\frac{x-1}{(x+2)(x+1)} > 0$$

Vertical asymptotes of $x=-2$ and $x=-1$, and horizontal asymptote of $y=0$ (i.e., the x-axis). The numerator tells us that the function becomes positive from $x=1$ -- the numerator and denominator are both positive thereafter as $x \to \infty$.

The solution is $(-2,-1)\cup(1,\infty)$.

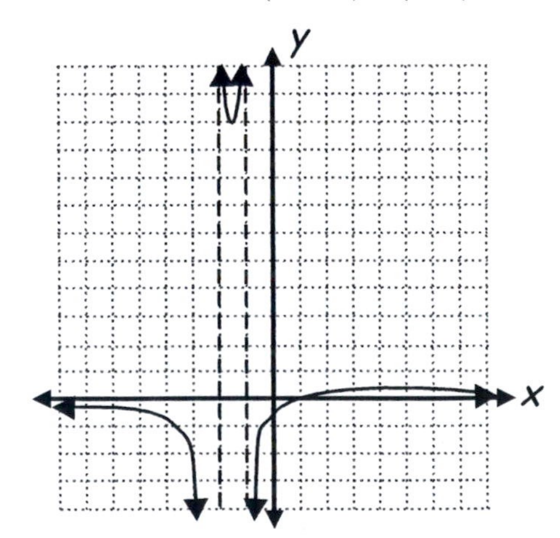

59. Solve $\frac{x}{x+1} \geq \frac{x+1}{x}$.

Re-write in standard form:

$$\frac{x}{x+1} - \frac{x+1}{x} \geq 0$$

$$\frac{x^2 - x^2 - 2x - 1}{x(x+1)} \geq 0$$

$$\frac{-2x-1}{x(x+1)} \geq 0$$

The factors of the denominator tells us that $x = 0$ and $x = -1$ are vertical asymptotes. The numerator tells us that the function has a zero at $x = -\frac{1}{2}$.

The graph indicates the solution set:

$(-\infty, -1) \cup \left[-\frac{1}{2}, 0\right)$.

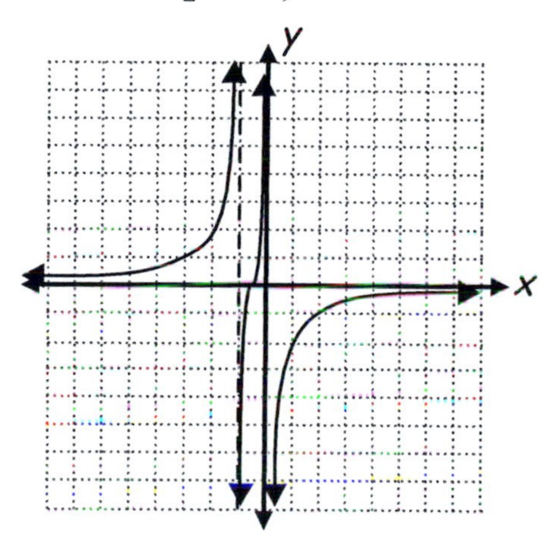

End of Section 4.5

Chapter Four Test
Solutions to Odd-Numbered Exercises

1. Substitute $x=2$:

$$4(2)^3-5(2)^2\stackrel{?}{=}-3(2)+18$$
$$32-20\stackrel{?}{=}-6+18$$
$$12=12$$

3. Substitute $x=5-4i$:

$$(5-4i)^3+5-4i\stackrel{?}{=}6(5-4i)^2-164$$
$$-115-236i+5-4i\stackrel{?}{=}-110-240i$$
$$-110-240i=-110-240i$$

5. Given: $f(x)=(x+1)^2(x-2)$.
The factors indicate the x-intercepts to be at $x=-1$ and $x=2$.
The y-intercept occurs at $f(0)=-2$.
The degree of the polynomial is 3.
$f(x)\to\infty$ as $x\to\infty$ and
$f(x)\to-\infty$ as $x\to-\infty$

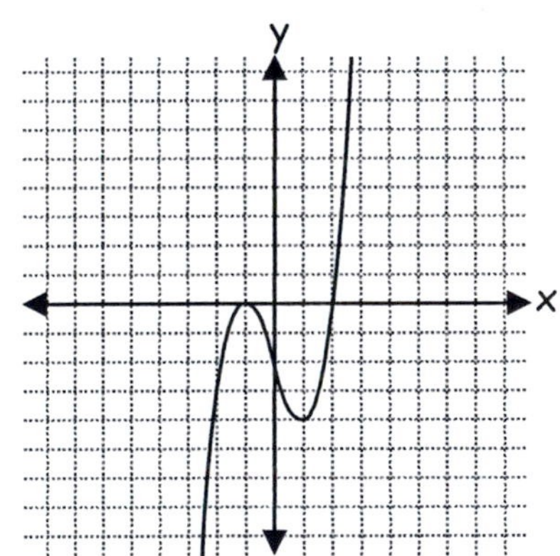

7. Given: $s(x)=x^3+5x^2+6x$.
Factor:
$x(x^2+5x+6)=x(x+3)(x+2)$
The factors indicate the x-intercepts to be at $x=0$, $x=-3$ and $x=-2$.
The y-intercept occurs at $s(0)=0$.
The degree of the polynomial is 3.

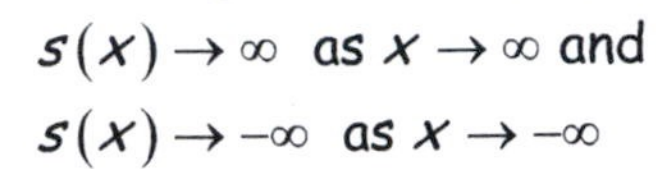

$s(x)\to\infty$ as $x\to\infty$ and
$s(x)\to-\infty$ as $x\to-\infty$

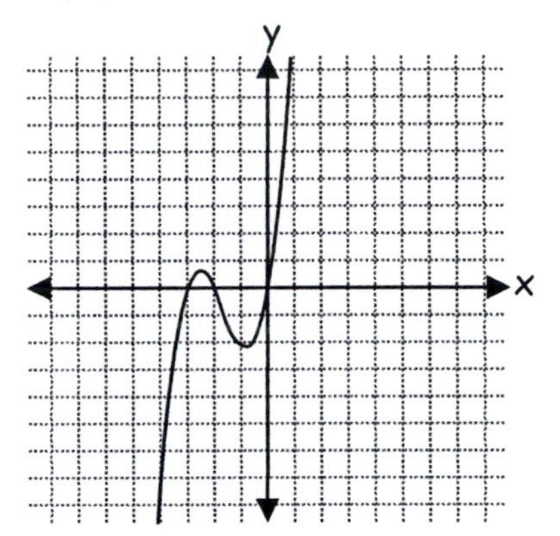

9. $(x-3)^2(x+1)^2>0$
The graph of the polynomial intersects the x-axis at 3 and -1. The intervals of interest are $(-\infty,-1)\cup(-1,3)\cup(3,\infty)$.

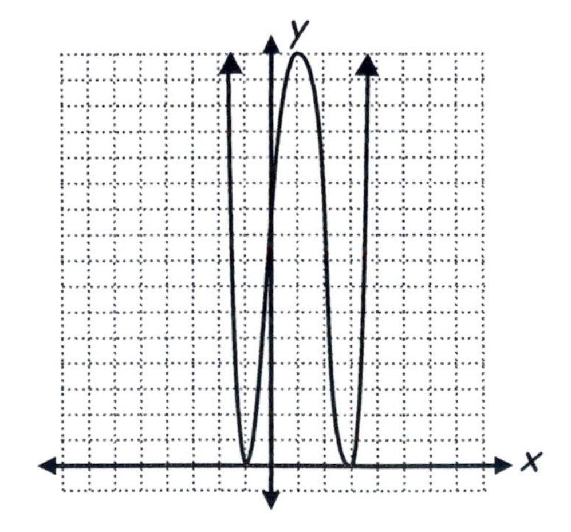

The polynomial is strictly greater than 0 on $(-\infty,-1)\cup(-1,3)\cup(3,\infty)$, or $\{x\mid x\in\mathbb{R}, x\neq-1 \text{ and } x\neq3\}$.

11.

$$4x^2-3x+3+\frac{7}{2x^2-1}$$

$$2x^2-1\overline{)8x^4-6x^3+2x^2+3x+4}$$

$$-(8x^4 \qquad -4x^2)$$

$$-6x^3+6x^2+3x+4$$

$$-(-6x^3 \qquad +3x)$$

$$6x^2 \qquad +4$$

$$-(6x^2 \qquad -3)$$

$$7$$

13.

$$x^2-3x+4-\frac{5x+16}{x^2+3x+2}$$

$$x^2+3x+2\overline{)x^4+0x^3-3x^2+x-8}$$

$$-(x^4+3x^3+2x^2)$$

$$-3x^3-5x^2+x-8$$

$$-(-3x^3-9x^2-6x)$$

$$4x^2+7x-8$$

$$-(4x^2+12x+8)$$

$$-5x-16$$

15.

$\frac{1}{6}$	48	10	−51	−10	3
		8	3	−8	−3
	48	18	−48	−18	0

$\frac{1}{6}$ is a zero. $p\left(\frac{1}{6}\right)=0$.

17. Given that $-1, 4$ and -5 are zeros of a 3rd degree polynomial $f(x)$ means that $x+1, x-4$ and $x+5$ are factors. In addition, given that $f(x)\to-\infty$ as $x\to\infty$, the lead coefficient is negative.

Let $f(x)=-(x+1)(x-4)(x+5)$

$f(x)=-x^3-2x^2+19x+20$

19. $f(x)=x^4+3x^3-3x^2-11x-6$

Factors of a_4: $\pm\{1\}$

Factors of a_0 : $\pm\{1, 2, 3, 6\}$

There are 8 potential rational zeros.

Potential zeros: $\pm\{1, 2, 3, 6\}$

Actual zeros:

Test $k=-3$:

−3	1	3	−3	−11	−6
		−3	0	9	6
	1	0	−3	−2	0

Therefore, $k=-3$ is a zero.

Test $k=-1$:

−1	1	0	−3	−2
		−1	1	2
	1	−1	−2	0

Therefore, $k=-1$ is a zero.

Test $k=2$:

2	1	−1	−2
		2	2
	1	1	0

Therefore, $k=2$ is a zero.

The last line indicates $(x+1)$ is another factor, so -1 is a repeated zero.

Actual zeros: $\{-3, -1, 2\}$

21. There are two variations in sign from left to right. According to Descartes' Rule, there are 2 or 0 possible real positive zeros.

Substituting $-x$ for x in $f(x)$:

$f(-x)=2(-x)^4-3(-x)^3$

$\qquad -(-x)^2+3(-x)+10$

$f(-x)=2x^4+3x^3-x^2-3x+10$

There are two variations in sign. According to Descartes' Rule, there are 2 or 0 possible negative real zeros.

23. Choose an integer to test as an upper or lower bound of the real zeros of the given polynomial,
$f(x) = 2x^3 - 11x^2 + 3x + 36$.
For example, test $x = 6$:

$$\begin{array}{r|rrrr} 6 & 2 & -11 & 3 & 36 \\ & & 12 & 6 & 54 \\ \hline & 2 & 1 & 9 & 90 \end{array}$$

Only positive coefficients in the last row satisfies the "upper bound test." I.e., 6 is an upper bound for the real zeros of f.
Test $x = 5$:

$$\begin{array}{r|rrrr} 5 & 2 & -11 & 3 & 36 \\ & & 10 & -5 & -10 \\ \hline & 2 & -1 & -2 & 26 \end{array}$$

5 does not satisfy the upper or lower bound test.
Test $x = -2$:

$$\begin{array}{r|rrrr} -2 & 2 & -11 & 3 & 36 \\ & & -4 & 30 & -66 \\ \hline & 2 & -15 & 33 & -30 \end{array}$$

The alternating signs in the last row of coefficients satisfies the "lower bound test." I.e., -2 is a lower bound of the real zeros of f.
The real zeros are bound by $[-2, 6]$.

25. Determine $f(1)$ and $f(2)$. If they differ in sign, then the Intermediate Value Theorem guarantees that $f(x)$ has a zero between 1 and 2:

$$f(1) = 1^3 + 3(1)^2 - 4(1) - 10 = -10$$

$$f(2) = 2^3 + 3(2)^2 - 4(2) - 10 = 2$$

Therefore, $f(c) = 0$ for some real $c, 1 < c < 2$.

27. Solve
$3x^5 + x^4 + 5x^3 = x^2 + 28x + 20$
Re-write:
$3x^5 + x^4 + 5x^3 - x^2 - 28x - 20 = 0$
Factors of a_5: $\pm\{1, 3\}$
Factors of a_0 : $\pm\{1, 2, 4, 5, 10, 20\}$
Potential rational zeros:

$$\pm\left\{\frac{1}{3}, \frac{2}{3}, 1, 2, 4, \frac{4}{3}, 5, \frac{5}{3}, 10, \frac{10}{3}, 20, \frac{20}{3}\right\}$$

Test the potential zeros by synthetic division. This could be a long process with so many possibilities. Consider also that a zero may have multiplicity as a zero:
Test $x = -1$:

$$\begin{array}{r|rrrrrr} -1 & 3 & 1 & 5 & -1 & -28 & -20 \\ & & -3 & 2 & -7 & 8 & 20 \\ \hline & 3 & -2 & 7 & -8 & -20 & 0 \end{array}$$

Therefore, $x = -1$ is a solution
Using the last row, continue the process
Test $x = \frac{5}{3}$:

$$\begin{array}{r|rrrrr} \frac{5}{3} & 3 & -2 & 7 & -8 & -20 \\ & & 5 & 5 & 20 & 20 \\ \hline & 3 & 3 & 12 & 12 & 0 \end{array}$$

Therefore, $x = \frac{5}{3}$ is a solution.
Using the last row, continue the process:
Test $x = -1$ (again):

$$\begin{array}{r|rrrr} -1 & 3 & 3 & 12 & 12 \\ & & -3 & 0 & -12 \\ \hline & 3 & 0 & 12 & 0 \end{array}$$

Therefore, $x = -1$ is a multiple solution.
The last row gives us a quadratic which is readily solvable:
$3x^2 + 12 = 0$, or $x = \pm 2i$
The solution set is $\left\{-1, \frac{5}{3}, -2i, 2i\right\}$

29. Given $2+i$ is a zero of

$f(x)=14x^4-109x^3+296x^2-321x+70$,

then, $2-i$ is also a zero, and

$x-2-i$ and $x-2+i$ are factors of f.

The strategy to factor f shown here is to multiply out and simplify the factor that results from these two known linear factors. Then, divide the original polynomial and factor the result (or else find its zeros which will give us the factors):

$$(x-(2+i))(x-(2-i))=x^2-4x+5$$

$$\begin{array}{r|l} & 14x^2-53x+14 \\ x^2-4x+5 & 14x^4-109x^3+296x^2-321x+70 \\ & -(14x^4-56x^3+70x^2) \\ \hline & -53x^3+226x^2-321x+70 \\ & -(-53x^3+212x^2-265x) \\ \hline & 14x^2-56x+70 \\ & -(14x^2-56x+70) \\ \hline & 0 \end{array}$$

$14x^2-53x+14$ factors as

$(7x-2)(2x-7)$.

The factorization of $f(x)$:

$(x-2+i)(x-2-i)(2x-7)(7x-2)$

31. $\frac{1}{2}$ and $1+2i$ are zeros of $f(x)$ means that $\left(x-\frac{1}{2}\right)$, $(x-1-2i)$ and $(x-1+2i)$ are factors of $f(x)$.

Given that $f(x)$ is a polynomial of the fourth degree with only real coefficients with lead coefficient 2, we know that for some real number c, $f(x)=$

$2\left(x-\frac{1}{2}\right)(x-1-2i)(x-1+2i)(x-c)$.

We can solve for c by setting $x=0$ and setting the product of the factors equal to the given y-intercept, -30:

$(-c)(-1)(5)=-30$. So, $c=-6$.

To find the polynomial explicitly, we need to multiply out the expression with the value of $c=-6$:

$2\left(x-\frac{1}{2}\right)(x-1-2i)(x-1+2i)(x+6)$

Rearranging the last factor with the first and multiplying out the first pair and the second pair we have:

$f(x)=(2x-1)(x+6)(x^2-2x+5)$

$f(x)=(2x^2+11x-6)(x^2-2x+5)$

$f(x)=2x^4+7x^3-18x^2+67x-30$

33. Given: $f(x)=\dfrac{4}{2x-5}$

The vertical asymptote is where

$2x-5=0$. I.e., at $x=\frac{5}{2}$.

35. Given: $f(x)=\dfrac{2x^3+5x^2-1}{x^2-2x}$

deg. of num. = deg. of den. +1.

This implies that there is an oblique asymptote. To find it, perform polynomial division:

$$\begin{array}{r|l} & 2x+9 \\ x^2-2x & 2x^3+5x^2+0x-1 \\ & -(2x^3-4x^2) \\ \hline & 9x^2 \qquad -1 \\ & -(9x^2-18x) \\ \hline & 18x-1 \end{array}$$

The oblique asymptote is $2x+9$.

37. Sketch the graph of $f(x)=\dfrac{x^2+5}{x+5}$.

Notice that the graph has a vertical asymptote at $x=-5$ and an oblique asymptote $y=x-5$. To derive the latter, use long division. The quotient is the oblique asymptote.

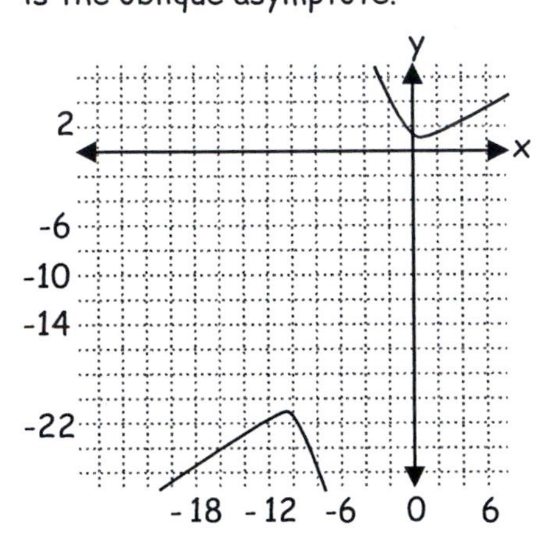

39. Solve $\dfrac{7}{x+3}\geq\dfrac{2x}{x+3}$.

Re-write as $\dfrac{7}{x+3}-\dfrac{2x}{x+3}\geq 0$

$$\frac{7-2x}{x+3}\geq 0.$$

Graph $\dfrac{7-2x}{x+3}=0$.

Line $x=-3$ is a vertical asymptote.

Line $y=-2$ is a horizontal asymptote.

The solution to the inequality can be determined from the graph: $\left(-3,\dfrac{7}{2}\right]$.

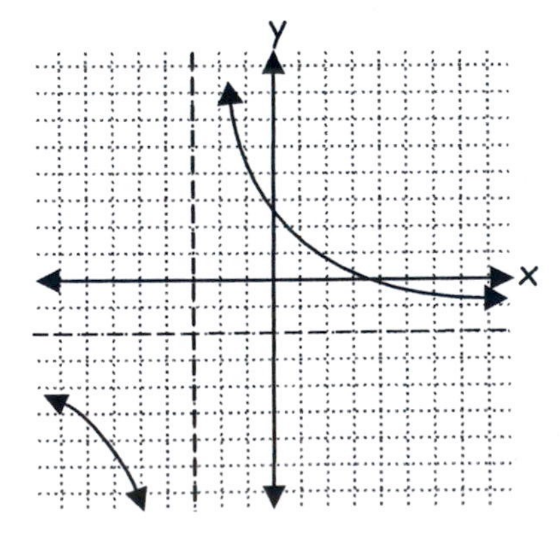

End of Chapter 4 Test

Chapter Five: Section 5.1
Solutions to Odd-numbered Exercises

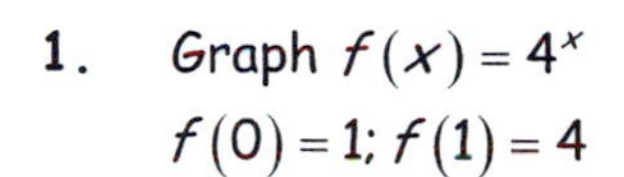

1. Graph $f(x)=4^x$

   $f(0)=1;\ f(1)=4$

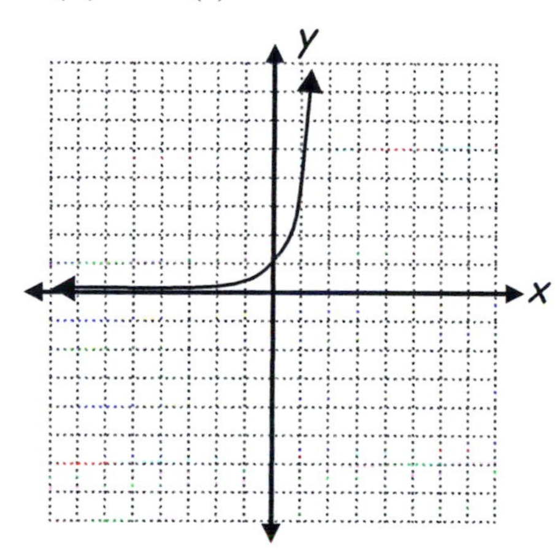

3. Graph $s(x)=3^{x-2}$

   $s(2)=1;\ s(3)=3$

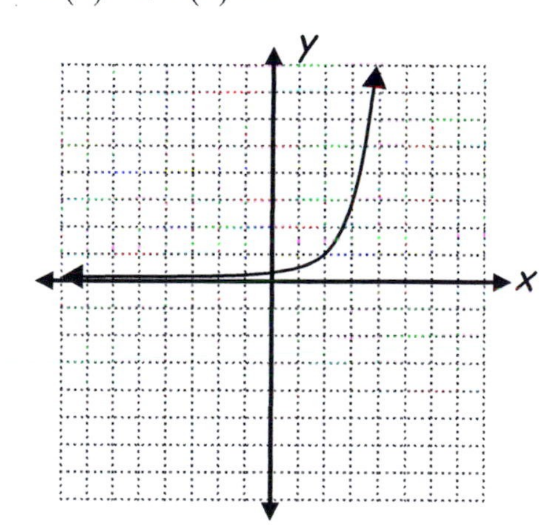

5. Graph $r(x)=5^{x-2}+3$

   $r(2)=4;\ r(3)=8$

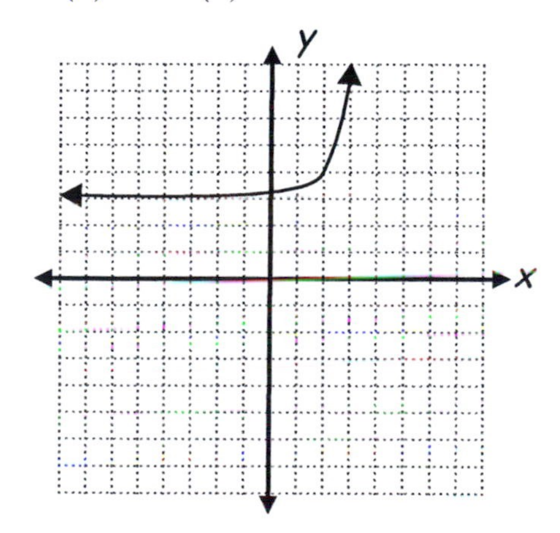

7. Graph $f(x)=2^{-x}$

   $f(0)=1;\ f(1)=\frac{1}{2};\ f(-1)=2$

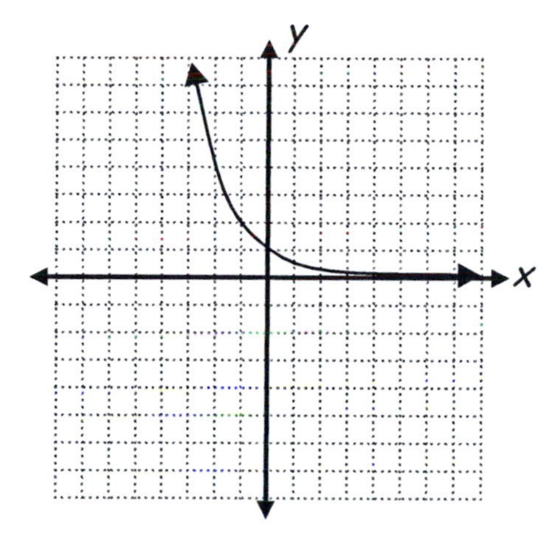

9. Graph $g(x)=3\left(2^{-x}\right)$

   $g(-1)=6;\ g(0)=3;\ g(1)=\frac{3}{2}$

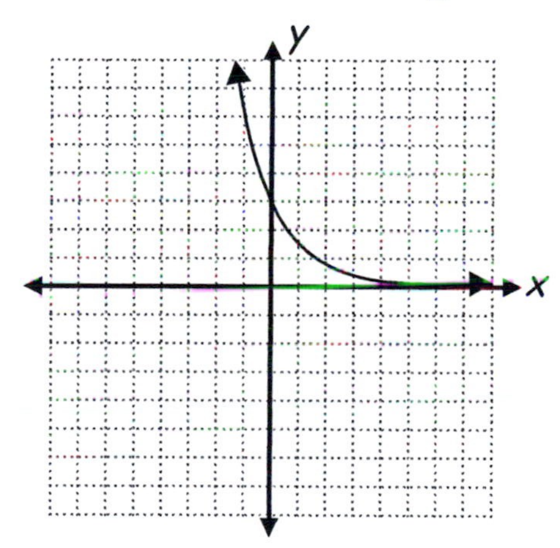

11. Graph $s(x)=(.2)^{-x}$

    $s(-1)=.2;\ s(0)=1;\ s(1)=5$

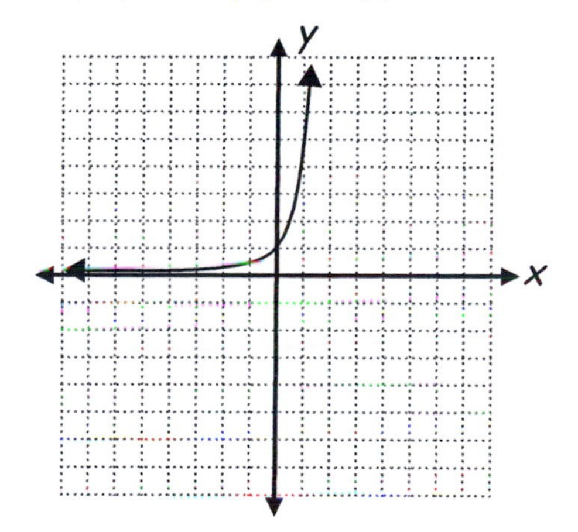

13. Graph $g(x) = 3 - 2^{-x}$

$g(-1) = 1;\ g(0) = 2;\ g(1) = \frac{5}{2}$

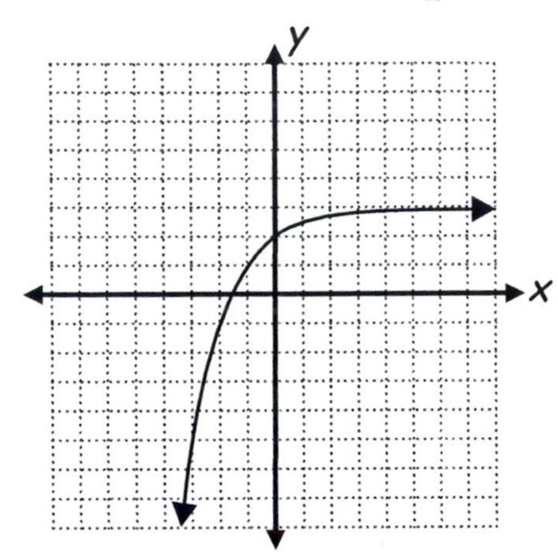

15. Graph $h(x) = \left(\frac{1}{2}\right)^{5-x}$

$h(5) = 1;\ h(6) = 2;\ h(0) = \frac{1}{32}$

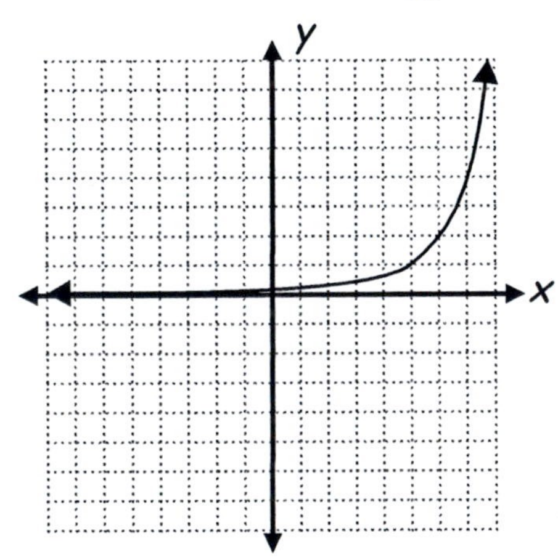

17. Graph $p(x) = 2 - 4^{2-x}$

$p(2) = 1;\ p(1) = -2$

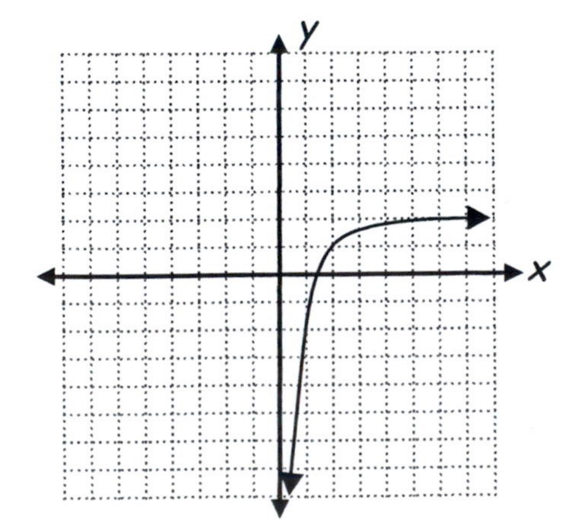

19. Solve $5^x = 125$:

$$5^x = 5^3$$

$$x = 3 \qquad \text{Solution Set: } \{3\}$$

21. Solve $9^{2x-5} = 27^{x-2}$:

$$3^{2(2x-5)} = 3^{3(x-2)}$$

$$4x - 10 = 3x - 6$$

$$x = 4 \qquad \text{Solution Set: } \{4\}$$

23. Solve $4^{-x} = 16$:

$$4^{-x} = 4^2$$

$$-x = 2$$

$$x = -2 \qquad \text{Solution Set: } \{-2\}$$

25. Solve $5^x = .2$:

$$\left(\frac{1}{5}\right)^{-x} = \frac{1}{5}$$

$$-x = 1$$

$$x = -1 \qquad \text{Solution Set: } \{-1\}$$

27. Solve $3^{x^2+4x} = 81^{-1}$:

$$3^{x^2+4x} = \left(3^4\right)^{-1}$$

$$3^{x^2+4x} = 3^{-4}$$

$$x^2 + 4x = -4$$

$$x^2 + 4x + 4 = 0$$

$$(x+2)^2 = 0$$

$$x = -2 \qquad \text{Solution Set: } \{-2\}$$

29. Solve $64^{x+\frac{7}{6}} = 2$:

$$\left(2^6\right)^{x+\frac{7}{6}} = 2^1$$

$$6x + 7 = 1$$

$$x = -1 \qquad \text{Solution Set: } \{-1\}$$

31. Solve $4^{2x-5} = 8^{\frac{x}{2}}$:

$$2^{2(2x-5)} = 2^{3\left(\frac{x}{2}\right)}$$

$$4x - 10 = \frac{3x}{2}$$

$$5x = 20$$

$$x = 4 \qquad \text{Solution Set: } \{4\}$$

33. Solve $4^{4x-7} = \frac{1}{64}$:

$$2^{2(4x-7)} = 2^{-6}$$
$$8x - 14 = -6$$
$$8x = 8$$
$$x = 1 \qquad \text{Solution Set: } \{1\}$$

35. Solve $3^x = 27^{x+4}$:

$$3^x = 3^{3(x+4)}$$
$$x = 3x + 12$$
$$2x = -12$$
$$x = -6 \qquad \text{Solution Set: } \{-6\}$$

37. Solve $2^x = \left(\frac{1}{2}\right)^{13}$:

$$2^x = 2^{-13}$$
$$x = -13 \qquad \text{Solution Set: } \{-13\}$$

39. Solve $5^{3x-1} = 625^x$:

$$5^{3x-1} = 5^{4x}$$
$$3x - 1 = 4x$$
$$x = -1 \qquad \text{Solution Set: } \{-1\}$$

41. Solve $\frac{1}{5}^{x-4} = 625^{\frac{1}{2}}$

The common base number is 5.

$\frac{1}{5} = 5^{-1}$ and $625 = 5^4$

Re-write with the base 5 subsitution:

$(5^{-1})^{x-4} = (5^4)^{\frac{1}{2}}$

By the power rule of exponents,

$$5^{-x+4} = 5^2$$
$$-x + 4 = 2$$
$$x = 2$$

Solution Set: $\{2\}$

43. Solve $\left(e^{x+2}\right)^3 = \left(e^x\right)\frac{1}{e^{3x}}$

The common base number is e.

$$e^{3x+6} = e^{-2x}$$
$$3x + 6 = -2x$$
$$5x = -6$$
$$x = -\frac{6}{5}$$

Solution Set: $\left\{-\frac{6}{5}\right\}$

45. Given $f(x) = 2^{3x}$

Notice that $f(0) = 1$ and $f(1) = 8$

Graph a satisfies these two conditions.

47. Given $g(x) = 2\left(4^{x-1}\right)$

Notice that $g(x) = 2\left(2^{2x-2}\right) = 2^{2x-1}$

$g(0) = \frac{1}{2}$ and $g(1) = 2$

Graph i satisfies these two conditions.

49. Given $f(x) = 6^{4-x}$

Notice that $f(4) = 1$ and $f(3) = 6$

Graph d satisfies these two conditions.

51. Given $m(x) = -2 + 2^{-3x}$

Notice that $m(0) = -1$ and $m(-1) = 6$

Graph e satisfies these two conditions.

53. Given $h(x) = 3^{\frac{1}{2}x}$

Notice that $h(0) = 1$ and $h(2) = 3$

Graph h satisfies these two conditions.

End of Section 5.1

Chapter Five: Section 5.2 Solutions to Odd-numbered Exercises

1. a. Using $A(t) = A_0 a^t$, we are given that

$A(1600) = A_0 a^{1600} = \frac{A_0}{2}$. Then,

$$a^{1600} = \frac{1}{2}$$
$$a \approx 0.999567$$

b. Given $A_0 = 1$ g, find $A(100)$:

$$A(100) \approx 1 \cdot (0.999567)^{100}$$
$$\approx 0.958 \text{ g}$$

c. Find $A(1000)$:

$$A(1000) \approx 1 \cdot (0.999567)^{1000}$$
$$\approx 0.648 \text{ g}$$

3. a. Given $P(t) = (1000)2^{\frac{t}{3}}$. Find t where $1000 \cdot 2^{\frac{t}{3}} = 2000$:

Solve $2^{\frac{t}{3}} = 2$

$$\frac{t}{3} = 1$$
$$t = 3 \text{ years}$$

b. Find t when $(1000) \cdot 2^{\frac{t}{3}} = 8000$:

Solve $2^{\frac{t}{3}} = 8$

$$\frac{t}{3} = 3$$
$$t = 9 \text{ years}$$

5. Using $A(t) = A_0 a^t$, we are given that $A_0 = 100$ and $A(1) = 500$.

So, $100a^1 = 500 \Rightarrow a = 5$

"Another 6 months" means $t = \frac{3}{2}$.

$$A\left(\frac{3}{2}\right) = 100(5)^{\frac{3}{2}}$$
$$\approx 1118 \text{ rabbits}$$

7. Determine the amount she would have after $2\frac{1}{2}$ years with each bank:

For $r = 2.75\%$ compounded monthly:

$$A\left(2\frac{1}{2}\right) = 3{,}500\left(1 + \frac{0.0275}{12}\right)^{12\left(\frac{5}{2}\right)}$$
$$\approx 3{,}500(1.0711)$$
$$\approx \$3{,}748.79$$

For $r = 2.7\%$ compounded daily:

$$A\left(2\frac{1}{2}\right) = 3{,}500\left(1 + \frac{0.027}{365}\right)^{365\left(\frac{5}{2}\right)}$$
$$\approx 3{,}500(1.0698)$$
$$\approx \$3{,}744.40$$

Therefore, the bank offering 2.75% and monthly compounding is better.

9. Using $A(t) = P\left(1 + \frac{r}{n}\right)^{nt}$, where $A(t) = 11{,}000$, $n = 12$ and $t = 3$, solve for r:

$$11{,}000 = 10{,}000\left(1 + \frac{r}{12}\right)^{12(3)}$$
$$1.1 = \left(1 + \frac{r}{12}\right)^{36}$$
$$1.002651013 \approx 1 + \frac{r}{12}$$
$$r \approx 0.03181215$$
$$r \approx 3.18\%$$

11.
$$C(10) = 100{,}000(1 + 0.03)^{10}$$
$$= 100{,}000(1.343916379)$$
$$= \$134{,}392 \text{ (rounded)}$$

13. a. Find $N(0)$.

$$N(0) = \frac{10{,}000}{1+999(e)^{0}}$$
$$= \frac{10{,}000}{1000} = 10 \text{ people}$$

b. $$N(8) = \frac{10{,}000}{1+999(e)^{-8}}$$
$$\approx \frac{10{,}000}{1+0.33513}$$
$$\approx 7{,}490 \text{ people}$$

c. As $t \to \infty$, $999e^{-t} \to 0$. This means that the denominator of $N(t) \to 1$ as $t \to \infty$. This means that $N(t) \to 10{,}000$ as $t \to \infty$.

15. Given that $A(t) = A_0 a^t$ and
$A(20) = \frac{A_0}{2} = A_0 a^{20}$:

a. $$a^{20} = \frac{1}{2}$$
$$a \approx 0.965936$$

b. $$A(30) = 2(0.965936)^{30}$$
$$= 0.707 \text{ kg}$$

c. Use the fact that 6 hrs. = 360 min.
$$A(360) = 2(0.965936)^{360}$$
$$\approx 7.6294 \times 10^{-6} \text{ kg, or}$$
7.628 mg

17. Use $A(t) = P\left(1+\frac{r}{2}\right)^{2t}$.

a. For $t = 3\frac{1}{2}$ years,

$$A\left(3\frac{1}{2}\right) = 1250\left(1+\frac{0.057}{2}\right)^{2\left(\frac{7}{2}\right)}$$
$$= 1250(1.0285)^{7}$$
$$= \$1521.74$$

b. $1521.74 - 1250 = \$271.74$

19. Given $V = P\left(1-e^{-0.18d}\right)$.
P is the number of people in village.
d is the number of days since the virus appeared.
V is the number of infected people in d days.
Set $V = 300\left(1-e^{-0.18(5)}\right)$
$V \approx 178$ or 179 people

21. Given $C = Ie^{0.08m}$.
I is the amount invested.
m is the number of months of the investment.
C is the amount accrued after m months.
Set $C = 1250\left(e^{0.08(24)}\right)$
$C \approx \$8{,}526.20$

23. Given $W = C\left(1-e^{-0.12h}\right)$.
C is the number of computers in the network.
h is the number of hours after discovery.
W is the number of infected computers after h hours.
Set $W = 150\left(1-e^{-0.12(8)}\right)$
$W \approx 93$ computers

End of Section 5.2

Chapter Five: Section 5.3
Solutions to Odd-numbered Exercises

1. Graph $f(x) = \log_3(x-1)$
$f(2) = 0;\ f(4) = 1$

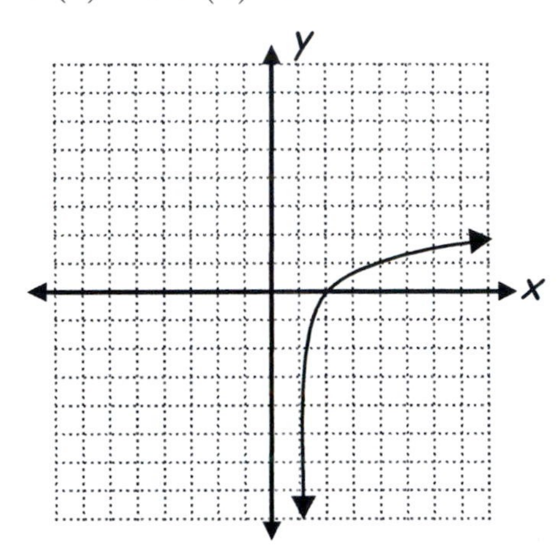

3. Graph $r(x) = \log_{\frac{1}{2}}(x-3)$

$r(4) = 0;\ r\left(3\frac{1}{2}\right) = 1$

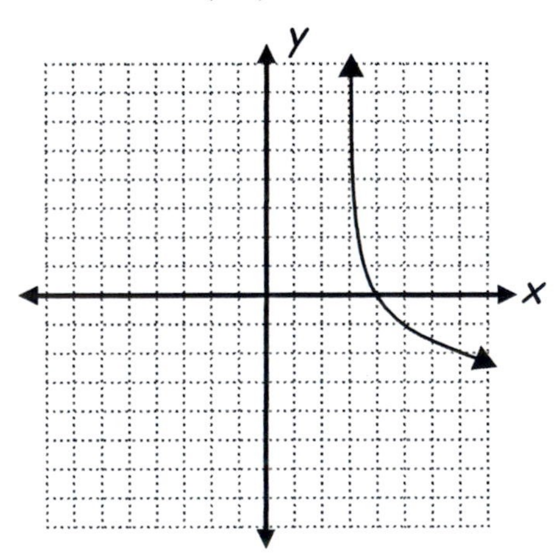

5. Graph $q(x) = \log_3(2-x)$
$q(1) = 0;\ q(-1) = 1$

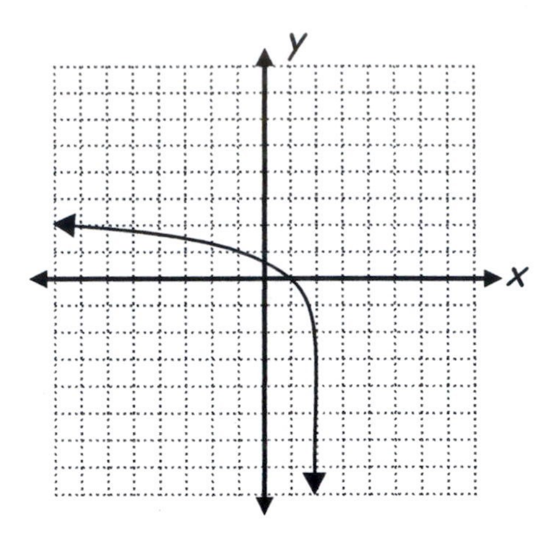

7. Graph $h(x) = \log_7(x-3)+3$
$h(4) = 3;\ h(10) = 4$

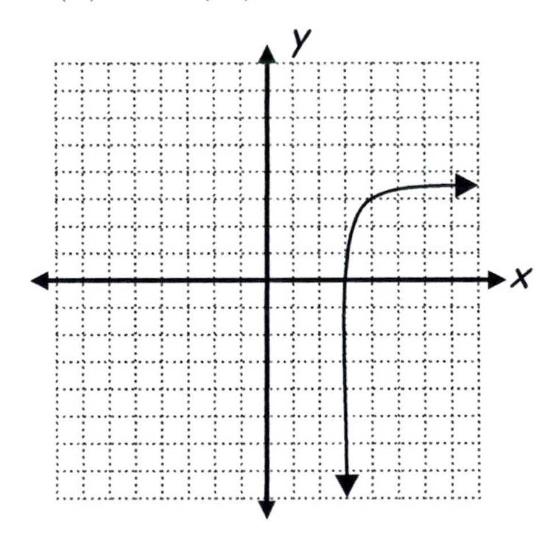

9. Graph $f(x) = \log_3(6-x)$
$f(5) = 0;\ f(3) = 1$

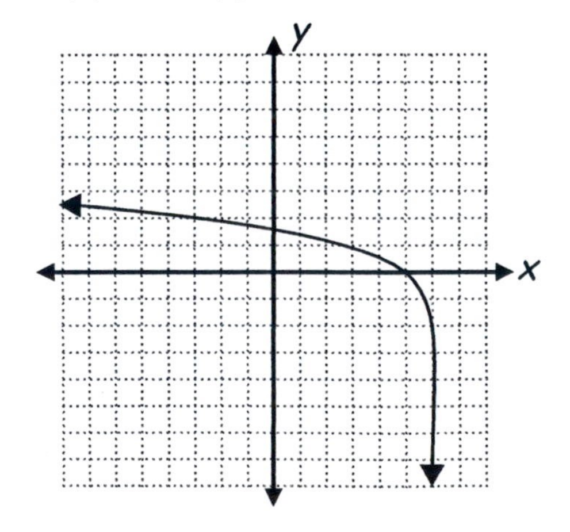

11. Graph $s(x) = -\log_{\frac{1}{3}}(-x)$

$s(-1) = 0;\ s\left(-\frac{1}{3}\right) = -1$

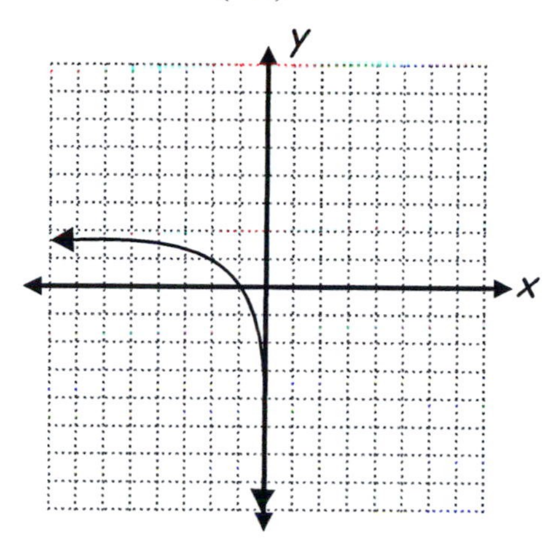

13. Evaluate $\log_7 \sqrt{7}$:

$$\log_7 \sqrt{7} = \log_7 7^{\frac{1}{2}} = \frac{1}{2}$$

15. Evaluate $\log_9\left(\frac{1}{81}\right)$:

$$\log_9\left(\frac{1}{81}\right) = \log_9\left(\frac{1}{9}\right)^2 = \log_9(9)^{-2} = -2$$

17. Evaluate $\log_{27} 3$:

$$\log_{27} 3 = \log_{27}(27)^{\frac{1}{3}} = \frac{1}{3}$$

19. Evaluate $\ln e^{2.89}$:

$$\ln e^{2.89} = \log_e e^{2.89} = 2.89$$

21. Evaluate: $\log_a a^{\frac{5}{3}} = \frac{5}{3}$

23. Evaluate $\log\left(\log(10^{10})\right)$:

$$\log\left(\log(10^{10})\right) = \log 10 = 1$$

25. Evaluate $\ln\sqrt[5]{e}$:

$$\ln\sqrt[5]{e} = \ln e^{\frac{1}{5}} = \frac{1}{5}$$

27. Evaluate $\log_8 4^{\log 1000}$:

$$\log_8 4^{\log 1000} = \log_8 4^3 = \log_8\left(2^2\right)^3 = \log_8\left(2^3\right)^2 = \log_8 8^2 = 2$$

29. Solve:

$$\log_{16} x^{\frac{1}{2}} = \frac{3}{4}$$

$$16^{\frac{3}{4}} = x^{\frac{1}{2}}$$

$$\left(16^{\frac{3}{4}}\right)^2 = x$$

$$x = 16^{\frac{6}{4}}$$

$$x = 64$$

Solution Set: $\{64\}$

31. Solve:

$$\log_5 5^{\log_3 x} = 2$$

$$\log_3 x = 2$$

$$x = 3^2$$

$$x = 9$$

Solution Set: $\{9\}$

33. Solve $\log_3 9^{2x} = -2$

$$9^{2x} = 3^{-2}$$

$$3^{4x} = 3^{-2}$$

$$4x = -2$$

$$x = -\frac{1}{2}$$

Solution Set: $\left\{-\frac{1}{2}\right\}$

35. Solve:

$$\log_7(3x) = -1$$
$$7^{-1} = 3x$$
$$\frac{1}{7} = 3x$$
$$x = \frac{1}{21}$$

Solution Set: $\left\{\frac{1}{21}\right\}$

37. Solve:

$$\log_9(2x-1) = 2$$
$$9^2 = 2x - 1$$
$$81 + 1 = 2x$$
$$x = 41$$

Solution Set: $\{41\}$

39. Solve:

$$6^{\log_x e^2} = e$$
$$\log_x e^2 = \log_6 e$$
$$x^{\log_6 e} = e^2$$

If we use the substitution $u = \log_6 e$ and note $6^u = e$, we have

$$x^u = (6^u)^2$$
$$x^u = (6^2)^u$$
$$x = 36$$

41. Solve:

$$\log x^2 = -2$$
$$10^{-2} = x^2$$
$$\left(\frac{1}{10}\right)^2 = x^2$$
$$x = \pm\frac{1}{10}$$

Solution Set: $\left\{\pm\frac{1}{10}\right\}$

43. Solve:

$$\ln 2x = -1$$
$$e^{-1} = 2x$$
$$x = \frac{1}{2e} \approx 0.1839$$

Solution Set: $\{0.184\}$

45. Solve:

$$\ln(\ln x^2) = 0$$
$$e^0 = \ln x^2$$
$$1 = \ln x^2$$
$$e^1 = x^2$$
$$x = \pm\sqrt{e}, \text{ or } x \approx \pm 1.65$$

Solution Set: $\approx \{-1.65, 1.65\}$

47. Solve:

$$\log e^x = 5.6$$
$$10^{5.6} = e^x$$
$$398{,}107.1706 = e^x$$
$$\ln 398{,}107.1706 = x$$
$$x \approx 12.8945$$

Solution Set: $\approx \{12.894\}$

49. Solve:

$$\log x^{10} = 10$$
$$10^{10} = x^{10}$$
$$x = 10$$

Solution Set: $\{10\}$

51. Given $f(x) = \log_2 x - 1$

$f(1) = \log_2(1) - 1 = -1$

$f(2) = \log_2(2) - 1 = 0$

Graph e satisfies these two conditions.

53. Given $f(x) = \log_2(-x)$

$f(-1) = \log_2 1 = 0$

$f(-2) = \log_2 2 = 1$

Graph b satisfies these two conditions.

55. Given $f(x) = 1 - \log_2 x$

$f(1) = 1 - \log_2 1 = 1$

$f(2) = 1 - \log_2 2 = 0$

Graph h satisfies these two conditions.

57. Given $f(x) = -\log_2(-x)$

$f(-1) = -\log_2 1 = 0$

$f(-2) = -\log_2 2 = -1$

Graph d satisfies these two conditions.

59. Given $f(x) = \log_2 x + 3$
$f(1) = \log_2(1) + 3 = 3$
$f(2) = \log_2(2) + 3 = 4$
Graph i satisfies these two conditions.

End of Section 5.3

61. Given $216 = 6^3$
$\log_6 216 = 3$

63. Given $b^2 = 3.2$
$\log_b 3.2 = 2$

65. Given $1.3^2 = V$
$\log_{1.3} V = 2$

67. Given $16^{2x} = 215$
$\log_{16} 215 = 2x$

69. Given $e^x = \pi$
$\log_e \pi = x$, or $\ln\pi = x$

71. Given $4^e = N$
$\log_4 N = e$

73. Given $\log_2 \frac{1}{8} = -3$
$2^{-3} = \frac{1}{8}$

75. Given $\log_y 9 = 2$
$y^2 = 9$

77. Given $\log_5 8 = d$
$5^d = 8$

79. Given $\log_7 T = 6$
$7^6 = T$

81. Given $\log_{\sqrt{3}} 2\pi = x$
$\sqrt{3}^x = 2\pi$

83. Given $\ln 5x = 3$
$e^3 = 5x$

Chapter Five: Section 5.4
Solutions to Odd-numbered Exercises

1. $\log_5(125x^3) = \log_5(5x)^3$
$= 3\log_5(5x)$
$= 3(\log_5 5 + \log_5 x)$
$= 3(1 + \log_5 x)$
$= 3 + 3\log_5 x$

3. $\ln\left(\frac{e^2p}{q^3}\right) = \ln e^2 p - \ln q^3$
$= 2(\ln e) + \ln p - 3\ln q$
$= 2 + \ln p - 3\ln q$

5. $\log_9 9xy^{-3} = \log_9 9 + \log_9 x + \log_9 y^{-3}$
$= 1 + \log_9 x - 3\log_9 y$

7. $\ln\left[\frac{\sqrt{x^3}pq^5}{e^7}\right] = \ln x^{\frac{3}{2}}pq^5 - \ln e^7$
$= \frac{3}{2}\ln x + \ln p + 5\ln q - 7$

9. $\log(\log(100x^3)) = \log(\log 100 + 3\log x)$
$= \log(2 + 3\log x)$

11. $\log\left[\frac{10}{\sqrt{x+y}}\right] = \log 10 - \log(x+y)^{\frac{1}{2}}$
$= 1 - \frac{1}{2}\log(x+y)$

13. $\log_2\left[\frac{y^2+z}{16x^4}\right] = \log_2(y^2+z) - \log_2(16x^4)$
$= \log_2(y^2+z) - \log_2(2x)^4$
$= \log_2(y^2+z) - 4(\log_2 2 + \log_2 x)$
$= \log_2(y^2+z) - 4(1 + \log_2 x)$
$= \log_2(y^2+z) - 4 - 4\log_2 x$

15. $\log_b\sqrt{\frac{x^4y}{z^2}} = \log_b\left(\frac{x^4y}{z^2}\right)^{\frac{1}{2}}$
$= \frac{1}{2}\left[\log_b x^4 + \log_b y - \log_b z^2\right]$
$= 2\log_b x + \frac{1}{2}\log_b y - \log_b z$

17. $\log_b ab^2c^b = \log_b a + 2\log_b b + \log_b c^b$
$= \log_b a + 2 + b\log_b c$

19. $\log_5 x - 2\log_5 y = \log_5 x - \log_5 y^2$
$= \log_5\left(\frac{x}{y^2}\right)$

21. $\ln(x^2y) - \ln y - \ln x = 2\ln x + \ln y - \ln y - \ln x$
$= \ln x$

23. $\frac{1}{5}(\log_7(x^2) - \log_7(pq)) = \frac{1}{5}\left(\log_7\left(\frac{x^2}{pq}\right)\right)$
$= \log_7\left(\frac{x^2}{pq}\right)^{\frac{1}{5}}$
$= \log_7\left(\frac{x^{\frac{2}{5}}}{p^{\frac{1}{5}}q^{\frac{1}{5}}}\right)$

25. $2(\log_5\sqrt{x} - \log_5 y) = 2\left(\log_5\frac{x^{\frac{1}{2}}}{y}\right)$
$= \log_5\left(\frac{x}{y^2}\right)$

27. $2\log a^2b - \log\frac{1}{b} + \log\frac{1}{a} =$
$2(2\log a + \log b) - \log 1 + \log b + \log 1 - \log a =$
$4\log a + 2\log b + \log b - \log a =$
$3\log a + 3\log b =$
$\log a^3b^3$

29. $\log x - \log y = \log\frac{x}{y}$

31. $\log_5 20 - \log_5 5 = \log_5\frac{20}{5}$
$= \log_5 4$

33. $\ln 15 + \ln 3 = \ln(15\cdot 3)$
$= \ln 45$

35. $0.5\log_3 16 - \log_3 4 = \log_3 16^{\frac{1}{2}} - \log_3 4$
$= \log_3 4 - \log_3 4$
$= 0 \ (\text{or } \log_3 1)$

37. $0.25\ln 81 + \ln 4 = \ln 81^{\frac{1}{4}} + \ln 4$
$= \ln 3 + \ln 4$
$= \ln(3\cdot 4)$
$= \ln 12$

39. $\log 11 + 0.5\log 9 - \log 3 =$
$\log 11 + \log 9^{0.5} - \log 3 =$
$\log 11 + \log 3 - \log 3 = \log 11$

41. $\log_8(2x^2 - 2y) - 0.25\log_8 16 =$
$\log_8(2x^2 - 2y) - \log_8 16^{\frac{1}{4}} =$
$\log_8\frac{2(x^2 - y)}{2} = \log_8(x^2 - y)$

43. $5^{2\log_5 x} = \left(5^{\log_5 x}\right)^2$
$= x^2$

45. $e^{2-\ln x+\ln p} = \frac{e^2 e^{\ln p}}{e^{\ln x}} = \frac{e^2 p}{x}$

47. $10^{\log x^3 - 4\log y} = \frac{10^{\log x^3}}{10^{\log y^4}} = \frac{x^3}{y^4}$

49. $10^{2\log x} = 10^{\log x^2}$
$= x^2$

51. $\log_4 16\cdot\log_x x^2 = 2\cdot 2\log_x x$
$= 2\cdot 2\cdot 1$
$= 4$

53. $4^{\log_4(3x)+0.5\log_4(16x^2)} = 4^{\log_4(3x)}\cdot 4^{\log_4(16x^2)^{\frac{1}{2}}}$
$= 3x\cdot(16x^2)^{\frac{1}{2}}$
$= 3x(4x)$
$= 12x^2$

For items 55-77: Select a convenient base, say e or 10, and convert by using $\log_b x = \frac{\log_a x}{\log_a b}$ when necessary.

55. $\log_4 17 = \frac{\ln 17}{\ln 4} \approx \frac{2.8332}{1.3863}$
≈ 2.04

57. $\log_9 8 = \frac{\ln 8}{\ln 9} \approx \frac{2.0794}{2.1972}$
≈ 0.95

59. $\log_{12} 10.5 = \frac{\ln 10.5}{\ln 12} \approx \frac{2.3514}{2.4849}$
≈ 0.95

61. $\log_6 3^4 = 4\left(\frac{\log 3}{\log 6}\right) \approx 4\left(\frac{0.4771}{0.7782}\right)$
≈ 2.45

63. $\log_{\frac{1}{2}} \pi^{-2} = -2\left(\frac{\log\pi}{\log\frac{1}{2}}\right) \approx -2\left(\frac{0.4971}{-0.3010}\right)$
≈ 3.30

65. $\ln(\log 123) = \ln(2.0899)$
≈ 0.74

67. $\log 16 \approx 1.20$

69. $\log_5(20) = \dfrac{\ln 20}{\ln 5} \approx 1.86$

71. $\log_4(0.25) = x$

$$4^x = \frac{1}{4}$$
$$x = -1$$

73. $\log_{2.5}(34) = \dfrac{\ln 34}{\ln 2.5} \approx 3.85$

75. $\log_4(2.9) = \dfrac{\ln 2.9}{\ln 4} \approx 0.77$

77. $\log_{0.2}(17) = \dfrac{\ln 17}{\ln 0.2} \approx -1.76$

79. $\log_4 16 = x$

$$4^x = 16$$
$$x = 2$$

81. $\ln e^4 + \ln e^3 = 4\ln e + 3\ln e$

$$= 4 + 3 = 7$$

83. $\ln e^{1.5} - \log_4 2 = 1.5\ln e - \dfrac{1}{2}$

$$= 1.5 - \frac{1}{2} = 1$$

85. Use $\text{pH} = -\log\left[H_3O^+\right]$:

$$\text{pH} = -\log\left(3.16 \times 10^{-6}\right)$$
$$= -\left[\log 3.16 - 6\log 10\right]$$
$$= -\left[\log 3.16 - 6\right] \approx 6 - 0.4997$$
$$\approx 5.5$$

87. Solve $3.2 = -\log\left[H_3O^+\right]$

$$3.2 = \log\left[\frac{1}{H_3O^+}\right]$$
$$10^{3.2} = \frac{1}{H_3O^+}$$
$$H_3O^+ = \frac{1}{10^{3.2}} \approx 6.31 \times 10^{-4} \text{ moles/liter}$$

89. Northridge quake was 6.7 on the Richter scale and Gujarat was 7.9--a difference of 1.2 units on the Richter.

$$1.2 = \log\left(\frac{I_G}{I_N}\right)$$
$$10^{1.2} = \frac{I_G}{I_N}$$
$$I_G = 10^{1.2}(I_N)$$
$$I_G = 15.8 I_N$$

The intensity of the quake at Gujarat was stronger by a factor of about 15.8 times that of Northridge.

91. Use $D = 10\log\left(\dfrac{I}{I_o}\right)$; $D = 105$; $I_o = 10^{-12}$

$$105 = 10\log\left(\frac{I}{10^{-12}}\right)$$
$$10.5 = \log I - \log 10^{-12}$$
$$10.5 = \log I + 12\log 10$$
$$10.5 = \log I + 12$$
$$-1.5 = \log I$$
$$10^{-1.5} = I$$
$$I \approx 0.03 \text{ watts/m}^2$$

93. Use $D = 10\log\left(\dfrac{I}{I_o}\right)$; $I = 2.19 \times 10^{-11}$; and $I_o = 10^{-12}$:

$$D = 10\log\left(\frac{2.19 \times 10^{-11}}{10^{-12}}\right)$$
$$D = 10\log(21.9)$$
$$D = 13.4 \text{ decibels}$$

End of Section 5.4

Chapter Five: Section 5.5
Solutions to Odd-numbered Exercises

1. Solve $3e^{5x}=11$

$$\ln 3e^{5x}=\ln 11$$
$$\ln 3+\ln e^{5x}=\ln 11$$
$$5x=\ln 11-\ln 3$$
$$x=\frac{1}{5}\ln\left(\frac{11}{3}\right)$$
$$x\approx 0.26$$

3. Solve $11^{\frac{3}{x}}=10$

$$\log 11^{\frac{3}{x}}=\log 10$$
$$\frac{3}{x}\log 11=1$$
$$x=3\log 11$$
$$x\approx 3.12$$

5. Solve $e^{15-3x}=28$

$$\ln e^{15-3x}=\ln 28$$
$$(15-3x)\ln e=\ln 28$$
$$3x=15-\ln 28$$
$$x=\frac{15-\ln 28}{3}$$
$$x\approx 3.89$$

7. Solve $10^{2x+5}=e$

$$\log 10^{2x+5}=\log e$$
$$2x+5=\log e$$
$$x=\frac{\log e-5}{2}$$
$$x\approx -2.28$$

9. Solve $6^{x-7}=7$

$$\log 6^{x-7}=\log 7$$
$$(x-7)\log 6=\log 7$$
$$(x-7)=\frac{\log 7}{\log 6}$$
$$x=7+\frac{\log 7}{\log 6}\text{ or }x=\frac{7\log 6+\log 7}{\log 6}$$
$$x\approx 8.09$$

11. Solve $2^{6-x}=10$

$$\log 2^{6-x}=\log 10$$
$$(6-x)\log 2=1$$
$$6-x=\frac{1}{\log 2}$$
$$x=6-\frac{1}{\log 2},\text{ or }x=\frac{6\log 2-1}{\log 2}$$
$$x\approx 2.68$$

13. Solve $e^{-4x-2}=12$

$$\ln e^{-4x-2}=\ln 12$$
$$(-4x-2)\ln e=\ln 12$$
$$-4x=2+\ln 12$$
$$x=-\frac{2+\ln 12}{4}$$
$$x\approx -1.12$$

15. Solve $e^{x-4}=4^{\frac{2x}{3}}$

$$\ln e^{x-4}=\ln 4^{\frac{2x}{3}}$$
$$(x-4)\ln e=\frac{2x}{3}\ln 4$$
$$3x-12=2x\ln 4$$
$$x(3-2\ln 4)=12$$
$$x=\frac{12}{3-2\ln 4}$$
$$x\approx 52.77$$

17. Solve $8^{x^2+1}=23$

$$\log 8^{x^2+1}=\log 23$$
$$x^2+1=\frac{\log 23}{\log 8}$$
$$x^2=\frac{\log 23}{\log 8}-1$$
$$x^2\approx 1.508-1$$
$$x\approx\sqrt{0.508}$$
$$x\approx\pm 0.71$$

19. Solve $6^{3x-4} = 36^{2x+4}$

$$\log 6^{3x-4} = \log 6^{2(2x+4)}$$
$$(3x-4)\log 6 = (4x+8)\log 6$$
$$3x - 4 = 4x + 8$$
$$x = -12$$

21. Solve $e^{2x} = 14$

$$\ln e^{2x} = \ln 14$$
$$2x \ln e = \ln 14$$
$$2x = \ln 14$$
$$x = \frac{\ln 14}{2}$$
$$x \approx 1.32$$

23. Solve $5^{5x-7} = 10^{2x}$

$$\log 5^{5x-7} = \log 10^{2x}$$
$$(5x-7)\log 5 = 2x \log 10$$
$$(5x-7)\log 5 = 2x \quad [\log 5 \approx 0.699]$$
$$3.495x - 4.893 \approx 2x$$
$$1.495x \approx 4.893$$
$$x \approx 3.27$$

25. Solve $\log_5 x = 3$

$$5^3 = x$$
$$x = 125$$

27. Solve $\log x + \log(4x) = 2$

$$\log 4x^2 = 2$$
$$10^2 = 4x^2$$
$$x^2 = 25$$
$$x = 5$$

29. Solve $\ln 2x - \ln 4 = 3$

$$\ln\left(\frac{2x}{4}\right) = 3$$
$$e^3 = \frac{x}{2}$$
$$x = 2e^3$$
$$x \approx 40.17$$

31. Solve $\log_4(x-3) + \log_4 2 = 3$

$$\log_4\left[(x-3)(2)\right] = 3$$
$$4^3 = 2x - 6$$
$$32 = x - 3$$
$$x = 35$$

33. Solve $\log_5(8x) - \log_5 3 = 2$

$$\log_5\left(\frac{8x}{3}\right) = 2$$
$$5^2 = \frac{8x}{3}$$
$$8x = 75$$
$$x = \frac{75}{8} \approx 9.38$$

35. Solve $9^{\log_3 x} = 16$

$$3^{2\log_3 x} = 16$$
$$x^2 = 16$$
$$x = 4$$

37. Solve $\log_3 6x - 2\log_3 6x = 3$

$$-\log_3 6x = 3$$
$$\log_3 6x = -3$$
$$3^{-3} = 6x$$
$$x = \frac{1}{27(6)}$$
$$x = \frac{1}{162} \approx 0.01$$

39. Solve $\ln 2^{4e^x} = \ln 16^e$

$$\ln 16^{e^x} = \ln 16^e$$
$$e^x = e$$
$$x = 1$$

41. Solve $\log(x-3) + \log(x+3) = 4$

$$\log\left[(x-3)(x+3)\right] = 4$$
$$x^2 - 9 = 10^4$$
$$x = \sqrt{10009}$$
$$x \approx 100.04$$

43. $\log_\pi(x-5)+\log_\pi(x+3)=\log_\pi(1-2x)$

$$\log_\pi[(x-5)(x+3)]=\log_\pi(1-2x)$$
$$\log_\pi[x^2-2x-15]=\log_\pi(1-2x)$$
$$x^2-2x-15=1-2x$$
$$x^2-16=0$$
$$x=\pm 4 \quad [\text{But wait!}]$$

Both 4 and -4 would render one or more of the terms in the original equation with an undefined expression (i.e., log(negative number)), so there is no solution.

45. Solve $\log x+\log(x-3)=1$

$$\log(x(x-3))=1$$
$$\log(x^2-3x)=1$$
$$10=x^2-3x$$
$$x^2-3x-10=0$$
$$(x-5)(x+2)=0$$

$x=5$; Solution Set: $\{5\}$

-2 is extraneous for original equation.

47. Solve $\log_2 x+\log_2(x-7)=3$

$$\log_2(x(x-7))=3$$
$$\log_2(x^2-7x)=3$$
$$x^2-7x=2^3$$
$$x^2-7x-8=0$$
$$(x-8)(x+1)=0$$

$x=8$; Solution Set: $\{8\}$

-1 is extraneous for original equation.

49. Solve $\log_3(x+1)-\log_3(x-4)=2$

$$\log_3\left(\frac{x+1}{x-4}\right)=2$$
$$\frac{x+1}{x-4}=3^2$$
$$9x-36=x+1$$
$$8x=37$$
$$x=\frac{37}{8}$$

Solution Set: $\left\{\frac{37}{8}\right\}$

51. Solve

$$\log_4(x-3)+\log_4(x-2)=\log_4(x+1)$$
$$\log_4[(x-3)(x-2)]=\log_4(x+1)$$
$$x^2-5x+6=x+1$$
$$x^2-6x+5=0$$
$$(x-5)(x-1)=0$$

$x=5$; Solution Set $\{5\}$

1 is extraneous for original equation.

53. $\log_5(x-1)+\log_5(x+4)=\log_5(x-5)$

$$\log_5[(x-1)(x+4)]=\log_5(x-5)$$
$$x^2+3x-4=x-5$$
$$x^2+2x+1=0$$
$$(x+1)^2=0$$

-1 is extraneous for the original equation. Therefore, the solution set is the empty set $\varnothing$.

55. Solve $\log_2(x-5)+\log_2(x+2)=3$

$$\log_2(x^2-3x-10)=3$$
$$x^2-3x-10=2^3$$
$$(x-6)(x+3)=0$$

$x=6$; Solution Set $\{6\}$

-3 is extraneous for original equation.

57. Solve

$$\ln(x+2)+\ln(x)=0$$
$$\ln(x^2+2x)=0$$
$$x^2+2x=e^0=1$$
$$x^2+2x-1=0$$
$$x=\frac{-2\pm\sqrt{8}}{2}$$
$$x=-1+\sqrt{2}$$

Solution Set: $\{-1+\sqrt{2}\}$

$-1-\sqrt{2}$ is extraneous, since $\ln(-1-\sqrt{2})$ is not defined.

59. Solve

$2^{2x}-12\left(2^{x}\right)+32=0$

Let $u=2^{x}$. First, solve for u:

$u^{2}-12u+32=0$

$(u-8)(u-4)=0$

$u=8$ or $u=4$. Substitute $u=2^{x}$:

$2^{x}=8$ or $2^{x}=4$

$x=3$ or $x=2$

Solution Set: $\{2,3\}$

61. Solve

$3^{2x}-12\left(3^{x}\right)+27=0$

Let $u=3^{x}$. First, solve for u:

$u^{2}-12u+27=0$

$(u-9)(u-3)=0$

$u=9$ or $u=3$. Substitute $u=3^{x}$:

$3^{x}=9$ or $3^{x}=3$

$x=2$ or $x=1$

Solution Set: $\{1,2\}$

63. Simplify $f(x)=0.25\log 16x^{8}$

$$f(x)=\log\left(16x^{8}\right)^{\frac{1}{4}}$$
$$f(x)=\log 2x^{2}$$

65. Simplify $f(x)=8\ln\sqrt[4]{3x}$

$$f(x)=8\ln(3x)^{\frac{1}{4}}$$
$$f(x)=\ln\left[(3x)^{\frac{1}{4}}\right]^{8}$$
$$f(x)=\ln\left(3^{2}x^{2}\right)$$
$$f(x)=\ln\left(9x^{2}\right)$$

67. Simplify $f(x)=10^{2x\log 16}$

$$f(x)=10^{\log 16^{(2x)}}$$
$$f(x)=16^{2x}$$
$$f(x)=256^{x}$$

69. Simplify $f(x)=2\ln x^{3}-\ln x^{6}$

$$f(x)=6\ln x-6\ln x$$
$$f(x)=0$$

71. Simplify $f(x)=2\ln 5^{x\log_{20}(2\sqrt{5})}$

$$f(x)=\ln 5^{2x\log_{20}(2\sqrt{5})}$$
$$f(x)=\ln 5^{x\log_{20}(2\sqrt{5})^{2}}$$
$$f(x)=\ln 5^{x\log_{20}20}$$
$$f(x)=\ln 5^{x(1)}$$
$$f(x)=\ln 5^{x}$$

73. Simplify $f(x)=2\ln 5^{\log_{4}2}$

$$f(x)=2\ln 5^{\frac{1}{2}}$$
$$f(x)=\ln 5^{2\left(\frac{1}{2}\right)}$$
$$f(x)=\ln 5$$

75. **a.** Let $A=$ the initial investment.

Use $2A=A\left(1+\frac{r}{n}\right)^{nt}$ and solve for t:

$$2A=A\left(1+\frac{0.04}{12}\right)^{12t}$$
$$2\approx(1.003333)^{12t}$$
$$\ln 2\approx\ln(1.003333)^{12t}$$
$$\ln 2\approx 12t\ln(1.003333)$$
$$t\approx\frac{\ln 2}{12\ln(1.003333)}$$
$$t\approx 17.36\ \text{years}$$

b. Use $2A=Ae^{rt}$

$$2=e^{0.07t}$$
$$\ln 2=0.07t\ln e$$
$$t=\frac{\ln 2}{0.07}$$
$$t\approx 9.90\ \text{years}$$

77. Use $P(t) = P_0 a^t$ and the given information that $P(0) = 10{,}000 = P_0 a^0$, and that $2P_0 = P_0 a^{1.5}$:

$$a^{1.5} = 2$$
$$1.5 \log a = \log 2$$
$$\log a = \frac{\log 2}{1.5}$$
$$\log a \approx 0.20069$$
$$a \approx 1.5874$$
$$10{,}000(1.5874)^t \approx 100{,}000$$
$$(1.5874)^t \approx 10$$
$$t \log(1.5874) \approx \log 10$$
$$t \approx \frac{1}{0.20069}$$
$$t \approx 4.98 \text{ hours}$$

79. Use $A = A_0\left(1 + \frac{r}{n}\right)^{nt}$ with $A_0 = \$12{,}500$ $r = 3.66\%$ and $n = 12$. Find t for

$$15{,}000 = 12{,}500\left(1 + \frac{0.0366}{12}\right)^{12t}$$
$$15{,}000 = 12{,}500(1.00305)^{12t}$$
$$6 = 5(1.00305)^{12t}$$
$$\ln\left(\frac{6}{5}\right) = 12t\ln(1.00305)$$
$$t = \frac{\ln(1.2)}{12\ln(1.00305)}$$
$$t \approx 4.99 \text{ years}$$

81. Use $A = A_0 a^t$. You are given that $\frac{1}{2}A_0 = A_0 a^{30}$, or $\frac{1}{2} = a^{30}$. First, find a:

$$\ln\left(\frac{1}{2}\right) = 30\ln a$$
$$\ln a = \frac{-\ln 2}{30}$$
$$\ln a \approx -0.023104906$$
$$a \approx e^{-0.023104906}$$
$$a \approx 0.9771599685$$

Now, solve for t:

$$159 \approx 160(0.9771599685)^t :$$
$$\ln\left(\frac{159}{160}\right) \approx t\ln(0.9771599685)$$
$$t \approx \frac{\ln(0.99375)}{\ln(0.97716)}$$
$$t \approx 0.271 \text{ years, or } 99 \text{ days}$$

End of Section 5.5

Chapter Five Test
Solutions to Odd-Numbered Exercises

1. Graph $f(x)=\left(\frac{1}{2}\right)^{x-1}+3$:

$f(1)=4;\ f(2)=3\frac{1}{2};\ f(-1)=7$

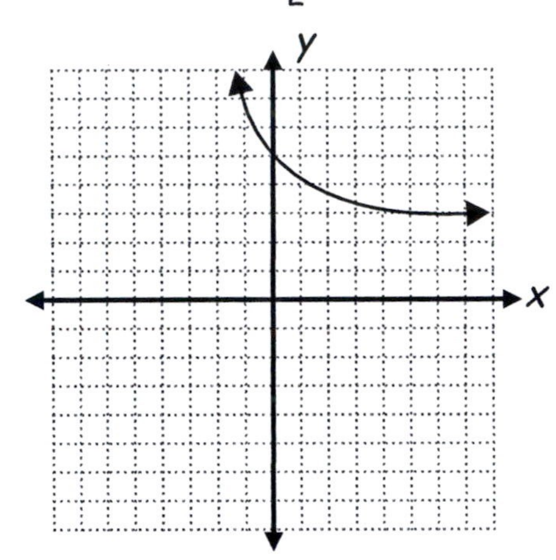

3. Solve $3^{3x-5}=81$:

$3^{3x-5}=3^4$

$3x-5=4$

$x=3;$ Solution Set: $\{3\}$

5. Solve $10{,}000^x=10^{-2x-12}$

$10^{4x}=10^{-2x-12}$

$4x=-2x-12$

$6x=-12$

$x=-2;$ Solution Set: $\{-2\}$

7. Use $A(t)=A_0a^t$ with $t=$ number of weeks. First, find a:

$11=37a^2$

$a^2=\frac{11}{37}$

$a=\sqrt{\frac{11}{37}}$

So $A(t)=37\left(\sqrt{\frac{11}{37}}\right)$

9. Sketch $f(x)=\log_4(3-x)+2$
The basic graph is that of $\log_4(-x)$. The graph of f is a shift right of 3 units and then up 2 units. Notice that $f(2)=2$:

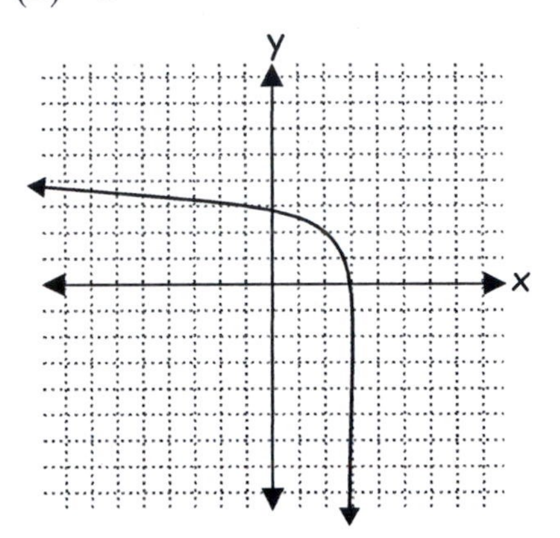

11. $\log_{27}9^{\log 1000}=\log_{27}9^3$

$=\log_{27}729$

$=\log_{27}(27)^2$

$=2$

13. $\log_4\left(\frac{1}{64}\right)=\log_4\left(\frac{1}{4}\right)^3$

$=\log_4(4)^{-3}$

$=-3$

15. Solve $\log_9 x^{\frac{1}{2}}=\frac{3}{4}$

$\frac{1}{2}\log_9 x=\frac{3}{4}$

$\log_9 x=\frac{3}{2}$

$x=9^{\frac{3}{2}}$

$x=27;$ Solution Set: $\{27\}$

17. Solve $\log 27=5x$

$x=\frac{\log 27}{5}$

$x\approx 0.2863$

19. $\log\sqrt{\dfrac{x^3}{4\pi^5}} = \dfrac{1}{2}\log\left(\dfrac{x^3}{4\pi^5}\right)$

$$= \frac{1}{2}\left[\log x^3 - \log 4\pi^5\right]$$

$$= \frac{1}{2}\left[3\log x - \log 2^2 - 5\log\pi\right]$$

$$= \frac{3}{2}\log x - \frac{5}{2}\log\pi - \log 2$$

21. $\dfrac{1}{3}\left(\log_2\left(a^5\right) - \log_2\left(bc^3\right)\right) = \dfrac{1}{3}\left(\log_2\left(\dfrac{a^5}{bc^3}\right)\right)$

$$= \log_2\left(\frac{a^{\frac{5}{3}}}{b^{\frac{1}{3}}c}\right)$$

23. Evaluate $\log_7 18 = \dfrac{\ln 18}{\ln 7}$

$$\approx 1.4854$$

25. Use $D = 10\log\left(\dfrac{I}{I_0}\right)$

$$= 10\log\left(\frac{10^{-1}}{10^{-12}}\right)$$

$$= 10\log 10^{11}$$

$$= 10(11) = 110 \text{ dB}$$

27. Solve $e^{8-5x} = 16$

$$\ln e^{8-5x} = \ln 16$$

$$8 - 5x = \ln 16$$

$$x = \frac{8 - \ln 16}{5}$$

$$x \approx 1.05$$

29. Solve $\ln(x+1) + \ln(x-1) = \ln(x+5)$

$$\ln((x+1)(x-1)) = \ln(x+5)$$

$$x^2 - 1 = x + 5$$

$$x^2 - x - 6 = 0$$

$$(x-3)(x+2) = 0$$

$$x = 3$$

Solution Set: $\{3\}$

Note: -2 is extraneous from the original equation.

31. Use $P(t) = P_0\left(1 + \dfrac{r}{n}\right)^{nt}$ and solve:

$$7000 = 6500\left(1 + \frac{0.0436}{12}\right)^{12t}$$

$$\frac{14}{13} \approx (1.0036333)^{12t}$$

$$1.076923 \approx (1.0036333)^{12t}$$

$$\ln(1.076923) \approx 12t\ln(1.0036333)$$

$$t \approx \frac{\ln(1.076923)}{12\ln(1.0036333)}$$

$t \approx 1.703$ years, or 20.4 mos

End of Chapter 5 Test

Chapter Six: Section 6.1
Solutions to Odd-numbered Exercises

1. $\frac{5\pi}{4}=\frac{5\pi}{4}\left(\frac{180}{\pi}\right)^\circ$
$=225^\circ$

3. $-\frac{3\pi}{8}=-\frac{3\pi}{8}\left(\frac{180}{\pi}\right)^\circ$
$=-\frac{135^\circ}{2}=-67.5^\circ$

5. $\frac{2\pi}{3}=\frac{2\pi}{3}\left(\frac{180}{\pi}\right)^\circ$
$=120^\circ$

7. $\frac{5\pi}{6}=\frac{5\pi}{6}\left(\frac{180}{\pi}\right)^\circ$
$=150^\circ$

9. $-\frac{9\pi}{4}=-\frac{9\pi}{4}\left(\frac{180}{\pi}\right)^\circ$
$=-405^\circ$

11. $47^\circ=47\left(\frac{\pi}{180}\right)$
$=\frac{47\pi}{180}$

13. $132^\circ=132\left(\frac{\pi}{180}\right)$
$=\frac{11\pi}{15}$

15. $148^\circ=148\left(\frac{\pi}{180}\right)$
$=\frac{37\pi}{45}$

17. $480^\circ=480\left(\frac{\pi}{180}\right)$
$=\frac{8\pi}{3}$

19. $125^\circ=125\left(\frac{\pi}{180}\right)$
$=\frac{25\pi}{36}$

21. $\frac{3\pi}{2}=\frac{3\pi}{2}\left(\frac{180}{\pi}\right)^\circ$
$=270^\circ$

23. $3\pi=3\pi\left(\frac{180}{\pi}\right)^\circ$
$=540^\circ$

25. $-\frac{2\pi}{5}=-\frac{2\pi}{5}\left(\frac{180}{\pi}\right)^\circ$
$=-72^\circ$

27. $20^\circ=20\left(\frac{\pi}{180}\right)$
$=\frac{\pi}{9}$

29. $-144^\circ=-144\left(\frac{\pi}{180}\right)$
$=-\frac{4\pi}{5}$

31. $30^\circ=30\left(\frac{\pi}{180}\right)$
$=\frac{\pi}{6}$

33. $166^\circ=166\left(\frac{\pi}{180}\right)$
$=\frac{83\pi}{90}$

35. $\frac{23\pi}{36}=\frac{23\pi}{36}\left(\frac{180}{\pi}\right)^\circ$
$=115^\circ$

37. $\frac{5\pi}{18}=\frac{5\pi}{18}\left(\frac{180}{\pi}\right)^\circ$
$=50^\circ$

39. $\frac{23\pi}{12} = \frac{23\pi}{12}\left(\frac{180}{\pi}\right)^{\circ}$
$= 345^{\circ}$

41. $\frac{59\pi}{36} = \frac{59\pi}{36}\left(\frac{180}{\pi}\right)^{\circ}$
$= 295^{\circ}$

43. $225^{\circ} = 225\left(\frac{\pi}{180}\right)$
$= \frac{5\pi}{4}$

45. The indicated right triangle is isosceles and therefore has congruent base angles of 45°. The indicated angle in question is a complementary angle to one of the base angles, so it is 45° or $45^{\circ} = 45\left(\frac{\pi}{180}\right) = \frac{\pi}{4}$.

47. The indicated angle is oriented with positive measure through a measure of π, or 180° with terminal ray along the hypotenuse of a $\frac{\pi}{6} - \frac{\pi}{3} - \frac{\pi}{2}$ right triangle. The measure of the given angle is $\pi + \frac{\pi}{3} = \frac{4\pi}{3}$.

49. The indicated angle is oriented with positive measure and is formed by one complete rotation (2π), and one semi-circle rotation (π) with terminal ray along the hypotenuse of a $\frac{\pi}{4} - \frac{\pi}{4} - \frac{\pi}{2}$ right triangle. The measure of the given angle is $2\pi + \pi + \frac{\pi}{4} = \frac{13\pi}{4}$.

51. Sketch $\frac{5\pi}{2} = 2\pi + \frac{\pi}{2}$ (one full rotation plus one quarter rotation in the positive (counterclockwise direction.)

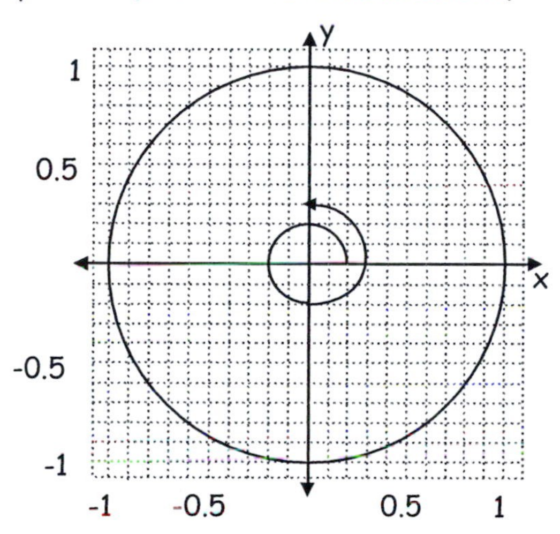

53. Sketch $210^{\circ} = 180^{\circ} + 30^{\circ}$ (one half circle rotation plus another 30° in the positive (counterclockwise direction.)

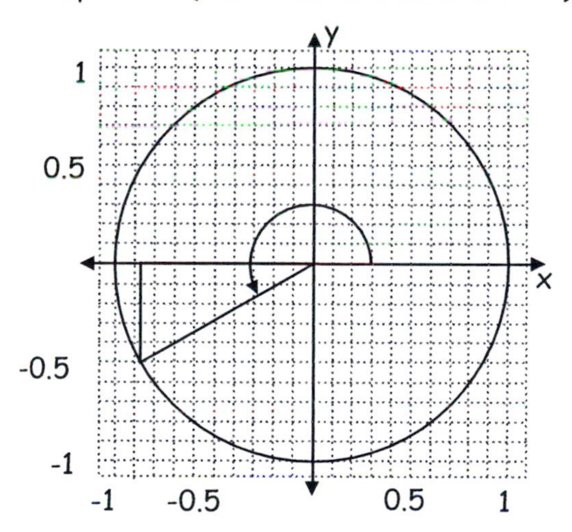

55. Sketch $\frac{7\pi}{4} = \pi + \frac{3\pi}{4}$ (one half circle rotation plus another $\frac{3\pi}{4}$ radians in the positive (counterclockwise direction.)

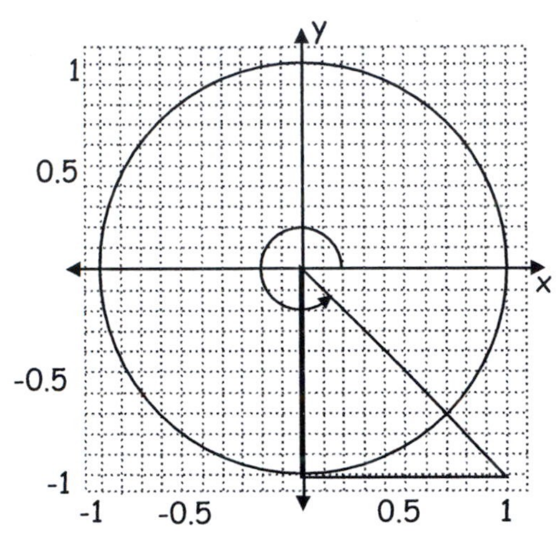

57. Given $r = 4$ in.; $\theta = 1$

$s = r\theta$

$s = 4(1) = 4$ in.

59. Given $r = 15$ ft.; $\theta = \frac{\pi}{4}$

$s = r\theta$

$s = 15\left(\frac{\pi}{4}\right) \approx 11.78$ ft.

61. Given $r = 16.5$ m; $\theta = 30°$
Change 30° to radians:

$30° = 30\left(\frac{\pi}{180}\right) = \frac{\pi}{6}$

$s = r\theta$

$s = 16.5\left(\frac{\pi}{6}\right) \approx 8.64$ m

63. $17° = 17\left(\frac{\pi}{180}\right) \approx 0.2967$

Use $s = r\theta$.

$s \approx 5(0.2967) \approx 1.48$ in.

65. $68° = 68\left(\frac{\pi}{180}\right) \approx 1.1868$

radius $= \frac{1}{2}$ of diameter $= 3$

Use $s = r\theta$.

$s \approx 3(1.1868) \approx 3.56$ ft.

67. $\theta = 34° - 29.25° = 4.75°$

Convert to radians:

$4.75° = 4.75\left(\frac{\pi}{180}\right) \approx 0.0829$

Use $s = r\theta$ and $r = 6370$ km

$s \approx 6370(0.0829) \approx 528.10$ km

69. Given $r = 1.2$ cm and $s = 9$ mm
(Note: $s = 0.9$ cm)
Use $s = r\theta$ to find θ.
$0.9 = 1.2\theta$

$\theta = \frac{0.9}{1.2} = 0.75 = 0.75\left(\frac{\pi}{180}\right)$

$\approx 43°$

71. $\theta = 39.17° - 33.67° = 5.5°$

Convert to radians:

$5.5° = 5.5\left(\frac{\pi}{180}\right) \approx 0.09599$

Use $s = r\theta$ and $r = 6370$ km

$s \approx 6370(0.096) \approx 611.48$ km

73. Add the two degree measures because one is north latitude and the other is south latitude:

$\theta = 16.5° + 10.52° = 27.02°$

$= 27.02\left(\frac{\pi}{180}\right) \approx 0.47159$

Use $s = r\theta$

$s = 6370(0.47159) \approx 3004.02$ km

75. Given $r=14$ ft.; $s=63$ ft.

$$\theta=\frac{s}{r}=\frac{63}{14}=\frac{9}{2}$$

77. Given $r=23.5$ dm; $s=10.5$ dm

$$\theta=\frac{s}{r}=\frac{10.5}{23.5}=\frac{105}{235}=\frac{21}{47}$$

79. Given $r=2$ km; $s=22.5$ km

$$\theta=\frac{s}{r}=\frac{22.5}{2}=\frac{45}{4}$$

81. Given $r=10$ in.; 1 revolution $=2\pi$ rad

a. $\omega=1000(2\pi)/\text{min.}$
$=2000\pi/\text{min.}$

b. Use 10 in $=\frac{5}{6}$ ft. and convert ω to radians per second (i.e., divide by 60):

$$\omega=\frac{2000\pi}{60}/\text{s}=\frac{100\pi}{3}/\text{s.}$$

$$v=r\omega$$

$$=\frac{5}{6}\left(\frac{100\pi}{3}\right)\text{ ft./s}$$

$$=\frac{250\pi}{9}\text{ ft./s}$$

$$\approx 87.27\text{ ft./s}$$

83. Given $r=14$ in.; 1 revolution $=2\pi$ rad

a. $\omega=50(2\pi)/\text{min.}$
$=100\pi/\text{min.}$

b. Use 14 in. $=\frac{14}{12}$ ft. $=\frac{7}{6}$ ft.

$$v=r\omega$$

$$=\frac{7}{6}\omega\text{ ft./min.}$$

$$=\frac{7(100\pi)}{6}\text{ ft./min.}$$

$$\approx 366.52\text{ ft./min.}$$

85. The indicated triangle is $\frac{\pi}{6}-\frac{\pi}{3}-\frac{\pi}{2}$

$$\theta=\pi-\frac{\pi}{6}=\frac{5\pi}{6}$$

Given $r=1$:

$$A=(1)^2\left(\frac{\frac{5\pi}{6}}{2}\right)=\frac{5\pi}{12}\text{ square units}$$

$A\approx 1.31$ square units

87. Given $r=7$ cm; $\theta=70°$

$$\theta=70°=70\left(\frac{\pi}{180}\right)=\frac{7\pi}{18}$$

$$A=7^2\left[\frac{\left(\frac{7\pi}{18}\right)}{2}\right]$$

$$=\frac{343\pi}{36}\text{ cm}^2$$

$$A\approx 29.93\text{ cm}^2$$

89. Given $r=4$ m; $\theta=\frac{3\pi}{5}$

$$A=\frac{4^2\left(\frac{3\pi}{5}\right)}{2}\approx 15.08\text{ m}^2$$

91. Given $r=20$ ft.; $\theta=\frac{\pi}{2}$

$$A=\frac{20^2\left(\frac{\pi}{2}\right)}{2}\approx 314.16\text{ ft.}^2$$

93. Given $r=5$ in.;

8 equal pieces $\Rightarrow$ central angles of

$$\frac{2\pi}{8}=\frac{\pi}{4}$$

$$A=\frac{5^2\left(\frac{\pi}{4}\right)}{2}=\frac{25\pi}{8}\approx 9.82\text{ in.}^2$$

95. Given $r = 5$ cm;

6 equal pieces $\Rightarrow$ central angles of

$\frac{2\pi}{6} = \frac{\pi}{3}$

Let $A =$ Area of 2 of the $\frac{\pi}{3}$ sectors

Then, $A = 2\left[\frac{5^2\left(\frac{\pi}{3}\right)}{2}\right] = 2\left[\frac{25\pi}{6}\right] \approx 26.18 \text{ cm}^2$

97. Given $r = 2.53$ in.^{2}

Number of rotations per minute =

360/min., $360(2\pi) = 720\pi$ radians/min.

a. Convert to seconds, i.e., divide by 60:

$\omega = \frac{720\pi}{60} = 12\pi$ rad/s

b. $v = r\omega$

$= 2.53(12\pi)$ in./s

≈ 95.38 in./s

99. The arc-lengths, s, of the two wheels are the same for any rotation. However, the central angles are different. Find s from the central angle measure (60°) of the smaller wheel. Then use the radius of the larger wheel to find its central angle:

Use $s = r\theta$. $s = 5.23\left(\frac{\pi}{3}\right) \approx 5.477$ ft.

Then, $5.477 = (8.16)\theta \Rightarrow \theta \approx 0.6712$

$\theta \approx 0.6712\left(\frac{180}{\pi}\right) \approx 38.5°$

End of Section 6.1

Chapter Six: Section 6.2
Solutions to Odd-numbered Exercises

1. Use Pythagoras' Formula to find the length of the side adjacent to θ:

$$\text{adj} = \sqrt{4^2 - \left(2\sqrt{2}\right)^2} = 2\sqrt{2}$$

$$\sin\theta = \frac{\text{opp}}{\text{hyp}} = \frac{2\sqrt{2}}{4} = \frac{\sqrt{2}}{2}$$

$$\csc\theta = \frac{1}{\sin\theta} = \frac{2}{\sqrt{2}} \cdot \frac{\sqrt{2}}{\sqrt{2}} = \sqrt{2}$$

$$\cos\theta = \frac{\text{adj}}{\text{hyp}} = \frac{2\sqrt{2}}{4} = \frac{\sqrt{2}}{2}$$

$$\sec\theta = \frac{1}{\cos\theta} = \frac{2}{\sqrt{2}} \cdot \frac{\sqrt{2}}{\sqrt{2}} = \sqrt{2}$$

$$\tan\theta = \frac{\text{opp}}{\text{adj}} = \frac{2\sqrt{2}}{2\sqrt{2}} = 1$$

$$\cot\theta = \frac{1}{\tan\theta} = 1$$

3. Use Pythagoras' Formula to find the length of the side opposite of θ:

$$\text{opp} = \sqrt{\left(6\sqrt{2}\right)^2 - 8^2} = \sqrt{72 - 64} = 2\sqrt{2}$$

$$\sin\theta = \frac{\text{opp}}{\text{hyp}} = \frac{2\sqrt{2}}{6\sqrt{2}} = \frac{1}{3}$$

$$\csc\theta = \frac{1}{\sin\theta} = 3$$

$$\cos\theta = \frac{\text{adj}}{\text{hyp}} = \frac{8}{6\sqrt{2}} = \frac{8\sqrt{2}}{6\sqrt{2} \cdot \sqrt{2}} = \frac{2\sqrt{2}}{3}$$

$$\sec\theta = \frac{1}{\cos\theta} = \frac{3}{2\sqrt{2}} = \frac{3\sqrt{2}}{4}$$

$$\tan\theta = \frac{\text{opp}}{\text{adj}} = \frac{2\sqrt{2}}{8} = \frac{\sqrt{2}}{4}$$

$$\cot\theta = \frac{1}{\tan\theta} = \frac{4}{\sqrt{2}} = 2\sqrt{2}$$

5. Use Pythagoras' Formula to find the length of the hypotenuse:

$$\text{hyp} = \sqrt{2(21)^2} = 21\sqrt{2}$$

$$\sin\theta = \frac{\text{opp}}{\text{hyp}} = \frac{21}{21\sqrt{2}} = \frac{1}{\sqrt{2}} = \frac{\sqrt{2}}{2}$$

$$\csc\theta = \frac{1}{\sin\theta} = \frac{2}{\sqrt{2}} = \sqrt{2}$$

$$\cos\theta = \frac{\text{adj}}{\text{hyp}} = \frac{21}{21\sqrt{2}} = \frac{1}{\sqrt{2}} = \frac{\sqrt{2}}{2}$$

$$\sec\theta = \frac{1}{\cos\theta} = \frac{2}{\sqrt{2}} = \sqrt{2}$$

$$\tan\theta = \frac{\text{opp}}{\text{adj}} = \frac{21}{21} = 1$$

$$\cot\theta = \frac{1}{\tan\theta} = 1$$

7. Use Pythagoras' Formula to find the length of the side adjacent to θ:

$$\text{adj} = \sqrt{7^2 - 5^2} = \sqrt{24} = 2\sqrt{6}$$

$$\sin\theta = \frac{\text{opp}}{\text{hyp}} = \frac{5}{7}$$

$$\csc\theta = \frac{1}{\sin\theta} = \frac{7}{5}$$

$$\cos\theta = \frac{\text{adj}}{\text{hyp}} = \frac{2\sqrt{6}}{7}$$

$$\sec\theta = \frac{1}{\cos\theta} = \frac{7}{2\sqrt{6}} \cdot \frac{\sqrt{6}}{\sqrt{6}} = \frac{7\sqrt{6}}{12}$$

$$\tan\theta = \frac{\text{opp}}{\text{adj}} = \frac{5}{2\sqrt{6}} = \frac{5\sqrt{6}}{12}$$

$$\cot\theta = \frac{1}{\tan\theta} = \frac{12}{5\sqrt{6}} = \frac{2\sqrt{6}}{5}$$

9. Use Pythagoras' Formula to find the length of the side opposite of θ:

$\text{opp} = \sqrt{13^2 - 5^2} = 12$

$\sin\theta = \frac{\text{opp}}{\text{hyp}} = \frac{12}{13}$

$\csc\theta = \frac{1}{\sin\theta} = \frac{13}{12}$

$\cos\theta = \frac{\text{adj}}{\text{hyp}} = \frac{5}{13}$

$\sec\theta = \frac{1}{\cos\theta} = \frac{13}{5}$

$\tan\theta = \frac{\text{opp}}{\text{adj}} = \frac{12}{5}$

$\cot\theta = \frac{1}{\tan\theta} = \frac{5}{12}$

11. Use Pythagoras' Formula to find the length of the side adjacent to θ:

$\text{adj} = \sqrt{65^2 - 33^2} = \sqrt{3136} = 56$

$\sin\theta = \frac{\text{opp}}{\text{hyp}} = \frac{33}{65}$

$\csc\theta = \frac{1}{\sin\theta} = \frac{65}{33}$

$\cos\theta = \frac{\text{adj}}{\text{hyp}} = \frac{56}{65}$

$\sec\theta = \frac{1}{\cos\theta} = \frac{65}{56}$

$\tan\theta = \frac{\text{opp}}{\text{adj}} = \frac{33}{56}$

$\cot\theta = \frac{1}{\tan\theta} = \frac{56}{33}$

13. $\theta = \frac{\pi}{2} - \frac{\pi}{3} = \frac{\pi}{6}$

$\sin\theta = \sin\frac{\pi}{6} = \frac{1}{2}$

$\csc\theta = \csc\frac{\pi}{6} = 2$

$\cos\theta = \cos\frac{\pi}{6} = \frac{\sqrt{3}}{2}$

$\sec\theta = \sec\frac{\pi}{6} = \frac{2}{\sqrt{3}} = \frac{2\sqrt{3}}{3}$

$\tan\theta = \tan\frac{\pi}{6} = \frac{\sqrt{3}}{3}$

$\cot\theta = \cot\frac{\pi}{6} = \sqrt{3}$

15. $\sin\frac{\pi}{4} = \frac{1}{\sqrt{2}} = \frac{\sqrt{2}}{2}$

$\csc\frac{\pi}{4} = \frac{2}{\sqrt{2}} = \sqrt{2}$

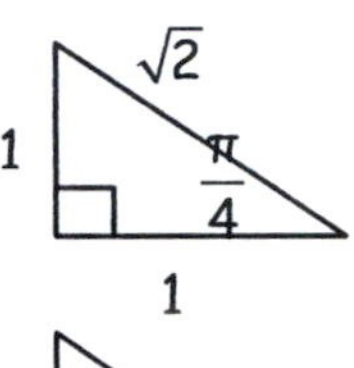

17. $\sec 60° = \sec\frac{\pi}{3} = \frac{2}{1}$

$= 2$

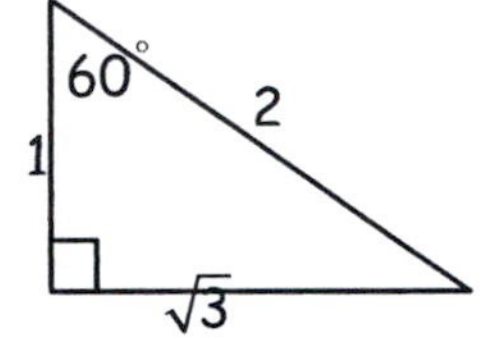

19. $\csc\frac{\pi}{6} = \frac{2}{1} = 2$

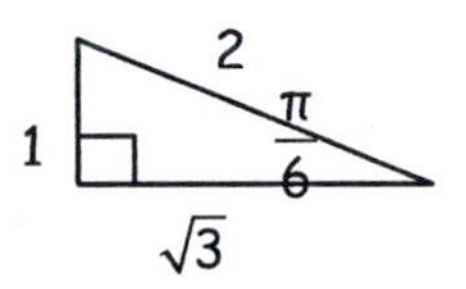

21. $\sec 5° \approx 1.0038$ (using a calculator)

$\tan 5° \approx 0.0875$ (using a calculator)

23. $\cot\frac{\pi}{3} = \frac{1}{\sqrt{3}} = \frac{\sqrt{3}}{3}$

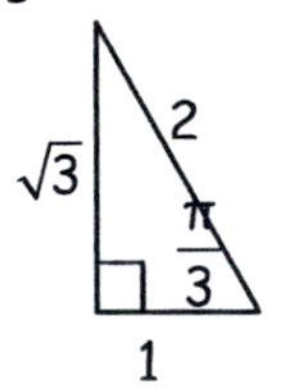

25. $\tan 87.2° \approx 20.4465$ (using a calculator)

For item 27-43, use a calculator. Be aware of the mode set for the calculator (i.e., radians or degrees).

27. $\sin 94° = 0.9976$ (be sure mode is set to degrees)

29. $\tan 146° \approx -0.6745$ (be sure mode is set to degrees)

31. $\sec 88° = \frac{1}{\cos 88°} \approx 28.6537$ (be sure mode is set to degrees)

33. $\tan\frac{4\pi}{5} = -0.7265$ (be sure mode is set to radians)

35. $\sin\frac{\pi}{8} = 0.3827$ (be sure mode is set to radians)

37. $\sec\frac{8\pi}{3} = -2$ $\left[\text{Use } \sec\frac{8\pi}{3} = \frac{1}{\cos\frac{8\pi}{3}}\right]$

(be sure mode is set to radians)

For items 39-43, use $1' = \frac{1}{60}^{\circ}$ and $1'' = \frac{1}{3600}^{\circ}$.

39. $38°\,54'\,19'' = 38 + \frac{54}{60} + \frac{19}{3600}$
$\approx 38.9053°$

41. $124°\,78'\,90'' = 124 + \frac{78}{60} + \frac{90}{3600}$
$\approx 125.325°$

43. $920°\,99'\,56'' = 920 + \frac{99}{60} + \frac{56}{3600}$
$\approx 921.6656°$

45. $\sin\theta = \frac{1}{\csc\theta}$
$= \frac{1}{8.7} \approx 0.1149$

47. $\tan\theta = \frac{1}{\cot\theta}$
$= \frac{1}{\frac{\sqrt{15}}{3}} = \frac{3}{\sqrt{15}}$
≈ 0.7746

49. $\sec\theta = \frac{1}{\cos\theta}$
$= \frac{1}{0.2} = 5$

51. If $\sin\theta = 0.8$, then, from $\csc\theta = \frac{1}{\sin\theta}$,
$\sin\theta = \frac{1}{0.8} = 1.25$.
[Answer: True]

53. If $\tan\theta = 4\frac{4}{9}$, then, from
$\cot\theta = \frac{1}{\tan\theta}$, $\cot\theta = \frac{1}{\frac{40}{9}} = \frac{9}{40} = 0.225$.
[Answer: True]

55. If $\cos\theta = 0.75$, then, from
$\sec\theta = \frac{1}{\cos\theta}$, $\sec\theta = \frac{4}{3}$ $\left(\text{not } \frac{8}{3}\right)$.
[Answer: False]

57. Let h = height in feet. Change the measure to decimal form and use the degree mode on your calculator.
$80°\,55'24'' \approx 80.9233°$

Use $\tan\theta = \frac{opp}{adj}$ and 40 yd. = 120 ft.:

$\tan 80.9233 = \frac{h}{120}$
$h = 120(6.2595) \approx 751.19$ ft.

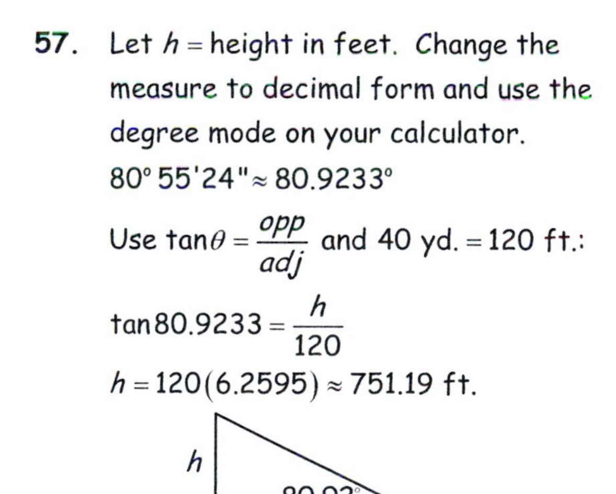

59. Using the given information, it is helpful to first draw a sketch of the described situation.

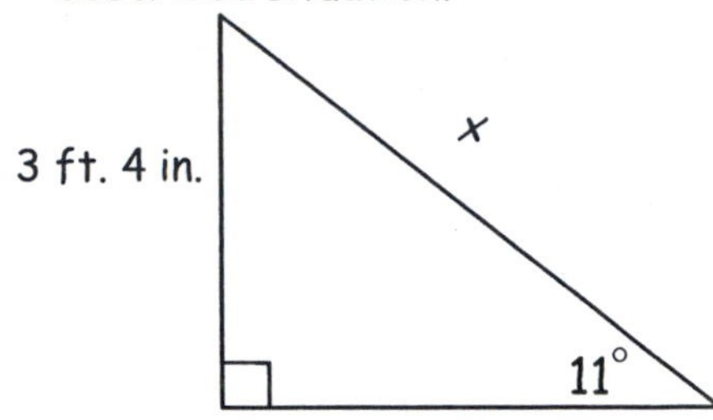

We can then set up the following equation, using $\sin\theta = \frac{\text{opp}}{\text{hyp}}$, and solve for x.

$$\sin 11^\circ = \frac{3\frac{1}{3}}{x}$$

$$x = \frac{3\frac{1}{3}}{\sin 11^\circ}$$

Using a calculator, we find $x \approx 17.47$ feet

61. Using the given information, it is helpful to first draw a sketch of the described situation.

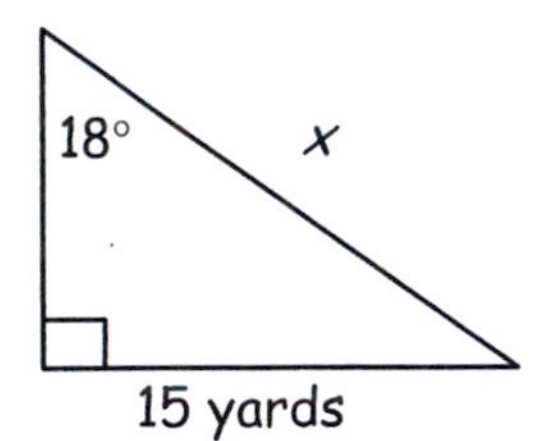

We can then set up the following equation, using $\sin\theta = \frac{\text{opp}}{\text{hyp}}$, and solve for x.

$$\sin 18^\circ = \frac{15\text{ yards}}{x}$$

$$= \frac{15\text{ yards}}{\sin 18^\circ}$$

Using a calculator, $x \approx 48.54$ yards.

63.

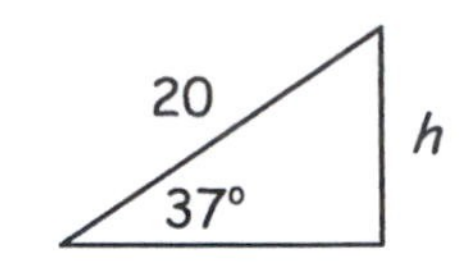

$$\sin 37^\circ = \frac{h}{20}$$

$$h = 20\sin 37^\circ$$

$$\approx 20(0.6018)$$

$$\approx 12.04\text{ m}$$

65. $\tan 78.5^\circ = \frac{h}{64}$

$$h = 64\tan 78.5^\circ$$

$$\approx 64(4.9152)$$

$$\approx 314.57\text{ ft.}$$

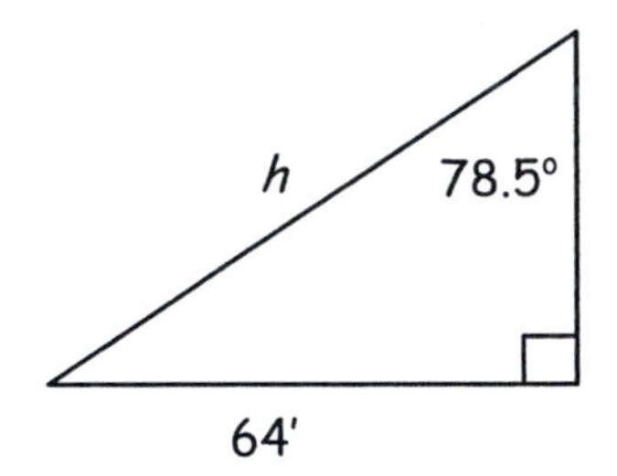

67. Using the given information, it is helpful to just draw a sketch of the described situation.

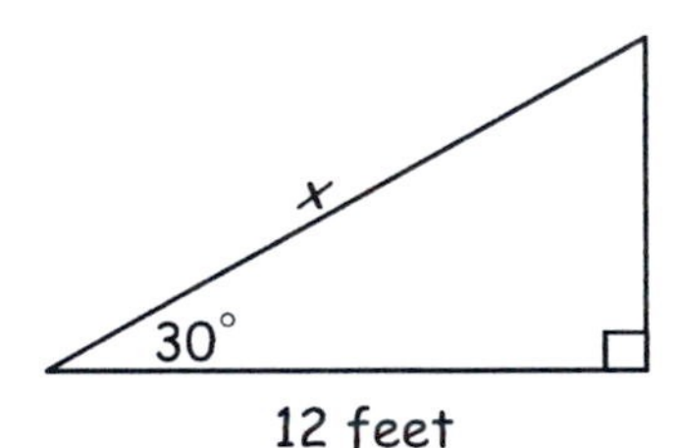

We can then set up the following equation, using $\cos\theta = \frac{\text{adj}}{\text{hyp}}$, and solve for x.

$$\cos 30^\circ = \frac{12\text{ feet}}{x}$$

$$x = \frac{12\text{ feet}}{\cos 30^\circ}$$

$$x = \frac{12}{\frac{\sqrt{3}}{2}}\text{ feet} \approx 13.86\text{ feet.}$$

69. Using the given information, it is helpful to just draw a diagram of the described situation.

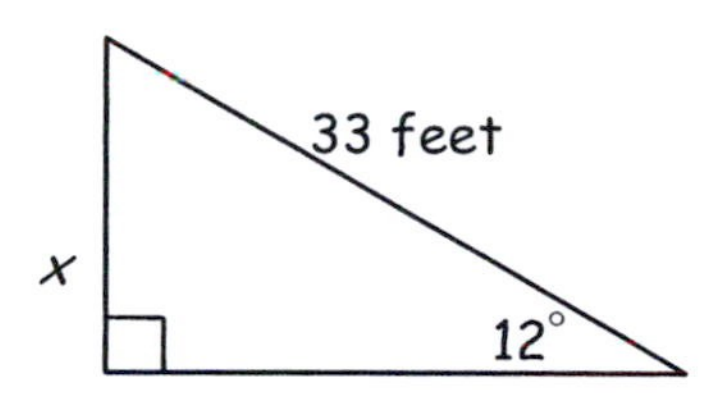

We can then set up the following equation, using $\sin\theta = \frac{\text{opp}}{\text{hyp}}$, and solve for x.

$$\sin 12^\circ = \frac{x}{33\,\text{feet}}$$

$$x = 33\sin 12^\circ\,\text{feet}$$

Using a calculator, $x \approx 6.86$ feet.

For 71 - 73, use the following:

$$h = \frac{d\tan\alpha\tan\beta}{\tan\alpha - \tan\beta}$$

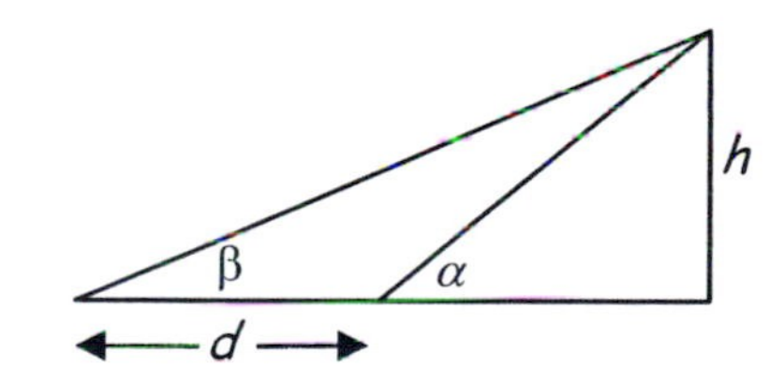

71.
$$h = \frac{68.2\tan 19^\circ \tan 9^\circ}{\tan 19^\circ - \tan 9^\circ}$$
$$\approx \frac{68.2(0.3443)(0.1584)}{0.3443 - 0.1584}$$
$$\approx \frac{3.7194}{0.1859} \approx 20\text{ m}$$

73. Before applying the formula, change the angle measure to decimal form:

$$\beta = 46 + \frac{57}{60} + \frac{12}{3600} \approx 46.9533^\circ$$

$$\alpha = 55 + \frac{37}{60} + \frac{70}{3600} \approx 55.6361^\circ$$

Also, given: $d = 800$

$$h \approx \frac{800\tan 55.6361^\circ \tan 46.9533^\circ}{\tan 55.6361^\circ - \tan 46.9533^\circ}$$
$$\approx 3196.80\text{ ft.}$$

End of Section 6.2

Chapter Seven: Section 6.3 Solutions to Odd-numbered Exercises

1. Given: $\theta = 45°$

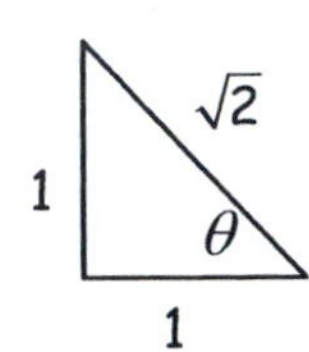

$\sin 45° = \frac{1}{\sqrt{2}} = \frac{\sqrt{2}}{2}$

$\cos 45° = \frac{1}{\sqrt{2}} = \frac{\sqrt{2}}{2}$

$\tan 45° = \frac{1}{1} = 1$

$\csc 45° = \frac{\sqrt{2}}{1} = \sqrt{2}$

$\sec 45° = \frac{\sqrt{2}}{1} = \sqrt{2}$

$\cot 45° = \frac{1}{1} = 1$

3. Given: $\theta = 60°$

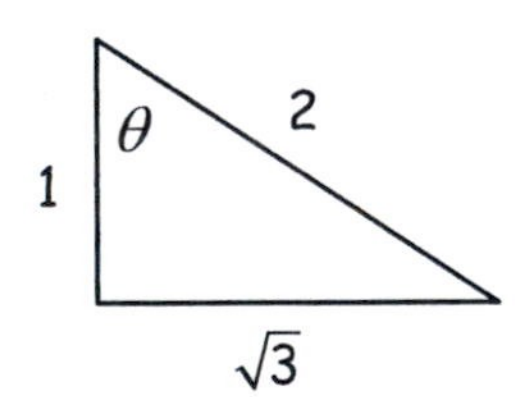

$\sin 60° = \frac{\sqrt{3}}{2}$

$\cos 60° = \frac{1}{2}$

$\tan 60° = \frac{\sqrt{3}}{1} = \sqrt{3}$

$\csc 60° = \frac{2}{\sqrt{3}} = \frac{2\sqrt{3}}{3}$

$\sec 60° = \frac{2}{1} = 2$

$\cot 60° = \frac{1}{\sqrt{3}} = \frac{\sqrt{3}}{3}$

5. Given: $\theta = \frac{5\pi}{2}$; then, $\theta' = \frac{\pi}{2}$

$\sin\frac{5\pi}{2} = \sin\frac{\pi}{2} = 1$

$\cos\frac{5\pi}{2} = \cos\frac{\pi}{2} = 0$

$\tan\frac{5\pi}{2} = \tan\frac{\pi}{2}$ is undefined

$\csc\frac{5\pi}{2} = \csc\frac{\pi}{2} = 1$

$\sec\frac{5\pi}{2} = \sec\frac{\pi}{2}$ is undefined

$\cot\frac{5\pi}{2} = \cot\frac{\pi}{2} = 0$

For items 7 - 15, first, find θ' for the given angle θ.

7. Given $\theta = 305°$; then $\theta' = -55°$.

$\sin 305° \approx -0.8192$

$\cos 305° \approx 0.5736$

$\tan 305° \approx -1.4281$

$\csc 305° = \frac{1}{\sin 305°} \approx -1.2208$

$\sec 305° = \frac{1}{\cos 305°} \approx 1.7434$

$\cot 305° = \frac{1}{\tan 305°} \approx -0.7002$

9. Given $\theta = 6\pi$; then $\theta' = 0$.

$\sin 6\pi = \sin 0 = 0$

$\cos 6\pi = \cos 0 = 1$

$\tan 6\pi = \tan 0 = 0$

$\csc 6\pi$ is undefined

$\sec 6\pi = \frac{1}{\cos 6\pi} = 1$

$\cot 6\pi$ is undefined

11. Given $\theta = \frac{3\pi}{2}$; then $\theta' = \frac{\pi}{2}$.

$\sin\frac{3\pi}{2} = -\sin\left(\frac{\pi}{2}\right) = -1$

$\cos\frac{3\pi}{2} = \cos\left(\frac{\pi}{2}\right) = 0$

$\tan\frac{3\pi}{2}$ is undefined

$\csc\frac{3\pi}{2} = \frac{1}{\sin\left(\frac{3\pi}{2}\right)} = \frac{1}{-\sin\left(\frac{\pi}{2}\right)} = -1$

$\sec\frac{3\pi}{2}$ is undefined

$\cot\frac{3\pi}{2} = \cot\left(\frac{\pi}{2}\right) = \frac{\cos\left(\frac{\pi}{2}\right)}{\sin\left(\frac{\pi}{2}\right)} = 0$

13. Given $\theta = \frac{5\pi}{4}$; then $\theta' = \frac{\pi}{4}$.

Note that θ is a third quadrant angle:

$\sin\frac{5\pi}{4} = -\sin\frac{\pi}{4} = -\frac{\sqrt{2}}{2}$

$\cos\frac{5\pi}{4} = -\cos\frac{\pi}{4} = -\frac{\sqrt{2}}{2}$

$\tan\frac{5\pi}{4} = \tan\frac{\pi}{4} = 1$

$\csc\frac{5\pi}{4} = \frac{1}{\sin\left(\frac{5\pi}{4}\right)} = \frac{1}{-\sin\frac{\pi}{4}} = -\sqrt{2}$

$\sec\frac{5\pi}{4} = \frac{1}{\cos\left(\frac{5\pi}{4}\right)} = \frac{1}{-\cos\frac{\pi}{4}} = -\sqrt{2}$

$\cot\frac{5\pi}{4} = \cot\frac{\pi}{4} = \frac{-\cos\frac{\pi}{4}}{-\sin\frac{\pi}{4}} = 1$

15. Given $\theta = -445°$; then $\theta' = 85°$ (from $445 - 360$).

Note: θ is a fourth quadrant angle:

$\sin(-445°) \approx -0.9962$

$\cos(-445°) \approx 0.0872$

$\tan(-445°) \approx -11.4301$

$\csc(-445°) \approx -1.0038$

$\sec(-445°) \approx 11.4737$

$\cot(-445°) \approx -0.0875$

17. Given $\theta = \frac{9\pi}{2}$. The terminal side is on the positive y-axis. Then, $\theta' = \frac{\pi}{2}$.

$\left(\text{from } \frac{9\pi}{2} - 4\pi\right)$

19. Given $\theta = \frac{5\pi}{4}$. The terminal side is in the third quadrant. Then, $\theta' = \frac{\pi}{4}$.

$\left(\text{from } \frac{5\pi}{4} - \pi\right)$

21. Given $\varphi = 313°$. The terminal side is in the fourth quadrant. Then, $\varphi' = 47°$. $(\text{from } 360° - 313°)$

23. Given $\varphi = -168°$. The terminal side is in the third quadrant. Then, $\varphi' = 12°$. $(\text{from } 180° - 168°)$

25. Given $\varphi = 216°$. The terminal side is in the third quadrant. Then, $\varphi' = 36°$. $(\text{from } 216° - 180°)$

27. Given $\varphi = -330°$. The terminal side is in the first quadrant. Then, $\varphi' = 30°$. $(\text{from } 360° - 330°)$

29. Given $\varphi = 718°$. The terminal side is in the first quadrant. Then, $\varphi' = 2°$. $\left(\text{from } 2(360°) - 718°\right)$

31. Given $\sin\theta > 0$ ($\Rightarrow y$-coordinate > 0) and $\tan\theta < 0$ ($\Rightarrow x$-coordinate < 0), then, the terminal side must lie in the second quadrant.

33. Given $\tan\theta > 0$ ($\Rightarrow (x, y)$ in I or III) and $\sec\theta > 0$ ($\Rightarrow x$-coordinate > 0), then, the terminal side must lie in quadrant I.

35. Given $\sec\theta < 0$ ($\Rightarrow x$-coordinate < 0) and $\csc\theta < 0$ ($\Rightarrow y$-coordinate < 0), then, (x, y) lies in the third quadrant. The terminal side must lie in the third quadrant.

37. Given $\cot\theta > 0$ ($\Rightarrow x$ and y same sign) and $\cos\theta < 0$ ($\Rightarrow x$-coordinate < 0), then, y-coordinate < 0. So, (x, y) must lie in quadrant III. The terminal side must lie in the third quadrant.

39. $\theta = 300°$ matches c; $\theta' = 60°$, since $360° - 300° = 60°$

41. $\theta = -135°$ matches b; $\theta' = 45°$, since $180° - 135° = 45°$

43. $\theta = -120°$ matches c; $\theta' = 60°$, since $180° - 120° = 60°$

45. $\theta = 510°$ matches a; $\theta' = 30°$, since $510° - 360° = 150°$ and $180° - 150° = 30°$

47. $\theta = 855°$ matches b; $\theta' = 45°$, since $855° - 2(360°) = 135°$ and $180° - 135° = 45°$

49. The terminal side of $98°$ lies in the second quadrant. $\theta' = 82°$.
$\tan 98° = -\tan 82° = -7.1154$

51. The terminal side of $-60°$ lies in the fourth quadrant. $\theta' = 60°$.
$\cos\text{-}60° = \cos 60° = \dfrac{1}{2}$

53. The terminal side of $\dfrac{5\pi}{2}$ lies along the positive y-axis. $\theta' = \dfrac{\pi}{2}$.
$\cos\dfrac{5\pi}{2} = \cos\dfrac{\pi}{2} = 0$

55. The terminal side of $\dfrac{7\pi}{6}$ lies in the third quadrant. $\theta' = \dfrac{\pi}{6}$.
$\cos\dfrac{7\pi}{6} = -\cos\dfrac{\pi}{6} = -\dfrac{\sqrt{3}}{2}$

57. The terminal side of $\dfrac{6\pi}{5}$ lies in the third quadrant. $\theta' = \dfrac{\pi}{5}$.
$\cos\dfrac{6\pi}{5} = -\cos\dfrac{\pi}{5} \approx -0.8090$

59. The terminal side of $\dfrac{3\pi}{2}$ lies along the negative y-axis. $\theta' = \dfrac{\pi}{2}$.
$\tan\dfrac{3\pi}{2}$ is undefined.

61. The terminal side of $\frac{7\pi}{4}$ lies in the fourth quadrant. $\theta' = \frac{\pi}{4}$.

$\sin\frac{7\pi}{4} = -\sin\frac{\pi}{4} = -\frac{\sqrt{2}}{2}$

63. The terminal side of 105° lies in the second quadrant. $\theta' = 75°$.

$\sin 105° = \sin 75° \approx 0.9659$

65. $\sec\frac{\pi}{6} = \csc\left(\frac{\pi}{2} - \frac{\pi}{6}\right)$

$= \csc\frac{\pi}{3}$ (Answer: a)

67. $\cos 87° = \sin(90° - 87°)$

$= \sin 3°$ (Answer: e)

69. $\sec\frac{\pi}{2} = \csc\left(\frac{\pi}{2} - \frac{\pi}{2}\right) = \csc(0) \Rightarrow$ undefined

71. $\cos\left(-\frac{3\pi}{4}\right) = \cos\left(\frac{\pi}{2} - \left(-\frac{3\pi}{4}\right)\right)$

$= \sin\frac{5\pi}{4}$

$= -\sin\frac{\pi}{4} = -\frac{\sqrt{2}}{2}$

73. $\cot 313° = \tan(90° - 313°)$

$= \tan(-223°)$

$= -\tan 43° \approx -0.9325$

75. $\csc(-168°) = \sec(90° - (-168°))$

$= \sec(258°)$

$= -\sec 78° \approx -4.8097$

77. $\sec 216° = \csc(90° - 216°)$

$= \csc(-126°)$

$= -\csc 54° \approx -1.2361$

79. $\cos(-15°) = \sin(90° - (-15°))$

$= \sin 105°$

$= \sin 75° \approx 0.9659$

81. $\tan(-105°) = \cot(90° - (-105°))$

$= \cot 195°$

$= \cot 15° \approx 3.7321$

83. $\tan\theta = \frac{\sin\theta}{\cos\theta} = \frac{0.978}{0.208} \approx 4.702$

$\cot\theta = \frac{1}{\tan\theta} \approx 0.213$

85. $\tan\theta = \frac{\sin\theta}{\cos\theta} = \frac{-0.966}{-0.259} \approx 3.730$

$\cot\theta = \frac{1}{\tan\theta} \approx 0.268$

87. $\tan\theta = \frac{\sin\theta}{\cos\theta} = \frac{-0.699}{-0.743} \approx -0.941$

$\cot\theta = \frac{1}{\tan\theta} \approx -1.063$

89. Prove: $\sin\theta = \cos\left(\frac{\pi}{2} - \theta\right)$

Using algebra,

$\theta = \theta + \frac{\pi}{2} - \frac{\pi}{2} = \frac{\pi}{2} - \left(\frac{\pi}{2} - \theta\right)$

Then, using the original cofunction identity $\cos\varphi = \sin\left(\frac{\pi}{2} - \varphi\right)$,

$\sin\theta = \sin\left(\frac{\pi}{2} - \left(\frac{\pi}{2} - \theta\right)\right)$

$= \cos\left(\frac{\pi}{2} - \theta\right)$

Prove: $\sec\theta = \csc\left(\frac{\pi}{2} - \theta\right)$

Using algebra,

$\theta = \theta + \frac{\pi}{2} - \frac{\pi}{2} = \frac{\pi}{2} - \left(\frac{\pi}{2} - \theta\right)$

Then, using the original cofunction identity $\csc\varphi = \sec\left(\frac{\pi}{2} - \varphi\right)$,

$$\sec\theta = \sec\left(\frac{\pi}{2}-\left(\frac{\pi}{2}-\theta\right)\right)$$
$$= \csc\left(\frac{\pi}{2}-\theta\right)$$

Prove: $\tan\theta = \cot\left(\frac{\pi}{2}-\theta\right)$

Again, using algebra,

$\theta = \theta + \frac{\pi}{2} - \frac{\pi}{2} = \frac{\pi}{2} - \left(\frac{\pi}{2}-\theta\right)$

Then, using the original cofunction identity $\cot\varphi = \tan\left(\frac{\pi}{2}-\varphi\right)$,

$$\tan\theta = \tan\left(\frac{\pi}{2}-\left(\frac{\pi}{2}-\theta\right)\right)$$
$$= \cot\left(\frac{\pi}{2}-\theta\right)$$

91. Given $\csc\theta = 0.6 = \frac{3}{5}$.

The implication of this ratio would render an associated right triangle with $hyp = 3$ and $opp = 5$. This presents an impossible situation because *hyp* cannot be less than either of the other two sides of the triangle.

93. Given $\sec\theta = 0.3 = \frac{3}{10}$ with terminal side in the fourth quadrant.

The implication of this ratio would render an associated right triangle with $hyp = 3$ and $adj = 10$. This presents an impossible situation because *hyp* cannot be less than either of the other two sides of the triangle.

95. $\csc\frac{8\pi}{3} = \sec\left(\frac{\pi}{2}-\frac{8\pi}{3}\right)$

$= \sec\left(-\frac{13\pi}{6}\right)$

≈ 1.1547

97. $\cos\left(-\frac{5\pi}{3}\right) = \sin\left(\frac{\pi}{2}-\left(-\frac{5\pi}{3}\right)\right)$

$= \sin\left(\frac{\pi}{2}+\frac{5\pi}{3}\right)$

$= \sin\left(\frac{13\pi}{6}\right)$

$= \frac{1}{2}$

99. $\sec(-315^\circ) = \csc(90^\circ - (-315^\circ))$

$= \csc(405^\circ)$

≈ 1.4142

End of Section 6.3

Chapter Six: Section 6.4
Solutions to Odd-numbered Exercises

1. Sketch the graph of cotangent.
Note that the cotangent is not defined where tangent is zero.
$\tan x = 0$ for $x = n\pi$, for any integer n.
The asymptotes for cotangent occur at $x = n\pi$. A plot for special x-values in the first and second quadrants will illustrate the pattern of the graph.

x	0	$\frac{\pi}{6}$	$\frac{\pi}{4}$	$\frac{\pi}{3}$	$\frac{\pi}{2}$	$\frac{2\pi}{3}$	$\frac{3\pi}{4}$	$\frac{5\pi}{6}$
$\cot x$	und	$\sqrt{3}$	1	$\frac{\sqrt{3}}{3}$	0	$-\frac{\sqrt{3}}{3}$	-1	$-\sqrt{3}$

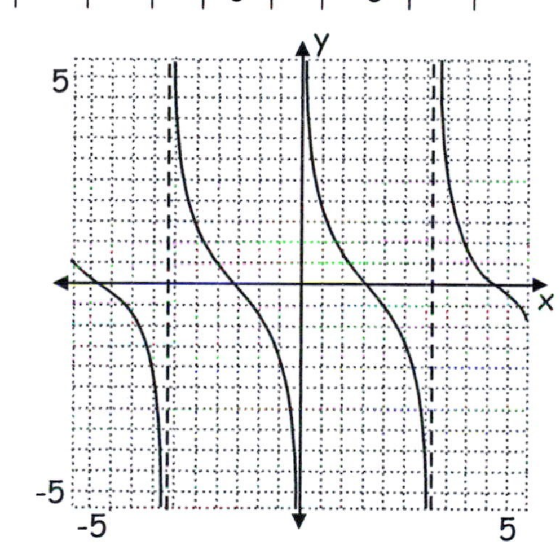

3. For simple harmonic motion, use $g(t) = a\cos bt$. When $t = 0$, the initial position is at -3; i.e., $g(0) = -3$. The amplitude is $|-3| = 3$. The period is the time for one oscillation, so period $P = 2$.
Using $P = \frac{2\pi}{b} \Rightarrow 2 = \frac{2\pi}{b} \Rightarrow b = \pi$.
Then, $g(t) = -3\cos \pi t$

5. Assuming that the ball is bounced from Marcel's waist each time, the distance traveled between "peaks" is 3 feet, so $a = \frac{3}{2} = 1.5$. The ball travels 6 ft. through one cycle, so at Marcel's rate of bounce (given as 10 ft./s), the frequency (number of cycles per second) is 10/6 or 5/3 cycles per second. Using $P = \frac{1}{f}$, the period $P = \frac{3}{5}$ s (the time for one cycle).
Now, the model $g(t) = a\cos bt$ has period $P = \frac{2\pi}{b}$. Applied to this problem,
$\frac{2\pi}{b} = \frac{3}{5} \Rightarrow b = \frac{10\pi}{3}$
Then, $g(t) = 1.5\cos\frac{10\pi}{3}t$

7. Given $y = \frac{3}{2}\sin x$:
Amplitude $A = \frac{3}{2}$; Period $P = \frac{2\pi}{1} = 2\pi$
There is no phase shift.

9. Given $y = \sin(x-5)$:
Amplitude $A = 1$; Period $P = 2\pi$.
The phase shift is 5 units to the right.

11. Given $2y = \cos x$:
Amplitude $A = \frac{1}{2}$; Period $P = 2\pi$
There is no phase shift.

13. Given $y = \frac{2}{3}\sin x$:

Amplitude $A = \frac{2}{3}$; Period $P = 2\pi$

There is no phase shift.

15. Given $y = -3\cos\left(\frac{1}{2}x\right)$:

Amplitude $A = |-3| = 3$;

Period $P = \frac{2\pi}{\frac{1}{2}} = 4\pi$

There is no phase shift.

17. Given $y = \cos(3\pi\theta - 2)$:

Rewrite: $y = \cos 3\pi\left(\theta - \frac{2}{3\pi}\right)$

Amplitude $A = 1$;

Period $P = \frac{2\pi}{3\pi} = \frac{2}{3}$

There is a $\frac{2}{3\pi}$ unit phase shift to the right.

19. Given $y = 7\cos\left(x \cdot \frac{\pi}{2} + \frac{3}{2}\right)$:

Rewrite: $y = 7\cos\frac{\pi}{2}\left(x + \frac{3}{\pi}\right)$

Amplitude $A = 7$;

Period $P = \frac{2\pi}{\pi/2} = 4$

There is a $\frac{3}{\pi}$ unit phase shift to the left.

21. Given $y = 2 - \frac{3}{4}\sin(-3 + x)$:

Amplitude $A = \left|-\frac{3}{4}\right| = \frac{3}{4}$; Period $P = 2\pi$.

There is a shift up of 2 units.

The phase shift is 3 units to the right.

23. Sketch $g(x) = -2\sin 5x$:

Sine curve amplitude $A = |-2| = 2$.

Period $P = \frac{2\pi}{5}$.

Cycle starting point: 0

Cycle ending point: $\frac{2\pi}{5}$

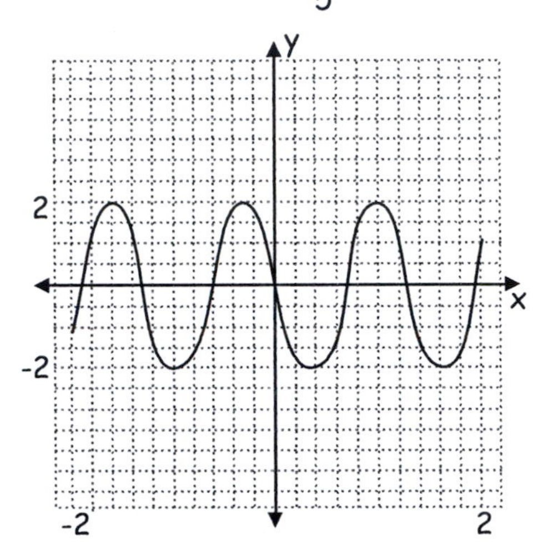

25. Sketch $g(x) = 3\sin(x - 2\pi)$:

Sine curve amplitude $A = 3$.

Period $P = 2\pi$.

Phase shift of 2π units to the right;

Cycle starting point: 0

Cycle ending point: 2π

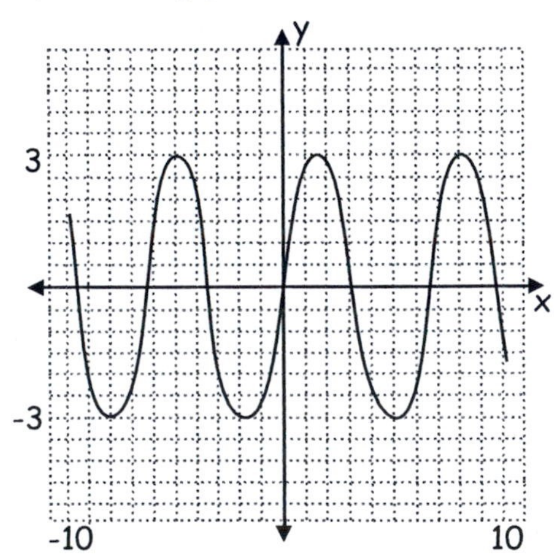

27. Sketch $f(x) = -5\cot \pi x$:
Start with the cotangent function.
Vertical asymptotes at ... $-\pi, 0, \pi, 2\pi, \ldots$.
The function $\cot \pi x$ has vertical asymptotes at ... $-2, -1, 0, 1, 2, 3, \ldots$.
The coefficient -5 "stretches" the graph by a factor of 5, and the negative sign reflects the cotangent graph through the x-axis.
A starting point of a cycle is at 0 with the ending point at 1.

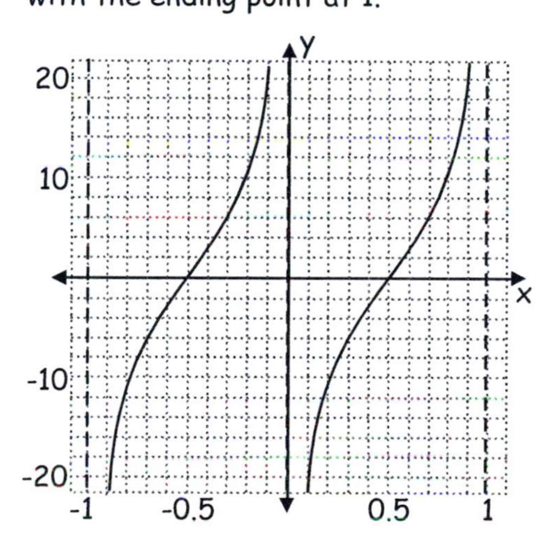

29. Sketch $f(x) = 4\cos\left(\frac{3x}{2} + \frac{\pi}{2}\right)$:

Rewrite as $f(x) = 4\cos\left(\frac{3}{2}\left(x + \frac{\pi}{3}\right)\right)$

Start with the cosine function.
The amplitude is 4.

The period $P = \dfrac{2\pi}{\frac{3}{2}} = \dfrac{4\pi}{3}$.

There is a phase shift of $\frac{\pi}{3}$ units to the left. A cycle starts at $-\frac{\pi}{3}$ and ends at $-\frac{\pi}{3} + \frac{4\pi}{3} = \pi$.

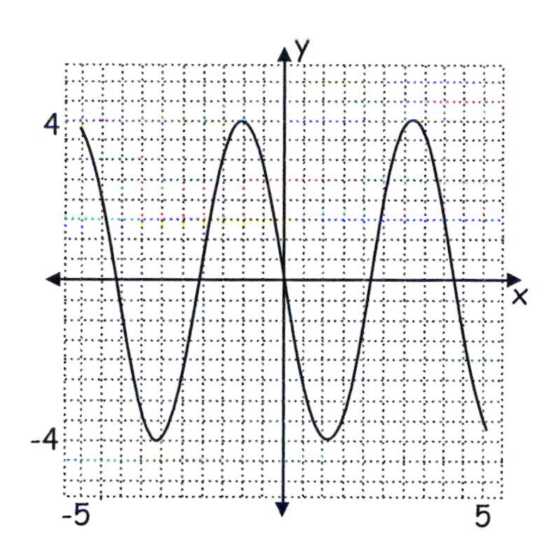

31. Sketch $g(x) = 2\cos(4x - 2)$:

Rewrite as $g(x) = 2\cos 4\left(x - \frac{1}{2}\right)$

Start with the cosine function.
The amplitude is 2.

The period $P = \frac{2\pi}{4} = \frac{\pi}{2}$.

There is a phase shift of $\frac{1}{2}$ unit to the right. A cycle starts at $\frac{1}{2}$ and ends at $\frac{1}{2} + \frac{\pi}{2} = \frac{1+\pi}{2} \approx 2.07$.

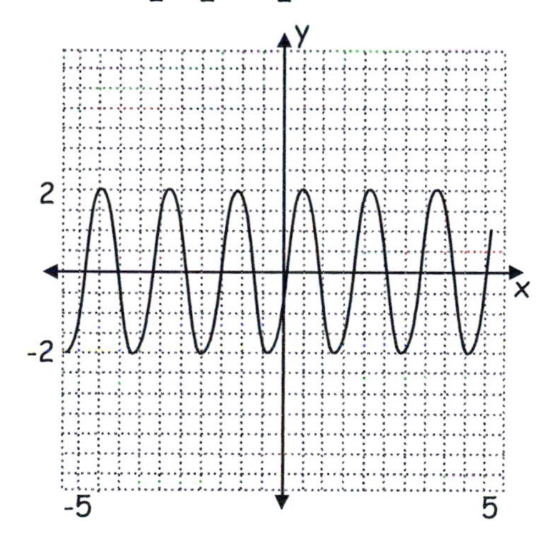

33. Sketch $g(x) = 3\sin 4x$:
Start with the sine function.
The amplitude $A = 3$.
Period $P = \frac{2\pi}{4} = \frac{\pi}{2}$.
There is no phase shift.
Cycle starting point: 0
Cycle ending point: $\frac{\pi}{2}$

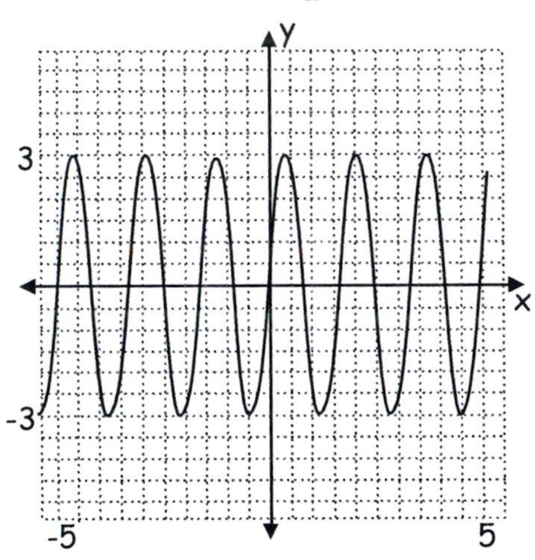

35. Sketch $f(x) = -\sin 2\pi x$:
Start with the sine function.
The amplitude $A = |-1| = 1$.
The graph is a reflection through the x-axis.
Period $P = \frac{2\pi}{2\pi} = 1$.
There is no phase shift.
Cycle starting point: 0
Cycle ending point: 1

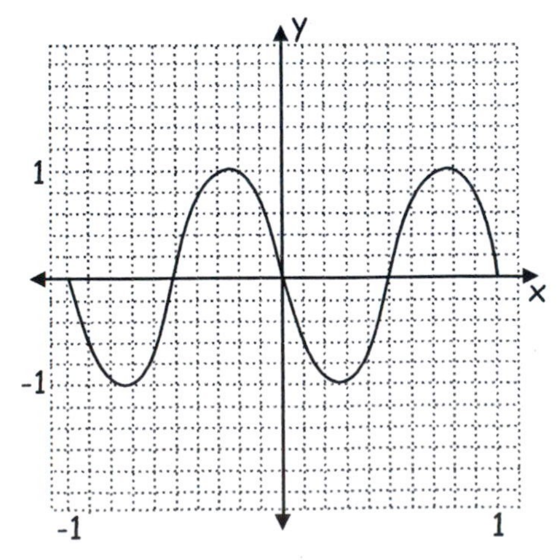

37. Sketch $g(x) = 1 + \sin(x - 2\pi)$:
Begin with the sine function.
The amplitude $A = 1$.
There is a shift of 1 unit up and a phase shift right of 2π units.
The period is 2π.
The left end-point of a cycle is 2π.
The right end-point of the cycle is 4π.

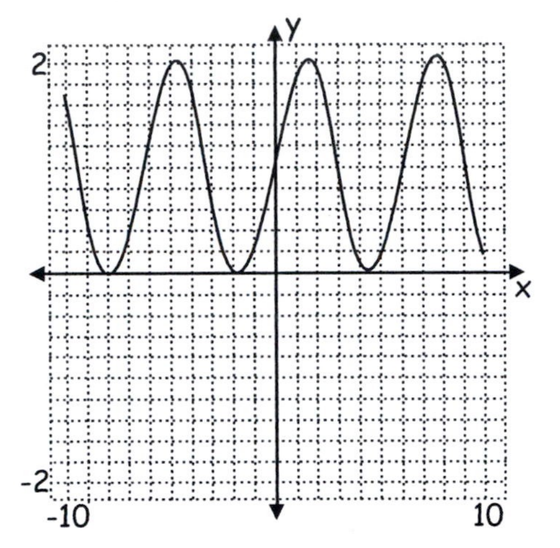

39. Sketch $f(x) = 4 + \csc\left(1 - \frac{5\pi}{4}x\right)$:

Begin with the graph of the cosecant function.
The graph is shifted up 4 units.
There are vertical asymptotes where $1 - \frac{5\pi}{4}x = n\pi$, where n is an integer,

i.e., where $\sin\left(1 - \frac{5\pi}{4}x\right) = 0$.

For example, setting $1 - \frac{5\pi}{4}x = 0 \Rightarrow$

$x = \frac{4}{5\pi} \approx 0.255$

The period is $\frac{2\pi}{\frac{5\pi}{4}} = \frac{8}{5} = 1.6$.

The left end-point of a cycle is at an asymptote:

$x = \frac{1}{\frac{5\pi}{4}} = \frac{4}{5\pi} \approx 0.255.$

The right end-point of the cycle is

$\frac{4}{5\pi} + \frac{8}{5} = \frac{4+8\pi}{5\pi} \approx 1.85.$

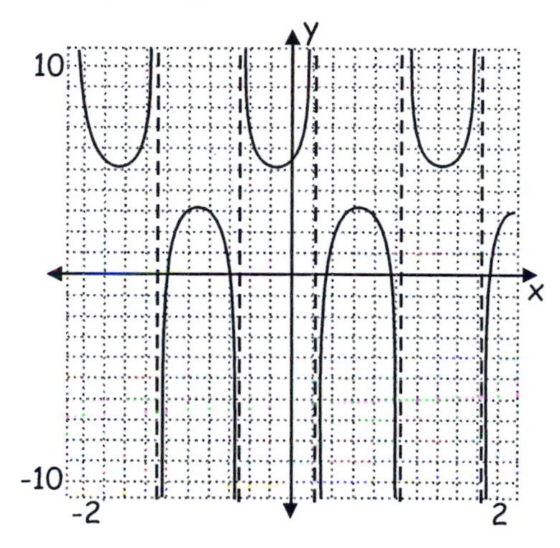

41. Sketch $g(x) = 1 + \tan\left(\pi x - \frac{\pi}{4}\right)$:

Begin with the graph of the tangent function. There is a shift up of one unit. Rewrite as $1 + \tan\left(\pi\left(x - \frac{1}{4}\right)\right)$

There is a phase shift of $\frac{1}{4}$ unit to the right.

Recall that $\tan\theta$ is not defined for $\theta = \frac{\pi}{2}$ and the odd multiples of $\frac{\pi}{2}$

i,e., $\pm\left\{\frac{\pi}{2}, \frac{3\pi}{2}, \frac{5\pi}{2}, \ldots\right\}$.

So, there are vertical asymptotes where $\pi x - \frac{\pi}{4} =$ any odd multiple of $\frac{\pi}{2}$.

For a left end-point, set $\pi x - \frac{\pi}{4} = \frac{\pi}{2}$ and solve for x: $x = \frac{3}{4}$

The left end-point of a cycle occurs at an asymptote.

For the right end-point, set $\pi x - \frac{\pi}{4} = \frac{3\pi}{2}$, and solve for x: $x = \frac{7}{4}$

The right end-point of the cycle occurs at an asymptote.

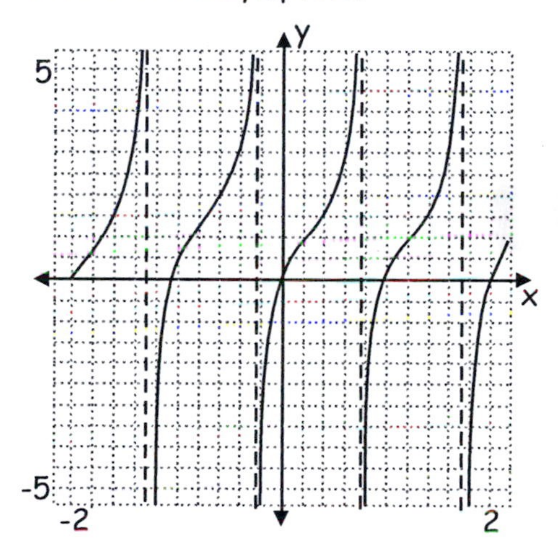

43. Sketch $g(x) = 2 - \sin\left(2x - \frac{\pi}{4}\right)$:

Rewrite as $g(x) = 2 - \sin\left(2\left(x - \frac{\pi}{8}\right)\right)$

Begin with the -sine function.

The amplitude is $|-1| = 1$.

The period $P = \frac{2\pi}{2} = \pi$.

There is a phase shift of $\frac{\pi}{8}$ units to the right. The left end-point of a cycle starts at $\frac{\pi}{8}$ and ends at $\frac{\pi}{8} + \pi = \frac{9\pi}{8}$.

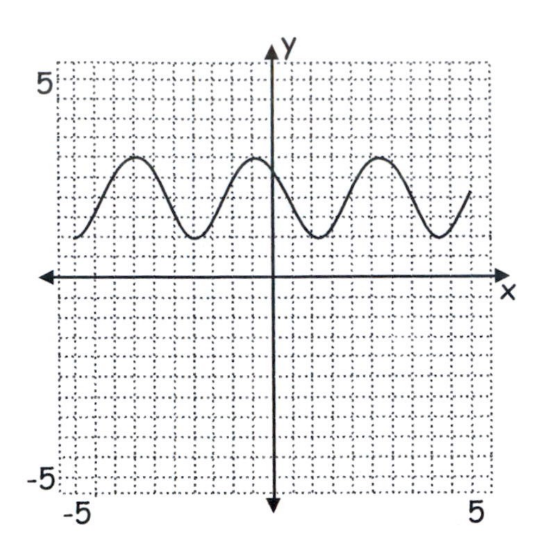

45. Sketch $f(x)=\frac{1}{2}-5\sin\left(\frac{1}{2}x-\frac{\pi}{2}\right)$:

Rewrite as $f(x)=\frac{1}{2}-5\sin\left(\frac{1}{2}(x-\pi)\right)$

Begin with the $-$sine function.

The amplitude is $|-5|=5$.

The period $P=\frac{2\pi}{\frac{1}{2}}=4\pi$.

There is a phase shift of $\frac{\frac{\pi}{2}}{\frac{1}{2}}=\pi$ units to the right. The left end-point of a cycle is at π with its right end-point at $\pi+4\pi=5\pi$.

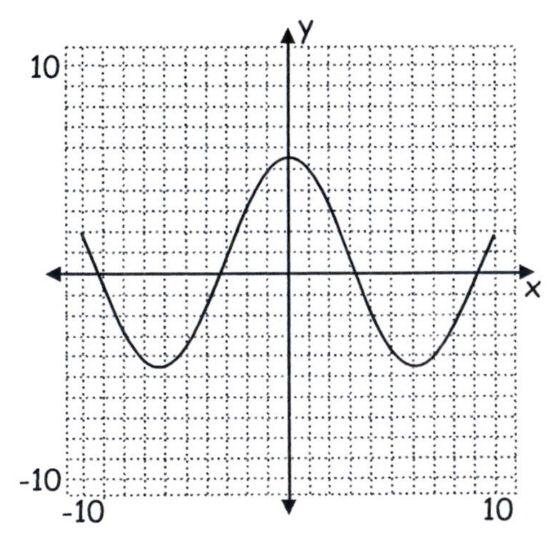

47. Sketch $g(x)=2+\frac{5}{6}\sec\left(\frac{1}{2}x-\pi x\right)$.

Rewrite as $g(x)=2+\frac{5}{6}\sec\left(\left(\frac{1}{2}-\pi\right)x\right)$.

Begin with the graph of the secant function, which has asymptotes where $\cos\theta=0$ $\left(\text{since } \sec\theta=\frac{1}{\cos\theta}\right)$.

The graph is shifted up 2 units and has a slight "stretch" factor of $\frac{5}{6}$.

There are vertical asymptotes at $x=\frac{n\pi}{1-2\pi}$, where n is an odd integer, i.e., where $\cos\left(\left(\frac{1}{2}-\pi\right)x\right)=0$.

So, to find an asymptote, set $\left(\frac{1}{2}-\pi\right)x=\frac{\pi}{2}$ and solve for x:

$x=\frac{\pi}{2\left(\frac{1}{2}-\pi\right)}=\frac{\pi}{1-2\pi}\approx-0.595$. There is another at $x=-\frac{\pi}{1-2\pi}\approx 0.595$

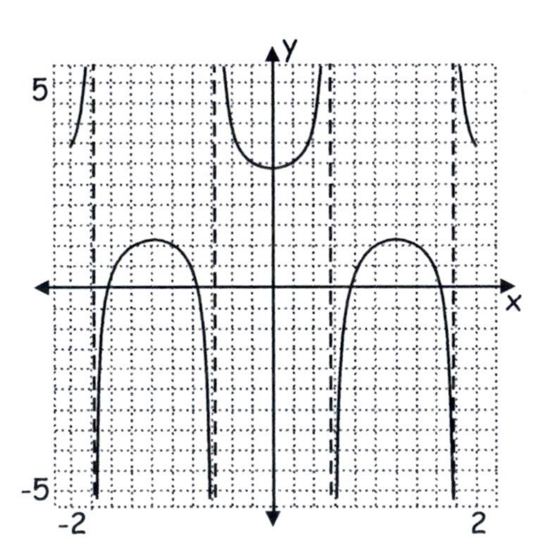

49. Sketch $g(t) = -2e^{-t}\cos 5\pi t$.
The graph is that of a damped harmonic motion function. First, graph $y = -2e^{-t}$ and $y = 2e^{-t}$. These graphs provide the envelope or "boundary" of the function $y = \cos 5\pi t$.
A table of selected values for the function $y = \cos 5\pi t$ will aid in graphing g manually, but a graphing calculator illustrates it best.

$\theta = 5\pi t$	$-\frac{3\pi}{2}$	$-\pi$	$-\frac{\pi}{2}$	0	$\frac{\pi}{2}$	π
t	$-\frac{3}{10}$	$-\frac{1}{5}$	$-\frac{1}{10}$	0	$\frac{1}{10}$	$\frac{1}{5}$
$\cos 5\pi t$	0	-1	0	1	0	-1

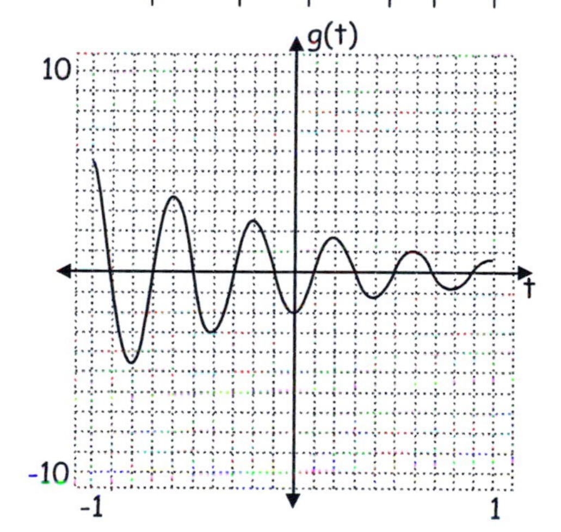

51. Sketch $g(t) = e^{t}\sin\left(3t - \frac{\pi}{2}\right)$.
The graph is that of a damped harmonic motion function. First, graph $y = -e^{t}$ and $y = e^{t}$. These graphs provide the envelope or "boundary" of the function $y = \sin\left(3t - \frac{\pi}{2}\right)$.
A table of selected values for the function $y = \sin\left(3t - \frac{\pi}{2}\right)$ will aid in graphing g manually, but a graphing calculator illustrates it best.

t	$3t - \frac{\pi}{2}$	$\sin\left(3t - \frac{\pi}{2}\right)$
$\frac{\pi}{6}$	0	0
$\frac{\pi}{3}$	$\frac{\pi}{2}$	1
$\frac{\pi}{2}$	π	0
$\frac{2\pi}{3}$	$\frac{3}{2}\pi$	-1
$\frac{5\pi}{6}$	2π	0
0	$-\frac{\pi}{2}$	-1

The envelope functions will dampen the amplitude of the function $y = \sin\left(3t - \frac{\pi}{2}\right)$, with selected values in the table.

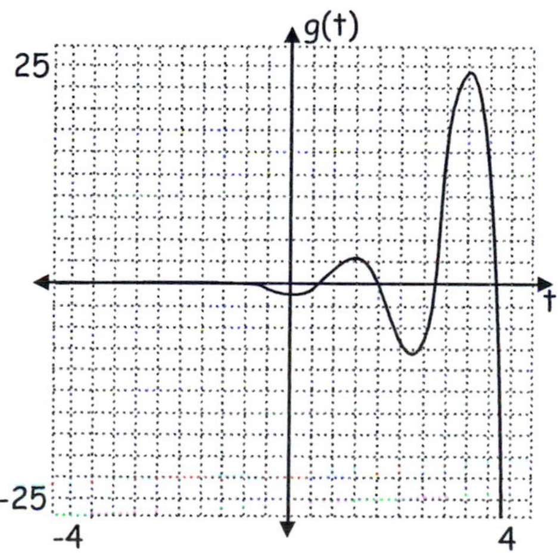

53. Sketch $f(t) = -3 + 5e^{-t}\cos t$.
The graph is that of a damped harmonic motion function. First, graph $y = -5e^{-t}$ and $y = 5e^{-t}$. These graphs provide the envelope or "boundary" of the function $y = \cos t$.
Note that the graph of the damped harmonic function $y = -3 + 5e^{-t}\cos t$ is a shift down of 3 units relative to the function $y = 5e^{-t}\cos t$:

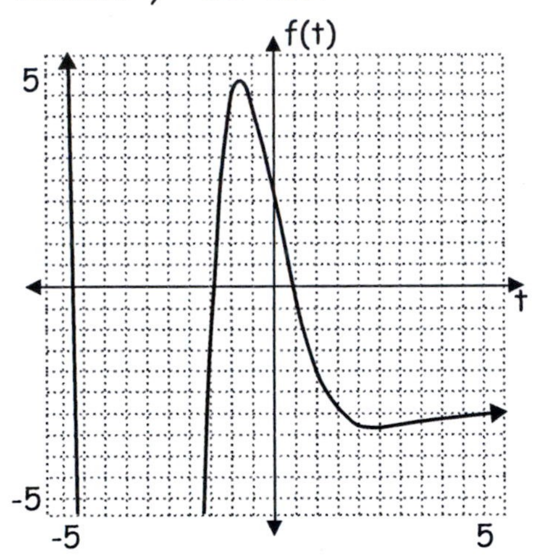

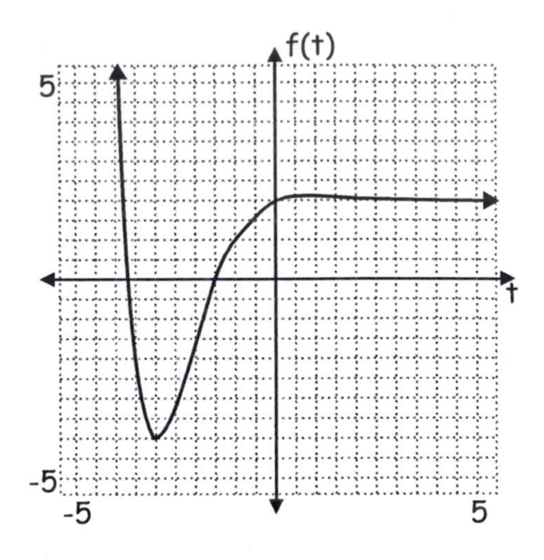

55. Sketch $f(t) = \frac{1}{2}e^{-t}\sin\left(\frac{5}{6}t - 4\pi\right) + 2$.
The graph of f is a 2-unit shift up of the graph of $g(t) = \frac{1}{2}e^{-t}\sin\left(\frac{5}{6}t - 4\pi\right)$.
The graphs of $y = -\frac{1}{2}e^{t}$ and $y = \frac{1}{2}e^{t}$ provide the envelope or "boundary" of the function $y = \sin\left(\frac{5}{6}t - 4\pi\right)$. We can also rewrite the latter form as and $y = \sin\frac{5}{6}\left(t - \frac{24}{5}\pi\right)$ to expose the period $P = \frac{2\pi}{\frac{5}{6}} = \frac{12\pi}{5}$ and phase shift to the right of $\frac{24}{5}\pi$.

End of Section 6.4

Chapter Six: Section 6.5
Solutions to Odd-numbered Exercises

1. Sketch arccosecant.
   First, consider the restricted domain of the cosecant function to provide a one-to-one function. A commonly selected domain is $\left[-\frac{\pi}{2}, 0\right) \cup \left(0, \frac{\pi}{2}\right]$ with a range of $(-\infty, -1] \cup [1, \infty)$.
   Graph of cosecant with restricted domain:

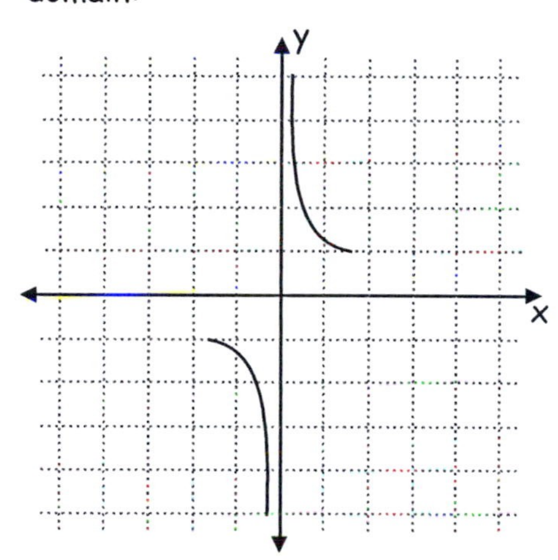

Reflect the graph of cosecant through $y = x$ to obtain the graph of the arccosecant function with domain $(-\infty, -1] \cup [1, \infty)$ and range $\left[-\frac{\pi}{2}, 0\right) \cup \left(0, \frac{\pi}{2}\right]$.

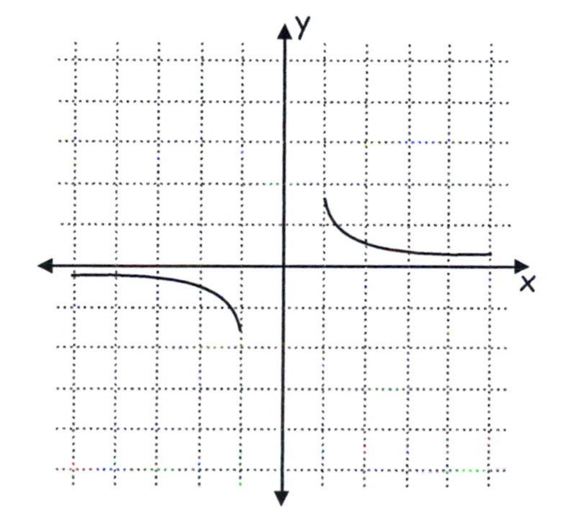

3. Sketch arccotangent.
   First, consider the restricted domain of the cotangent function to provide a one-to-one function. A commonly selected domain is $\left[-\frac{\pi}{2}, 0\right) \cup \left(0, \frac{\pi}{2}\right]$ with a range of $(-\infty, \infty)$.
   Graph of cotangent with restricted domain:

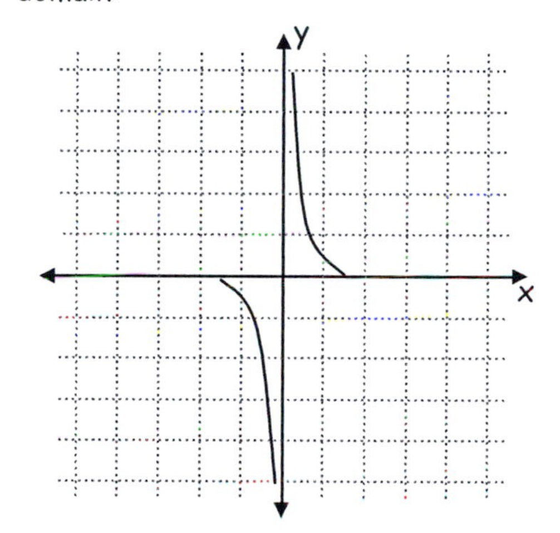

Reflect the graph of cotangent through $y = x$ to obtain the graph of the arccotangent function with domain $(-\infty, \infty)$ and range $\left[-\frac{\pi}{2}, 0\right) \cup \left(0, \frac{\pi}{2}\right]$.

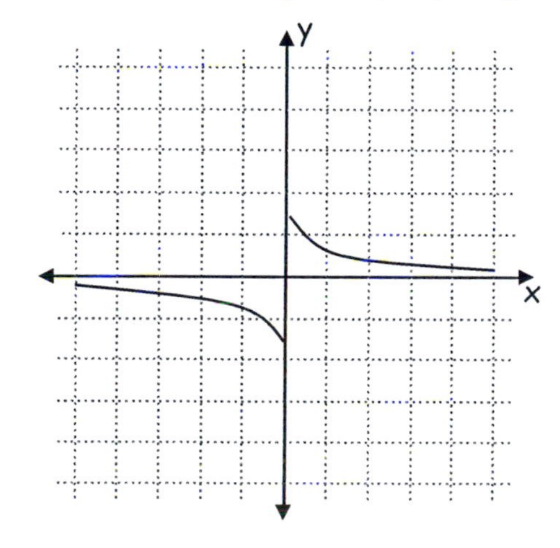

5. Given $y = \cos^{-1}\left(\frac{\sqrt{2}}{2}\right)$:

$\cos y = \frac{\sqrt{2}}{2}$

$y = \frac{\pi}{4}$

7. Given $y = \cot^{-1}\left(-\frac{\sqrt{3}}{3}\right)$:

$\cot y = -\frac{\sqrt{3}}{3}$

or

Use the triangle and take into consideration that $\cot y < 0$. The triangle is $\frac{\pi}{6}-\frac{\pi}{3}-\frac{\pi}{2}$, and the associated angle has measure $\frac{\pi}{3}$.

$-\frac{\sqrt{3}}{3} < 0 \Rightarrow y$ is a fourth quadrant angle $\Rightarrow y = -\frac{\pi}{3}$.

9. Given $y = \csc^{-1}(-2)$:

$\csc y = -2$

Use the triangle and take into consideration that $\csc y < 0$. The triangle is $\frac{\pi}{6}-\frac{\pi}{3}-\frac{\pi}{2}$, and the associated angle has measure $\frac{\pi}{6}$.

$\csc y = -2 < 0 \Rightarrow y$ is a fourth quadrant angle $\Rightarrow y = -\frac{\pi}{6}$.

11. Given $y = \arccos(-1)$.

$\cos y = -1$

Over the restricted domain of the cosine function, namely $[0, \pi]$,

$\cos y = -1 \Rightarrow y = \pi$.

13. Given $y = \operatorname{arccot}(-\sqrt{3})$.

$\cot y = -\sqrt{3}$

Over the restricted domain of the cotangent function, namely $(0, \pi)$,

$\cot y = -\sqrt{3} \Rightarrow y$ is the measure of an angle in the second quadrant,

so, $y = \frac{5\pi}{6}$.

15. Given $y = \operatorname{arccsc}(\sqrt{2})$.

$\csc y = \sqrt{2}$

Over the restricted domain of the cosecant function, namely

$\left[-\frac{\pi}{2}, 0\right) \cup \left(0, \frac{\pi}{2}\right]$,

$\csc y = \sqrt{2} \Rightarrow y = \frac{\pi}{4}$

17. Given $y = \tan^{-1}\left(\frac{\sqrt{3}}{3}\right)$.

$\tan y = \frac{\sqrt{3}}{3}$

Over the restricted domain of the cosecant function, namely $\left(-\frac{\pi}{2}, \frac{\pi}{2}\right)$

$\tan y = \frac{\sqrt{3}}{3} \Rightarrow y = \frac{\pi}{6}$

19. Given $y = \csc^{-1} 2$. Solve for y.
$\csc y = 2$
Over the restricted domain of the cosecant function, namely
$\left[-\frac{\pi}{2}, 0\right) \cup \left(0, \frac{\pi}{2}\right]$,
$\csc y = 2 \Rightarrow y = \frac{\pi}{6}$.

21. Given $y = \sec^{-1}(-1)$. Solve for y.
$\sec y = -1$
Over the restricted domain of the secant function, namely
$\left[0, \frac{\pi}{2}\right) \cup \left(\frac{\pi}{2}, \pi\right]$,
$\sec y = -1 \Rightarrow y = \pi$.

23. Given $y = \arctan 0$. Solve for y.
$\tan y = 0$
Over the restricted domain of the tangent function, namely $\left(-\frac{\pi}{2}, \frac{\pi}{2}\right)$
$\tan y = 0 \Rightarrow y = 0$.

25. Given $y = \arccos\left(-\frac{\sqrt{2}}{2}\right)$. Solve for y.
$\cos y = -\frac{\sqrt{2}}{2}$
Over the restricted domain of the cosine function, namely $[0, \pi]$,
$\cos y = -\frac{\sqrt{2}}{2} \Rightarrow y = \frac{3\pi}{4}$.

27. Given $y = \cot^{-1}\left(-\sqrt{3}\right)$. Solve for y.
$\cot y = -\sqrt{3}$
Over the restricted domain of the cotangent function, namely $\left[-\frac{\pi}{2}, \frac{\pi}{2}\right]$,
$\cot y = -\sqrt{3} \Rightarrow y$ is the measure of an angle in the fourth quadrant,
so, $y = -\frac{\pi}{6}$.

29. $\cos^{-1} 4$ is not defined since the domain of the $\cos^{-1}$ function is $[-1, 1]$.

31. $\tan^{-1} 5 \approx 1.3734$

33. $\tan^{-1} 0.8 \approx 0.6747$

35. Let $\theta = \sec^{-1}(-0.5)$.
Then, $\sec\theta = -0.5$

$$\frac{1}{\cos\theta} = -0.5 \;\Rightarrow\; \cos\theta = \frac{1}{-0.5}$$

Then, $\theta = \cos^{-1}\left(\frac{1}{-0.5}\right) = \cos^{-1}(-2)$;
Since -2 is not in the domain of $\cos^{-1}$ (domain is $[-1, 1]$), θ does not exist.

37. Let $\theta = \cot^{-1}(-0.2)$
$\cot\theta = -0.2$
$\frac{1}{\tan\theta} = -0.2$
$\tan\theta = \frac{1}{-0.2} \Rightarrow \theta = \tan^{-1}\left(\frac{1}{-0.2}\right)$
$\theta \approx -1.3734$

39. Let $\theta = \sec^{-1} 2$
$\sec\theta = 2$
$\frac{1}{\cos\theta} = 2$
$\cos\theta = \frac{1}{2} \Rightarrow \theta = \cos^{-1}\left(\frac{1}{2}\right)$
$\theta = \frac{\pi}{3} \approx 1.0472$

41. $\sin^{-1}\left(\sin\frac{3\pi}{2}\right) = \sin^{-1}(-1)$
$= -\frac{\pi}{2}$
$\left[-\frac{\pi}{2} \in \text{Range of } \sin^{-1}\right]$

43. $\sin^{-1}\left(\sin\frac{7\pi}{6}\right) = \sin^{-1}\left(-\frac{1}{2}\right)$

$= -\frac{\pi}{6}$

$\left[-\frac{\pi}{6} \in \text{Range of } \sin^{-1}\right]$

45. $\tan^{-1}\left(\tan\frac{5\pi}{4}\right) = \tan^{-1}(1)$

$= \frac{\pi}{4}$

$\left[\frac{\pi}{4} \in \text{Range of } \tan^{-1}\right]$

47. $\sin^{-1}\left(\cos\frac{3\pi}{2}\right) = \sin^{-1} 0$

$= 0 \quad (\text{since } \sin 0 = 0)$

49. $\arcsin(\tan 1) \approx \arcsin(1.5574)$

But, this value cannot exist because $1.5574 \notin$ domain of arcsine function.

51. $\tan^{-1}(\cos 5) \approx \tan^{-1}(0.2837)$

≈ 0.2764

53. $\cos\left(\sec^{-1}(-2)\right) = \cos\theta,$

with $\theta = \sec^{-1}(-2)$. Then, $\sec\theta = -2$.

Rewrite as $\frac{1}{\cos\theta} = -2 \Rightarrow \cos\theta = -\frac{1}{2}$

$\theta' = \frac{\pi}{3}$, so $\theta = \frac{2\pi}{3}$. Substitute in the original expression:

$\cos\left(\sec^{-1}(-2)\right) = \cos\frac{2\pi}{3} = -\frac{1}{2}$

55. $\csc\left(\arccos\left(-\frac{\sqrt{3}}{2}\right)\right) = \csc\theta,$

with $\theta = \arccos\left(-\frac{\sqrt{3}}{2}\right)$.

Then, $\cos\theta = -\frac{\sqrt{3}}{2}$.

$\theta = \frac{5\pi}{6}$ and

$\csc\left(\arccos\left(-\frac{\sqrt{3}}{2}\right)\right) = \csc\frac{5\pi}{6} = 2$

57. $\sec\left(\csc^{-1}\frac{2\sqrt{3}}{3}\right) = \sec\theta,$

with $\theta = \csc^{-1}\frac{2\sqrt{3}}{3}$.

Then, $\csc\theta = \frac{2\sqrt{3}}{3}$ or, $\frac{1}{\sin\theta} = \frac{2\sqrt{3}}{3}$

$\Rightarrow \sin\theta = \frac{3}{2\sqrt{3}} = \frac{\sqrt{3}}{2} \Rightarrow \theta = \frac{\pi}{3}$ and

$\sec\left(\csc^{-1}\frac{2\sqrt{3}}{3}\right) = \sec\frac{\pi}{3}$

$= \frac{1}{\cos\frac{\pi}{3}} = \frac{1}{\frac{1}{2}} = 2$

59. $\sec\left(\arcsin\left(-\frac{1}{2}\right)\right) = \sec\theta,$

with $\theta = \arcsin\left(-\frac{1}{2}\right)$.

Then, $\sin\theta = -\frac{1}{2}$ and, $\theta = -\frac{\pi}{6}$.

$\sec\left(\arcsin\left(-\frac{1}{2}\right)\right) = \sec\left(-\frac{\pi}{6}\right)$

$= \frac{1}{\cos\left(-\frac{\pi}{6}\right)}$

$= \frac{1}{\frac{\sqrt{3}}{2}}$

$= \frac{2}{\sqrt{3}} = \frac{2\sqrt{3}}{3}$

61. $\cot(\operatorname{arccsc}(-2)) = \cot\theta$,
with $\theta = \operatorname{arccsc}(-2)$.

Then, $\csc\theta = -2$ or, $\dfrac{1}{\sin\theta} = -2$

Rewrite as $\sin\theta = -\dfrac{1}{2}$.

Then, $\theta = -\dfrac{\pi}{6}$

$$\cot(\operatorname{arccsc}(-2)) = \cot\left(-\frac{\pi}{6}\right) = -\sqrt{3}$$

63. $\sec\left(\tan^{-1}\left(-\dfrac{\sqrt{3}}{3}\right)\right) = \sec\theta$,

with $\theta = \tan^{-1}\left(-\dfrac{\sqrt{3}}{3}\right)$.

Then, $\tan\theta = -\dfrac{\sqrt{3}}{3} \Rightarrow \theta = -\dfrac{\pi}{6}$ and

$$\sec\left(\tan^{-1}\left(-\frac{\sqrt{3}}{3}\right)\right) = \sec\left(-\frac{\pi}{6}\right) = \frac{1}{\cos\left(-\frac{\pi}{6}\right)} = \frac{1}{\frac{\sqrt{3}}{2}} = \frac{2\sqrt{3}}{3}$$

65. $\tan(\csc^{-1}(-2)) = \tan\theta$,
with $\theta = \csc^{-1}(-2)$. Then, $\csc\theta = -2$

Rewrite as $\sin\theta = -\dfrac{1}{2} \Rightarrow \theta = -\dfrac{\pi}{6}$

$$\text{Then, } \tan(\csc^{-1}(-2)) = \tan\left(-\frac{\pi}{6}\right) = \frac{\sin\left(-\frac{\pi}{6}\right)}{\cos\left(-\frac{\pi}{6}\right)} = \frac{-\frac{1}{2}}{\frac{\sqrt{3}}{2}} = -\frac{1}{\sqrt{3}} = -\frac{\sqrt{3}}{3}$$

67. $\csc\left(\cot^{-1}\sqrt{3}\right) = \csc\theta$
with $\theta = \cot^{-1}\sqrt{3}$

Then, $\cot\theta = \sqrt{3} \Rightarrow \theta = \dfrac{\pi}{6}$

$$\text{Then, } \csc\left(\cot^{-1}\sqrt{3}\right) = \csc\frac{\pi}{6} = \frac{1}{\sin\frac{\pi}{6}} = \frac{1}{\frac{1}{2}} = 2$$

69. $\cos\left(\arctan\left(-\dfrac{\sqrt{3}}{3}\right)\right) = \cos\theta$

with $\theta = \arctan\left(-\dfrac{\sqrt{3}}{3}\right)$

Then, $\tan\theta = -\dfrac{\sqrt{3}}{3} \Rightarrow \theta = -\dfrac{\pi}{6}$

$$\text{Then, } \cos\left(\arctan\left(-\frac{\sqrt{3}}{3}\right)\right) = \cos\left(-\frac{\pi}{6}\right) = \frac{\sqrt{3}}{2}$$

71.

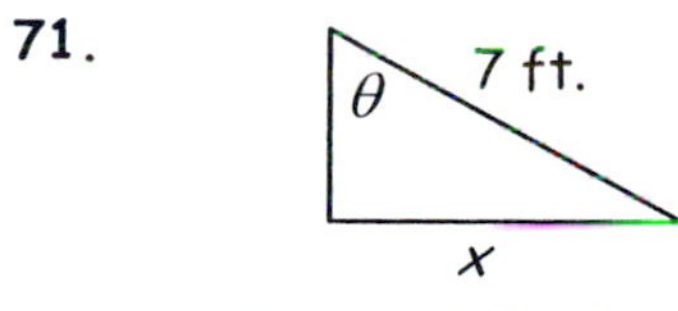

a. For $x = 2$ ft., find θ:

$\sin\theta = \dfrac{2}{7}$

$\theta = \sin^{-1}\left(\dfrac{2}{7}\right) \approx 0.2898$

b. For $x = 3$ ft., find θ:

$\sin\theta = \dfrac{3}{7}$

$\theta = \sin^{-1}\left(\dfrac{3}{7}\right) \approx 0.4429$

c. For $x = 5$ ft., find θ:

$\sin\theta = \dfrac{5}{7}$

$\theta = \sin^{-1}\left(\dfrac{5}{7}\right) \approx 0.7956$

73. Given $y=\cot\left(\sin^{-1}\frac{2}{x}\right)$. Rewrite as a purely algebraic function:

Let $\theta=\sin^{-1}\frac{2}{x}$; then, $\sin\theta=\frac{2}{x}$.

$y=\cot\left(\sin^{-1}\frac{2}{x}\right)=\cot\theta$

$y=\frac{\sqrt{x^2-4}}{2}$

75. Given $y=\tan\left(\sin^{-1}\frac{x}{\sqrt{x^2+3}}\right)$. Rewrite as a purely algebraic function:

Let $\theta=\sin^{-1}\frac{x}{\sqrt{x^2+3}}$; then,

$\sin\theta=\frac{x}{\sqrt{x^2+3}}$.

$y=\tan\left(\sin^{-1}\frac{x}{\sqrt{x^2+3}}\right)=\tan\theta$

$y=\frac{x}{\sqrt{3}}=\frac{\sqrt{3}x}{3}$

77. Given $y=\cos\left(\tan^{-1}\frac{x}{4}\right)$. Rewrite as a purely algebraic function:

Let $\theta=\tan^{-1}\frac{x}{4}$; then,

$\tan\theta=\frac{x}{4}$.

$y=\cos\left(\tan^{-1}\frac{x}{4}\right)=\cos\theta$

$y=\frac{4}{\sqrt{x^2+16}}$

79. $\theta=\arctan(-0.258416)$
$=-14.48917361°$

81. $\theta=\sec^{-1}(-1.1224539)\Rightarrow$
$\sec\theta=-1.1224539\Rightarrow$
$\frac{1}{\cos\theta}=-1.1224539\Rightarrow$
$\cos\theta=\frac{1}{-1.1224539}\approx-0.8909051855$
$\theta\approx\cos^{-1}(-0.8909051855)\approx152.987°$

83. $\theta=\arccos(-0.1115598)$
≈1.682588837

85. $\theta=\tan^{-1}5.999999$
≈1.405647622

87. $\theta=\arcsin(0.65937229)$
≈0.7199835305

89. The graph of $f(x)=\sec^{-1}(2x)$ will be a horizontal compression of $f(x)=\sec^{-1}(x)$ by a factor of 2.

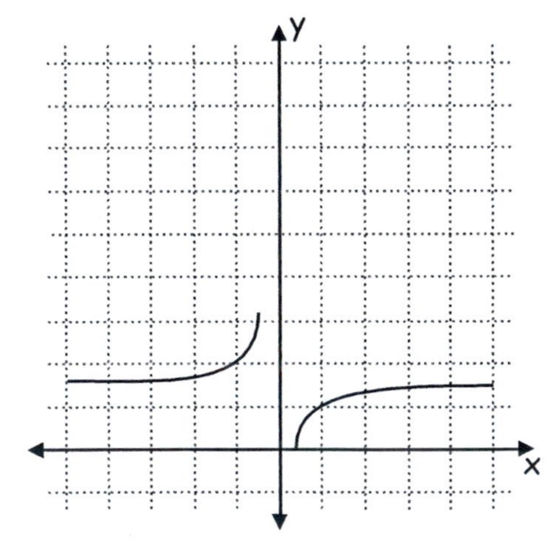

91. Sketch $f(x) = 2\arccos x$:

Domain $= [-1, 1]$. Range $= [0, 2\pi]$.

x	0	1	-1	$\frac{1}{2}$	$\frac{\sqrt{3}}{2}$
$2\arccos x$	π	0	2π	$\frac{2\pi}{3}$	$\frac{\pi}{3}$

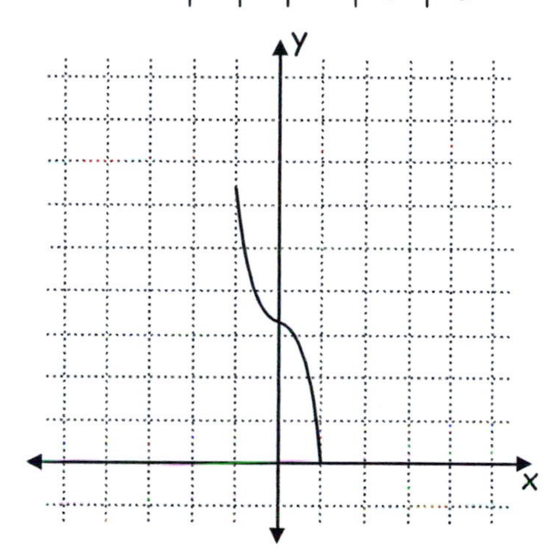

End of Section 6.5

Chapter Six Test
Solutions to Odd-Numbered Exercises

1. $\frac{4\pi}{3} = \frac{4\pi}{3} \cdot \frac{180}{\pi} = 240°$

3. $\frac{\pi}{2} + \frac{\pi}{4} = \frac{3\pi}{4}$

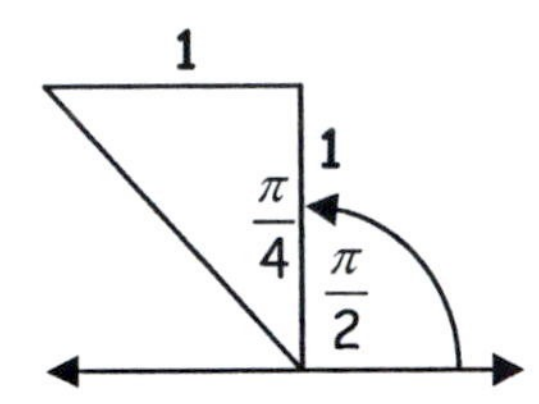

5. Sketch of $-\frac{5\pi}{3}$:

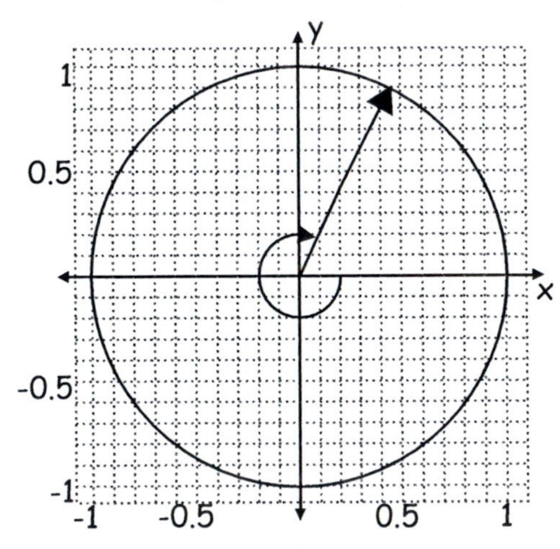

7. Given $d = 1.5$ $(r = 0.75)$
Central angle $\theta = 92°$
Convert to radians:
$92° = 92\left(\frac{\pi}{180}\right) \approx 1.605702912$
The arclength $s = r\theta$
$s \approx 0.75(1.605702912)$
≈ 1.204 m

9. Given: $r = 2$ in.
600 rpm $= 600(2\pi)$ rad/min.

a. Use $\omega = \frac{\theta}{t}$.
$\omega = 1200\pi$ rad/min.,

b. Use $v = r\omega$ and convert the answer in (a) to inches per second by multiplying by $r = 2$ inches and dividing by 60:
$\omega = 1200\pi$ rad/60 per second
$\omega = 20$rad/second
$v = 2(20\pi)$ in./s ≈ 125.66 in./s

11. Use $A = \frac{r^2\theta}{2}$. $r = 1$; $\theta = \frac{3\pi}{4}$

$$A = \frac{1^2\left(\frac{3\pi}{4}\right)}{2} = \frac{3\pi}{8} \text{ units}^2$$

13. $\text{opp} = \sqrt{\left(17\sqrt{2}\right)^2 - 17^2} = \sqrt{289} = 17$
The given triangle is similar to the $\frac{\pi}{4} - \frac{\pi}{4} - \frac{\pi}{2}$ triangle. So, $\theta = \frac{\pi}{4}$.

$\sin\frac{\pi}{4} = \frac{1}{\sqrt{2}} = \frac{\sqrt{2}}{2}$

$\cos\frac{\pi}{4} = \frac{1}{\sqrt{2}} = \frac{\sqrt{2}}{2}$

$\tan\frac{\pi}{4} = \frac{17}{17} = 1$

$\csc\frac{\pi}{4} = \frac{17\sqrt{2}}{17} = \sqrt{2}$

$\sec\frac{\pi}{4} = \frac{17\sqrt{2}}{17} = \sqrt{2}$

$\cot\frac{\pi}{4} = \frac{17}{17} = 1$

15. $\tan 63° \approx 1.96$
$\cot 63° = \frac{1}{\tan 63°} \approx \frac{1}{1.96} \approx 0.51$

17. Change 55° 33'20" to decimal notation:

$$55°\,33'20" = 55 + \frac{33}{60} + \frac{20}{3600} \approx 55.5556°$$

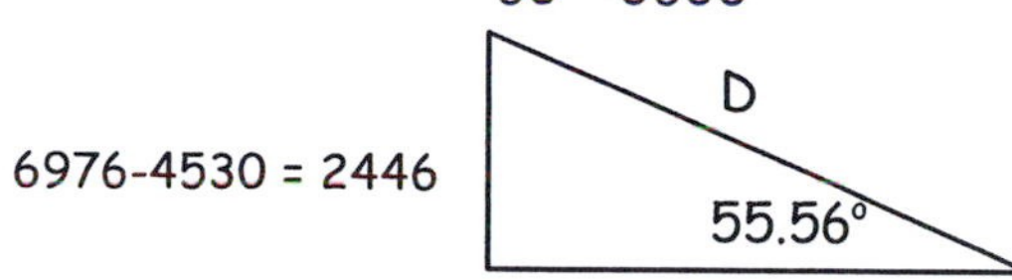

Set mode on calculator to degrees (or else convert to radians and use radian mode):

$\sin 55.5556° = \frac{2446}{D}$; Solve for D:

$$D = \frac{2446}{\sin 55.5556}$$
$$\approx \frac{2446}{0.824675}$$
$$\approx 2966.02 \text{ ft.}$$

19. Given $\beta = 44°$; $\alpha = 52°$; $d = 45$ ft.

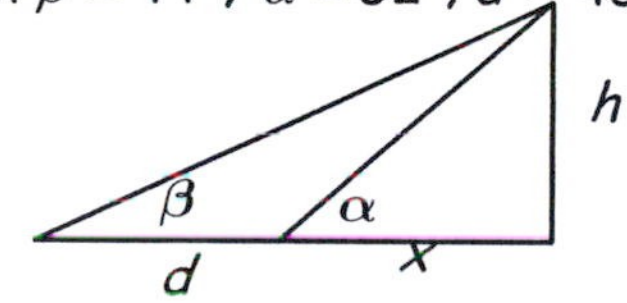

Use $h = \frac{d}{\cot\beta - \cot\alpha}$

$$h = \frac{45}{\cot 44° - \cot 52°}$$
$$h \approx \frac{45}{1.03553 - 0.78129} \approx 176.99 \text{ ft.}$$

21. $\sin(-125°) \approx -0.82$

$\cos(-125°) \approx -0.57$

$\tan(-125°) \approx 1.43$

$\csc(-125°) \approx \frac{1}{\sin(-125°)} \approx -1.22$

$\sec(-125°) \approx \frac{1}{\cos(-125°)} \approx -1.74$

$\cot(-125°) \approx \frac{1}{\tan(-125°)} \approx 0.7$

23. For $\varphi = 86°$, the reference angle is $\varphi' = 86°$.

25. **a.** $\sin 179° = \sin 1°$

b. $\sin 179° \approx 0.02$

27. $\tan\left(-\frac{\pi}{5}\right) = \cot\left(\frac{\pi}{2} - \left(-\frac{\pi}{5}\right)\right)$

$$= \cot\left(\frac{7\pi}{10}\right)$$

Both ≈ -0.73.

29. $\left\{\tan\theta = \frac{1}{2} (\text{i.e., } > 0) \text{ and } \sin\theta < 0\right\} \Rightarrow$

$\cos\theta < 0 \Rightarrow \theta$ is in the third quadrant.

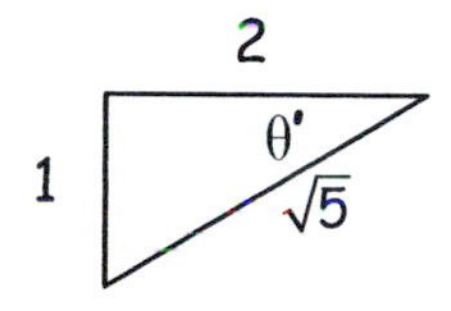

For the reference angle θ',

$\sin\theta' = \frac{1}{\sqrt{5}}$ or $\frac{\sqrt{5}}{5}$

$\sin\theta = -\sin\theta' = -\frac{\sqrt{5}}{5}$

31. Graph of the sine function

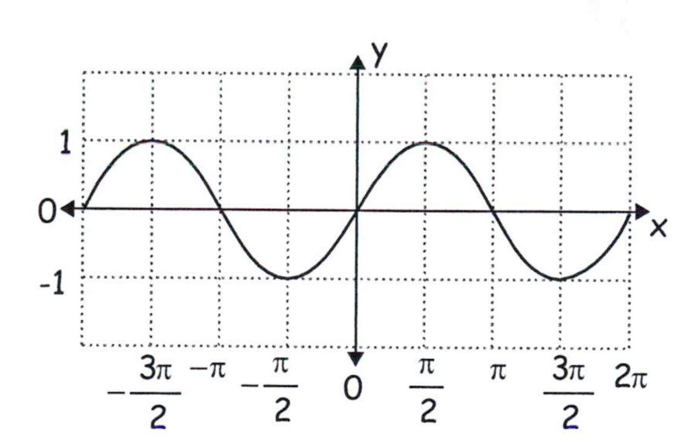

33. To sketch $f(x) = \csc 2\pi x$, begin with the graph of the cosecant function.

Period $= \frac{2\pi}{2\pi} = 1$. The vertical asymptotes occur whenever $2\pi x = n\pi$, or whenever $x = \frac{n}{2}$, for n any integer;

i.e., for $x \in \left\{\ldots, -\frac{3}{2}, -1, -\frac{1}{2}, 0, \frac{1}{2}, 1, \frac{3}{2}, \ldots\right\}$.

The left end-point of one cycle is 0.
The right end-point of the cycle is 1.

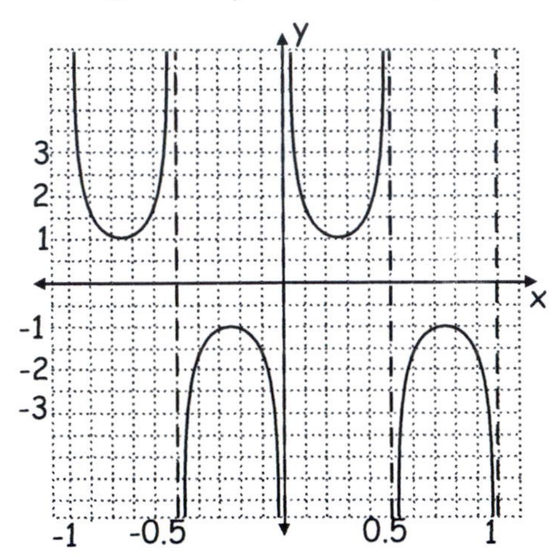

35. To sketch $g(x) = 2 + \sin(x - \pi)$, begin with the graph of the sine function. The sine function is shifted 2 units up, with a phase shift of π units to the right:

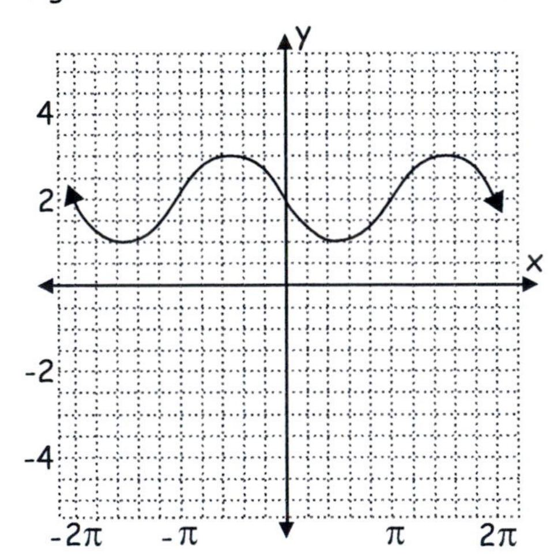

37. To sketch $g(t) = -e^{-t+2}\sin 3\pi t$, first sketch the "envelope" functions $\pm e^{-t+2}$. To graph $\sin 3\pi t$ manually, use a table of key values for t:

t	0	$\frac{1}{6}$	$\frac{1}{3}$	$\frac{1}{2}$
$3\pi t$	0	$\frac{\pi}{2}$	π	$\frac{3\pi}{2}$
$\sin 3\pi t$	0	1	0	-1

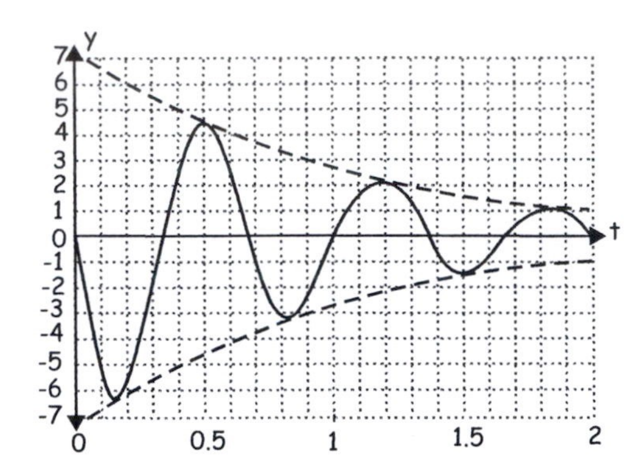

39.

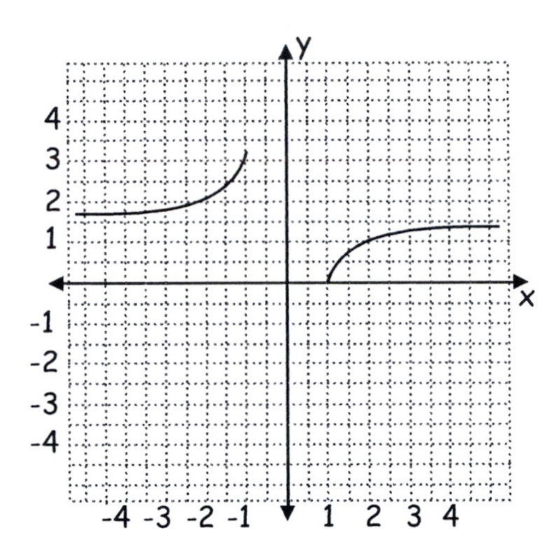

41. $\cos^{-1}(-2)$ cannot be evaluated because -2 is not in the domain of the $\cos^{-1}$ function. In other words, there is no angle whose cosine value is -2.

43. $\cot^{-1} 127 = \theta \Rightarrow \cot\theta = 127$

$$\Rightarrow \frac{1}{\tan\theta} = 127$$

$$\Rightarrow \tan\theta = \frac{1}{127}$$

$$\Rightarrow \tan^{-1}\left(\frac{1}{127}\right) = \theta$$

Then, $\cot^{-1} 127 = \tan^{-1}\left(\frac{1}{127}\right)$

$$\approx 0.0079$$

45. $\tan\left(\tan^{-1} 0.75\right) = 0.75$

(0.75 is in the restricted domain of the $\tan^{-1}$ function--$\left(-\frac{\pi}{2}, \frac{\pi}{2}\right)$.

47. $\sin^{-1}\left(\cos\frac{3\pi}{4}\right) = \sin^{-1}\left(-\frac{\sqrt{2}}{2}\right)$.

Use the associated angle value in the fourth quadrant to stay within the domain of the $\sin^{-1}$ function. The reference angle in the fourth quadrant is $-\frac{\pi}{4}$.

$\sin^{-1}\left(\cos\frac{3\pi}{4}\right) = \sin^{-1}\left(-\frac{\sqrt{2}}{2}\right) = \theta \Rightarrow$

$\sin\theta = -\frac{\sqrt{2}}{2}$ (which is a special sine value): $\theta = -\frac{\pi}{4}$.

49. Set $\sin^{-1}\frac{x}{\sqrt{x^2+4}} = \theta$

Then, $\sin\theta = \frac{x}{\sqrt{x^2+4}}$.

The associated right triangle:

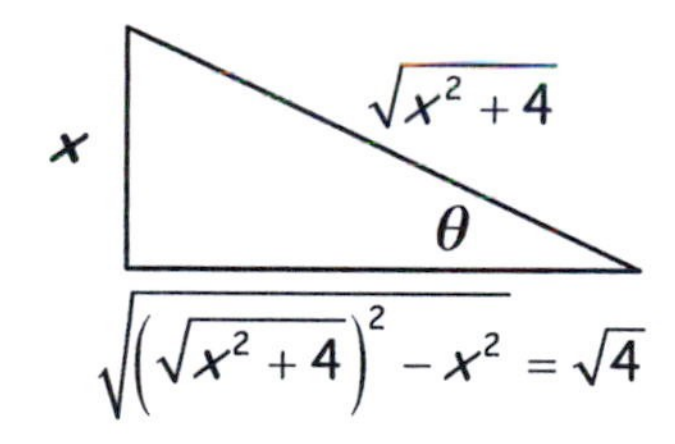

$\tan\left(\sin^{-1}\frac{x}{\sqrt{x^2-4}}\right) = \tan\theta$

$= \frac{x}{\sqrt{4}} = \frac{x}{2}$

End of Chapter 6 Test

Chapter 7: Section 7.1
Solutions to Odd-numbered Exercises

1. $\tan x \csc x = \dfrac{\sin x}{\cos x} \cdot \dfrac{1}{\sin x}$
$= \dfrac{1}{\cos x}$
$= \sec x$

3. $\dfrac{\tan t}{\sec t} = \dfrac{\frac{\sin t}{\cos t}}{\frac{1}{\cos t}}$
$= \dfrac{\sin t}{\cos t} \cdot \dfrac{\cos t}{1}$
$= \sin t$

5. $\sin(-x)\tan x = -\sin x\left(\dfrac{\sin x}{\cos x}\right)$
$= -\dfrac{\sin^2 x}{\cos x}$
$= -\dfrac{1-\cos^2 x}{\cos x}$
$= -\dfrac{1}{\cos x} + \dfrac{\cos^2 x}{\cos x}$
$= -\sec x + \cos x$, or
$\cos x - \sec x$

7. $\sin(\alpha + 2\pi)\sec\alpha = \sin\alpha \cdot \dfrac{1}{\cos\alpha}$
$= \dfrac{\sin\alpha}{\cos\alpha}$
$= \tan\alpha$

9. $\cos y\left(1+\tan^2 y\right) = \cos y\left(1 + \dfrac{\sin^2 y}{\cos^2 y}\right)$
$= \cos y + \dfrac{1-\cos^2 y}{\cos y}$
$= \cos y + \dfrac{1}{\cos y} - \cos y$
$= \sec y$

11. $\dfrac{1-\tan^2 x}{\cot^2 x - 1} = \dfrac{1 - \frac{\sin^2 x}{\cos^2 x}}{\frac{\cos^2 x}{\sin^2 x} - 1}$
$= \dfrac{\frac{\cos^2 x - \sin^2 x}{\cos^2 x}}{\frac{\cos^2 x - \sin^2 x}{\sin^2 x}}$
$= \dfrac{\cos^2 x - \sin^2 x}{\cos^2 x} \cdot \dfrac{\sin^2 x}{\cos^2 x - \sin^2 x}$
$= \dfrac{\sin^2 x}{\cos^2 x}$
$= \tan^2 x$

13. Verify $(1-\cos\theta)(1+\cos\theta) = \sin^2\theta$:
$(1-\cos\theta)(1+\cos\theta) = 1 - \cos^2\theta$
$= \sin^2\theta$

15. Verify $\sec^2 y - \tan^2 y = \sec y \cos(-y)$:
$\sec y \cos(-y) = \sec y \cos y$
$= \sec y \cdot \dfrac{1}{\sec y}$
$= 1$
$= \sec^2 y - \tan^2 y$

17. Verify $\dfrac{\sin\left[\frac{\pi}{2} - x\right]}{\cos\left[\frac{\pi}{2} - x\right]} = \cot x$:
$\dfrac{\sin\left[\frac{\pi}{2} - x\right]}{\cos\left[\frac{\pi}{2} - x\right]} = \dfrac{\cos x}{\sin x}$
$= \cot x$

19. $\dfrac{1}{\tan x} + \tan x = \dfrac{1+\tan^2 x}{\tan x}$
$= \dfrac{\sec^2 x}{\tan x}$

21. $$\frac{1}{\sin(\theta+2\pi)+1}+\frac{1}{\csc(\theta+2\pi)+1}=1$$
$$\frac{1}{\sin\theta+1}+\frac{1}{\csc\theta+1}=1$$
$$\frac{\frac{1}{\sin\theta}+1+\sin\theta+1}{(\sin\theta+1)\left(\frac{1}{\sin\theta}+1\right)}=1$$
$$\frac{\frac{1}{\sin\theta}+1+\sin\theta+1}{\frac{1}{\sin\theta}+1+\sin\theta+1}=1$$

23. Verify:

$\sin^2 x-\sin^4 x=\cos^2(-x)-\cos^4(-x)$

Simplify the right-hand side:

$$\cos^2(-x)-\cos^4(-x)=$$
$$\cos^2 x-\cos^4 x=$$
$$1-\sin^2 x-\left(1-\sin^2 x\right)^2=$$
$$1-\sin^2 x-\left(1-2\sin^2+\sin^4 x\right)=$$
$$1-\sin^2 x-1+2\sin^2-\sin^4 x=$$

The last expression simplifies to

$\sin^2 x-\sin^4 x$

25. $$\frac{\cos\left(\frac{\pi}{2}-\alpha\right)}{\csc\alpha}-1=\frac{\sin\alpha}{\csc\alpha}-1$$
$$=\frac{\sin\alpha}{\frac{1}{\sin\alpha}}-1$$
$$=\sin^2\alpha-1$$

Now, rewriting the right side of the given equation, we have

$$\sin\alpha\cot(-\alpha)\cos(-\alpha)=$$
$$\sin\alpha(-\cot\alpha)\cos\alpha=$$
$$\sin\alpha\left(-\frac{\cos\alpha}{\sin\alpha}\right)\cos\alpha=-\cos^2\alpha$$
$$=-\left(1-\sin^2\alpha\right)$$
$$=\sin^2\alpha-1$$

27. Rewrite $\sqrt{x^2-16}$ with $x=4\sec\theta$.

$$\sqrt{x^2-16}=\sqrt{16\sec^2\theta-16}$$
$$=4\sqrt{\sec^2-1}$$
$$=4\sqrt{\tan^2\theta+1-1}$$
$$=4\tan\theta$$

29. Rewrite $\sqrt{4x^2+100}$ with $\cot\theta=\frac{x}{5}$.

$$\sqrt{4x^2+100}=\sqrt{4(5\cot\theta)^2+100}$$
$$=2\sqrt{25\cot^2\theta+25}$$
$$=10\sqrt{\cot^2\theta+1}$$
$$=10\sqrt{\csc^2\theta}$$
$$=10\csc\theta$$

31. Rewrite $\sqrt{x^2-4}$ with $x=2\csc\theta$.

$$\sqrt{x^2-4}=\sqrt{(2\csc\theta)^2-4}$$
$$=\sqrt{4\csc^2\theta-4}$$
$$=2\sqrt{\csc^2\theta-1}$$
$$=2\sqrt{1+\cot^2\theta-1}$$
$$=2\sqrt{\cot^2\theta}$$
$$=2\cot\theta$$

33. Rewrite $\sqrt{144-9x^2}$ with $x=4\cos\theta$.

$$\sqrt{144-9x^2}=\sqrt{9\left(16-x^2\right)}$$
$$=3\sqrt{16-(4\cos\theta)^2}$$
$$=3\sqrt{16\left(1-\cos^2\theta\right)}$$
$$=12\sqrt{\sin^2\theta}$$
$$=12\sin\theta$$

35. Show $1+\cot^2 x=\csc^2 x$

Use $\sin^2 x+\cos^2 x=1$

$$1+\cot^2 x=1+\left(\frac{\cos x}{\sin x}\right)^2$$
$$=\frac{\sin^2 x+\cos^2 x}{\sin^2 x}$$
$$=\frac{1}{\sin^2 x}$$
$$=\csc^2 x$$

End of Section 7.1

Chapter Seven: Section 7.2
Solutions to Odd-numbered Exercises

1. $\cos\left[\frac{\pi}{4}+\frac{\pi}{3}\right]=\cos\frac{\pi}{4}\cos\frac{\pi}{3}-\sin\frac{\pi}{4}\sin\frac{\pi}{3}$
$=\left(\frac{\sqrt{2}}{2}\right)\left(\frac{1}{2}\right)-\left(\frac{\sqrt{2}}{2}\right)\left(\frac{\sqrt{3}}{2}\right)$
$=\frac{\sqrt{2}}{4}-\frac{\sqrt{6}}{4}$
$=\frac{\sqrt{2}-\sqrt{6}}{4}$

3. $\tan\left(\frac{4\pi}{3}+\frac{5\pi}{4}\right)=\frac{\tan\frac{4\pi}{3}+\tan\frac{5\pi}{4}}{1-\tan\frac{4\pi}{3}\tan\frac{5\pi}{4}}$
$=\frac{\sqrt{3}+1}{1-\sqrt{3}(1)}$
$=\frac{1+\sqrt{3}}{1-\sqrt{3}}\cdot\frac{1+\sqrt{3}}{1+\sqrt{3}}$
$=\frac{1+2\sqrt{3}+3}{1-3}$
$=-2-\sqrt{3}$

5. $\cos\left(\frac{7\pi}{6}-\frac{\pi}{6}\right)=\cos\frac{7\pi}{6}\cos\frac{\pi}{6}+\sin\frac{7\pi}{6}\sin\frac{\pi}{6}$
$=\left(-\frac{\sqrt{3}}{2}\right)\left(\frac{\sqrt{3}}{2}\right)+\left(-\frac{1}{2}\right)\left(\frac{1}{2}\right)$
$=\frac{-3}{4}-\frac{1}{4}$
$=-1$

7. $\cos\left(\frac{4\pi}{3}+\frac{5\pi}{3}\right)=$
$\cos\frac{4\pi}{3}\cos\frac{5\pi}{3}-\sin\frac{4\pi}{3}\sin\frac{5\pi}{3}=$
$\left(-\frac{1}{2}\right)\left(\frac{1}{2}\right)-\left(-\frac{\sqrt{3}}{2}\right)\left(-\frac{\sqrt{3}}{2}\right)=$
$\frac{-1}{4}-\frac{3}{4}=-1$

9. $\cos\left(\frac{7\pi}{6}-\frac{5\pi}{3}\right)=\cos\frac{7\pi}{6}\cos\frac{5\pi}{3}+\sin\frac{7\pi}{6}\sin\frac{5\pi}{3}$
$=\left(-\frac{\sqrt{3}}{2}\right)\left(\frac{1}{2}\right)+\left(\frac{1}{2}\right)\left(-\frac{\sqrt{3}}{2}\right)$
$=\frac{-\sqrt{3}}{4}-\frac{\sqrt{3}}{4}$
$=0$

11. $\sin\left(\frac{7\pi}{4}+\frac{5\pi}{4}\right)=\sin\frac{7\pi}{4}\cos\frac{5\pi}{4}+\cos\frac{7\pi}{4}\sin\frac{5\pi}{4}$
$=\left(-\frac{1}{\sqrt{2}}\right)\left(-\frac{1}{\sqrt{2}}\right)+\left(\frac{1}{\sqrt{2}}\right)\left(-\frac{1}{\sqrt{2}}\right)$
$=\frac{1}{2}-\frac{1}{2}$
$=0$

13. $\cos\left(\frac{\pi}{4}-\frac{\pi}{6}\right)=\cos\frac{\pi}{4}\cos\frac{\pi}{6}+\sin\frac{\pi}{4}\sin\frac{\pi}{6}$
$=\frac{\sqrt{2}}{2}\cdot\frac{\sqrt{3}}{2}+\frac{\sqrt{2}}{2}\cdot\frac{1}{2}$
$=\frac{\sqrt{6}+\sqrt{2}}{4}$

15. $\sin\left(\frac{7\pi}{4}+\frac{2\pi}{3}\right)=\sin\frac{7\pi}{4}\cos\frac{2\pi}{3}+\cos\frac{7\pi}{4}\sin\frac{2\pi}{3}$
$=-\frac{\sqrt{2}}{2}\cdot-\frac{1}{2}+\frac{\sqrt{2}}{2}\cdot\frac{\sqrt{3}}{2}$
$=\frac{\sqrt{2}+\sqrt{6}}{4}$

17. $\tan 75^\circ=\tan(30^\circ+45^\circ)$
$=\frac{\tan 30^\circ+\tan 45^\circ}{1-\tan 30^\circ\tan 45^\circ}$
$=\frac{\frac{\sqrt{3}}{3}+1}{1-\frac{\sqrt{3}}{3}(1)}$
$=\frac{\sqrt{3}+3}{3-\sqrt{3}}$
$=\frac{\sqrt{3}+3}{3-\sqrt{3}}\cdot\frac{3+\sqrt{3}}{3+\sqrt{3}}$
$=\frac{9+6\sqrt{3}+3}{9-3}$
$=2+\sqrt{3}$

19. $\sin 165° = \sin(120° + 45°)$

Use a sum formula to rewrite:

$$\sin 120° \cos 45° + \cos 120° \sin 45° =$$
$$= \left(\frac{\sqrt{3}}{2}\right)\left(\frac{\sqrt{2}}{2}\right) + \left(-\frac{1}{2}\right)\left(\frac{\sqrt{2}}{2}\right)$$
$$= \frac{\sqrt{6}}{4} - \frac{\sqrt{2}}{4}$$
$$= \frac{\sqrt{6} - \sqrt{2}}{4}$$

21. $\cos -15° = \cos(-45° + 30°)$

Use a sum formula to rewrite:

$$= \cos(-45°)\cos 30° - \sin(-45°)\sin 30°$$
$$= \left(\frac{\sqrt{2}}{2}\right)\left(\frac{\sqrt{3}}{2}\right) - \left(-\frac{\sqrt{2}}{2}\right)\left(\frac{1}{2}\right)$$
$$= \frac{\sqrt{6}}{4} + \frac{\sqrt{2}}{4}$$
$$= \frac{\sqrt{6} + \sqrt{2}}{4}$$

23. $\tan 255° = \tan(210° + 45°)$

$$= \frac{\tan 210° + \tan 45°}{1 - \tan 210° \tan 45°}$$
$$= \frac{\tan 30° + \tan 45°}{1 - \tan 30° \tan 45°}$$
$$= \frac{\frac{\sqrt{3}}{3} + 1}{1 - \frac{\sqrt{3}}{3}(1)}$$
$$= \frac{\sqrt{3} + 3}{3 - \sqrt{3}}$$
$$= \frac{\sqrt{3} + 3}{3 - \sqrt{3}} \cdot \frac{3 + \sqrt{3}}{3 + \sqrt{3}}$$
$$= \frac{6\sqrt{3} + 12}{6}$$
$$= \sqrt{3} + 2$$

25. $\cos 195° = \cos(135° + 60°)$

$$= \cos 135° \cos 60° - \sin 135° \sin 60°$$
$$= -\cos 45° \cos 60° - \sin 45° \sin 60°$$
$$= \left(-\frac{\sqrt{2}}{2}\right)\left(\frac{1}{2}\right) - \left(\frac{\sqrt{2}}{2}\right)\left(\frac{\sqrt{3}}{2}\right)$$
$$= -\frac{\sqrt{2}}{4} - -\frac{\sqrt{6}}{4}$$
$$= \frac{-\sqrt{2} - \sqrt{6}}{4}$$

27. $\cos -165° = \cos(120° + 45°)$

$$= \cos 120° \cos 45° - \sin 120° \sin 145°$$
$$= \left(-\frac{1}{2} \cdot \frac{\sqrt{2}}{2}\right) - \left(\frac{\sqrt{3}}{2} \cdot \frac{\sqrt{2}}{2}\right)$$

29. $\sin\frac{\pi}{12} = \sin\left(\frac{\pi}{3} - \frac{\pi}{4}\right)$

$$= \sin\frac{\pi}{3}\cos\frac{\pi}{4} - \cos\frac{\pi}{3}\sin\frac{\pi}{4}$$
$$= \left(\frac{\sqrt{3}}{2}\right)\left(\frac{\sqrt{2}}{2}\right) - \left(\frac{1}{2}\right)\left(\frac{\sqrt{2}}{2}\right)$$
$$= \frac{\sqrt{6}}{4} - \frac{\sqrt{2}}{4}$$
$$= \frac{\sqrt{6} - \sqrt{2}}{4}$$

31. $\sin\frac{-5\pi}{6} = \sin\left(\frac{\pi}{6} - \pi\right)$

$$= \sin\frac{\pi}{6}\cos\pi - \cos\frac{\pi}{6}\sin\pi$$
$$= \frac{1}{2}(-1) - \frac{\sqrt{3}}{2}(0)$$
$$= -\frac{1}{2}$$

33. $\cos\frac{25\pi}{12}=\cos\left(\frac{3\pi}{4}+\frac{4\pi}{3}\right)$

$=\cos\frac{3\pi}{4}\cos\frac{4\pi}{3}-\sin\frac{3\pi}{4}\sin\frac{4\pi}{3}$

$=\left(-\frac{\sqrt{2}}{2}\right)\left(-\frac{1}{2}\right)-\left(\frac{\sqrt{2}}{2}\right)\left(-\frac{\sqrt{3}}{2}\right)$

$=\frac{\sqrt{2}}{4}+\frac{\sqrt{6}}{4}$

$=\frac{\sqrt{2}+\sqrt{6}}{4}$

35. $\sin\frac{11\pi}{12}=\sin\left(\frac{\pi}{4}+\frac{2\pi}{3}\right)$

$=\sin\frac{\pi}{4}\cos\frac{2\pi}{3}+\cos\frac{\pi}{4}\sin\frac{2\pi}{3}$

$=\frac{\sqrt{2}}{2}\left(-\frac{1}{2}\right)+\frac{\sqrt{2}}{2}\left(\frac{\sqrt{3}}{2}\right)$

$=-\frac{\sqrt{2}}{4}+\frac{\sqrt{6}}{4}$

$=\frac{\sqrt{6}-\sqrt{2}}{4}$

37. $\cos\frac{-7\pi}{6}=\cos\left(\frac{-\pi}{2}-\frac{2\pi}{3}\right)$

$=\cos\frac{-\pi}{2}\cos\frac{2\pi}{3}+\sin\frac{-\pi}{2}\sin\frac{2\pi}{3}$

$=\cos\frac{\pi}{2}\cos\frac{2\pi}{3}-\sin\frac{\pi}{2}\sin\frac{\pi}{3}$

$=0\left(-\frac{1}{2}\right)-(1)\frac{\sqrt{3}}{2}$

$=-\frac{\sqrt{3}}{2}$

39. $\sin\frac{5\pi}{12}=\sin\left(\frac{\pi}{6}+\frac{\pi}{4}\right)$

$=\sin\frac{\pi}{6}\cos\frac{\pi}{4}+\cos\frac{\pi}{6}\sin\frac{\pi}{4}$

$=\left(\frac{1}{2}\right)\left(\frac{\sqrt{2}}{2}\right)+\left(\frac{\sqrt{3}}{2}\right)\left(\frac{\sqrt{2}}{2}\right)$

$=\frac{\sqrt{2}}{4}+\frac{\sqrt{6}}{4}$

$=\frac{\sqrt{2}+\sqrt{6}}{4}$

41. $\cos(\alpha-\beta)=\cos\alpha\cos\beta+\sin\alpha\sin\beta$

$\sin\alpha=\frac{4}{5}\Rightarrow\cos\alpha=\frac{3}{5}$, and

$\sin\beta=\frac{5}{13}\Rightarrow\cos\beta=\frac{12}{13}$

Then, $\cos(\alpha-\beta)=\left(\frac{3}{5}\right)\left(\frac{12}{13}\right)+\left(\frac{4}{5}\right)\left(\frac{5}{13}\right)$

$=\frac{36}{65}+\frac{20}{65}$

$=\frac{56}{65}$

43. $\sin(\alpha+\beta)=\sin\alpha\cos\beta+\cos\alpha\sin\beta$

$\cos\alpha=\frac{-15}{17}\Rightarrow\sin\alpha=\frac{8}{17}$, and

$\cos\beta=\frac{-3}{5}\Rightarrow\sin\beta=\frac{-4}{5}$

$\sin(\alpha+\beta)=\frac{8}{17}\left(\frac{-3}{5}\right)+\left(\frac{-15}{17}\right)\left(\frac{-4}{5}\right)$

$=\frac{36}{85}$

45. Use $\sin(\beta-\alpha)=\sin\beta\cos\alpha-\cos\beta\sin\alpha$

$\cos\alpha=\frac{2}{5}$ and α in quadrant I $\Rightarrow$

$\sin\alpha=\frac{\sqrt{21}}{5}$

$\cos\beta=\frac{1}{5}$ and β in quadrant I $\Rightarrow$

$\sin\beta=\frac{2\sqrt{6}}{5}$

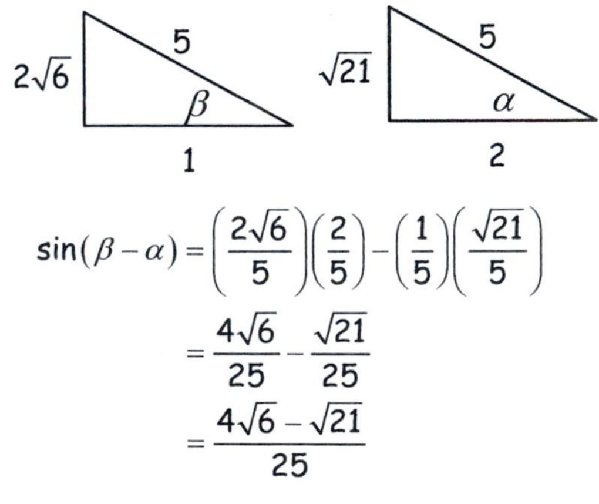

$\sin(\beta-\alpha)=\left(\frac{2\sqrt{6}}{5}\right)\left(\frac{2}{5}\right)-\left(\frac{1}{5}\right)\left(\frac{\sqrt{21}}{5}\right)$

$=\frac{4\sqrt{6}}{25}-\frac{\sqrt{21}}{25}$

$=\frac{4\sqrt{6}-\sqrt{21}}{25}$

47. $\sin 15^\circ \cos 30^\circ + \cos 15^\circ \sin 30^\circ =$

$\sin(15^\circ + 30^\circ) =$

$\sin 45^\circ = \dfrac{\sqrt{2}}{2}$

49. $\dfrac{\tan 100^\circ + \tan 35^\circ}{1 - \tan 100^\circ \tan 35^\circ} = \tan(100^\circ + 35^\circ)$

$= \tan 135^\circ$

$= -\tan 45^\circ$

$= -1$

51. $\dfrac{\tan\frac{5\pi}{16} - \tan\frac{\pi}{16}}{1 + \tan\frac{5\pi}{16}\tan\frac{\pi}{16}} = \tan\left(\frac{5\pi}{16} - \frac{\pi}{16}\right)$

$= \tan\frac{\pi}{4}$

$= 1$

53. $\sin 70^\circ \cos 80^\circ + \cos 70^\circ \sin 80^\circ =$

$\sin(70^\circ + 80^\circ) =$

$\sin 150^\circ =$

$\sin 30^\circ = \dfrac{1}{2}$

55. $\cos 182^\circ \cos 47^\circ + \sin 182^\circ \sin 47^\circ =$

$\cos(182^\circ - 47^\circ) =$

$\cos 135^\circ =$

$-\cos 45^\circ = -\dfrac{\sqrt{2}}{2}$

57. $\dfrac{\tan 70^\circ - \tan 10^\circ}{1 + \tan 70^\circ \tan 10^\circ} = \tan(70^\circ - 10^\circ)$

$= \tan 60^\circ$

$= \sqrt{3}$

59. $\tan\left(\frac{\pi}{2} - \theta\right) = \dfrac{\sin\left(\frac{\pi}{2} - \theta\right)}{\cos\left(\frac{\pi}{2} - \theta\right)}$

$= \dfrac{\sin\frac{\pi}{2}\cos\theta - \cos\frac{\pi}{2}\sin\theta}{\cos\frac{\pi}{2}\cos\theta + \sin\frac{\pi}{2}\sin\theta}$

$= \dfrac{(1)\cos\theta - (0)(\sin\theta)}{(0)\cos\theta + (1)\sin\theta}$

$= \dfrac{\cos\theta}{\sin\theta}$

$= \cot\theta$

61. $\cos\left(\frac{3\pi}{2} - \alpha\right) = \cos\frac{3\pi}{2}\cos\alpha + \sin\frac{3\pi}{2}\sin\alpha$

$= (0)\cos\alpha + -1\sin\alpha$

$= -\sin\alpha$

63. $\tan\left(\alpha - \frac{5\pi}{4}\right) = \dfrac{\tan\alpha - \tan\frac{5\pi}{4}}{1 + \tan\alpha\tan\frac{5\pi}{4}}$

$= \dfrac{\tan\alpha - \tan\frac{\pi}{4}}{1 + \tan\alpha\tan\frac{\pi}{4}}$

$= \dfrac{\tan\alpha - 1}{1 + \tan\alpha}$

65. $\tan(\pi + 2\pi) = \dfrac{\tan\pi + \tan 2\pi}{1 - \tan\pi\tan 2\pi}$

$= \dfrac{0 + 0}{1 - (0)(0)}$

$= 0$

67.

$\sin(u+v)\sin(u-v) =$

$(\sin u\cos v + \cos u\sin v)(\sin u\cos v - \cos u\sin v) =$

$\sin^2 u\cos^2 v - \cos^2 u\sin^2 v =$

$(\sin^2 u)(1 - \sin^2 v) - \sin^2 v(1 - \sin^2 u) =$

$\sin^2 u - \sin^2 u\sin^2 v - \sin^2 v + \sin^2 v\sin^2 u =$

$\sin^2 u - \sin^2 v$

For items 69 - 73, use a graphing calculator to test whether the given identities are true or false.

69. Use the $\langle y = \rangle$ to input the two expressions. You should observe that the two graphs coincide. The identity is true.

71. Use the $\langle y = \rangle$ to input the two expressions. To do this use the identity $\cot\theta = \dfrac{1}{\tan\theta}$.
You should observe that the two graphs do not coincide. The identity is false.

73. Use the $\langle y = \rangle$ to input the two expressions. You should observe that the two graphs coincide. The identity is true.

75. Given: $\sin\left(\sin^{-1}2x + \cos^{-1}2x\right)$

Let $u = \sin^{-1}2x$ and $v = \cos^{-1}2x$
Then, $\sin u = 2x$

$$\cos v = 2x$$
$$\cos u = \sqrt{1-4x^2}$$
$$\sin v = \sqrt{1-4x^2}$$

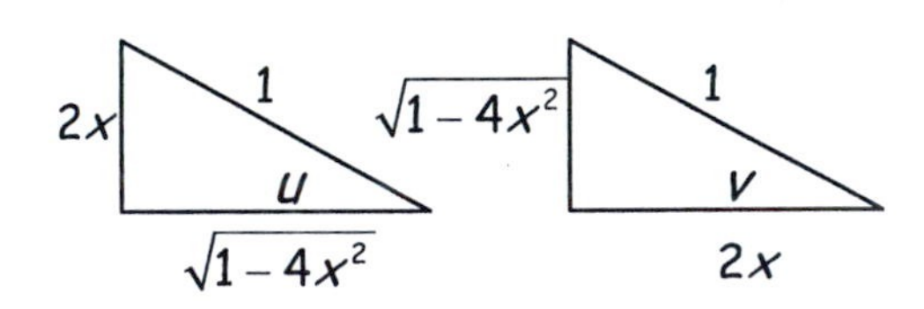

$$\sin\left(\sin^{-1}2x + \cos^{-1}2x\right) =$$
$$\sin(u+v) =$$
$$\sin u \cos v + \cos u \sin v =$$
$$2x(2x) + \sqrt{1-4x^2}\cdot\sqrt{1-4x^2} =$$
$$4x^2 + 1 - 4x^2 = 1$$

77. Given: $\cos(\arctan 2x - \arcsin x)$

Let $u = \arctan 2x$ and $v = \arcsin x$
Then, $\tan u = 2x$ and $\sin v = x$

$$\cos v = \sqrt{1-x^2}$$
$$\cos u = \frac{1}{\sqrt{4x^2+1}}$$
$$\sin u = \frac{2x}{\sqrt{4x^2+1}}$$

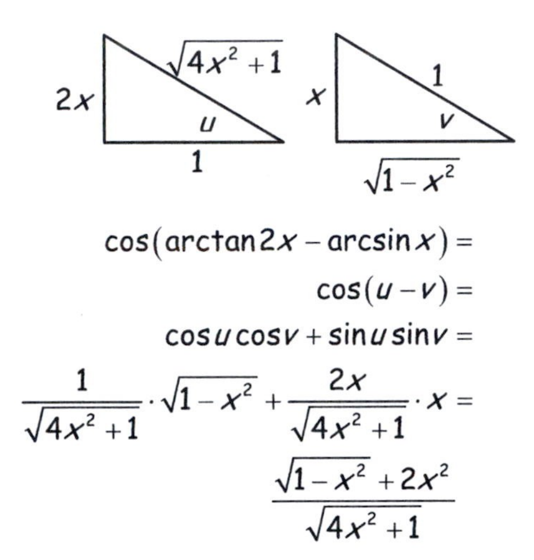

$$\cos(\arctan 2x - \arcsin x) =$$
$$\cos(u-v) =$$
$$\cos u \cos v + \sin u \sin v =$$
$$\frac{1}{\sqrt{4x^2+1}}\cdot\sqrt{1-x^2} + \frac{2x}{\sqrt{4x^2+1}}\cdot x =$$
$$\frac{\sqrt{1-x^2}+2x^2}{\sqrt{4x^2+1}}$$

79. Given: $\cos(\arccos x + \arcsin 2x)$

Let $u = \arccos x$ and $v = \arcsin 2x$
Then, $\cos u = x$

$$\sin u = \sqrt{1-x^2}$$
$$\cos v = \sqrt{1-4x^2}$$
$$\sin v = 2x$$

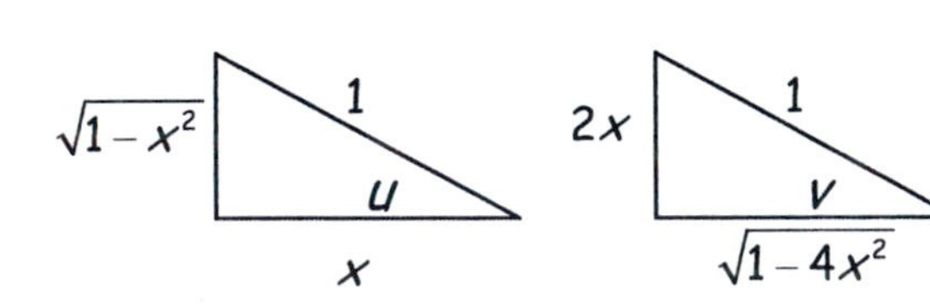

$$\cos(\arccos x + \arcsin 2x) =$$
$$\cos(u+v) =$$
$$\cos u \cos v - \sin u \sin v =$$
$$x\sqrt{1-4x^2} - 2x\sqrt{1-x^2}$$

For items 81 - 85, use the formula from The text:

$$A\sin x + B\cos x = \sqrt{A^2+B^2}\sin(x+\varphi),$$

where $\cos\varphi = \dfrac{A}{\sqrt{A^2+B^2}}$; $\sin\varphi = \dfrac{B}{\sqrt{A^2+B^2}}$

81. $f(x) = \sin x + \cos x$

$A = 1$ and $B = 1$

$\cos\varphi = \dfrac{1}{\sqrt{2}}$; $\sin\varphi = \dfrac{1}{\sqrt{2}} \Rightarrow \varphi = \dfrac{\pi}{4}$

Substituting in the given formula:

$$f(x) = \sqrt{1^2+1^2}\sin\left(x+\frac{\pi}{4}\right)$$

$$f(x) = \sqrt{2}\sin\left(x+\frac{\pi}{4}\right)$$

In graphing f, think first of the sine function. The sine function is shifted $\dfrac{\pi}{4}$ unit to the left and has an amplitude of $\sqrt{2}$. The period is 2π.

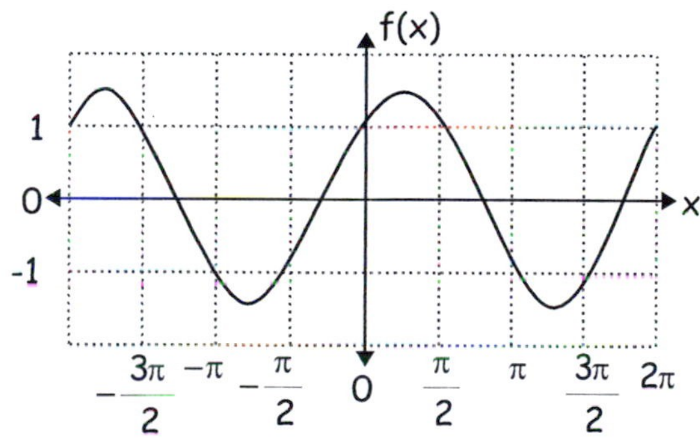

83. $h(\beta) = \sin 2\beta - \cos 2\beta$

Rewrite as $h(\beta) = (1)\sin 2\beta + (-1)\cos 2\beta$

Set $A = 1$ and $B = -1$

$\cos\varphi = \dfrac{1}{\sqrt{2}}$; $\sin\varphi = \dfrac{-1}{\sqrt{2}} \Rightarrow \varphi$ lies in quadrant IV $\Rightarrow \varphi = -\dfrac{\pi}{4}$

Substituting in the given formula:

$$h(\beta) = \sqrt{1^2+(-1)^2}\sin\left(2\beta-\frac{\pi}{4}\right)$$

$$h(\beta) = \sqrt{2}\sin\left(2\beta-\frac{\pi}{4}\right), \text{ or}$$

$$h(\beta) = \sqrt{2}\sin\left[2\left(\beta-\frac{\pi}{8}\right)\right]$$

In graphing h, think first of the sine function. The sine function is shifted $\dfrac{\pi}{8}$ unit to the right and has an amplitude of $\sqrt{2}$. The period is π.

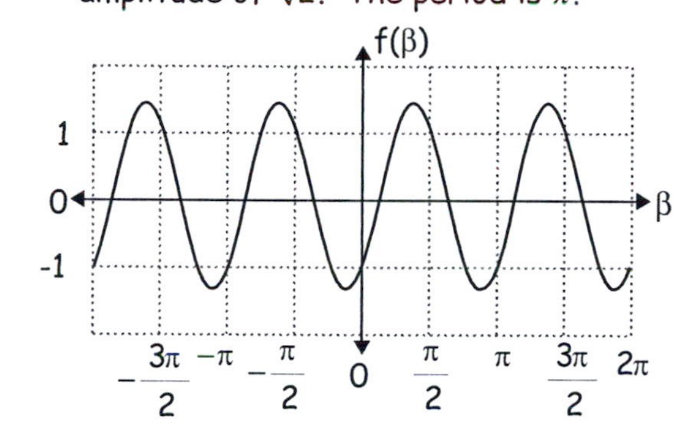

85. $g(u) = 5\sin 5u + 12\cos 5u$

Set $A = 5$ and $B = 12$

$\cos\varphi = \dfrac{5}{\sqrt{5^2+12^2}} = \dfrac{5}{13}$;

$\sin\varphi = \dfrac{12}{13} \Rightarrow \varphi = \sin^{-1}\dfrac{12}{13}$

Use a calulator to approximate φ:

$\varphi \approx 1.1760$

Substituting in the given formula:

$g(u) = 13\sin(5u + 1.1760)$, or

$g(u) = 13\sin 5(u + 0.2352)$,

In graphing g, think first of the sine function. The sine function is shifted 0.2352 unit to the left and has an amplitude of 13. The period is $\dfrac{2\pi}{5}$.

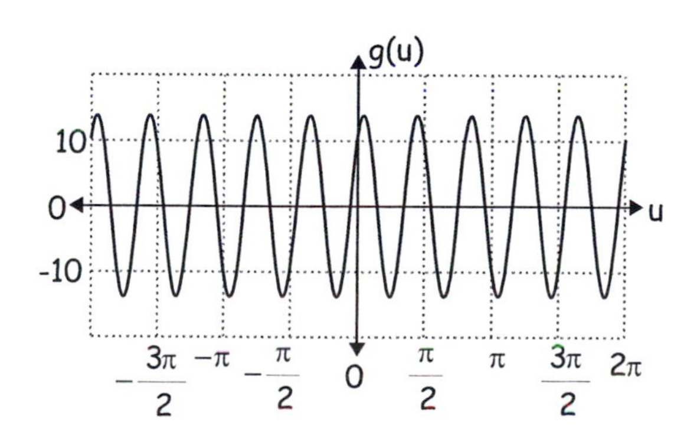

87. Prove: $\sin(u+v)=\sin u\cos v+\cos u\sin v$

Use $\sin\theta=\cos\left(\frac{\pi}{2}-\theta\right)$:

$$\sin(u+v)=\cos\left(\frac{\pi}{2}-(u+v)\right)$$
$$=\cos\left(\left(\frac{\pi}{2}-u\right)-v\right)$$

Then, by the difference identity for cosine:

$$\sin(u+v)=\cos\left(\frac{\pi}{2}-u\right)\cos v+\sin\left(\frac{\pi}{2}-u\right)\sin v$$
$$=\sin u\cos v+\cos u\sin v$$

Prove: $\sin(u-v)=\sin u\cos v-\cos u\sin v$

Use $\sin\theta=\cos\left(\frac{\pi}{2}-\theta\right)$:

$$\sin(u-v)=\cos\left(\frac{\pi}{2}-(u-v)\right)$$
$$=\cos\left(\left(\frac{\pi}{2}-u\right)+v\right)$$

Then, by the sum identity for cosine:

$$\sin(u-v)=\cos\left(\frac{\pi}{2}-u\right)\cos v-\sin\left(\frac{\pi}{2}-u\right)\sin v$$
$$=\sin u\cos v-\cos u\sin v$$

89. The identity is true:

$\sin(u+v)+\sin(u-v)=$
$\sin u\cos v+\cos u\sin v+$
$\sin u\cos v-\cos u\sin v=2\sin u\cos v$

91. This proposition is false:

$$\frac{\cos(u-v)}{\cos(u+v)}=\frac{\cos u\cos v+\sin u\sin v}{\cos u\cos v-\sin u\sin v}$$

Divide numerator and denominator by $\cos u\cos v$:

$$=\frac{\dfrac{\cos u\cos v+\sin u\sin v}{\cos u\cos v}}{\dfrac{\cos u\cos v-\sin u\sin v}{\cos u\cos v}}$$
$$=\frac{1+\tan u\tan v}{1-\tan u\tan v}$$
$$\neq 2\tan u\tan v$$

Another approach to disprove this proposition is to select values for u and v, for which $\cos(u+v)\neq 0$, e.g., let $u=30^\circ$ and $v=45^\circ$. Use the values to test the proposition by substitution. Note that one can disprove a propostion in this manner by showing a counter-example, but one cannot prove an identity with a finite number of positive examples.

93. Prove:

$$\tan(u-v)=\frac{\tan u-\tan v}{1+\tan u\tan v}$$

First, use the sine and cosine identities for tangent. Then, divide numerator and denominator by $\cos u\cos v$:

$$\tan(u-v)=\frac{\sin(u-v)}{\cos(u-v)}$$
$$=\frac{\sin u\cos v-\cos u\sin v}{\cos u\cos v+\sin u\sin v}$$
$$=\frac{\dfrac{\sin u\cos v-\cos u\sin v}{\cos u\cos v}}{\dfrac{\cos u\cos v+\sin u\sin v}{\cos u\cos v}}$$
$$=\frac{\tan u-\tan v}{1+\tan u\tan v}$$

End of Section 7.2

Chapter Seven: Section 7.3
Solutions to Odd-numbered Exercises

1. Given $\sin x = \frac{3}{5}$; $\cos x > 0 \Rightarrow x$ lies in quadrant I

5 3 x

$\sqrt{25-9} = 4$

Then, $\cos x = \frac{4}{5}$; $\tan x = \frac{3}{4}$

a. $\cos 2x = 1 - 2\sin^2 x$

$$= 1 - 2\left(\frac{3}{5}\right)^2$$

$$= 1 - \frac{18}{25}$$

$$= \frac{7}{25}$$

b. $\sin 2x = 2\sin x \cos x$

$$= 2\left(\frac{3}{5}\right)\left(\frac{4}{5}\right)$$

$$= \frac{24}{25}$$

c. $\tan 2x = \frac{2\tan x}{1 - \tan^2 x}$

$$= \frac{2\left(\frac{3}{4}\right)}{1 - \left(\frac{3}{4}\right)^2}$$

$$= \frac{\frac{3}{2}}{1 - \frac{9}{16}}$$

$$= \frac{3}{2} \cdot \frac{16}{7}$$

$$= \frac{24}{7}$$

3. Given $\cos x = -\frac{2}{\sqrt{6}}$; $\sin x > 0 \Rightarrow x$ lies in quadrant II.

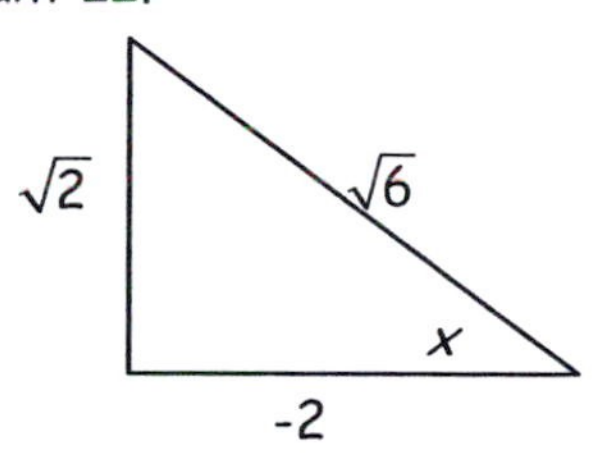

Then, $\sin x = \frac{\sqrt{2}}{\sqrt{6}} = \frac{2\sqrt{3}}{6} = \frac{\sqrt{3}}{3}$;

$\tan x = \frac{\sqrt{2}}{-2} = -\frac{\sqrt{2}}{2}$

a. $\cos 2x = 1 - 2\sin^2 x$

$$= 1 - 2\left(\frac{\sqrt{3}}{3}\right)^2$$

$$= 1 - \frac{6}{9}$$

$$= 1 - \frac{2}{3}$$

$$= \frac{1}{3}$$

b. $\sin 2x = 2\sin x \cos x$

$$= 2\left(\frac{\sqrt{3}}{3}\right)\left(\frac{-2}{\sqrt{6}}\right)$$

$$= -\frac{4}{3}\sqrt{\frac{1}{2} \cdot \frac{2}{2}}$$

$$= -\frac{2\sqrt{2}}{3}$$

c. $\tan 2x = \frac{2\tan x}{1 - \tan^2 x}$

$$= \frac{2\left(-\frac{\sqrt{2}}{2}\right)}{1 - \left(-\frac{\sqrt{2}}{2}\right)^2}$$

$$= -\frac{\sqrt{2}}{1 - \frac{2}{4}}$$

$$= -2\sqrt{2}$$

5. Given $\tan x = \frac{1}{\sqrt{3}}$; $\cos x < 0 \Rightarrow x$ lies in quadrant III.

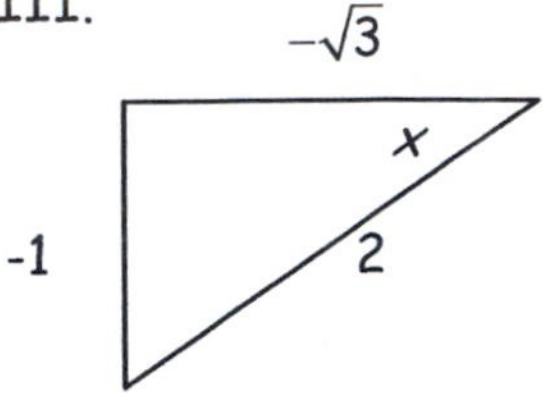

Then, $\sin x = -\frac{1}{2}$; $\cos x = \frac{-\sqrt{3}}{2}$

$\tan x = \frac{-1}{-\sqrt{3}} = \frac{\sqrt{3}}{3}$

a. $\cos 2x = 1 - 2\sin^2 x$

$$= 1 - 2\left(-\frac{1}{2}\right)^2$$

$$= 1 - \frac{1}{2}$$

$$= \frac{1}{2}$$

b. $\sin 2x = 2\sin x \cos x$

$$= 2\left(-\frac{1}{2}\right)\left(-\frac{\sqrt{3}}{2}\right)$$

$$= \frac{\sqrt{3}}{2}$$

c. $\tan 2x = \frac{2\tan x}{1 - \tan^2 x}$

$$= \frac{2\left(\frac{\sqrt{3}}{3}\right)}{1 - \left(\frac{\sqrt{3}}{3}\right)^2}$$

$$= \left(\frac{2\sqrt{3}}{3}\right)\left(\frac{3}{2}\right)$$

$$= \sqrt{3}$$

7. Show: $\tan 3x = \frac{3\tan x - \tan^3 x}{1 - 3\tan^2 x}$

$\tan 3x = \tan(2x + x)$

$$= \frac{\tan 2x + \tan x}{1 - \tan 2x \tan x}$$

$$= \frac{\frac{2\tan x}{1 - \tan^2 x} + \tan x}{1 - \frac{2\tan x}{1 - \tan^2 x}\tan x}$$

$$= \frac{\frac{2\tan x + \tan x(1 - \tan^2 x)}{\cancel{1 - \tan^2 x}}}{\frac{1 - \tan^2 x - 2\tan^2 x}{\cancel{1 - \tan^2 x}}}$$

$$= \frac{2\tan x + \tan x - \tan^3 x}{1 - 3\tan^2 x}$$

$$= \frac{3\tan x - \tan^3 x}{1 - 3\tan^2 x}$$

9. Show: $\frac{\sin 4x - \sin 2x}{\cos 4x + \cos 2x} = \tan x$

Let $u = 2x$.

$$\frac{\sin 4x - \sin 2x}{\cos 4x + \cos 2x} = \frac{\sin 2u - \sin u}{\cos 2u + \cos u}$$

$$= \frac{2\sin u \cos u - \sin u}{1 - 2\sin^2 u + \cos u}$$

$$= \frac{(2\cos u - 1)\sin u}{1 - 2(1 - \cos^2 u) + \cos u}$$

$$= \frac{(2\cos u - 1)\sin u}{-1 + 2\cos^2 u + \cos u}$$

$$= \frac{(2\cos u - 1)\sin u}{(2\cos u - 1)(\cos u + 1)}$$

$$= \frac{\sin u}{\cos u + 1}$$

$$= \tan\frac{1}{2}u$$

$$= \tan\frac{1}{2}(2x) = \tan x$$

11. Show: $2\sin^2 3x = 1-\cos 6x$

Let $u = 3x$.

$$1-\cos 6x = 1-\cos 2u$$
$$= 1-\left(\cos^2 u - \sin^2 u\right)$$
$$= 1-\left(1-\sin^2 u - \sin^2 u\right)$$
$$= 2\sin^2 u$$
$$= 2\sin^2 3x$$

13. $\sin^3 x = \sin^2 x(\sin x)$

$$= \left(1-\cos^2 x\right)(\sin x)$$
$$= \left(1-\left(\frac{1+\cos 2x}{2}\right)\right)\sin x$$
$$= \sin x - \frac{\sin x \cos 2x + \sin x}{2}$$
$$= \frac{2\sin x - \sin x - \sin x\cos 2x}{2}$$
$$= \frac{\sin x - \sin x \cos 2x}{2}$$

15.

$$\sin^4 x \cos^2 x =$$
$$\sin^2 x \sin^2 x \cos^2 x =$$
$$\left(\frac{1-\cos 2x}{2}\right)\left(\frac{1-\cos 2x}{2}\right)\left(\frac{1+\cos 2x}{2}\right) =$$
$$\frac{1}{8}(1-\cos 2x)\left(1-\cos^2 2x\right) =$$
$$\frac{1}{8}(1-\cos 2x)\left(1-\frac{1+\cos 4x}{2}\right) =$$
$$\frac{1}{8}(1-\cos 2x)\left(\frac{2-1-\cos 4x}{2}\right) =$$
$$\frac{1}{8}(1-\cos 2x)\left(\frac{1-\cos 4x}{2}\right) =$$
$$\frac{1}{16}(1-\cos 2x - \cos 4x + \cos 2x \cos 4x)$$

17.

$$\tan^4 x \sin x =$$
$$\tan^2 x \tan^2 x \sin x =$$
$$\left(\frac{1-\cos 2x}{1+\cos 2x}\right)\left(\frac{1-\cos 2x}{1+\cos 2x}\right)\sin x =$$
$$\left(\frac{1-2\cos 2x+\cos^2 2x}{1+2\cos 2x+\cos^2 2x}\right)\sin x =$$
$$\left(\frac{1-2\cos 2x+\frac{1+\cos 4x}{2}}{1+2\cos 2x+\frac{1+\cos 4x}{2}}\right)\sin x =$$
$$\left(\frac{2-4\cos 2x+1+\cos 4x}{2+4\cos 2x+1+\cos 4x}\right)\sin x =$$
$$\left(\frac{3-4\cos 2x+\cos 4x}{3+4\cos 2x+\cos 4x}\right)\sin x =$$
$$\frac{3\sin x - 4\sin x\cos 2x + \sin x \cos 4x}{3+4\cos 2x+\cos 4x}$$

19. Exact value of $\sin\left(\frac{3\pi}{8}\right)$:

$$\sin\left(\frac{3\pi}{8}\right) = \sin\frac{1}{2}\left(\frac{3\pi}{4}\right)$$
$$= \sqrt{\frac{1-\cos\frac{3\pi}{4}}{2}}$$
$$= \sqrt{\frac{1-\left(-\cos\frac{\pi}{4}\right)}{2}}$$
$$= \sqrt{\frac{1+\frac{\sqrt{2}}{2}}{2}}$$
$$= \sqrt{\frac{2+\sqrt{2}}{4}}$$
$$= \frac{\sqrt{2+\sqrt{2}}}{2}$$

21. Exact value for $\cos\left(-\frac{\pi}{12}\right)$:

$$\cos\left(-\frac{\pi}{12}\right)=\cos\frac{1}{2}\left(-\frac{\pi}{6}\right)$$
$$=\sqrt{\frac{1+\cos\left(-\frac{\pi}{6}\right)}{2}}$$
$$=\sqrt{\frac{1+\cos\frac{\pi}{6}}{2}}$$
$$=\sqrt{\frac{1+\frac{\sqrt{3}}{2}}{2}}$$
$$=\sqrt{\frac{2+\sqrt{3}}{4}}$$
$$=\frac{1}{2}\sqrt{2+\sqrt{3}}$$

23. Exact value for $\sin 75°$:

$$\sin 75° = \sin\left[\frac{1}{2}(150°)\right]$$
$$=\sqrt{\frac{1-\cos 150°}{2}}$$
$$=\sqrt{\frac{1-(-\cos 30°)}{2}}$$
$$=\sqrt{\frac{1+\frac{\sqrt{3}}{2}}{2}}$$
$$=\sqrt{\frac{2+\sqrt{3}}{4}}$$
$$=\frac{\sqrt{2+\sqrt{3}}}{2}$$

25. Use the product-to-sum identity:

$$\sin 3x\cos 3x =$$
$$\frac{1}{2}[\sin(3x+3x)+\sin(3x-3x)]=$$
$$\frac{1}{2}[\sin 6x+\sin 0]=$$
$$\frac{1}{2}\sin 6x+\sin 0=\frac{\sin 6x+\sin 0}{2}$$

27. Use the product-to-sum identity:

$$5\cos 105°\sin 15° = \frac{5}{2}[\sin 120° - \sin 90°]$$

29. Use the product-to-sum identity:

$$\sin(x+y)\sin(x-y)=$$
$$\frac{1}{2}[\cos(x+y-x+y)-\cos(x+y+x-y)]=$$
$$\frac{1}{2}[\cos 2y-\cos 2x]$$

31. Use the product-to-sum identity:

$$\sin\frac{5\pi}{4}\sin\frac{2\pi}{3}=$$
$$\frac{1}{2}\left[\cos\left(\frac{5\pi}{4}-\frac{2\pi}{3}\right)-\cos\left(\frac{5\pi}{4}+\frac{2\pi}{3}\right)\right]=$$
$$\frac{1}{2}\left[\cos\frac{7\pi}{12}-\cos\frac{23\pi}{12}\right]$$

33. Use the product-to-sum identity:

$$2\cos\frac{\pi}{3}\sin\frac{\pi}{6}=$$
$$2\left(\frac{1}{2}\right)\left[\sin\left(\frac{\pi}{3}+\frac{\pi}{6}\right)-\sin\left(\frac{\pi}{3}-\frac{\pi}{6}\right)\right]=$$
$$\sin\frac{\pi}{2}-\sin\frac{\pi}{6}$$

35. Use the sum-to-product identity:

$$\sin 6x+\sin 2x=$$
$$2\sin\left(\frac{6x+2x}{2}\right)\cos\left(\frac{6x-2x}{2}\right)=$$
$$2\sin 4x\cos 2x$$

37. Use the sum-to-product identity:

$$\cos 3\beta-\cos\beta=$$
$$-2\sin\left(\frac{3\beta+\beta}{2}\right)\sin\left(\frac{3\beta-\beta}{2}\right)=$$
$$-2\sin 2\beta\sin\beta$$

39. Use the sum-to-product identity:

$$\sin 135^\circ - \sin 15^\circ =$$
$$2\cos\left(\frac{135^\circ + 15^\circ}{2}\right)\sin\left(\frac{135^\circ - 15^\circ}{2}\right) =$$
$$2\cos\left(\frac{150^\circ}{2}\right)\sin\left(\frac{120^\circ}{2}\right) =$$
$$2\cos 75^\circ \sin 60^\circ$$

41. Use the sum-to-product identity:

$$\cos\frac{7\pi}{6} - \cos\frac{\pi}{4} =$$
$$-2\sin\left(\frac{\frac{7\pi}{6} + \frac{\pi}{4}}{2}\right)\sin\left(\frac{\frac{7\pi}{6} - \frac{\pi}{4}}{2}\right) =$$
$$-2\sin\frac{17\pi}{24}\sin\frac{11\pi}{24}$$

43. a. Prove: $\sin 2u = 2\sin u \cos u$

Use the sum identity for sine:

$$\begin{aligned} \sin 2u &= \sin(u+u) \\ &= \sin u \cos u + \cos u \sin u \\ &= 2\sin u \cos u \end{aligned}$$

b. Prove: $\cos 2u = 1 - 2\sin^2 u$

Use the sum identity for cosine and the Pythagorean identity:

$$\begin{aligned} \cos 2u &= \cos(u+u) \\ &= \cos u \cos u - \sin u \sin u \\ &= \cos^2 u - \sin^2 u \\ &= \left(1 - \sin^2 u\right) - \sin^2 u \\ &= 1 - 2\sin^2 u \end{aligned}$$

c. Prove: $\tan 2u = \dfrac{2\tan u}{1 - \tan^2 u}$

Use the sum identity for tangent:

$$\begin{aligned} \tan 2u &= \tan(u+u) \\ &= \frac{\tan u + \tan u}{1 - \tan u \tan u} \\ &= \frac{2\tan u}{1 - \tan^2 u} \end{aligned}$$

45. a. Prove: $\sin\frac{x}{2} = \pm\sqrt{\frac{1-\cos x}{2}}$

Use the power-reducing identity:

$$\sin^2\frac{x}{2} = \frac{1-\cos x}{2}$$

Then, $\sin\frac{x}{2} = \pm\sqrt{\frac{1-\cos x}{2}}$

b. Prove: $\cos\frac{x}{2} = \pm\sqrt{\frac{1+\cos x}{2}}$

Use the power-reducing identity:

$$\cos^2\frac{x}{2} = \frac{1+\cos x}{2}$$

Then, $\cos\frac{x}{2} = \pm\sqrt{\frac{1+\cos x}{2}}$

End of Section 7.3

Chapter Seven: Section 7.4
Solutions to Odd-numbered Exercises

1. Solve $2\sin x + 1 = 0$:

$\sin x = -\frac{1}{2}$

Note: $\sin x < 0 \Rightarrow x$ lies in the third or fourth quadrant. The reference angle is $\frac{\pi}{6}$. In other words, $x = \frac{7\pi}{6}$ or $x = -\frac{\pi}{6}$. Using a period of 2π, the solution set is

$\left\{x : x = \frac{7\pi}{6} + 2n\pi\right\} \cup \left\{x : x = -\frac{\pi}{6} + 2n\pi\right\}$,

where n is an integer.

3. Solve $\sqrt{2} - 2\cos x = 0$:

$\cos x = \frac{\sqrt{2}}{2}$

Note: $\cos x > 0 \Rightarrow x$ lies in the first or fourth quadrant. The reference angle is $\frac{\pi}{4}$. In other words, $x = \frac{\pi}{4}$ or $x = -\frac{\pi}{4}$. Using a period of 2π, the solution set is

$\left\{x : x = \frac{\pi}{4} + 2n\pi\right\} \cup \left\{x : x = -\frac{\pi}{4} + 2n\pi\right\}$,

where n is an integer.

5. Solve $2\cos x - \sqrt{3} = 0$:

$\cos x = \frac{\sqrt{3}}{2}$

Note: $\cos x > 0 \Rightarrow x$ lies in the first or fourth quadrant. The reference angle is $\frac{\pi}{6}$. In other words, $x = \frac{\pi}{6}$ or $x = \frac{11\pi}{6}$. Using a period of 2π, the solution set is

$\left\{x : x = \frac{\pi}{6} + 2n\pi\right\} \cup \left\{x : x = \frac{11\pi}{6} + 2n\pi\right\}$,

where n is an integer.

7. Solve: $-\frac{1}{\sqrt{48}\sin x} = \frac{1}{8}$

$-8 = \sqrt{48}\sin x$

$\sin x = \frac{-8}{\sqrt{48}}$

$\sin x = \frac{-8}{4\sqrt{3}}$

$\sin x = -\frac{2}{\sqrt{3}} = -\frac{2\sqrt{3}}{3} < -1$

Therefore, since $-\frac{2\sqrt{3}}{3}$ is outside the range of sine, the solution set is the empty set, symbolized by $\varnothing$.

9. Solve $\sqrt{3}\tan x + 1 = -2$:

$\sqrt{3}\tan x = -3$

$\tan x = \frac{-3}{\sqrt{3}} = -\sqrt{3}$

Note: $\tan x < 0 \Rightarrow x$ lies in the second or fourth quadrant. The reference angle is $\frac{\pi}{3}$. In other words, $x = \frac{2\pi}{3}$ or $x = -\frac{\pi}{3}$. Using a period of π, the solution set is $\left\{x : x = \frac{2\pi}{3} + n\pi\right\}$, where n is an integer. Also, note that $-\frac{\pi}{3}$ would yield the same solution set as $\frac{2\pi}{3}$.

11. Solve $\sec^2 x - 1 = 0$:

$\sec^2 x = 1$

$\sec x = \pm 1$

Note: $\sec x = \pm 1 \Rightarrow \cos x = \pm 1$.

The reference angle is 0 or π.

Using the period of $\sec^2$, π, the solution set is $\{x : x = n\pi\}$, where n is an integer.

13. Solve: $\sec x + \tan x = 1$

Use the sine and cosine identities:

$$\frac{1}{\cos x} + \frac{\sin x}{\cos x} = 1$$

(Multiply both sides by $\cos x$)

$$1 + \sin x = \cos x$$

$$\sin x = \cos x - 1$$

Square both sides (being wary that we may introduce extraneous solutions):

$$\sin^2 x = (\cos x - 1)^2$$

$$\sin^2 x = \cos^2 x - 2\cos x + 1$$

Use a Pythagorean identity:

$$1 - \cos^2 x = \cos^2 x - 2\cos x + 1$$

$$2\cos^2 x - 2\cos x = 0$$

$$2\cos x(\cos x - 1) = 0$$

Set each factor equal to zero and solve:

$\cos x = 0$	$\cos x = 1$
$\Rightarrow$	$\Rightarrow$
$x = \frac{\pi}{2}$	$x = 0$
Discard	

$\frac{\pi}{2}$ is discarded because secant and tangent are both not defined at $x = \frac{\pi}{2}$.

The solution set is $\{x : x = 0 + 2n\pi\} = \{x : x = 2n\pi\}$, for any integer n.

15. Solve: $\sin^2 x + \cos^2 x + \tan^2 x = 0$

Notice that each of the three terms on the left is always non-negative. For any value of x for which $\cos x = 0$; (i.e., for $x = \frac{\pi n}{2}$, where n = any odd integer), $\tan x$ is not a defined term. For all other values of x, the value of the left hand side is positive. Therefore, we can conclude that there are no real number solutions.

17. Solve: $2\cos^2 x - 3 = 5\cos x$

Treat as a quadratic equation in $\cos x$. Factor the quadratic, set each factor equal to zero and solve. Examine each of the potential solutions for its applicability to the original problem:

$$2\cos^2 x - 3 = 5\cos x$$

$$2\cos^2 x - 5\cos x - 3 = 0$$

$$(2\cos x + 1)(\cos x - 3) = 0$$

$2\cos x + 1 = 0$	$\cos x - 3 = 0$
$\cos x = -\frac{1}{2}$	$\cos x = 3$
	Discard

$\cos x = 3$ is discarded because 3 is not in the range of cosine.

$\cos x = -\frac{1}{2} \Rightarrow x$ lies in the second or third quadrants $\Rightarrow$ the basic angles are $\frac{2\pi}{3}$ and $\frac{4\pi}{3}$. With a period of 2π for cosine, the solution set is

$$\left\{x : x = \frac{2\pi}{3} + 2n\pi\right\} \cup \left\{x : x = \frac{4\pi}{3} + 2n\pi\right\}$$

for any integer n.

19. Solve: $\frac{\cot 2x}{\sqrt{3}} = -1$

$$\cot 2x = -\sqrt{3}$$

The negative value indicates that $2x$ lies in the second or fourth quadrants.

$$\cot 2x = -\sqrt{3} \Rightarrow 2x = -\frac{\pi}{6} \text{ or } \frac{5\pi}{6}$$

Recall that the period of cotangent is π. Then, the equation is true for

$2x = -\frac{\pi}{6} + n\pi \Rightarrow x = -\frac{\pi}{12} + \frac{n\pi}{2}$, or

$2x = \frac{5\pi}{6} + n\pi \Rightarrow x = \frac{5\pi}{12} + \frac{n\pi}{2}$, for any integer n. These implications describe the same solution set.

The solution set could be represented as $\left\{x : x = -\frac{\pi}{12} + \frac{n\pi}{2}\right\}$ or equivalently as $\left\{x : x = \frac{5\pi}{12} + \frac{n\pi}{2}\right\}$.

21. Solve: $2\sin^2 x + 7\sin x = 4$ on $[0, 2\pi]$
Treat as a quadratic equation in $\sin x$. Factor the quadratic, set each factor equal to zero and solve. Examine each of the potential solutions for its applicability to the range of sine:

$$2\sin^2 x + 7\sin x - 4 = 0$$
$$(2\sin x - 1)(\sin x + 4) = 0$$

$2\sin x - 1 = 0$	$\sin x + 4 = 0$
$\sin x = \frac{1}{2}$	$\sin x = -4$
	Discard

$\sin x = -4$ is discarded because -4 is not in the range of sine.

$\sin x = \frac{1}{2} \Rightarrow x$ lies in the first or second quadrants $\Rightarrow$ the angles are $\frac{\pi}{6}$ and $\frac{5\pi}{6}$.

23. Solve: $2\cos^2 x - 1 = 0$

$$\cos^2 x = \frac{1}{2}$$
$$\cos x = \pm\sqrt{\frac{1}{2}}$$
$$\cos x = \pm\frac{\sqrt{2}}{2}$$
$$x = \frac{\pi}{4}, \frac{3\pi}{4}, \frac{5\pi}{4}, \frac{7\pi}{4}$$

25. Solve: $0.05\sin^3 x = 0.1\sin x$

$$5\sin^3 x = 10\sin x$$
$$\sin^3 x = 2\sin x$$

This equation can be true only for $\sin x = 0$. This implies that $x = 0,\ \pi$ and 2π on the interval $[0,\ 2\pi]$.

For other possible values of $\sin x \neq 0$, dividing both sides by $\sin x$ implies that $\sin^2 x = 2$. But, this implies that $\sin x > 1$, which is outside the range of sine.

27. Solve: $2\cos^2 x + 11\cos x = -5$
Treat as a quadratic equation in $\cos x$. Factor the quadratic, set each factor equal to zero and solve. Examine each of the potential solutions for its applicability to the original problem:

$$2\cos^2 x + 11\cos x + 5 = 0$$
$$(2\cos x + 1)(\cos x + 5) = 0$$

$2\cos x + 1 = 0$	$\cos x + 5 = 0$
$\cos x = -\frac{1}{2}$	$\cos x = -5$
	Discard

$\cos x = -5$ is discarded because -5 is not in the range of cosine.

$\cos x = -\frac{1}{2} \Rightarrow x$ lies in the second or third quadrants.

$$x = \frac{2\pi}{3}, \frac{4\pi}{3}.$$

29. Solve: $\sec^2 x - 2 + 8\tan x = 42$
Use the $\sec^2 x = \tan^2 x + 1$ identity.

$$\tan^2 x + 1 - 2 + 8\tan x = 42$$

Rewrite and use the quadratic formula to solve the equation in $\tan x$:

$$\tan^2 x + 8\tan x - 43 = 0$$
$$\tan x = \frac{-8 \pm \sqrt{64 - 4(-43)}}{2}$$
$$= \frac{-8 \pm \sqrt{4(16 + 43)}}{2}$$
$$= -4 \pm \sqrt{59}$$

Then, $x = \tan^{-1}\left(-4 + \sqrt{59}\right)$

$x = \tan^{-1}\left(-4 - \sqrt{59}\right)$

31. Verify solutions for $2\cos x+1=0$:

a. $$2\cos\frac{2\pi}{3}+1=2\left(-\frac{1}{2}\right)+1$$
$$=-1+1$$
$$=0$$

b. $$2\cos\frac{4\pi}{3}+1=2\left(-\frac{1}{2}\right)+1$$
$$=-1+1$$
$$=0$$

33. Verify solutions for $2\sin^2 x-\sin x-1=0$:

a. $$2\sin^2\frac{\pi}{2}-\sin\frac{\pi}{2}-1=2(1)^2-1-1$$
$$=2-1-1$$
$$=0$$

b.
$$2\sin^2\frac{7\pi}{6}-\sin\frac{7\pi}{6}-1=2\left(-\frac{1}{2}\right)^2-\left(-\frac{1}{2}\right)-1$$
$$=\frac{1}{2}+\frac{1}{2}-1$$
$$=0$$

35. Verify solutions for
$\csc^4 x-4\csc^2 x=0$:

a. $$\csc^4\frac{\pi}{6}-4\csc^2\frac{\pi}{6}=2^4-4(2)^2$$
$$=16-16$$
$$=0$$

b. $$\csc^4\frac{5\pi}{6}-4\csc^2\frac{5\pi}{6}=2^4-4(2)^2$$
$$=16-16$$
$$=0$$

37. Verify solutions for $2\cot x+1=-1$:

a. $$2\cot\frac{3\pi}{4}+1=2(-1)+1$$
$$=-2+1$$
$$=-1$$

b. $$2\cot\frac{7\pi}{4}+1=2(-1)+1$$
$$=-2+1$$
$$=-1$$

39. Verify solutions for
$2\sec x+1=\sec x+3$:

a. $$2\sec\frac{\pi}{3}+1=2(2)+1$$
$$=5$$
$$\sec\frac{\pi}{3}+3=2+3$$
$$=5$$

b. $$2\sec\frac{5\pi}{3}+1=2(2)+1$$
$$=5$$
$$\sec\frac{5\pi}{3}+3=2+3$$
$$=5$$

41. Use a graphing calculator to solve
$x\tan x-3=0$
Use the equation solver or the <math> zero features of the calculator:
$x\approx 1.1925$ and $x\approx 3.8088$

43. Use a graphing calculator to solve
$2\cos^2 x-\sin x=0$
Use the equation solver or the <math> zero features of the calculator:
$x\approx 0.8959$ and $x\approx 2.2457$

45. Use a graphing calculator to solve
$2\sin x-\csc^2 x=0$

Rewrite as $2\sin x-\left(\frac{1}{\sin x}\right)^2 x=0$

Use the equation solver or the <math> zero features of the calculator:
$x\approx 0.9169$ and $x\approx 2.2247$

47. Use a graphing calculator to solve
$\log x=-\sin x$
Use the equation solver or the <math> zero features of the calculator:
$x\approx 0.4043$, $x\approx 3.7535$ and $x\approx 5.4549$

49. Test: $x = \frac{4\pi}{3} + 2n\pi$

$$2\cos\left(\frac{4\pi}{3}\right) = 2\left(-\frac{1}{2}\right)$$
$$= -1$$

The period of cosine is 2π.

Therefore, $\cos\left(\frac{4\pi}{3} + 2n\pi\right) = -\frac{1}{2}$ is true for any integer n.

51. Test: $x = \frac{\pi}{6} + n\pi$

$$3\sec^2\frac{\pi}{6} = 3\left(\frac{2}{\sqrt{3}}\right)^2$$
$$= 3\left(\frac{4}{3}\right)$$
$$= 4$$

The period of $\sec^2$ is π.

Therefore, $\sec^2\frac{\pi}{6} = \sec^2\left(\frac{\pi}{6} + n\pi\right)$ is true for any integer n.

Then, $3\sec^2\left(\frac{\pi}{6} + n\pi\right) = 4$ is also true for any integer n.

53. Test: $x = \frac{2\pi}{3} + 2n\pi$

$$\sqrt{3}\csc\frac{2\pi}{3} = \sqrt{3}\left(\frac{2}{\sqrt{3}}\right)$$
$$= 2$$

The period of cosecant is 2π.

Therefore, $\sqrt{3}\csc\left(\frac{2\pi}{3} + 2n\pi\right) = 2$ is true for any integer n.

55. Test: $x = \frac{\pi}{6} + n\pi$

$$\tan\frac{\pi}{6} \neq -\sqrt{3}$$

Rather, $\tan\frac{2\pi}{3} = -\sqrt{3}$.

The period of tangent is π.

The solution set is

$x = \frac{2\pi}{3} + n\pi$, for any integer n.

57. Test: $x = \frac{\pi}{3} + n\pi$

$$3\cot^2\frac{\pi}{3} = 3\left(\frac{1}{\sqrt{3}}\right)^2$$
$$= 1$$

The period of $\cot^2$ is π.

Therefore, $\cot^2\frac{\pi}{3} = \cot^2\left(\frac{\pi}{3} + n\pi\right)$ is true for any integer n.

Then, $3\cot^2\left(\frac{\pi}{3} + n\pi\right) = 1$ is also true for any integer n.

59. Solve: $2\sin^2 x - \sin x - 1 = 0$ on $[0, 2\pi]$

Factor the left side and set each factor equal to zero:

$$(2\sin x + 1)(\sin x - 1) = 0$$

$2\sin x + 1 = 0$	$\sin x - 1 = 0$
$\sin x = -\frac{1}{2}$	$\sin x = 1$

The left-hand solution yields two values of x, one in the third quadrant and one in the fourth quadrant, or $x = \frac{7\pi}{6}$ and $\frac{11\pi}{6}$. The right-hand side yields $x = \frac{\pi}{2}$. The solution set is $\left\{\frac{\pi}{2}, \frac{7\pi}{6}, \frac{11\pi}{6}\right\}$.

61. Solve: $\sin x - \cos x - 1 = 0$ on $[0, 2\pi]$
Rewrite the equation with just $\sin x$ on the left side and square each side.

$\sin x = \cos x + 1$

$\sin^2 x = \cos^2 x + 2\cos x + 1$

Use a Pythagorean identity for $\sin^2 x$:

$$1 - \cos^2 x = \cos^2 x + 2\cos x + 1$$

$2\cos^2 x + 2\cos x = 0$

Set each factor equal to zero and solve:

$\cos x = 0$	$\cos x + 1 = 0$
$\Rightarrow x = \dfrac{\pi}{2}$	$\cos x = -1$
or $x = \dfrac{3\pi}{2}$ *	$\Rightarrow x = \pi$

* Extraneous solution-discard

The solution set is $\left\{\frac{\pi}{2}, \pi\right\}$

63. Solve: $\cos 2x - \cos x = 0$ on $[0, 2\pi]$
Rewrite the equation using a $\cos 2x$ identity:

$2\cos^2 x - 1 - \cos x = 0$, or

$2\cos^2 x - \cos x - 1 = 0$

Now, factor:

$(2\cos x + 1)(\cos x - 1) = 0$

Set each factor equal to zero and solve:

$2\cos x + 1 = 0$	$\cos x - 1 = 0$
$\cos x = -\dfrac{1}{2}$	$\cos x = 1$
$\Rightarrow x = \dfrac{2\pi}{3}$	$\Rightarrow x = 0$
or $x = \dfrac{4\pi}{3}$	

All solutions check. The solution set is $\left\{0, \frac{2\pi}{3}, \frac{4\pi}{3}\right\}$.

65. Solve: $\csc^2 x - 2\cot x = 0$ on $[0, 2\pi]$
Rewrite the equation using the $\csc^2 x$ identity:

$1 + \cot^2 x - 2\cot x = 0$, or

$\cot^2 x - 2\cot x + 1 = 0$

Now, factor:

$(\cot x - 1)^2 = 0$

Set the factor equal to zero and solve:

$\cot x - 1 = 0$

$\cot x = 1 \Rightarrow x = \frac{\pi}{4}, \frac{5\pi}{4}$

All solutions check. The solution set is $\left\{\frac{\pi}{4}, \frac{5\pi}{4}\right\}$.

67. Solve: $\cos^2 x = \sin^2 x$
Use a Pythagorean identity and simplify: $1 - \sin^2 x = \sin^2 x$

$$2\sin^2 x = 1$$

$$\sin^2 x = \frac{1}{2}$$

$$\sin x = \pm\sqrt{\frac{1}{2}} = \pm\frac{\sqrt{2}}{2}$$

On the interval $[0°, 360°)$,

$x = 45°, 135°, 225°, 315°$

69. Solve: $2\sin x = \csc x + 1$ on $[0, 360°)$
Rewrite the equation using the $\csc x = \dfrac{1}{\sin x}$ identity:

$2\sin x = \dfrac{1}{\sin x} + 1$

Multiply both side by $\sin x$:

$$2\sin^2 x = 1 + \sin x$$

$2\sin^2 x - \sin x - 1 = 0$ Now, factor:

$(2\sin x + 1)(\sin x - 1) = 0$

$2\sin x + 1 = 0$	$\sin x - 1 = 0$
$\sin x = -\dfrac{1}{2}$	$\sin x = 1$
$\Rightarrow x = 210°, 330°$	$\Rightarrow x = 90°$

The solution set is $\{90°, 210°, 330°\}$.

71. Solve: $\sin^2 x = 2\sin x - 3$

Rewrite: $\sin^2 x - 2\sin x + 3 = 0$

Use the quadratic formula:

$$\sin x = \frac{2 \pm \sqrt{4 - 4(3)}}{2}$$

$$= 1 \pm \sqrt{-2}, \text{ or } 1 \pm i\sqrt{2}$$

These solutions for $\sin x$ do not lead to a real number solution for x.

The solution set is the empty set, $\varnothing$.

73. Solve:

$2\sin x \cot x + \sqrt{3}\cot x - 2\sqrt{3}\sin x - 3 = 0$

Factor by grouping:

$$(2\sin x + \sqrt{3})\cot x - \sqrt{3}(2\sin x + \sqrt{3}) = 0$$

$$(2\sin x + \sqrt{3})(\cot x - \sqrt{3}) = 0$$

Set each factor equal to zero and solve:

$2\sin x + \sqrt{3} = 0$	$\cot x - \sqrt{3} = 0$
$2\sin x = -\sqrt{3}$	$\cot x = \sqrt{3}$
$\sin x = -\frac{\sqrt{3}}{2}$	$\Rightarrow x = 30°, 210°$
$\Rightarrow x = 240°, 300°$	

The solution set is

$\{30°, 210°, 240°, 300°\}$.

75. Solve: $s^2 + s - 12 = 0$

Factor: $(s + 4)(s - 3) = 0$

Set each factor equal to zero and solve:

$s + 4 = 0$	$s - 3 = 0$
$s = -4$	$s = 3$

Solve: $\sin^2 t + \sin t - 12 = 0$

Factor: $(\sin t + 4)(\sin t - 3) = 0$

Set each factor equal to zero and solve:

$\sin t + 4 = 0$	$\sin t - 3 = 0$
$\sin t = -4$	$\sin t = 3$

Both values are outside the range of the sine function. No solution exists. The solution set is the empty set $\varnothing$.

77. Solve: $4s^2 - 4s - 1 = 0$

Use the quadratic formula:

$$s = \frac{-(-4) \pm \sqrt{16 - 4(4)(-1)}}{2(4)}$$

$$s = \frac{4 \pm \sqrt{32}}{8} = \frac{1 \pm \sqrt{2}}{2}$$

Solve: $4\cos^2 t - 4\cos t - 1 = 0$

Use the quadratic formula:

$$\cos t = \frac{-(-4) \pm \sqrt{16 - 4(4)(-1)}}{2(4)}$$

$$\cos t = \frac{4 \pm \sqrt{32}}{8} = \frac{1 \pm \sqrt{2}}{2}$$

Note: $\cos t = \frac{1 + \sqrt{2}}{2} \approx 1.2071$. This value is outside the range of cosine, so no solution is offered by this choice.

$\cos t = \frac{1 - \sqrt{2}}{2} \approx -0.2071 \Rightarrow x$ lies in the second or third quadrant.

$$t = \cos^{-1}\left(\frac{1 - \sqrt{2}}{2}\right)$$

≈ 1.7794 (quadrant II)

and $t \approx \pi + (\pi - 1.7794)$

≈ 4.5038 (quadrant III))

79. For the items 74 - 77, the minimum number of solutions of the quadratic trigonometric equations is zero and the maximum is four solutions.

81. Set the mode of a graphing calculator to "degrees" and solve the equation $\sin 3x - \frac{1}{2} = 0$ on $[0°, 360°)$.

Solution set:

$\{10°, 50°, 130°, 170°, 250°, 290°\}$

83. The solution set of trigonometric equations of the form $y = \sin ax$ may have as many as $2a$ solutions over the interval $[0°, 360°)$.

85. Given $v_0 = 95\text{ft}/\text{sec}$

$$r = \frac{1}{32}v_0^2 \sin 2\theta$$

$$160 = \frac{1}{32}(95)^2 \sin 2\theta$$

$$\sin 2\theta \approx 0.56731$$

$$2\theta \approx \sin^{-1}(0.56731) \approx 34.56°$$

$$\theta \approx 17.3°$$

End of Section 7.4

Chapter Seven Test
Solutions to Odd-Numbered Exercises

1. Simplify:

$$\sin^2(-x) - 5\sec^2 x \cot^2 x =$$
$$(-\sin x)^2 - 5\left(\frac{1}{\cos^2 x}\right)\left(\frac{\cos^2 x}{\sin^2 x}\right) =$$
$$\sin^2 x - \frac{5}{\sin^2 x} =$$
$$\sin^2 x - 5\csc^2 x$$

3. Verify: Substitute a Pythagorean identity.

$$\csc^2\theta - \cot^2\theta + \sin^2\theta =$$
$$1 + \cot^2\theta - \cot^2\theta + \sin^2\theta =$$
$$1 + \sin^2\theta$$

5. Given: $\frac{x}{13} = \tan\theta$. Rewrite as $x = 13\tan\theta$.

$$\sqrt{x^2 + 169} = \sqrt{(13\tan\theta)^2 + 169}$$
$$= \sqrt{13^2(\tan^2\theta + 1)}$$
$$= 13\sqrt{\sec^2\theta}$$
$$= 13\sec\theta$$

7. The first Pythagorean identity is $\sin^2 x + \cos^2 x = 1$. From this identity

$$\tan^2 x + 1 = \frac{\sin^2 x}{\cos^2 x} + 1$$
$$= \frac{\sin^2 x + \cos^2 x}{\cos^2 x}$$
$$= \frac{1}{\cos^2 x}$$
$$= \sec^2 x$$

9. Evaluate $\tan 165^\circ$:

$$\tan 165^\circ = \tan(135^\circ + 30^\circ)$$
$$= \frac{\tan 135^\circ + \tan 30^\circ}{1 - \tan 135^\circ \tan 30^\circ}$$
$$= \frac{-\tan 45^\circ + \tan 30^\circ}{1 + \tan 45^\circ \tan 30^\circ}$$
$$= \frac{-1 + \frac{\sqrt{3}}{3}}{1 + (1)\left(\frac{\sqrt{3}}{3}\right)}$$
$$= \frac{-3 + \sqrt{3}}{3 + \sqrt{3}}$$
$$= \frac{-3 + \sqrt{3}}{3 + \sqrt{3}} \cdot \frac{3 - \sqrt{3}}{3 - \sqrt{3}}$$
$$= \frac{-9 + 6\sqrt{3} - 3}{9 - 3}$$
$$= \frac{-12 + 6\sqrt{3}}{6} = -2 + \sqrt{3}$$

11.

$$\frac{\tan\frac{5\pi}{8} + \tan\frac{\pi}{4}}{1 - \tan\frac{5\pi}{8}\tan\frac{\pi}{4}} = \tan\left(\frac{5\pi}{8} + \frac{\pi}{4}\right)$$
$$= \tan\frac{7\pi}{8}$$
$$\approx -0.4142$$

13. First, note that $\cos\frac{5\pi}{4} = \cos\left(\frac{\pi}{2} + \frac{3\pi}{4}\right)$ and that $\cos\left(-\frac{\pi}{4}\right) = \cos\left(\frac{\pi}{2} - \frac{3\pi}{4}\right)$.

$$\cos\frac{5\pi}{4}\cos\left(-\frac{\pi}{4}\right) =$$
$$\cos\left(\frac{\pi}{2} + \frac{3\pi}{4}\right)\cos\left(\frac{\pi}{2} - \frac{3\pi}{4}\right) =$$
$$\left[\cos\frac{\pi}{2}\cos\frac{3\pi}{4} - \sin\frac{\pi}{2}\sin\frac{3\pi}{4}\right] \cdot$$
$$\left[\cos\frac{\pi}{2}\cos\frac{3\pi}{4} + \sin\frac{\pi}{2}\sin\frac{3\pi}{4}\right] =$$

$$\cos^2\frac{\pi}{2}\cos^2\frac{3\pi}{4}-\sin^2\frac{\pi}{2}\sin^2\frac{3\pi}{4}=$$
$$\left[(0)\left(-\frac{\sqrt{2}}{2}\right)^2-(1)\left(\frac{\sqrt{2}}{2}\right)^2\right]=$$
$$0-\frac{1}{2}=-\frac{1}{2}$$

Also, $\cos^2\frac{\pi}{2}-\sin^2\frac{3\pi}{4}=0-\left(\frac{\sqrt{2}}{2}\right)^2$
$$=-\frac{2}{4}=-\frac{1}{2}$$

15. Given: $h(\beta)=2\sqrt{3}\sin 4\beta+2\cos 4\beta$
Using the formula for sum of sines and cosines, we have:

$A=2\sqrt{3}$, $B=2$ and

$\sqrt{A^2+B^2}=\sqrt{12+4}=4$.

So, $\cos\phi=\frac{2\sqrt{3}}{4}=\frac{\sqrt{3}}{2}$ and

$\sin\phi=\frac{2}{4}=\frac{1}{2}$.

So, $\phi=\frac{\pi}{6}$.

Thus $h(\beta)=4\sin\left(4\beta+\frac{\pi}{6}\right)$.

17. Show: $\tan 4x=\frac{4\tan x-4\tan^3 x}{1-6\tan^2 x+\tan^4 x}$

$$\tan 4x=\tan(2x+2x)$$
$$=\frac{\tan 2x+\tan 2x}{1-\tan 2x\cdot\tan 2x}$$
$$=\frac{2\tan 2x}{1-\tan^2 2x}$$
$$=\frac{2\left(\frac{2\tan x}{1-\tan^2 x}\right)}{1-\left(\frac{2\tan x}{1-\tan^2 x}\right)\left(\frac{2\tan x}{1-\tan^2 x}\right)}$$
$$=\frac{4\tan x}{1-\tan^2 x-\frac{4\tan^2 x}{1-\tan^2 x}}$$
$$=\frac{(1-\tan^2 x)4\tan x}{(1-\tan^2 x)(1-\tan^2 x)-4\tan^2\theta}$$
$$=\frac{4\tan x-4\tan^3 x}{1-6\tan^2 x+\tan^4 x}$$

19. Use a half-angle formula:

$$\sin\frac{11\pi}{12}=\sin\left(\frac{\frac{11\pi}{6}}{2}\right)$$
$$=\sqrt{\frac{1-\cos\left(\frac{11\pi}{6}\right)}{2}}$$
$$=\sqrt{\frac{1-\frac{\sqrt{3}}{2}}{2}}$$
$$=\sqrt{\frac{2-\sqrt{3}}{4}}$$
$$=\frac{\sqrt{2-\sqrt{3}}}{2}$$

21. $\sin 5x\cos 5x=\frac{1}{2}[\sin 10x+\sin 0]$
$$=\frac{1}{2}\sin 10x,\text{ or }\frac{\sin 10x}{2}$$

23. $\sin\frac{5x}{6}+\sin\frac{x}{6}=2\sin\left(\frac{\frac{5x}{6}+\frac{x}{6}}{2}\right)\cos\left(\frac{\frac{5x}{6}-\frac{x}{6}}{2}\right)$
$$=2\sin\frac{x}{2}\cos\frac{x}{3}$$

25. Solve: $4\sin x\cos x=\sqrt{3}$. Divide both sides by 2 and use a double angle identity:

$$2\sin x\cos x=\frac{\sqrt{3}}{2}$$
$$\sin 2x=\frac{\sqrt{3}}{2}\Rightarrow$$
$$2x=\sin^{-1}\frac{\sqrt{3}}{2}\Rightarrow$$
$$2x=\frac{\pi}{3}\text{ or }\frac{2\pi}{3}$$
$$x=\frac{\pi}{6}\text{ or }\frac{\pi}{3}$$

The period of $\sin 2x$ is $\frac{2\pi}{2}=\pi$, so the solution set is

$$\left\{x: x=\frac{\pi}{6}+n\pi\right\}\cup\left\{x: x=\frac{\pi}{3}+n\pi\right\},$$

for any integer n.

27. Solve: $2\cos^8 x = 4\cos^6 x$
The equation is equivalent to
$\cos^8 x = 2\cos^6 x$, by dividing by 2.
Notice that if $\cos x = 0$, then the equation is satisfied. For $x = \frac{\pi}{2}$, $\cos x = 0$ and the equation is true.
In general, the solution set is
$\left\{x : x = \frac{n\pi}{2}\right\}$, for any odd integer n.
Also, it is important to note that for $\cos x \neq 0$, you can divide both sides by $\cos^6 x$, and have

$$\frac{\cos^8 x}{\cos^6 x} = 2 \Rightarrow \cos^2 x = 2!$$

But, this result is impossible and indicates that there are no other solutions, since this result suggests that $\cos x > 1$.

29. Solve: $5\sin^2 x + 4\sin x = 6$
Use the quadratic formula. Solve for $\sin x$ and then express the solution for x in terms of the inverse of sine:

$$\sin x = \frac{-4 \pm \sqrt{16 - 4(5)(-6)}}{10}$$
$$= \frac{-4 \pm \sqrt{136}}{10}$$
$$= \frac{-2 \pm \sqrt{34}}{5}$$

Then, $x = \sin^{-1}\left(\frac{-2+\sqrt{34}}{5}\right)$ or
$x = \sin^{-1}\left(\frac{-2-\sqrt{34}}{5}\right)$. (Discard since not in the range of sine.)
Since $\sin x$ is 2π periodic, we have

$$x = \sin^{-1}\left(\frac{-2+\sqrt{34}}{5}\right) + 2n\pi.$$

31. Solve: $4\sec^2 x - 4 + 16\tan x = 24$
Substitute a Pythagorean identity for $\sec^2 x$ and form a quadratic equation in terms of $\tan x$. Then, the quadratic formula can be applied:

$$4(\tan^2 x + 1) + 16\tan x - 28 = 0$$
$$\tan^2 x + 1 + 4\tan x - 7 = 0$$
$$\tan^2 x + 4\tan x - 6 = 0$$

$$\tan x = \frac{-4 \pm \sqrt{16 - 4(-6)}}{2}$$
$$\tan x = \frac{-4 \pm 2\sqrt{10}}{2}$$
$$\tan x = -2 \pm \sqrt{10}$$

$x = \tan^{-1}\left(-2+\sqrt{10}\right)$, or
$x = \tan^{-1}\left(-2-\sqrt{10}\right)$
Since $\tan x$ is π periodic, we have
$x = \tan^{-1}\left(-2+\sqrt{10}\right) + n\pi$,
$\tan^{-1}\left(-2-\sqrt{10}\right) + n\pi$.

End of Chapter 7 Test

Chapter Eight: Section 8.1
Solutions to Odd-numbered Exercises

1. Given:

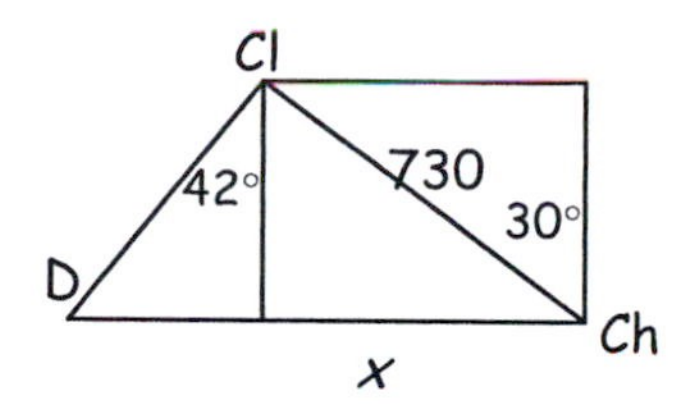

We can determine the values of the angles from what we know about right triangles and about parallel lines (e.g., alternate interior angles being congruent):

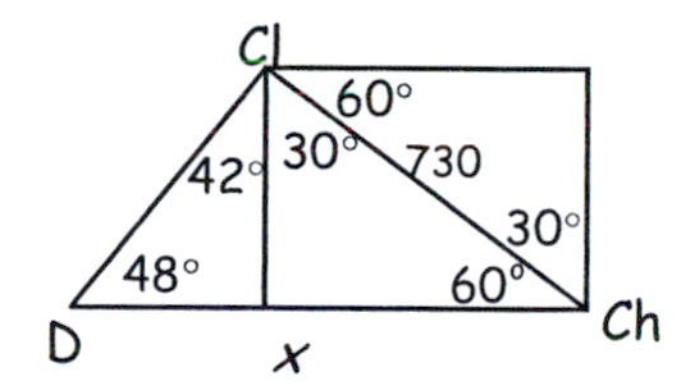

$$\frac{\sin 48°}{730} = \frac{\sin 72°}{x}$$

$$x = \frac{730 \sin 72°}{\sin 48°}$$

$$x \approx \frac{730(0.9511)}{0.7431}$$

$$x \approx 934.2 \text{ miles}$$

3. Given:

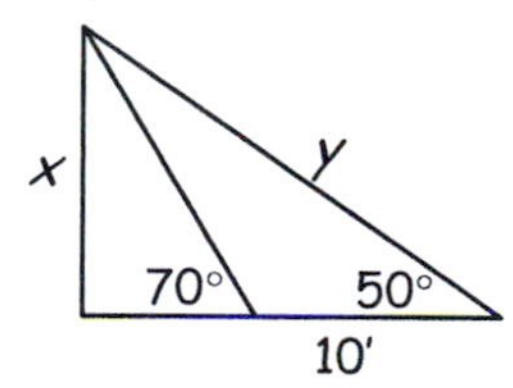

We can determine the values of the angles from what we know about triangles and about supplementary angles (e.g., their sum is 180°):

$$\frac{y}{\sin 110°} = \frac{10}{\sin 20°}$$

$$y = \frac{10 \sin 110°}{\sin 20°}$$

$$y \approx \frac{9.3969}{0.3420}$$

$$y \approx 27.48 \text{ ft.}$$

$$\frac{x}{\sin 50°} \approx \frac{27.48}{\sin 90°}$$

$$x \approx 27.48 \sin 50°$$

$$x \approx 27.48(0.7660)$$

$$x \approx 21 \text{ ft.}$$

5. Construct a triangle and label it according to the given description. This will take some experimentation on your part.

$$\frac{20}{\sin 40°} = \frac{30}{\sin H}$$

$$\sin H = \frac{30 \sin 40°}{20}$$

$$\sin H \approx \frac{3(0.6428)}{2}$$

$$\sin H \approx 0.9642$$

$$H \approx \sin^{-1}(0.9642)$$

$$H \approx 74.62°$$

$$R \approx 180° - 74.62° - 40°$$

$$R \approx 65.38°$$

$$\frac{x}{\sin 65.38°} \approx \frac{20}{\sin 40°}$$

$$x \approx \frac{20 \sin 65.38°}{\sin 40°}$$

$$x \approx \frac{20(0.9091)}{0.6428}$$

$$x \approx 28.3 \text{ ft.}$$

7. Given:

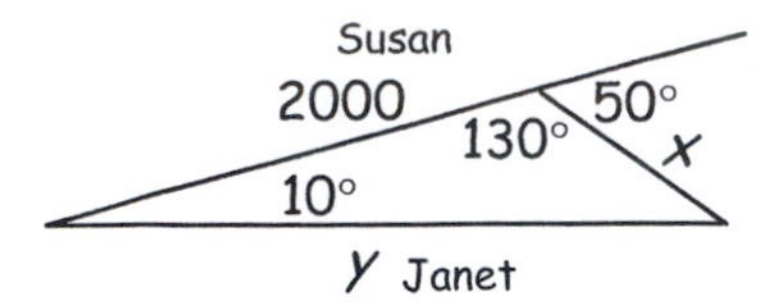

$$\frac{x}{\sin 10°} = \frac{2000}{\sin 40°}$$
$$x = \frac{2000 \sin 10°}{\sin 40°}$$
$$x \approx \frac{2000(0.1736)}{0.6428}$$
$$x \approx 540.1 \text{ ft.}$$

$$\frac{y}{\sin 130°} = \frac{2000}{\sin 40°}$$
$$y = \frac{2000 \sin 130°}{\sin 40°}$$
$$y \approx \frac{2000(0.7660)}{0.6428}$$
$$y \approx 2383.3 \text{ ft.}$$

$2000 + 540.1 - 2383.3 \approx 156.8$ ft.

Janet ran about 156.8 ft. less.

9. Find the depth, d:

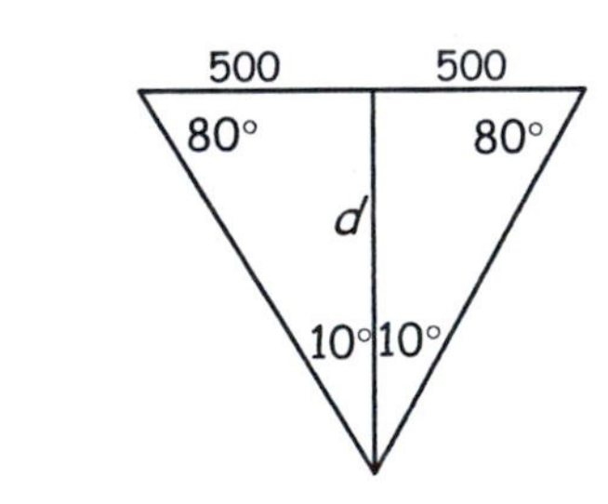

$$\frac{d}{\sin 80°} = \frac{500}{\sin 10°}$$
$$d = \frac{500 \sin 80°}{\sin 10°}$$
$$d \approx \frac{500(0.9848)}{0.1736}$$
$$d \approx 2835.6 \text{ ft.}$$

11. Find the distance, d:

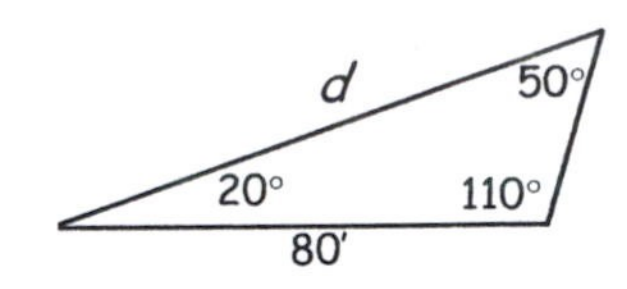

$$\frac{d}{\sin 110°} = \frac{80}{\sin 50°}$$
$$d = \frac{80 \sin 110°}{\sin 50°}$$
$$d \approx \frac{80(0.9397)}{0.7660}$$
$$d \approx 98.1 \text{ ft.}$$

13. Given:

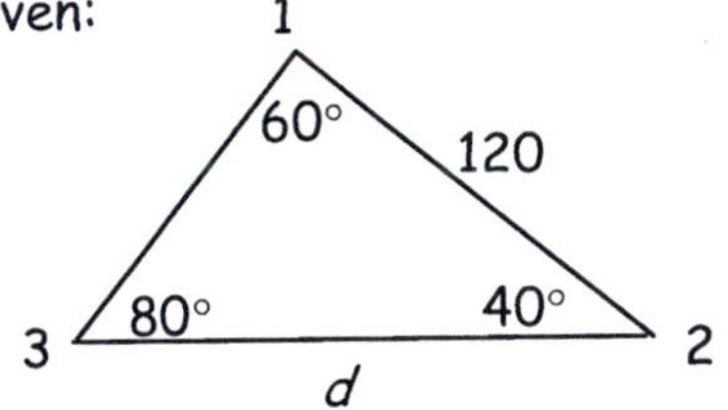

Find the distance, d:

$$\frac{d}{\sin 60°} = \frac{120}{\sin 80°}$$
$$d = \frac{120 \sin 60°}{\sin 80°}$$
$$d \approx \frac{120(0.8660)}{0.9848}$$
$$d \approx 105.5 \text{ miles}$$

15. Given: $A = 30°, B = 45°, a = 3$

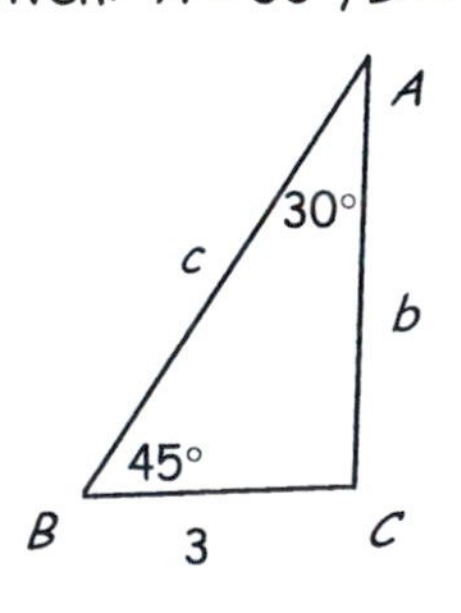

$$C = 180° - 45° - 30° = 105°$$

$$\frac{b}{\sin 45°} = \frac{3}{\sin 30°}$$
$$b = \frac{3 \sin 45°}{\sin 30°}$$

$$b = \frac{3\left(\frac{\sqrt{2}}{2}\right)}{\frac{1}{2}}$$

$$b = 3\sqrt{2} \approx 4.2426$$

$$\frac{c}{\sin 105^\circ} = \frac{3}{\sin 30^\circ}$$

$$c = \frac{3\sin 105^\circ}{\sin 30^\circ}$$

$$c \approx \frac{3(0.9659)}{0.5}$$

$$c \approx 5.7956$$

17. Given: $A = 70^\circ$, $B = 50^\circ$, $b = 4$

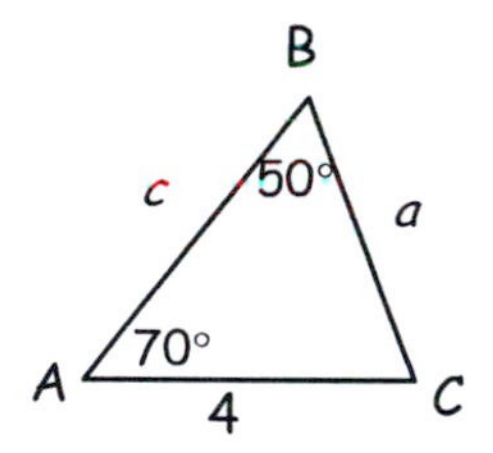

$$C = 180^\circ - 70^\circ - 50^\circ = 60^\circ$$

$$\frac{a}{\sin 70^\circ} = \frac{4}{\sin 50^\circ}$$

$$a = \frac{4\sin 70^\circ}{\sin 50^\circ}$$

$$a \approx \frac{4(0.9397)}{0.7660}$$

$$a \approx 4.9067$$

$$\frac{c}{\sin 60^\circ} = \frac{4}{\sin 50^\circ}$$

$$c = \frac{4\sin 60^\circ}{\sin 50^\circ}$$

$$c \approx \frac{4(0.8660)}{0.7660}$$

$$c \approx 4.5221$$

19. Given: $B = 70^\circ$, $C = 30^\circ$, $c = 2$

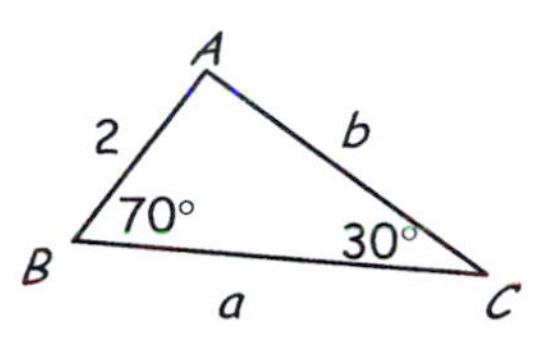

$$A = 180^\circ - 70^\circ - 30^\circ = 80^\circ$$

$$\frac{a}{\sin 80^\circ} = \frac{2}{\sin 30^\circ}$$

$$a = \frac{2\sin 80^\circ}{\sin 30^\circ}$$

$$a \approx \frac{2(0.9848)}{0.5}$$

$$a \approx 3.9392$$

$$\frac{b}{\sin 70^\circ} = \frac{2}{\sin 30^\circ}$$

$$b = \frac{2\sin 70^\circ}{\sin 30^\circ}$$

$$b \approx \frac{3(0.9397)}{0.5}$$

$$b \approx 3.7588$$

21. Given: $A = 20^\circ$, $B = 10^\circ$, $a = 2$

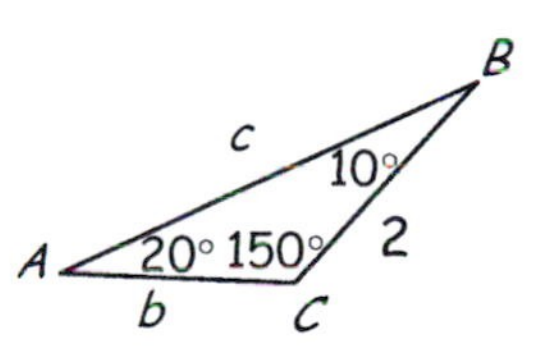

$$C = 180^\circ - 20^\circ - 10^\circ = 150^\circ$$

$$\frac{b}{\sin 10^\circ} = \frac{2}{\sin 20^\circ}$$

$$b = \frac{2\sin 10^\circ}{\sin 20^\circ}$$

$$b \approx \frac{2(0.1736)}{0.3420}$$

$$b \approx 1.0154$$

$$\frac{c}{\sin 150^\circ} = \frac{2}{\sin 20^\circ}$$

$$c = \frac{2\sin 150^\circ}{\sin 20^\circ}$$

$$c \approx \frac{2(0.5)}{0.3420}$$

$$c \approx 2.9238$$

23. Given: $A = 40°$, $a = 2$, $b = 4$

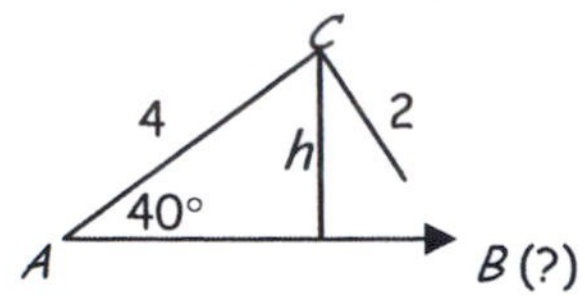

Test the height, h, for the existence of the triangle (i.e., Are the sides long enough to reach, given the angle and length of the two sides?):

$h = 4\sin 40°$

$h \approx 2.5712$

$a < h(\text{height})$ Therefore the triangle does not exist.

25. Given: $C = 45°$, $a = 2$, $c = 4$

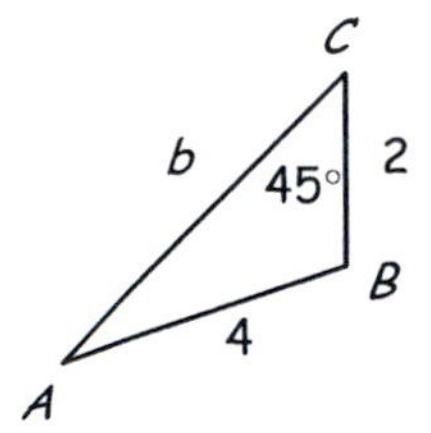

$$\frac{\sin A}{2} = \frac{\sin 45°}{4}$$

$$\sin A = \frac{2\left(\frac{\sqrt{2}}{2}\right)}{4}$$

$$\sin A = \frac{\sqrt{2}}{4} \approx 0.3536$$

$$A \approx \sin^{-1}(0.3536)$$

$$A \approx 20.7048°$$

$$B \approx 180° - 20.7048° - 45° \approx 114.2952°$$

$$\frac{b}{\sin 114.2952°} = \frac{4}{\sin 45°}$$

$$b \approx \frac{4\sin 114.2952°}{\sin 45°}$$

$$b \approx 5.1559$$

27. Given: $A = 60°$, $a = 5$, $c = 6$

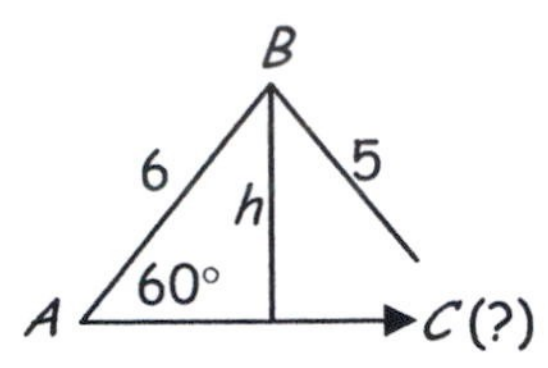

Check for triangle existence by determining whether side a is at least as long as the height h as shown in the figure.

$$h = 6\sin 60° = 6\left(\frac{\sqrt{3}}{2}\right) \approx 5.196$$

$a = 5 < h = 5.196 \Rightarrow$ a is not long enough. Therefore the triangle does not exist.

29. Given: $B = 50°$, $b = 2$, $c = 5$

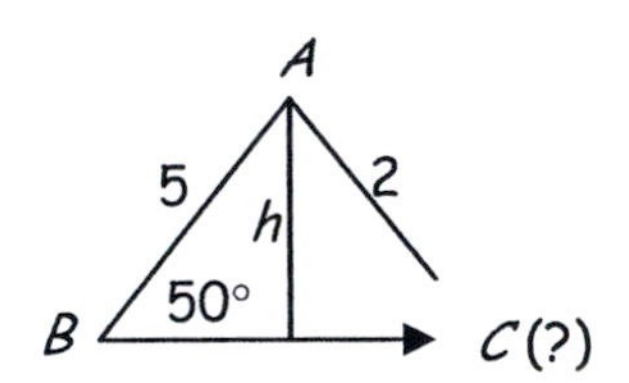

$h = 5\sin 50° \approx 5(0.7660) \approx 3.83$

$b = 2 < h = 3.83 \Rightarrow$ b is not long enough. Therefore the triangle does not exist.

31. Given: $A = 60°$, $a = 10$, $b = 6$

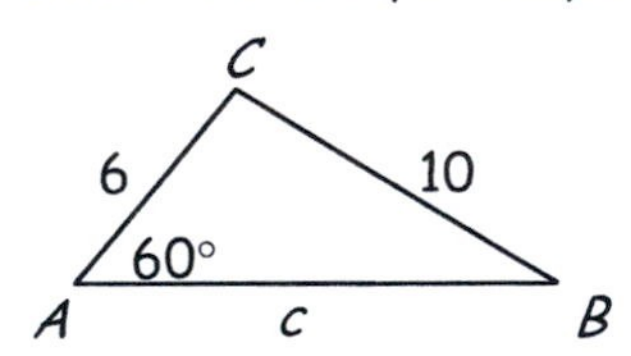

$$\frac{\sin B}{6} = \frac{\sin 60°}{10}$$

$$\sin B = \frac{6\left(\frac{\sqrt{3}}{2}\right)}{10}$$

$$\sin B \approx 0.5196$$

$$B \approx \sin^{-1}(0.5196)$$

$$B \approx 31.31°$$

$$C = 180° - 60° - 31.31° \approx 88.69°$$

$$\frac{c}{\sin 88.69^\circ} \approx \frac{10}{\sin 60^\circ}$$

$$c \approx \frac{10\sin 88.69^\circ}{\sin 60^\circ}$$

$$c \approx \frac{10(0.9997)}{0.8660}$$

$$c \approx 11.54$$

33. Given: $B = 13.2^\circ$, $A = 63.7^\circ$, $b = 21.2$

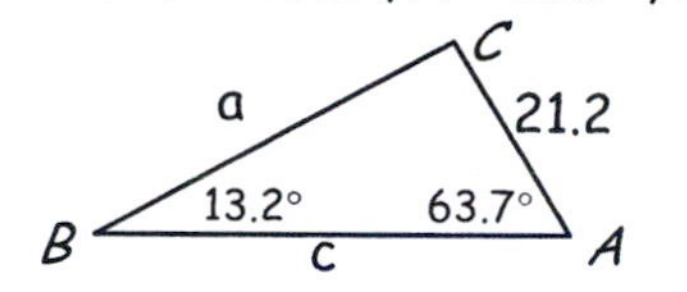

$C = 180^\circ - 13.2^\circ - 63.7^\circ = 103.1^\circ$

$$\frac{a}{\sin 63.7^\circ} = \frac{21.2}{\sin 13.2^\circ}$$

$$a = \frac{21.2\sin 63.7^\circ}{\sin 13.2^\circ}$$

$$a \approx 83.23$$

$$\frac{c}{\sin 103.1^\circ} = \frac{21.2}{\sin 13.2^\circ}$$

$$c = \frac{21.2\sin 103.1^\circ}{\sin 13.2^\circ}$$

$$c \approx 90.42$$

35. Given: $C = 100^\circ$, $a = 18.1$, $c = 20.4$

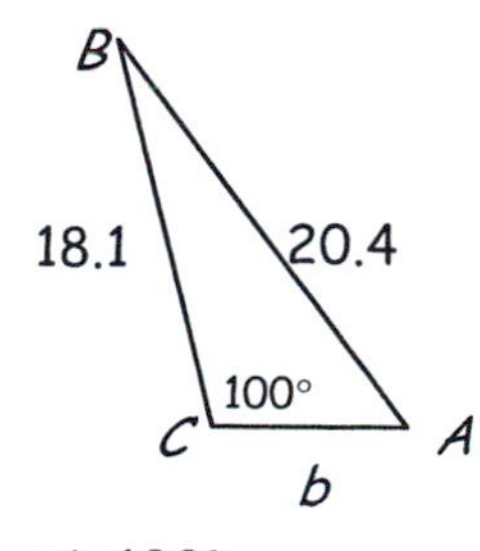

$$\frac{\sin A}{18.1} = \frac{\sin 100^\circ}{20.4}$$

$$\sin A = \frac{18.1\sin 100^\circ}{20.4}$$

$$\sin A \approx 0.8738$$

$$A \approx \sin^{-1}(0.8738)$$

$$A \approx 60.9^\circ$$

$$B \approx 180^\circ - 100^\circ - 60.9^\circ \approx 19.1^\circ$$

$$\frac{b}{\sin 19.1^\circ} = \frac{20.4}{\sin 100^\circ}$$

$$b = \frac{20.4\sin 19.1^\circ}{\sin 100^\circ}$$

$$b \approx 6.78$$

37. Given: $C = 24^\circ$, $b = 2.4$, $c = 1.5$

First, test the length of c relative to the length of h:

Find h:

$h = 2.4\sin 24^\circ$

$h \approx 0.9762$

With $h \approx 0.9762 < c = 1.5 < b = 2.4$, there are two possible triangles (see figure):

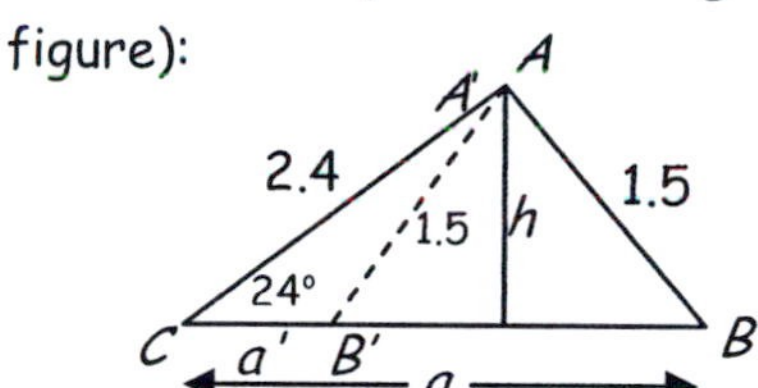

Triangle 1:

$$\frac{\sin B}{2.4} = \frac{\sin 24^\circ}{1.5}$$

$$\sin B = \frac{2.4\sin 24^\circ}{1.5}$$

$$\sin B \approx 0.6508$$

$$B \approx \sin^{-1}(0.6508)$$

$$B \approx 40.60^\circ$$

$$A \approx 180^\circ - 40.60^\circ - 24^\circ \approx 115.40$$

$$\frac{a}{\sin 115.4^\circ} \approx \frac{1.5}{\sin 24^\circ}$$

$$a \approx \frac{1.5\sin 115.4^\circ}{\sin 24^\circ}$$

$$a \approx 3.33$$

Triangle 2:

$$B' = 180^\circ - B \approx 139.40^\circ$$

$$A' \approx 180^\circ - 24^\circ - 139.4^\circ \approx 16.60^\circ$$

$$\frac{a'}{\sin 16.6^\circ} \approx \frac{1.5}{\sin 24^\circ}$$

$$a' \approx \frac{1.5\sin 16.6^\circ}{\sin 24^\circ}$$

$$a' \approx 1.05$$

39. Given: $A = 46°53'$, $B = 74°13'$, $c = 3.1$

$C = 180° - 46°53' - 74°13'$
$= 180° - 121°6'$
$= 58°54'$

Change the degrees to decimal form to compute a and b.

Convert A to decimal form:

$A = 46\frac{53}{60} \approx 46.8833°$

Convert B to decimal form:

$B = 74\frac{13}{60} \approx 74.2167°$

Convert C to decimal form:

$C = 58\frac{54}{60} \approx 58.90°$

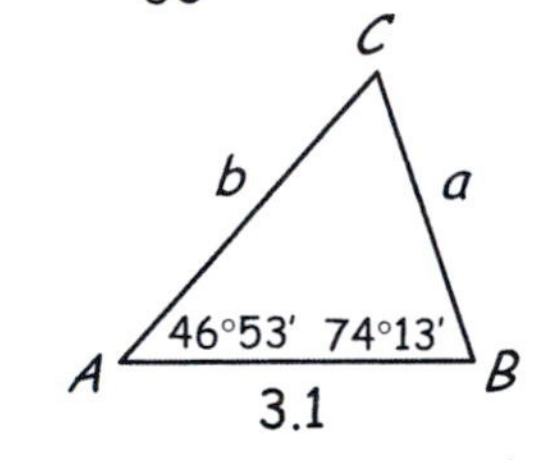

$\frac{a}{\sin 46.8833°} \approx \frac{3.1}{\sin 58.90°}$

$a \approx \frac{3.1 \sin 46.8833°}{\sin 58.90°}$

$a \approx 2.64$

$\frac{b}{\sin 74.2167°} \approx \frac{2.64}{\sin 46.8833°}$

$b \approx \frac{2.64 \sin 74.2167°}{\sin 46.8833°}$

$b \approx 3.48$

41. Given: $A = 30°$, $a = 15$, $b = 13$

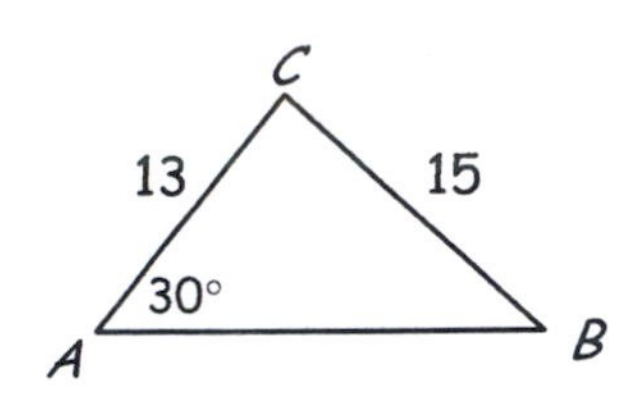

$\frac{\sin B}{13} = \frac{\sin 30°}{15}$

$\sin B = \frac{13 \sin 30°}{15}$

$\sin B \approx 0.4333$

$B \approx \sin^{-1}(0.4333)$

$B \approx 25.68°$

$C = 180° - 30° - 25.68° = 124.32°$

$\frac{c}{\sin 124.32°} \approx \frac{15}{\sin 30°}$

$c \approx \frac{15 \sin 124.32°}{.5}$

$c \approx 24.78$

43. Given: $A = 60°$, $b = 3$, $c = 7$

Use the Law of Cosines to find a:

$a^2 = 3^2 + 7^2 - 2(3)(7)\cos 60°$

$a = \sqrt{9 + 49 - 2(3)7\left(\frac{1}{2}\right)}$

$a = \sqrt{37} \approx 6.0828$

Use the Law of Sines to find B:

$\frac{\sin B}{3} \approx \frac{\sin 60°}{6.0828}$

$\sin B \approx 0.4271$

$B \approx \sin^{-1}(0.4271)$

$B \approx 25.28°$

$C \approx 180° - 60° - 25.2850° \approx 94.7150°$

45. Given: $B = 50°, a = 4, c = 6$

Use the Law of Cosines to find b:

$$b^2 = 4^2 + 6^2 - 2(4)(6)\cos 50°$$
$$b = \sqrt{16 + 36 - 48(0.6428)}$$
$$b \approx 4.5985$$

Use the Law of Sines to find A:

$$\frac{\sin A}{4} \approx \frac{\sin 50°}{4.5985}$$
$$\sin A \approx \frac{4(0.7660)}{4.5985}$$
$$\sin A \approx 0.6663$$
$$A \approx \sin^{-1}(0.6663)$$
$$A \approx 41.7854°$$

$$C \approx 180° - 50° - 41.7854° \approx 88.2146°$$

47. Given: $C = 30°, a = 8, b = 6$

Use the Law of Cosines to find c:

$$c^2 = 8^2 + 6^2 - 2(8)(6)\cos 30°$$
$$c \approx \sqrt{64 + 36 - 96(0.8660)}$$
$$c \approx 4.1063$$

Use the Law of Sines to find A:

$$\frac{\sin A}{8} \approx \frac{\sin 30°}{4.1063}$$
$$\sin A \approx \frac{8(0.5)}{4.1063}$$
$$\sin A \approx 0.9741$$
$$A \approx \sin^{-1}(0.9741)$$
$$A \approx 76.9357°\text{, or}$$
$$A \approx 180° - 76.9357° = 103.0643°$$

The latter value is selected because angle A is an obtuse angle, since it is opposite the longest side.

$$B \approx 180° - 30° - 103.0643° \approx 46.9357°$$

49. Given: $C = 70°, a = 5, b = 7$

Use the Law of Cosines to find c:

$$c^2 = 5^2 + 7^2 - 2(5)(7)\cos 70°$$
$$c \approx \sqrt{25 + 49 - 70(0.3420)}$$
$$c \approx 7.0752$$

Use the Law of Sines to find A:

$$\frac{\sin A}{5} \approx \frac{\sin 70°}{7.0752}$$
$$\sin A \approx \frac{5(0.9397)}{7.0752}$$
$$\sin A \approx 0.6641$$
$$A \approx \sin^{-1}(0.6641)$$
$$A \approx 41.6113°$$

$$B \approx 180° - 70° - 41.6113° \approx 68.3887°$$

51. Given: $a = 3, b = 4, c = 2$

Use the Law of Cosines to solve for A and B:

$$\cos A = \frac{a^2 - b^2 - c^2}{-2bc}$$
$$= \frac{9 - 16 - 4}{-2(4)(2)}$$
$$= 0.6875$$
$$A = \cos^{-1}(0.6875)$$
$$A \approx 46.5675°$$

$$\cos B = \frac{b^2 - a^2 - c^2}{-2ac}$$
$$= \frac{16 - 9 - 4}{-2(3)(2)}$$
$$= -0.25$$
$$B = \cos^{-1}(0.25)$$
$$B \approx 104.4775°$$

$$C \approx 180° - 46.5675° - 104.4775°$$
$$C \approx 28.9550°$$

53. Given: $a = 8, b = 6, c = 3$
Use the Law of Cosines to solve for A and B:

$$\cos A = \frac{a^2 - b^2 - c^2}{-2bc} = \frac{64 - 36 - 9}{-2(6)(3)} \approx -0.5278$$
$$A \approx \cos^{-1}(-0.5278)$$
$$A \approx 121.8554^\circ$$

$$\cos B = \frac{b^2 - a^2 - c^2}{-2ac} = \frac{36 - 64 - 9}{-2(8)(3)} \approx 0.7708$$
$$B \approx \cos^{-1}(0.7708)$$
$$B \approx 39.5712^\circ$$

$$C \approx 180^\circ - 121.8554^\circ - 39.5712^\circ$$
$$C \approx 18.5734^\circ$$

55. Given: $a = 5, b = 5, c = 5$
This is an equilateral triangle. Therefore, all three angles are congruent; i.e., $A = B = C = 60^\circ$.

57. Given: $a = 5, b = 3, c = 4$
You may recognized these values as describing a 3-4-5 right triangle. The angle opposite the longest side is a right angle. $A = 90^\circ$.
Otherwise, you may use the Law of Cosines to confirm it. Note: You could also use the fact that we have a right triangle and $\cos B = \frac{4}{5}$ to solve for B.

$$\cos B = \frac{b^2 - a^2 - c^2}{-2ac} = \frac{9 - 25 - 16}{-2(5)(4)} = 0.8$$
$$B = \cos^{-1}(0.8)$$
$$B \approx 36.8699^\circ$$

$$C \approx 180^\circ - 90^\circ - 36.8699^\circ \approx 53.1301^\circ$$

59. Given:
Find d:

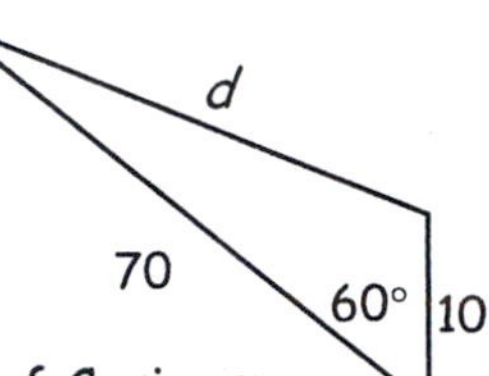

Use the Law of Cosines:
$$d^2 = 70^2 + 10^2 - 2(70)(10)\cos 60^\circ$$
$$d^2 = 4300$$
$$d \approx 65.5744 \text{ ft.}$$

61. Given:

Find d:

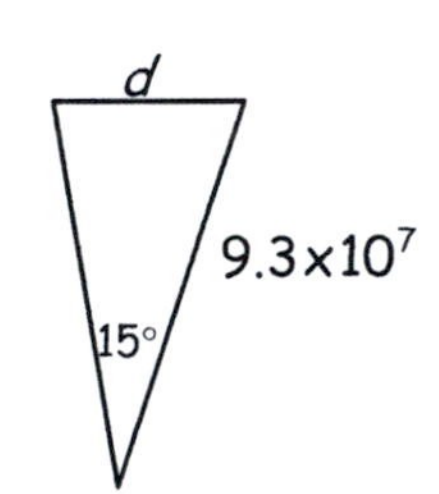

Use the Law of Cosines:
$$d^2 = (9.3\times10^7)^2 + (9.3\times10^7)^2 - 2(9.3)(9.3\times10^{14})\cos 15^\circ$$
$$\approx 2(86.49)\times10^{14} - 2(86.49)\times10^{14}(0.96593)$$
$$d^2 \approx 1.7298\times10^{16} - 1.670858\times10^{16}$$
$$d^2 \approx 5.8941506\times10^{14}$$
$$d \approx 24{,}277{,}871 \text{ miles, or}$$
$$d \approx 2.4\times10^7 \text{ miles}$$

63. Given:

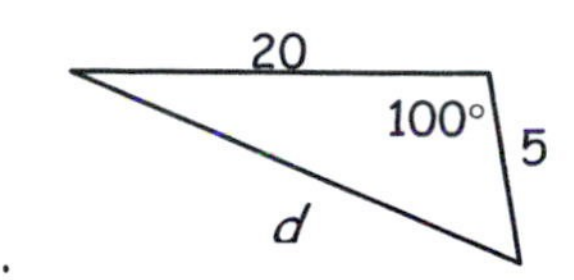

Find d:

Use the Law of Cosines:

$d^2 = 20^2 + 5^2 - 2(20)(5)\cos 100°$

$d^2 \approx 459.7296$

$d \approx 21.4413$ ft

65. Given:

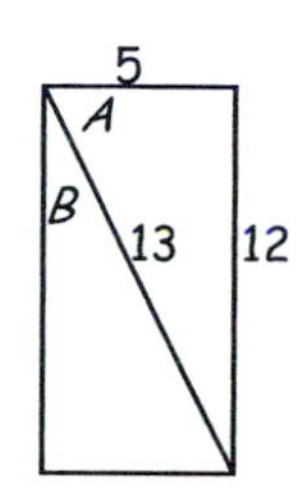

Find A and B:

Use the Law of Cosines to find A:

$\cos A = \dfrac{12^2 - 13^2 - 5^2}{-2(5)(13)}$

$\cos A \approx 0.3846$

$A \approx \cos^{-1}(0.3846)$

$A \approx 67.3801°$

$B \approx 90° - 67.3801° \approx 22.6199°$

67. Given: Find $\alpha + \beta$:

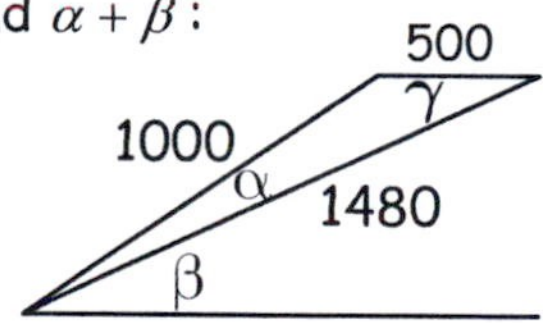

Use the Law of Cosines to find α:

$\cos\alpha = \dfrac{500^2 - 1000^2 - 1480^2}{-2(1000)(1480)}$

$\approx \dfrac{-2940400}{-2960000}$

≈ 0.993378

$\alpha \approx 6.5972°$

Note: In the figure the angle denoted by γ has the same measure as the angle denoted by β, by property of alternate interior angles. So, we can find β by finding γ with the use of the Law of Cosines:

$\cos\gamma = \dfrac{1000^2 - 500^2 - 1480^2}{-2(500)(1480)}$

$\approx \dfrac{-1440400}{-1480000}$

≈ 0.973243

$\gamma \approx \cos^{-1}(0.973243) \approx 13.2840°$

Then, $\alpha + \beta \approx 6.5972 + 13.2840 \approx 19.8812°$

69. Given: $C = 35°$, $b = 12$, $a = 14$

Use the Law of Cosines to find c:

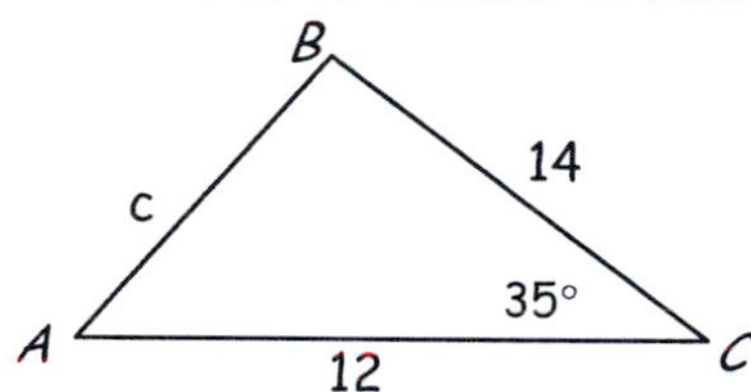

$c^2 = 12^2 + 14^2 - 2(12)(14)\cos 35°$

$c^2 \approx 64.7649$

$c \approx 8.05$

Use the Law of Sines to find A:

$\dfrac{\sin A}{14} \approx \dfrac{\sin 35°}{8.05}$

$\sin A \approx \dfrac{14\sin 35°}{8.05}$

$\sin A \approx 0.9978$

$A \approx \sin^{-1}(0.9978)$

$A \approx 86.21°$

$B \approx 180° - 86.21° - 35° \approx 58.79°$

71. Given: $C = 46°7'$, $a = 27.8$, $b = 19.4$

Use Law of Cosines to find c:

$$c^2 = 27.8^2 + 19.4^2 - 2(27.8)(19.4)\cos 46°7'$$

$$c^2 = 401.4952$$

$$c \approx 20.04$$

Use Law of Sines to find A:

$$\frac{\sin A}{27.8} \approx \frac{\sin 46°7'}{20.04}$$

$$\sin A \approx \frac{27.8 \sin 46°7'}{20.04}$$

$$\sin A \approx 0.999976973$$

$$A \approx \sin^{-1}(0.999976973)$$

$$A \approx 89°38'$$

$$B \approx 180° - 89°38' - 46°7' \approx 44°15'$$

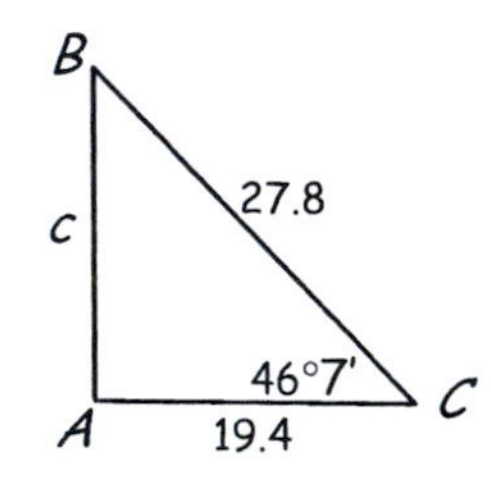

73. Given: $C = 75°4'$, $b = 15.4$, $a = 16.8$

Use the Law of Cosines to find c:

$$c^2 = 15.4^2 + 16.8^2 - 2(15.4)(16.8)\cos 75°4'$$

$$c^2 \approx 386.0586$$

$$c \approx 19.65$$

Use Law of Sines to find A:

$$\frac{\sin A}{16.8} \approx \frac{\sin 75°4'}{19.65}$$

$$\sin A \approx \frac{16.8 \sin 75°4'}{19.65}$$

$$\sin A \approx 0.82615$$

$$A \approx \sin^{-1}(0.82615)$$

$$A \approx 55.71 \approx 55°42'$$

$$B \approx 180° - 55°42' - 75°4' \approx 49°14'$$

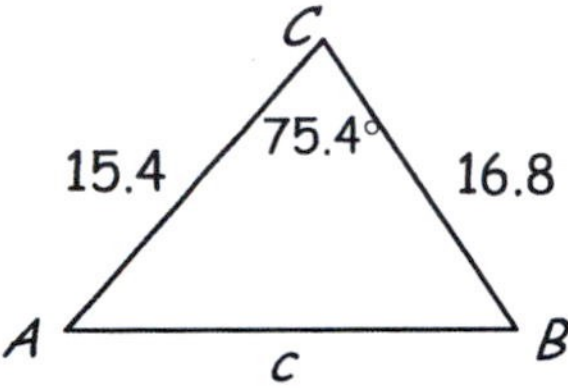

75. Given: $c = 4.78$, $b = 16.46$, $a = 16.54$

Use the Law of Cosines to find A:

$$\cos A = \frac{(16.54)^2 - (16.46)^2 - (4.78)^2}{-2(16.46)(4.78)}$$

$$\approx \frac{-20.2084}{-157.3576}$$

$$\approx 0.128423$$

$$A \approx \cos^{-1}(0.128423)$$

$$A \approx 82.62°$$

Use the Law of Sines to find B:

$$\frac{\sin B}{16.46} \approx \frac{\sin 82.62°}{16.54}$$

$$\sin B \approx \frac{16.46 \sin 82.62°}{16.54}$$

$$\sin B \approx 0.9869$$

$$B \approx \sin^{-1}(0.9869)$$

$$B \approx 80.72°$$

$$C \approx 180° - 80.72° - 82.62° \approx 16.66°$$

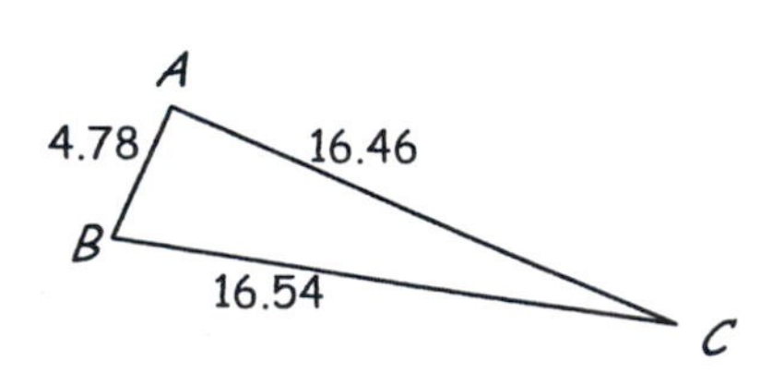

77. Given: $b = 6.84$, $c = 10.87$, $a = 7.37$

Use the Law of Cosines to find A:

$$\cos A = \frac{(7.37)^2 - (6.84)^2 - (10.87)^2}{-2(6.84)(10.87)}$$

$$\approx \frac{-110.6256}{-148.7016}$$

$$\approx 0.74394$$

$$A \approx \cos^{-1}(0.7439)$$

$$A \approx 41.93^\circ$$

Use the Law of Sines to find B:

$$\frac{\sin B}{6.84} \approx \frac{\sin 41.93^\circ}{7.37}$$

$$\sin B \approx \frac{6.84 \sin 41.93^\circ}{7.37}$$

$$\sin B \approx 0.6202$$

$$B \approx \sin^{-1}(0.6202)$$

$$B \approx 38.33^\circ$$

$$C \approx 180^\circ - 41.93^\circ - 38.33^\circ \approx 99.74^\circ$$

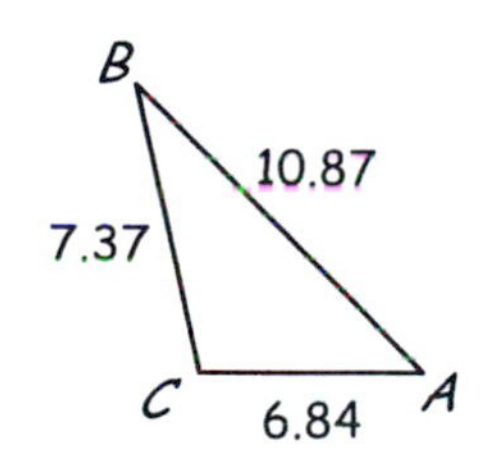

79. Given: $a = 5$, $b = 10$, $c = 7$

Use the Law of Cosines to find A:

$$\cos A = \frac{5^2 - 10^2 - 7^2}{-2(10)(7)}$$

$$\approx \frac{-124}{-140}$$

$$\approx 0.885714$$

$$A \approx \cos^{-1}(0.885714)$$

$$A \approx 27.66^\circ$$

Use the Law of Sines to find B:

$$\frac{\sin B}{10} \approx \frac{\sin 27.66^\circ}{5}$$

$$\sin B \approx \frac{10 \sin 27.66^\circ}{5}$$

$$\sin B \approx 0.9285$$

$$B \approx \sin^{-1}(0.9285)$$

$$B \approx 68.20^\circ \quad \text{or} \quad 111.80^\circ$$

Now, consider the fact that b is the longest side, so $B > 90^\circ$.

Therefore $B \approx 111.80^\circ$ is the appropriate choice.

$$C \approx 180^\circ - 111.80^\circ - 27.66^\circ \approx 40.54^\circ$$

81. Given: $B = 60^\circ 7'$, $c = 18$, $a = 6$

Use $\text{Area} = \frac{1}{2} ac \sin B$:

$$\text{Area} \approx \frac{1}{2}(6)(18)(0.8670)$$

$$\approx 46.82 \text{ sq. units}$$

83. Given: $B = 54^\circ$, $a = 10$, $c = 7$

Use $\text{Area} = \frac{1}{2} ac \sin B$:

$$\text{Area} \approx \frac{1}{2}(10)(7)(0.80902)$$

$$\approx 28.32 \text{ sq. units}$$

85. Given: $C = 46^\circ$, $b = 20$, $a = 19$

Use $\text{Area} = \frac{1}{2} ab \sin C$:

$$\text{Area} \approx \frac{1}{2}(20)(19)(0.7193)$$

$$\approx 136.67 \text{ sq. units}$$

87. Given: $b = 12$, $c = 18$, $a = 15$

Use $\text{Area} = \sqrt{s(s-a)(s-b)(s-c)}$

where $s = \frac{a+b+c}{2}$:

Here $s = \frac{15+12+18}{2} = 22.5$

$$\text{Area} = \sqrt{22.5(7.5)(10.5)(4.5)}$$

$$\approx \sqrt{7973.4375}$$

$$\approx 89.29 \text{ sq. units}$$

89. The area being planted is represented by the figure:

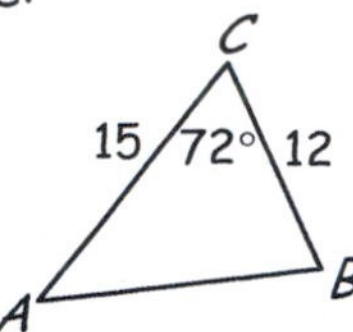

Given: $C = 72°$, $b = 15$, $a = 12$

Use Area $= \frac{1}{2}ab\sin C$:

$$\text{Area} \approx \frac{1}{2}(15)(12)(0.95106)$$
$$\approx 85.5951 \text{ ft.}^2$$

91. The area of the windows is represented by the figure:

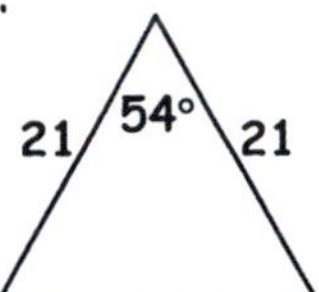

$$\text{Area} = \frac{1}{2}(21)(21)\sin 54°$$
$$\approx 178.3882 \text{ ft.}^2$$

93. a. For the pentagon:

The central angles $= 72°$ (from $360° \div 5$)

The base angles $= \dfrac{180° - 72°}{2} = 54°$

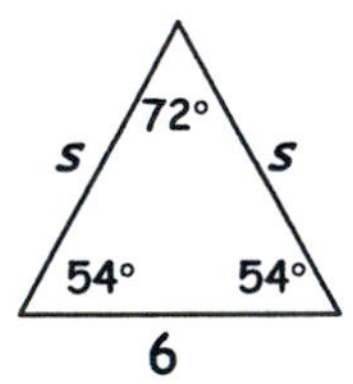

Let s = length of the legs of the five isosceles triangles formed within the pentagon. See the figure of one of the triangles. Find s:

$$\frac{s}{\sin 54°} = \frac{6}{\sin 72°}$$
$$s = \frac{6\sin 54°}{\sin 72°}$$
$$s \approx 5.104$$

$$\text{Area of one triangle} = \frac{1}{2}(5.104)^2 \sin 72°$$
$$\approx 12.3874$$
$$\text{Area of Pentagon} \approx 5(12.3874)$$
$$\approx 61.9372 \text{ in.}^2$$

b. For the octagon:

The central angles $= 45°$ (from $360° \div 8$)

The base angles $= \dfrac{180° - 45°}{2} = 67.5°$

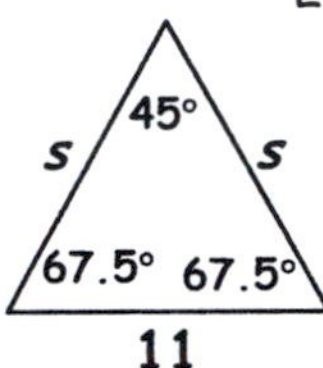

Let s = length of the legs of the eight isosceles triangles formed within the octagon. See the figure of one of the triangles. Find s:

$$\frac{s}{\sin 67.5°} = \frac{11}{\sin 45°}$$
$$s = \frac{11\sin 67.5°}{\sin 45°}$$
$$s \approx 14.3722$$

$$\text{Area of 1 triangle} = \frac{1}{2}(14.3722)^2 \sin 45°$$
$$\approx 73.03$$
$$\text{Area of Pentagon} \approx 8(73.03)$$
$$\approx 584.2397 \text{ in.}^2$$

c. There are several ways to find the area of the given five pointed star. The way shown below uses some of the base angle information from part a. First, we find the length of a in the figure to enable us to find the area of the central pentagon. Then, by determining the angles of the points (triangles), we can find the area of each of the triangular points. Finally, we will add the area of the pentagon and that of the five triangular points to obtain the area of the star:

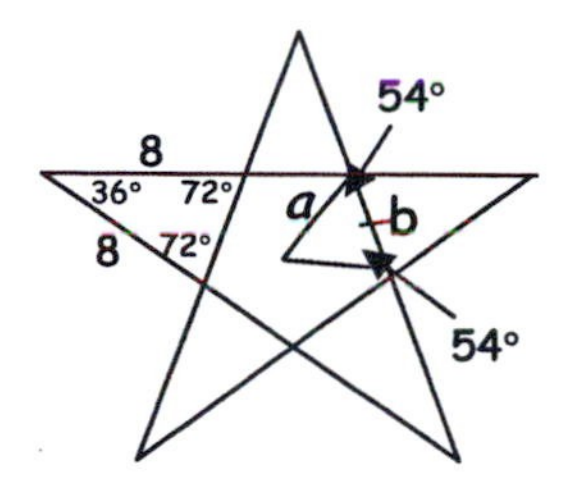

To find the area of the central pentagon, first find b:

Use the Law of Sines:

$$\frac{b}{\sin 36^\circ} = \frac{8}{\sin 72^\circ}$$

$$b = \frac{8\sin 36^\circ}{\sin 72^\circ} \approx 4.9443$$

Similarly, find a:

$$\frac{a}{\sin 54^\circ} \approx \frac{4.9443}{\sin 72^\circ}$$

$$a \approx \frac{4.9443\sin 54^\circ}{\sin 72^\circ} \approx 4.2058$$

Using Heron's Formula:

$$s \approx \frac{4.2058 + 4.2058 + 4.9443}{2}$$

$$\text{Area of Pentagon} \approx 5\sqrt{6.6780(6.6780-4.2058)(6.6780-4.2058)(6.6780-4.9443)}$$

$$\approx 42.0585$$

Area of the Star point triangles:

$$5\left[\frac{1}{2}8^2 \sin 36^\circ\right] \approx 94.0456$$

Area of the five pointed star:

$42.0585 + 94.0456 \approx 136.1041 \text{ in.}^2$

End of Section 8.1

Chapter Eight: Section 8.2
Solutions to Odd-numbered Exercises

1. Plot $\left(-1, \frac{5\pi}{4}\right)$:

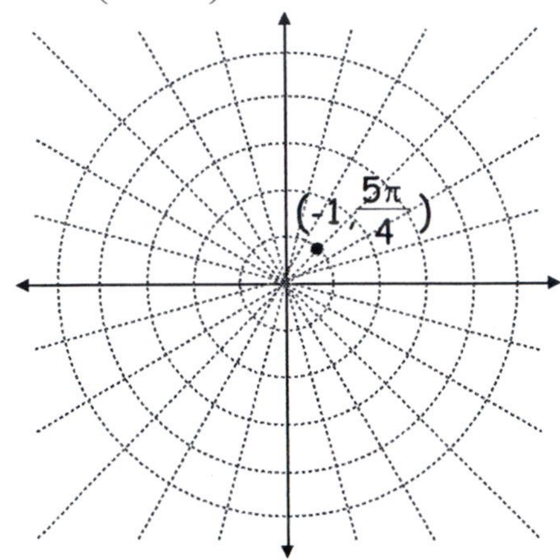

3. Plot $\left(\frac{1}{4}, \frac{-7\pi}{6}\right)$:

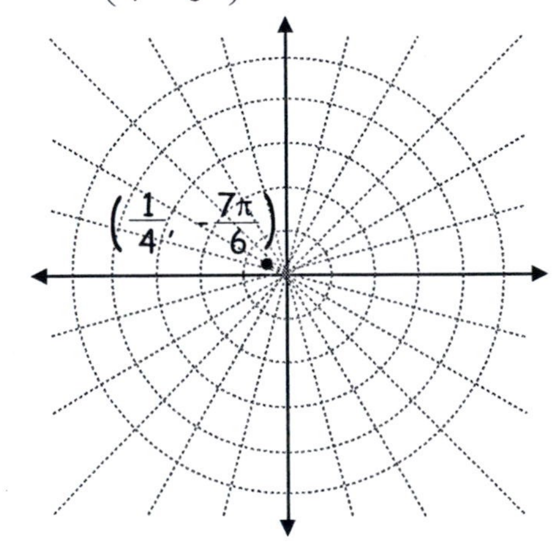

5. Plot $\left(4\frac{8}{9}, -\pi\right)$:

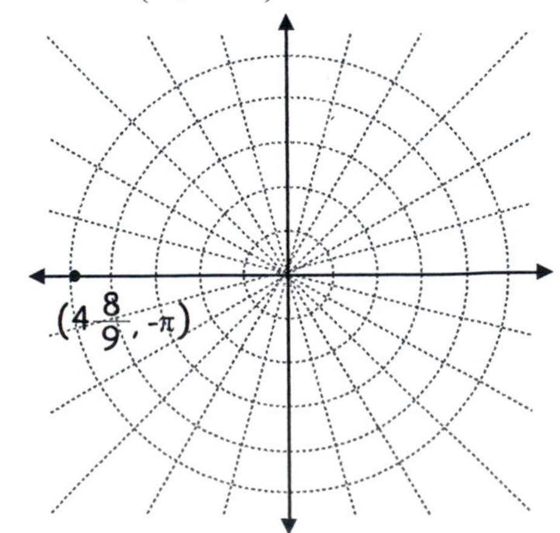

7. Convert Polar $\left(5, \frac{7\pi}{4}\right)$ to Cartesian coordinate:

Use $x = r\cos\theta;\ y = r\sin\theta$

$$x = 5\cos\frac{7\pi}{4};\ y = 5\sin\frac{7\pi}{4}$$
$$x \approx 3.54; \quad y \approx -3.54$$

9. Convert Polar $\left(6.25, \frac{-3\pi}{4}\right)$ to Cartesian coordinate:
Use $x = r\cos\theta;\ y = r\sin\theta$

$$x = 6.25\cos\frac{-3\pi}{4};\ y = 6.25\sin\frac{-3\pi}{4}$$
$$x \approx -4.42; \quad y \approx -4.42$$

11. Convert Polar $\left(3, \frac{-5\pi}{6}\right)$ to Cartesian coordinate:
Use $x = r\cos\theta;\ y = r\sin\theta$

$$x = 3\cos\frac{-5\pi}{6};\ y = 3\sin\frac{-5\pi}{6}$$
$$x \approx -2.6; \quad y \approx -1.5$$

13. Convert Cartesian $(-3,0)$ to Polar coordinate:

$$r = \pm\sqrt{0^2 + (-3)^2} = \pm 3$$
$$\tan\theta = \frac{0}{-3} = 0$$
$$\Rightarrow \theta = 0 \text{ with } -3 \text{ and } \pi \text{ with } 3.$$

Multiple Polar coordinate equivalencies exist: $(3, \pi)$ and $(-3, 0)$ are two.

15. Convert Cartesian $(12, -1)$ to Polar coordinate:

$$r = \pm\sqrt{12^2 + (-1)^2} = \pm\sqrt{145}$$
$$\tan\theta = \frac{-1}{12} = -0.8\overline{3} \Rightarrow$$
$$\theta = \tan^{-1}\left(-0.8\overline{3}\right) \approx -0.0831 \text{ with } r = \sqrt{145}$$
$$\theta = -0.0831 + \pi = 3.058 \text{ with } r = -\sqrt{145}$$

Multiple Polar coordinate equivalencies exist: $\left(\sqrt{145},-0.08\right)$ and $\left(-\sqrt{145},3.06\right)$ are two.

17. Convert Cartesian $\left(-\sqrt{3},9\right)$ to Polar coordinate: Note that the point lies in quadrant 2.

$$r=\pm\sqrt{\left(-\sqrt{3}\right)^2+9^2}=\pm\sqrt{84}=\pm2\sqrt{21}$$

$$\tan\theta=\frac{9}{-\sqrt{3}}\Rightarrow$$

$$\theta=\tan^{-1}\left(\frac{9}{-\sqrt{3}}\right)\approx-1.381 \text{ with } r=-2\sqrt{21}.$$

$\theta=\pi-1.381=1.761$ with $r=2\sqrt{21}$.
Multiple Polar coordinate equivalencies exist: $\left(2\sqrt{21},1.76\right)$ and $\left(-2\sqrt{21},-1.38\right)$ are two.

19. Rewrite $x^2+y^2=25$ in Polar form:

$$(r\cos\theta)^2+(r\sin\theta)^2=25$$
$$r^2\cos^2\theta+r^2\sin^2\theta=25$$
$$r^2\left(\cos^2\theta+\sin^2\theta\right)=25$$
$$r^2=25$$

21. Rewrite $x=12$ in Polar form:

$$r\cos\theta=12$$

23. Rewrite $y=x$ in Polar form:

$$r\sin\theta=r\cos\theta$$

$\sin\theta=\cos\theta$, or $\dfrac{\sin\theta}{\cos\theta}=1$, or

$$\tan\theta=1$$

25. Rewrite $x=16a$ in Polar form:

$$r\cos\theta=16a$$

27. Rewrite $x^2+y^2=4ax$ in Polar form:

$$(r\cos\theta)^2+(r\sin\theta)^2=4a(r\cos\theta)$$
$$r^2\cos^2\theta+r^2\sin^2\theta=4ar\cos\theta$$
$$r^2\left(\cos^2\theta+\sin^2\theta\right)=4ar\cos\theta$$
$$r^2-4ar\cos\theta=0$$

29. Rewrite $y^2-4=4x$ in Polar form:

$$(r\sin\theta)^2-4=4(r\cos\theta)$$
$$r^2\sin^2\theta-4ar\cos\theta-4=0$$

31. Rewrite the Polar equation $r=5\cos\theta$ in rectangular form:

$$\cos\theta=\frac{r}{5}$$
$$r\cos\theta=\frac{r^2}{5}$$
$$x=\frac{r^2}{5}$$
$$r^2=5x$$
$$x^2+y^2=5x$$

33. Rewrite the Polar equation $r=7$ in rectangular form:

$$x^2+y^2=49$$

35. Rewrite the Polar equation $18r=9\csc\theta$ in rectangular form:

$$\csc\theta=2r$$
$$\frac{1}{\sin\theta}=2r$$
$$2r\sin\theta=1$$
$$r\sin\theta=\frac{1}{2}$$
$$y=\frac{1}{2}$$

37. Rewrite the Polar equation $r^2=\sin2\theta$ in rectangular form:

$$r^2\left(r^2\right)=r^2 2\sin\theta\cos\theta$$
$$\left(x^2+y^2\right)\left(x^2+y^2\right)=r2\sin\theta r\cos\theta$$
$$x^4+y^4+2x^2y^2=2xy$$

39. Rewrite the Polar equation $r=\dfrac{12}{4\sin\theta+7\cos\theta}$ in rectangular form:

$$r(4\sin\theta+7\cos\theta)=12$$
$$4r\sin\theta+7r\cos\theta=12$$
$$4y+7x=12$$

41. Rewrite the Polar equation $r = 3$ in rectangular form:

$$r^2 = 9$$
$$x^2 + y^2 = 9$$

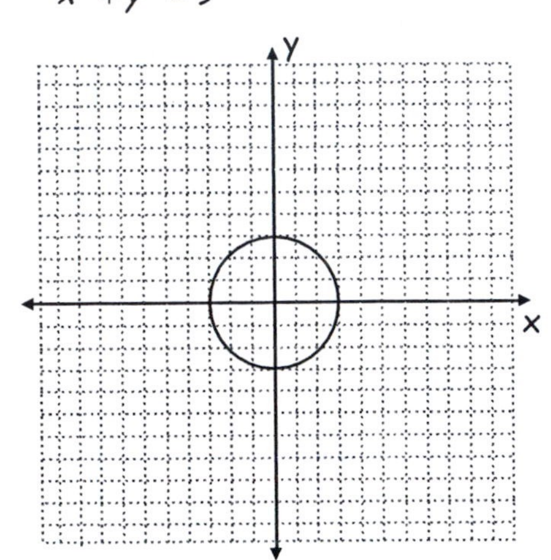

43. Rewrite the Polar equation $\theta = \frac{5\pi}{6}$ in rectangular form:

$$\frac{y}{x} = \tan\theta$$
$$\frac{y}{x} = \tan\left(\frac{5\pi}{6}\right) = -\frac{1}{\sqrt{3}}$$
$$y = -\frac{x}{\sqrt{3}}$$

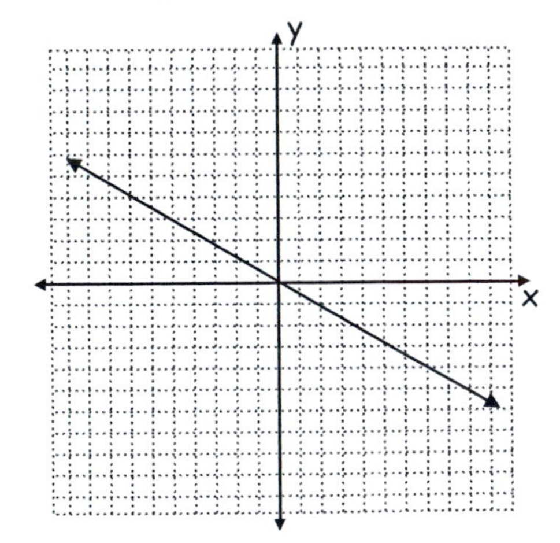

45. Rewrite the Polar equation $r = 7\sec\theta$ in rectangular form:

$$r = 7\left(\frac{1}{\cos\theta}\right)$$
$$r\cos\theta = 7$$
$$x = 7$$

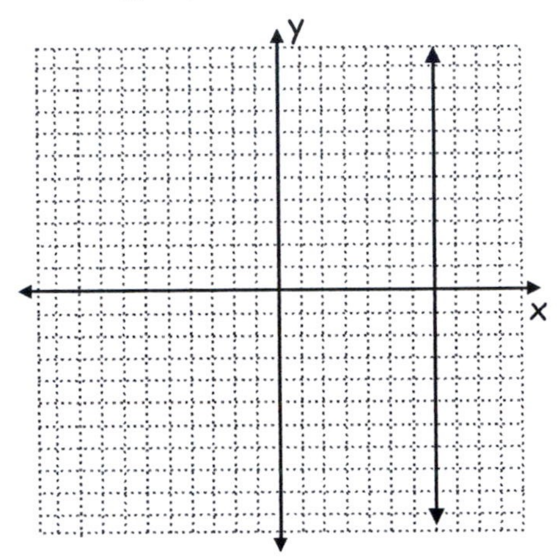

47. Graph of $r = 4$:

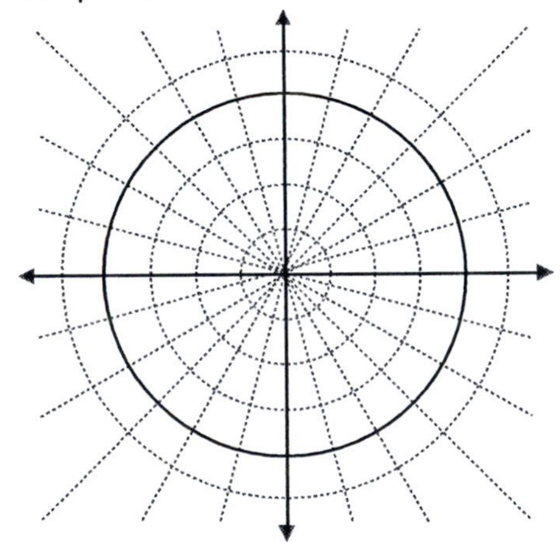

49. Graph of $\theta = \frac{4\pi}{3}$:

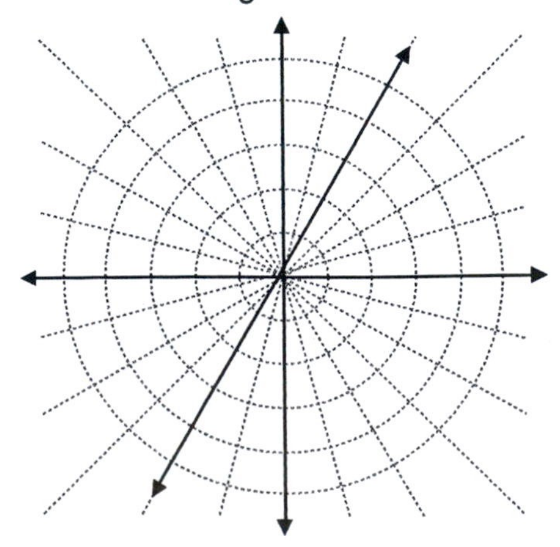

51. Graph of $r = 6\cos\theta$:

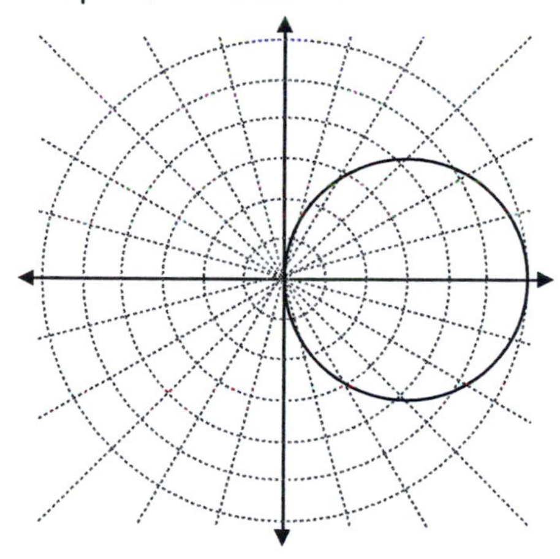

53. Graph of $r = 3 - 3\sin\theta$:
The form of the equation indicates the graph will be a cardiod. Plot some points to find the graph of the equation.

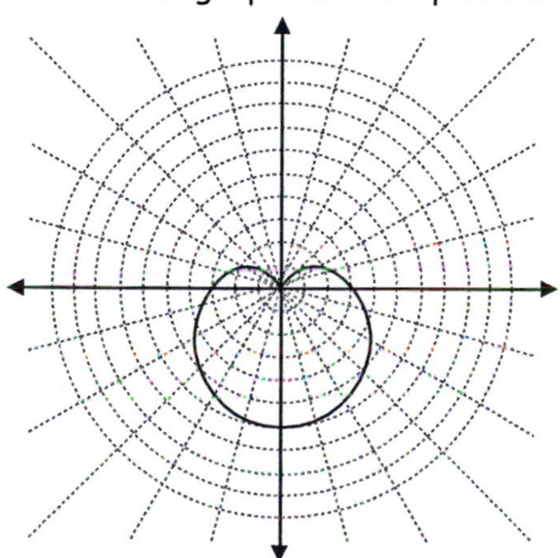

55. Graph of $r = 7(1 + \cos\theta)$:
The form of the equation indicates the graph will be a cardiod. Plotting some points will help sketch the graph.

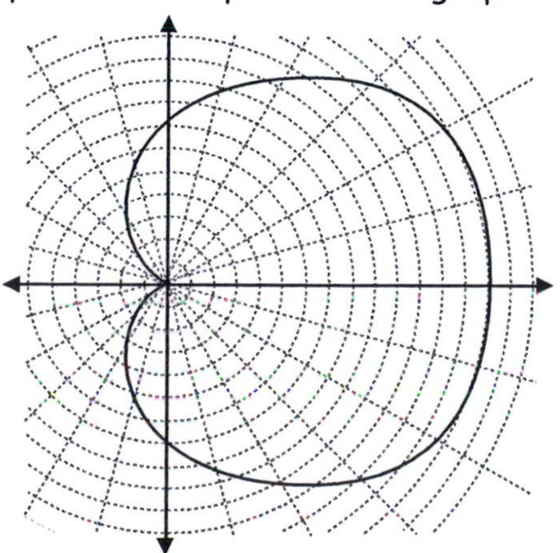

57. Graph of $r = 4 - 3\sin\theta$:
The form of the equation indicates the graph will be a dimpled limacon. Plotting some points will help in sketching the graph.

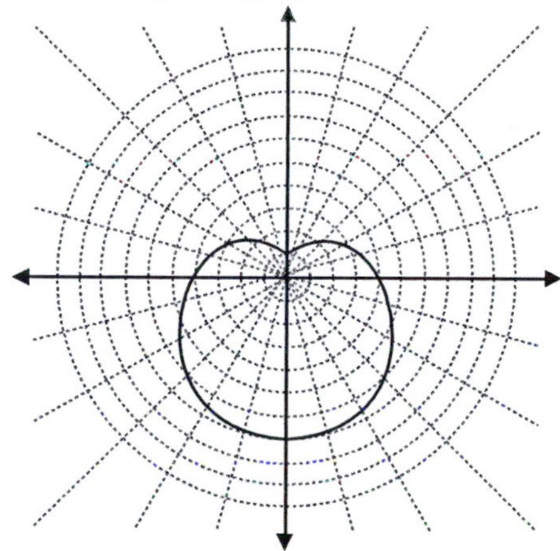

59. Graph of $r = 3\sin 3\theta$:
The form of the equation indicates the graph will be a three-leaved rose. Plotting some points will help in sketching the graph.

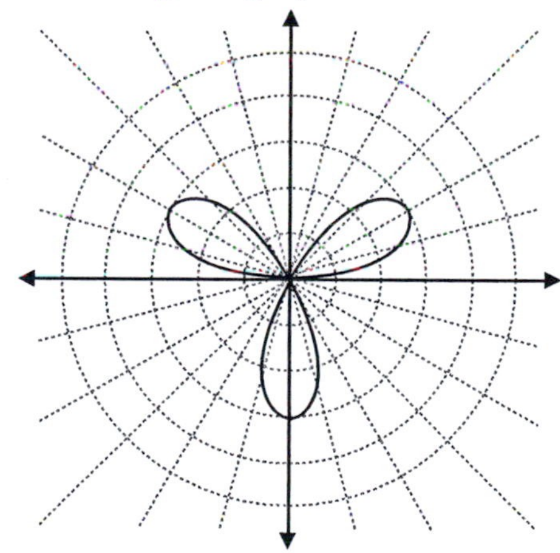

61. Graph of $r = 2\sin 2\theta$:
The form of the equation indicates the graph will be a four-leaved rose. Plotting some points will help in sketching the graph.

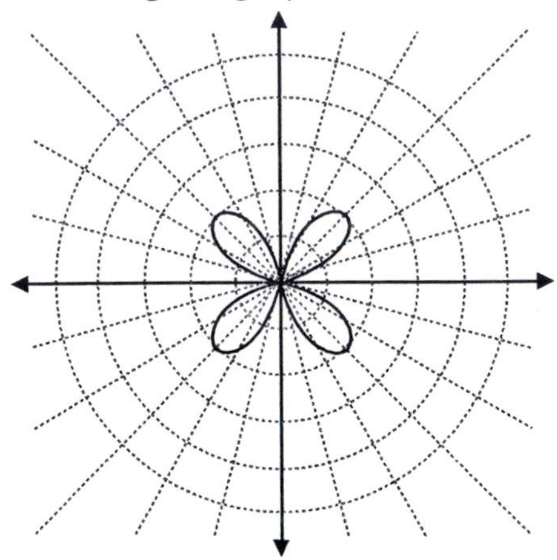

63. Graph of $r = 5\cos 5\theta$:
The form of the equation indicates the graph will be a five-leaved rose. Plotting some points will help in sketching the graph.

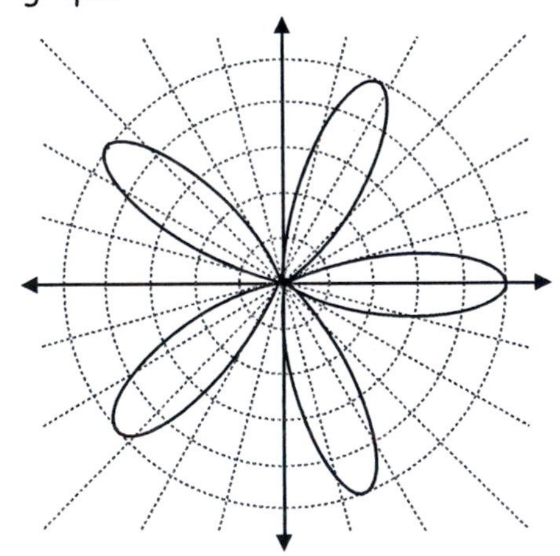

65. Graph of $r = 4\cos 4\theta$:
The form of the equation indicates the graph will be a eight-leaved rose. Plotting some points will help in sketching the graph.

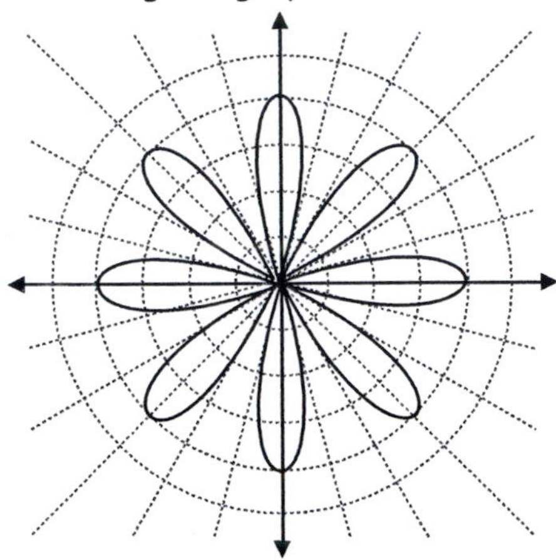

67. Graph of $r^2 = 16\sin 2\theta$:
The form of the equation indicates the graph will be a lemniscate. Plotting some points will help in sketching the graph.

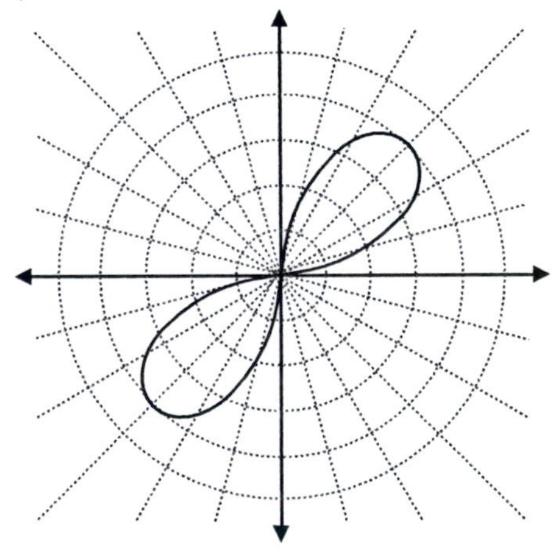

End of Section 8.2

Chapter Eight: Section 8.3
Solutions to Odd-numbered Exercises

1. Given: $x = 5 + t;\ y = \dfrac{\sqrt{t}}{t-2}$

t	0	1	2
(x,y)	$(5,0)$	$(6,-1)$	$(7,\text{undefined})$

t	3	4	5	6
(x,y)	$(8,\sqrt{3})$	$(9,1)$	$\left(10,\dfrac{\sqrt{5}}{3}\right)$	$\left(11,\dfrac{\sqrt{6}}{4}\right)$

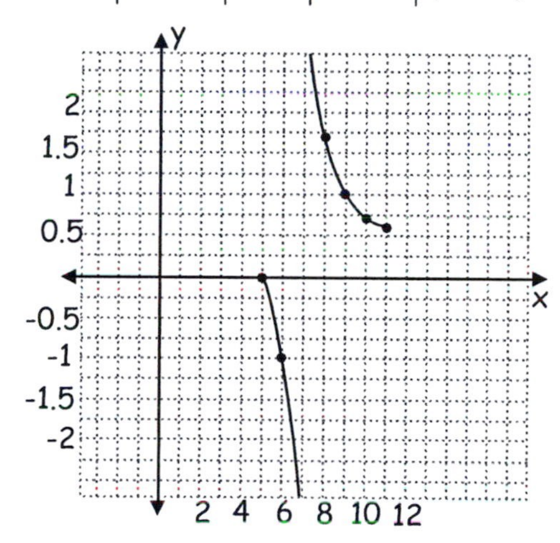

3. Use $y = -\dfrac{1}{2}gt^2 + v_0(\sin\theta)t + h_0$

$x = (v_0\cos\theta)t$

Given: $h_0 = 10$ ft; $v_0 = 80$ mph; $\theta = 60°$; $g = 32\ \text{ft/s}^2$. Convert v_0 to ft/s

$$v_0 = \frac{80(5280)}{3600} \approx 117.3 \text{ ft/s}$$

a. Parametric equations:

$x \approx (117.3\cos 60°)t$

$x \approx 58.67t$

$y = -\dfrac{1}{2}32t^2 + 117.3(\sin 60°)t + 10$

$y \approx -16t^2 + 101.61t + 10$

b. Sketch graph of flight:

$x = 58.67t$

$y = -16t^2 + 101.61t + 10$

t	0	1	2
(x,y)	$(0,10)$	$(59,96)$	$(117,149)$

3	4	5
$(176,174)$	$(235,160)$	$(293,118.05)$

6	7
$(352,44)$	$(411,-63)$

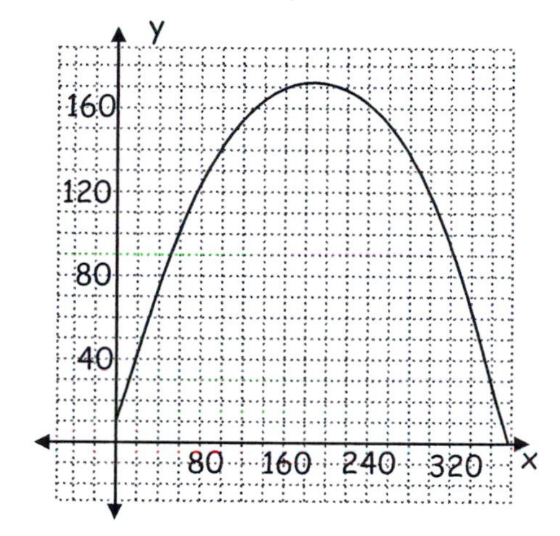

c. For $t = 1.5$

$y = -16(1.5)^2 + 101.61(1.5) + 10$

$y \approx 126.42$ ft.

d. Find t for $y = 0$:

Set $-16t^2 + 101.6t + 10 = 0$.

Solve for t:

(use the quadratic formula):

$$t = \frac{-101.6 \pm \sqrt{101.6^2 - 4(10)(-16)}}{2(-16)}$$

$$t \approx \frac{-101.6 - 104.7}{-32}$$

$t \approx 6.45$ s

Then, $x \approx 58.67(6.45) \approx 378.42$ ft.

(the point at which to place the net)

e. From d above, the performer would land in the net at $t \approx 6.45$ s.

f. At $x = 70$ ft., $t = \dfrac{70}{58.67} \approx 1.193$ s.

Then,

$y = -16(1.193)^2 + 101.61(1.193) + 10$

≈ 108.45 ft.

Yes, the performer should clear the 12 ft. high wall of flames unharmed.

5. Given: $y = -\frac{1}{2}t^2 + v_0 \sin\theta + h_0$,

$g = 32$ ft./s^2, $\theta = 48°$, $v_0 = 21$ ft./s

a. Parametric equations:

$x = 21\cos 48° t$

$x = 14.05t$

$y = -\frac{1}{2}(32)t^2 + 21(\sin 48°)t + 7$

$y = -16t^2 + 15.61t + 7$

b. Sketch the graph:

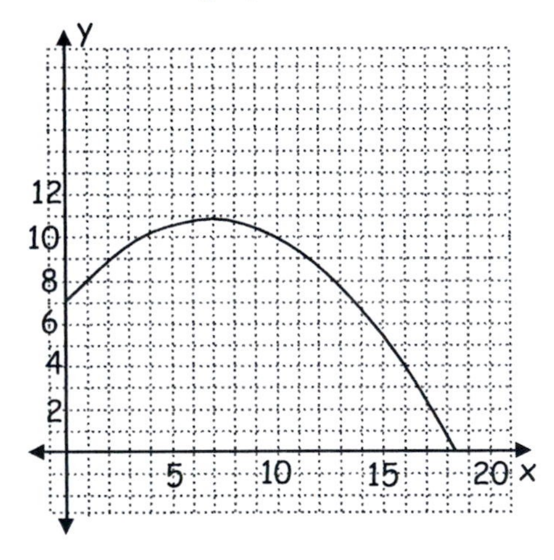

t	0	0.25	0.5
(x, y)	$(0, 7)$	$(3.5, 9.9)$	$(7, 10.8)$

0.75	1
$(10.5, 9.7)$	$(14.1, 6.6)$

c. For $x = 15$, $t \approx \frac{15}{14.05} \approx 1.07$

Find t for $y = 11$:

$-16t^2 + 15.61t + 7 = 11$, or

$-16t^2 + 15.61t - 4 = 0$

Use quadratic formula:

$$t = \frac{-15.61 \pm \sqrt{15.61^2 - 4(-16)(-4)}}{2(-16)}$$

$$t \approx \frac{-15.61 \pm \sqrt{-12.34}}{-32} \Rightarrow$$

t is not a real number

No, he will not make the shot.

7. Sketch: $x = \sqrt{t-2}$; $y = 3t - 2$

Eliminate the parameter t.

First, solve for t in one equation, and substitute in the other equation:

From $y = 3t - 2$, $t = \frac{y+2}{3}$. Then,

$x = \sqrt{\frac{y+2}{3} - 2}$.

Note that for $\frac{y+2}{3} - 2 \geq 0$,

$y + 2 - 6 \geq 0$

$y \geq 4$,

$t \geq 2$ and $x \geq 0$.

Then,

$x = \sqrt{\frac{y+2}{3} - 2}$

$x^2 = \frac{y+2}{3} - 2$, or $y = 3x^2 + 4$

$(x \geq 0)$ as $\sqrt{t-2}$.

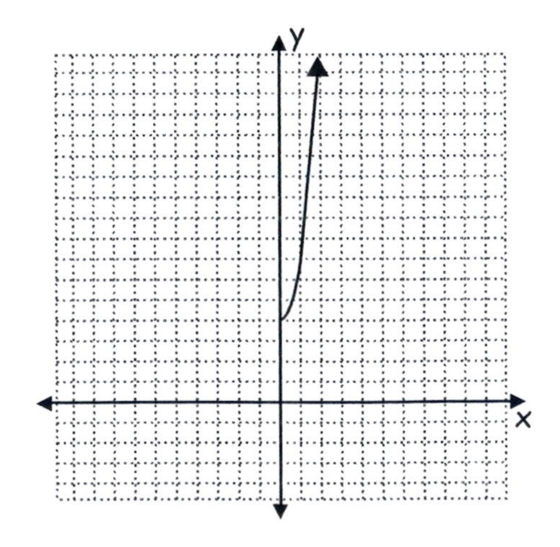

9. Sketch: $x = |t+3|;\ y = t-5$

Eliminate the parameter t.

First, solve for t in one equation, and substitute in the other equation:

From $y = t-5,\ t = y+5$. Then,

$x = |y+8|$.

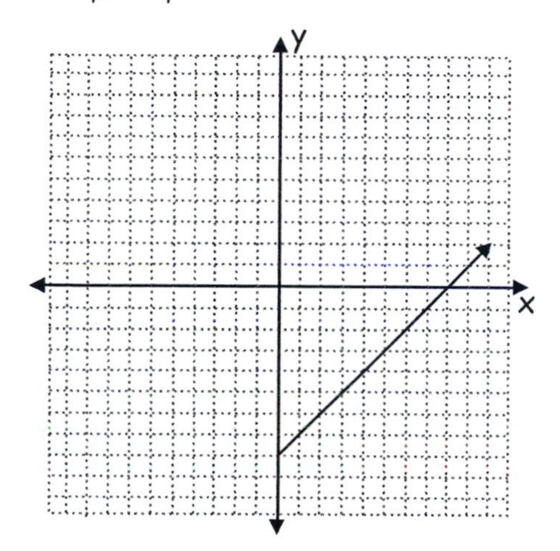

11. Sketch: $x = \dfrac{t}{t+2};\ y = \sqrt{t}$

Note: $t \geq 0$

Eliminate the parameter t.

First, solve for t in one equation, and substitute in the other equation:

From $y = \sqrt{t},\ t = y^2$. Then,

$$x = \frac{y^2}{y^2+2}$$

$$x(y^2+2) = y^2$$

$$xy^2 + 2x = y^2$$

$$xy^2 - y^2 = -2x$$

$$(x-1)y^2 = -2x$$

$$y^2 = \frac{-2x}{x-1}$$

$$y = \sqrt{\frac{2x}{1-x}},\ \text{with } 0 \leq x < 1$$

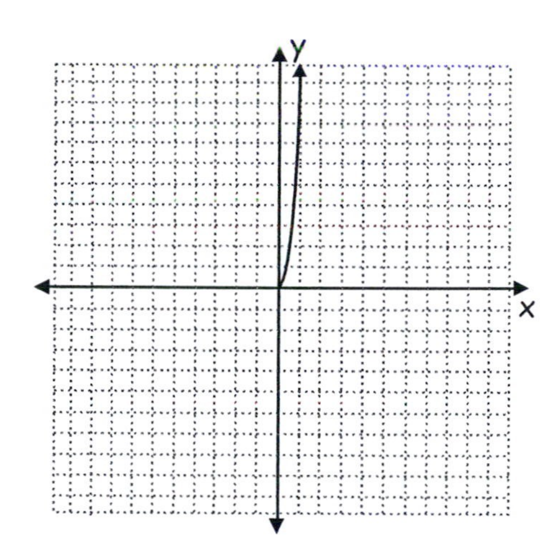

13. Sketch: $x = \dfrac{2}{|t-3|};\ y = 2t-1$

Note: $t \neq 3$

Eliminate the parameter t.

First, solve for t in one equation, and substitute in the other equation:

From $y = 2t-1,\ t = \dfrac{y+1}{2}$. Then,

$$x = \frac{2}{\left|\frac{y+1}{2} - 3\right|}$$

$$x = \frac{2}{\left|\frac{y-5}{2}\right|}$$

$$x = \frac{2}{\frac{|y-5|}{2}} = \frac{4}{|y-5|}\ \text{ with } t \neq 3,\ y \neq 5$$

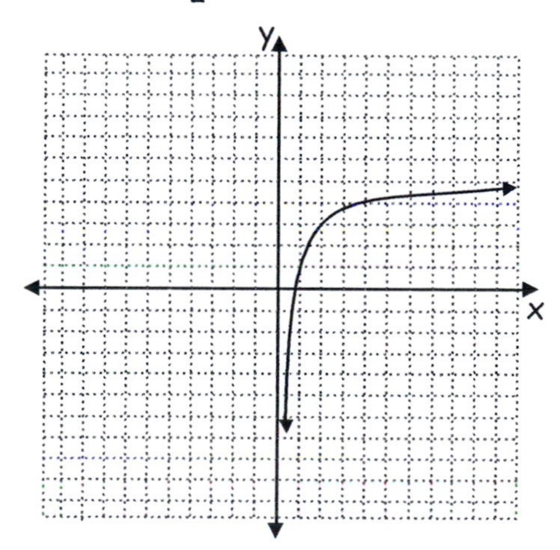

15. Sketch: $x = 3\sin\theta - 1,\ y = \dfrac{\cos\theta}{2}$

$x = 3\sin\theta - 1 \Rightarrow \sin\theta = \dfrac{x+1}{3}$

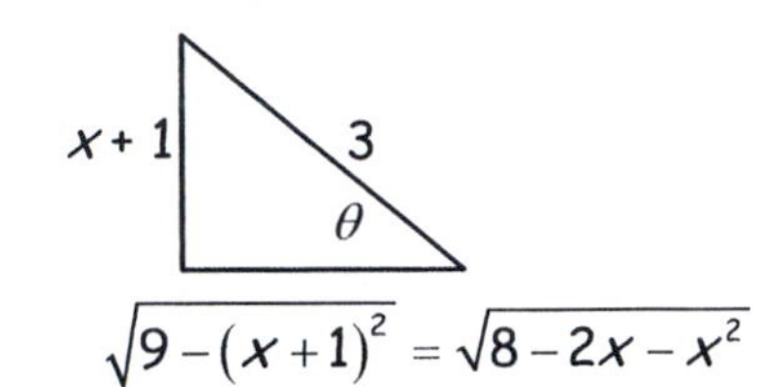

$\sqrt{9-(x+1)^2} = \sqrt{8-2x-x^2}$

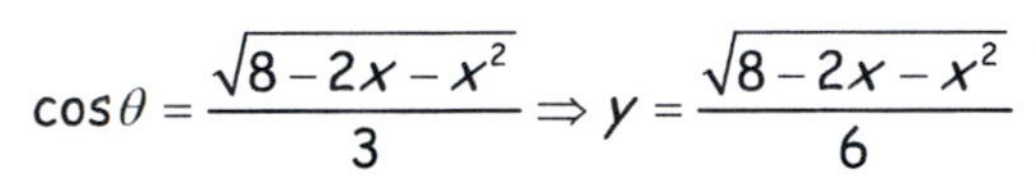

$\cos\theta = \dfrac{\sqrt{8-2x-x^2}}{3} \Rightarrow y = \dfrac{\sqrt{8-2x-x^2}}{6}$

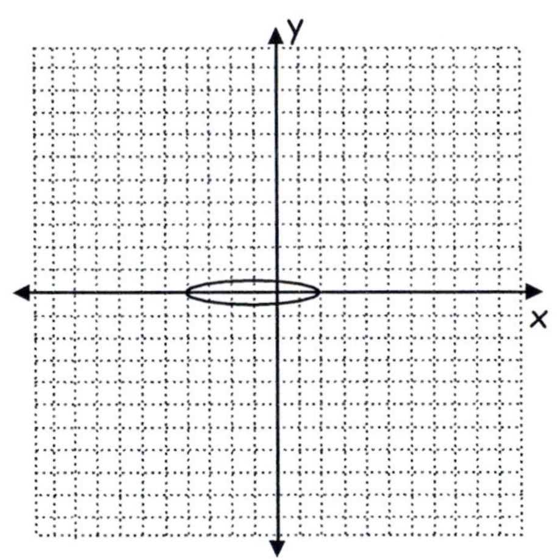

17. Sketch: $x = 2\cos\theta,\ y = 3\cos\theta$

$x = 2\cos\theta \Rightarrow \cos\theta = \dfrac{x}{2}$

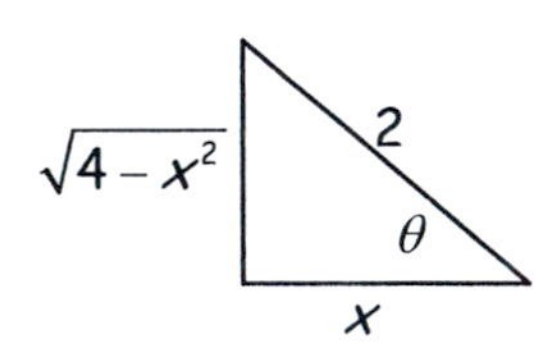

With $\cos\theta = \dfrac{x}{2}, y = 3\cos\theta \Rightarrow y = \dfrac{3x}{2}$

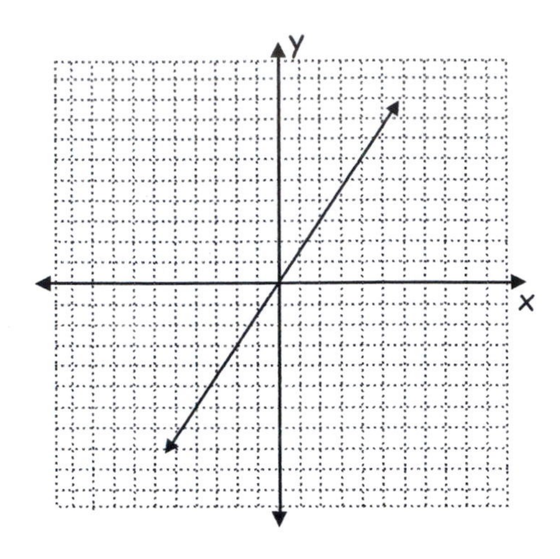

19. Sketch: $x = \sin\theta,\ y = 4 - 3\cos\theta$

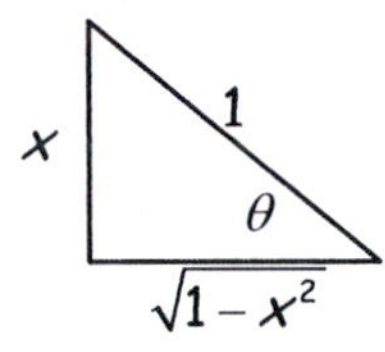

$\cos\theta = \sqrt{1-x^2}$

$y = 4 - 3\sqrt{1-x^2}$

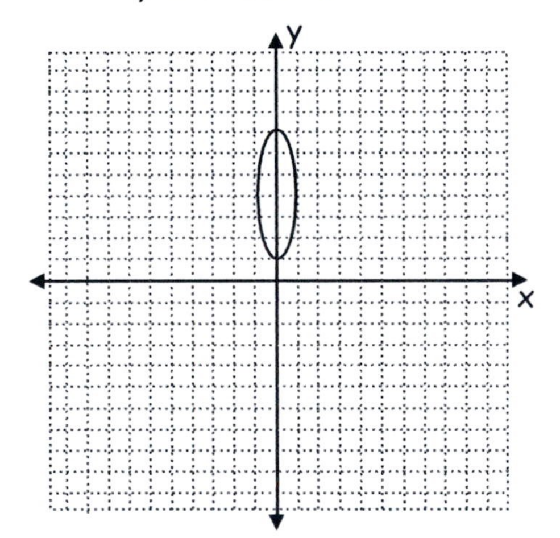

Questions 21-51: Note answers may vary.

21. Given: $y = 5x - 2$; Let $x = t$
Then, $y = 5t - 2$

23. Given: $x^2 + \frac{y^2}{4} = 1$; Let $x = t$
Then, $t^2 + \frac{y^2}{4} = 1$
$y^2 = 4(1 - t^2)$
$y = \pm\sqrt{4 - 4t^2} = \pm 2\sqrt{1 - t^2}$

25. Given: $y = x^2 + 1$; Let $x = t$
Then, $y = t^2 + 1$

27. Given: $x = 4y - 6$; Let $x = t$
Then, $t = 4y - 6 \Rightarrow y = \frac{t+6}{4}$

29. Given: $x = 2(y - 3)$; Let $x = t$
Then, $t = 2y - 6 \Rightarrow y = \frac{t+6}{2}$

31. Given: $x = \frac{1}{3y}$; Let $x = t$
Then, $t = \frac{1}{3y} \Rightarrow y = \frac{1}{3t}$

33. Given: Line with Slope -2, passing through $(-5, -2)$:
Then, $y + 2 = -2(x + 5)$.
Let $x = t$
Then, $y + 2 = -2(t + 5)$
$y = -2t - 12$

35. Given: Line with Slope 3, passing through $(7, 2)$:
Then, $y - 2 = 3(x - 7)$.
Let $x = t$
Then, $y - 2 = 3(t - 7)$
$y = 3t - 19$

37. Given: Line passing through $(6, -3)$ and $(2, 3)$: Find the Slope:
Slope $= \frac{3-(-3)}{2-6} = \frac{6}{-4} = -\frac{3}{2}$
Then,
$y - 3 = -\frac{3}{2}(x - 2)$.
Let $x = t$
Then, $y - 3 = -\frac{3}{2}(t - 2)$
$y = -\frac{3}{2}t + 6$

39. Given: Circle with Center $(0, 0)$ and radius $r = 1$:
$x^2 + y^2 = 1$
Let $x = \cos\theta$
$\cos^2\theta + y^2 = 1$
$y = \sin\theta$

41. Given: Circle with Center $(7, -5)$ and radius $r = 4$:
$(x-7)^2 + (y+5)^2 = 16$
Let $x = 7 + 4\cos\theta$
$(x-7)^2 + (y+5)^2 = 16$
$(7 + 4\cos\theta - 7)^2 + (y+5)^2 = 16$
$(4\cos\theta)^2 + (y+5)^2 = 16$
$(y+5)^2 = 16(1 - \cos^2\theta)$
$(y+5)^2 = 16(\sin^2\theta)$
$y + 5 = 4\sin\theta$
$y = 4\sin\theta - 5$

43. Given: Ellipse with Vertices $(\pm 3, 0)$ and Foci $(\pm 2, 0)$. Use the general equation of an ellipse and the fact that the Center is $(0, 0)$:
$\frac{x^2}{a^2} + \frac{y^2}{b^2} = 1$; $c^2 = a^2 - b^2$, where c is the distance of one of the foci to the Center, and $a = 3$ and $c = 2$.

From $c^2 = a^2 - b^2$

$4 = 9 - b^2$

$b^2 = 5$

$b = \sqrt{5}$

So $\frac{x^2}{9} + \frac{y^2}{5} = 1$

Let $x = 3\cos\theta$, then

$\frac{(3\cos\theta)^2}{9} + \frac{y^2}{5} = 1$

$\cos^2\theta + \frac{y^2}{5} = 1$

$5\cos^2\theta + y^2 = 5$

$y^2 = 5 - 5\cos^2\theta$

$y^2 = 5(1 - \cos^2\theta)$

$y^2 = 5(\sin^2\theta)$

$y = \sqrt{5}\sin\theta$

45. Given: Ellipse with Vertices $(5, 2)$ and $(5, -4)$ and Foci $(5, 0)$ and $(5, -2)$. Use the general equation of an ellipse and the fact that the Center is $(5, -1)$:

$\frac{(x-h)^2}{b^2} + \frac{(y-k)^2}{a^2} = 1$; $c^2 = a^2 - b^2$, where c is the distance of one of the foci to the Center, and $a = 3$ and $c = 1$.

From $c^2 = a^2 - b^2$

$1 = 9 - b^2$

$b^2 = 8$

The ellipse is $\frac{(x-5)^2}{8} + \frac{(y+1)^2}{9} = 1$

Let $x = 5 + 2\sqrt{2}\cos\theta$, then

$\frac{(5+2\sqrt{2}\cos\theta-5)^2}{8} + \frac{(y+1)^2}{9} = 1$

$\frac{8\cos^2\theta}{8} + \frac{(y+1)^2}{9} = 1$

$(y+1)^2 = 9(1-\cos^2\theta)$

$y+1 = 3\sin\theta$

$y = -1+3\sin\theta$

47. Given: Hyperbola with Vertices $(\pm 3, 0)$ and Foci $(\pm 4, 0)$. Use the general equation of a hyperbola and the fact that the Center (h, k) is $(0, 0)$:

$\frac{(x-h)^2}{a^2} - \frac{(y-k)^2}{b^2} = 1$;

$c^2 = a^2 + b^2$, where $a = 3$ and $c = 4$. Then, $b^2 = 16 - 9 = 7$, or $b = \sqrt{7}$.

The hyperbola is $\frac{x^2}{9} - \frac{y^2}{7} = 1$

Let $x = t$. Then,

$\frac{t^2}{9} - \frac{y^2}{7} = 1$

$\frac{y^2}{7} = \frac{t^2}{9} - 1$

$y^2 = \frac{7t^2}{9} - 7$

$y = \pm\sqrt{\frac{7t^2}{9} - 7}$

$y = \pm\sqrt{7\left(\frac{t^2}{9} - 1\right)}$

49. Given: Hyperbola with Vertices $(0, \pm 3)$ and Foci $(0, \pm 5)$. Use the general equation of a hyperbola and the fact that the Center (h, k) is $(0, 0)$:
First, note that the hyperbola is vertically oriented.

$$\frac{(y-k)^2}{a^2} - \frac{(x-h)^2}{b^2} = 1;$$

$c^2 = a^2 + b^2$, where $a = 3$ and $c = 5$. Then, $b^2 = 25 - 9 = 16$, or $b = 4$

The hyperbola is $\dfrac{y^2}{9} - \dfrac{x^2}{16} = 1$

Let $x = t$. Then,

$$\frac{y^2}{9} - \frac{t^2}{16} = 1$$

$$\frac{y^2}{9} = \frac{t^2}{16} + 1$$

$$y^2 = 9\left(\frac{t^2}{16} + 1\right)$$

$$y = \pm 3\sqrt{\frac{t^2}{16} + 1}$$

51. Using the equations derived in example 6 with $a = 12$,
$x = 12(\theta - \sin\theta)$; $y = 12(1 - \cos\theta)$

End of Section 8.3

Chapter Eight: Section 8.4
Solutions to Odd-numbered Exercises

1. Graph $3+5i$: Plot (3, 5) in the complex plane.

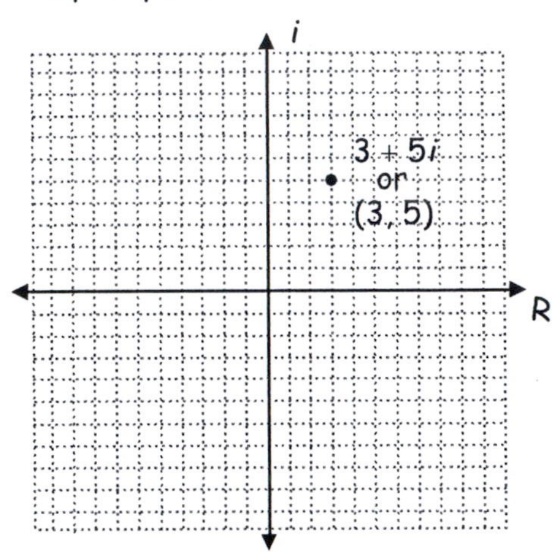

$$|3+5i| = \sqrt{3^2+5^2}$$
$$= \sqrt{34}$$

3. Graph $2-4i$: Plot (2, -4) in the complex plane.

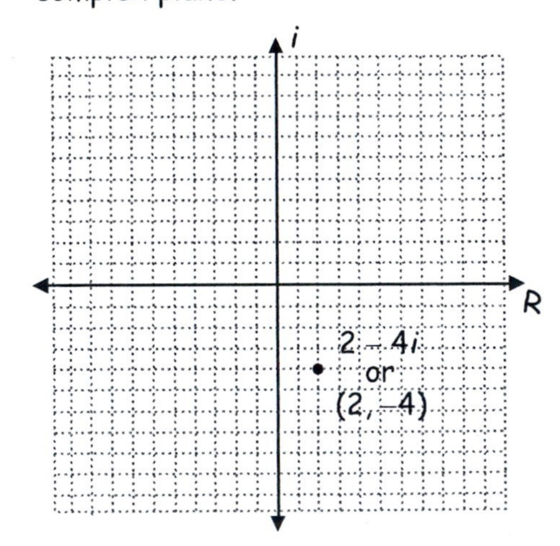

$$|2-4i| = \sqrt{2^2+(-4)^2}$$
$$= 2\sqrt{5}$$

5. Graph $4+4i$: Plot (4, 4) in the complex plane.

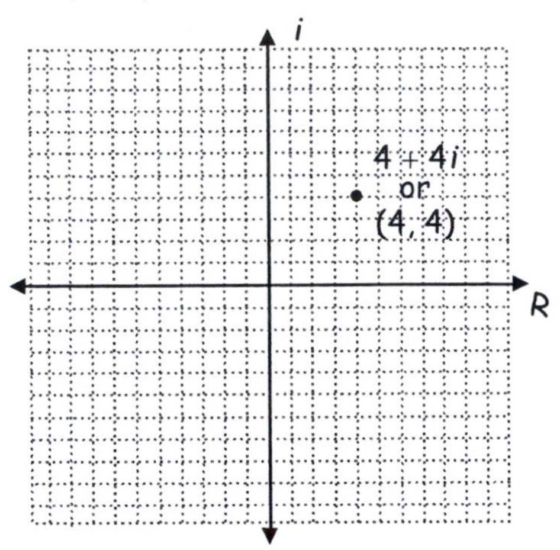

$$|4+4i| = \sqrt{4^2+4^2}$$
$$= 4\sqrt{2}$$

7. Given $z_1 = 7+2i$ and $z_2 = -2+3i$;
Find $z_1 + z_2$ and $z_1 z_2$:
$z_1 + z_2 = 7+2i+(-2+3i) = 5+5i$, which is $(5, 5)$ in the complex plane, and
$z_1 z_2 = (7+2i)(-2+3i) = -20+17i$, which is $(-20, 17)$ in the complex plane

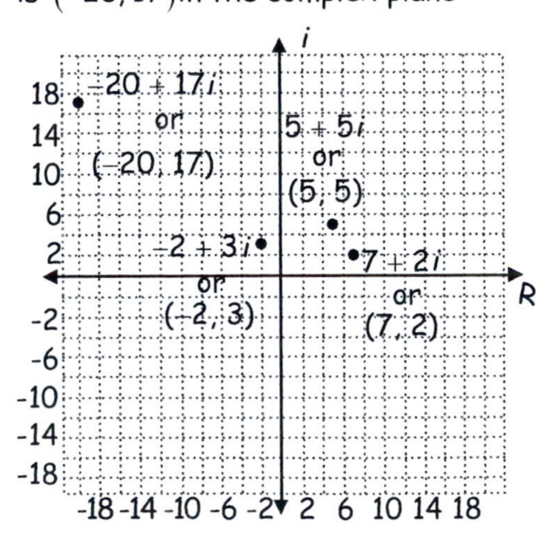

9. Given $z_1 = 3 + i$ and $z_2 = 5 - i$;
Find $z_1 + z_2$ and $z_1 z_2$:
$z_1 + z_2 = 3 + i + (5 - i) = 8 + 0i$, which is $(8,0)$ in the complex plane, and
$z_1 z_2 = (3 + i)(5 - i) = 16 + 2i$, which is $(16, 2)$ in the complex plane

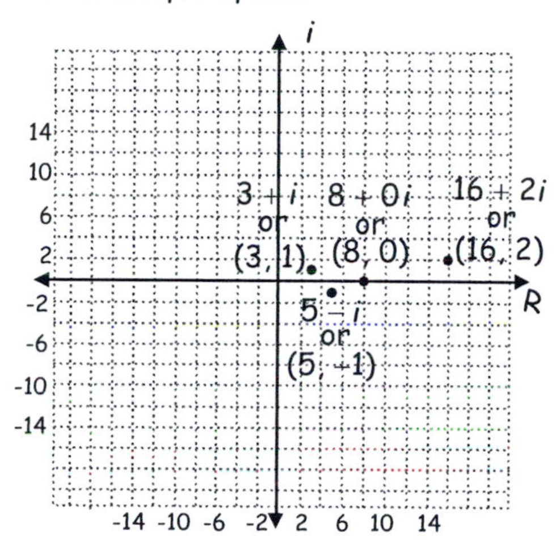

11. Graph $\{z \| z| < 3\}$. This is the set of complex numbers with magnitude less than 3. In the complex plane, this is the interior of a circle of radius 3 centered at $(0, 0)$.

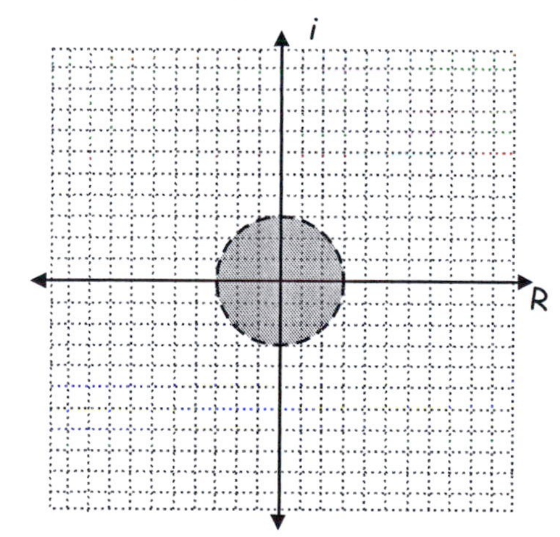

13. Graph $\{z \| z| \geq 1\}$. This is the set of complex numbers with magnitude greater than or equal to 1. In the complex plane this is a circle of radius 1 cenetered at $(0, 0)$ and all the points, exterior to the circle.

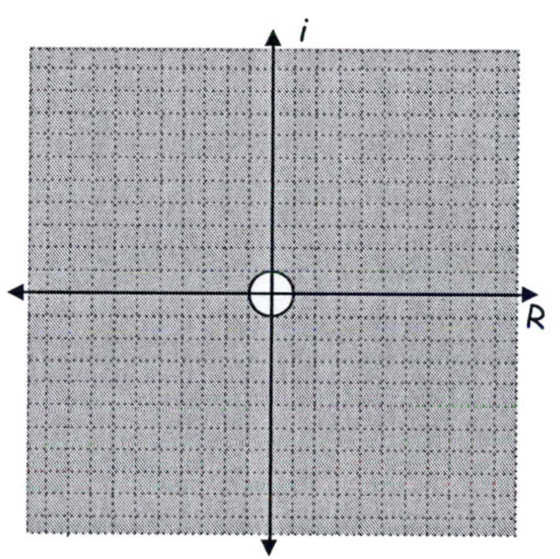

15. Graph $\{z = a + bi | a \geq b\}$.
This set is the same as
$\{z = a + bi | a - b \geq 0\}$.
First, graph the line $a - b = 0$.
Using a test point we see $a \geq b$ is the region to the right of the line.

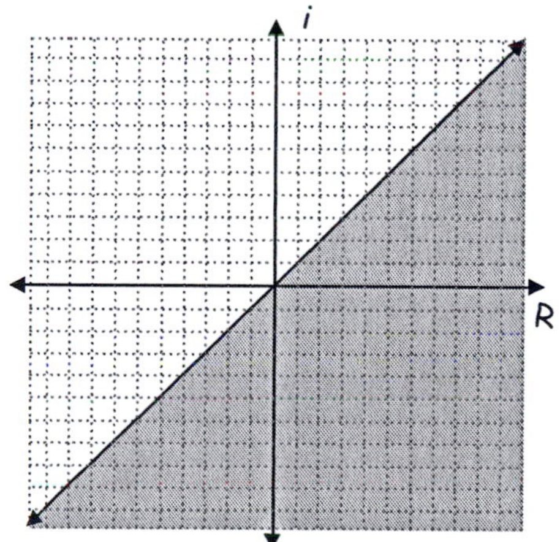

17. Given $z = -3 - i$

In trigonometric form:

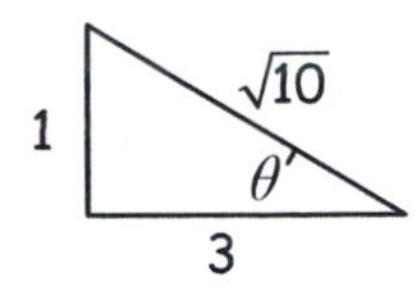

Use $z = |z|(\cos\theta + i\sin\theta)$, where

$\tan\theta = \frac{-1}{-3} = \frac{1}{3}$, and θ lies in quadrant 3.

Reference angle $\theta' \approx 0.3218$.

Then, $\theta \approx \pi + \theta' \approx 3.46$

$z = \sqrt{10}\left(\cos(3.46) + i\sin(3.46)\right)$

19. Given $z = 1 + 2i$

In trigonometric form:

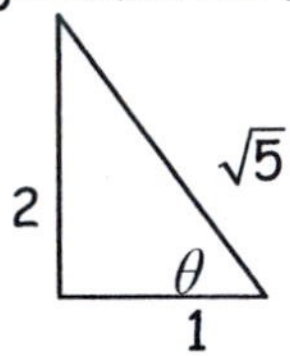

Use $z = |z|(\cos\theta + i\sin\theta)$, where

$\tan\theta = \frac{2}{1} = 2$, and θ lies in quadrant 1.

Then, $\theta = \tan^{-1} 2 \approx 1.11$

$z \approx \sqrt{5}\left(\cos(1.11) + i\sin(1.11)\right)$

21. Given $z = 4 + 2i$

In trigonometric form:

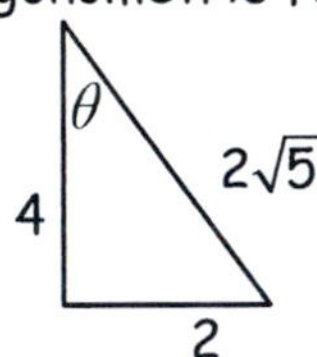

Use $z = |z|(\cos\theta + i\sin\theta)$, where

$\tan\theta = \frac{2}{4} = \frac{1}{2}$, and θ lies in quadrant 1.

Then, $\theta = \tan^{-1}\left(\frac{1}{2}\right) \approx 0.46$

$z \approx 2\sqrt{5}\left(\cos(0.46) + i\sin(0.46)\right)$

23. Given $z = \sqrt{2} - \sqrt{2}i$

In trigonometric form:

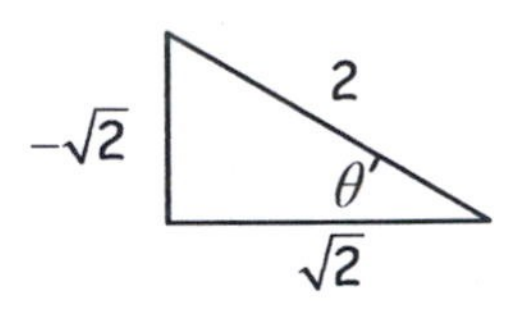

Use $z = |z|(\cos\theta + i\sin\theta)$, where

$\tan\theta = \frac{-\sqrt{2}}{\sqrt{2}} = -1$, and θ lies in quadrant 4.

Reference angle (a special angle)

$\theta' = \frac{\pi}{4}$.

Then, $\theta = -\theta' = -\frac{\pi}{4}$, or $\frac{7\pi}{4}$

$z = 2\left(\cos\left(-\frac{\pi}{4}\right) + i\sin\left(-\frac{\pi}{4}\right)\right)$

25. Given $z = 3 + 4i$

In trigonometric form:

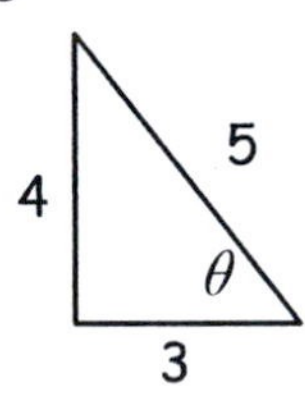

Use $z = |z|(\cos\theta + i\sin\theta)$, where

$\tan\theta = \frac{4}{3}$, and θ lies in quadrant 1.

$\theta = \tan^{-1}\frac{4}{3} \approx 0.9273$

$z \approx 5\left(\cos(0.93) + i\sin(0.93)\right)$

27. Given $z = 4 - 4\sqrt{3}i$

In trigonometric form:

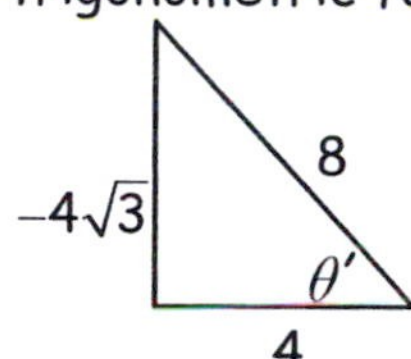

Use $z = |z|(\cos\theta + i\sin\theta)$, where $\tan\theta = \dfrac{-4\sqrt{3}}{4} = -\sqrt{3}$, and θ lies in quadrant 4. Reference angle (a special angle); $\theta' = \dfrac{\pi}{3}$.

Then, $\theta = -\theta' = -\dfrac{\pi}{3}$, or $\dfrac{5\pi}{3}$

$$|z| = \sqrt{4^2 + \left(4\sqrt{3}\right)^2} = \sqrt{64} = 8$$

$$z = 8\left(\cos\left(-\frac{\pi}{3}\right) + i\sin\left(-\frac{\pi}{3}\right)\right)$$

29. Write $z = 3\left(\cos\dfrac{5\pi}{6} + i\sin\dfrac{5\pi}{6}\right)$ in standard form:

$$z = 3\left(-\frac{\sqrt{3}}{2} + \frac{1}{2}i\right)$$

$$z = -\frac{3\sqrt{3}}{2} + \frac{3}{2}i$$

31. Write $z = 2\left(\cos\dfrac{4\pi}{3} + i\sin\dfrac{4\pi}{3}\right)$ in standard form:

$$z = 2\left(-\frac{1}{2} - \frac{\sqrt{3}}{2}i\right)$$

$$z = -i - \sqrt{3}i$$

33. Write $z = 5\left(\cos\dfrac{3\pi}{4} + i\sin\dfrac{3\pi}{4}\right)$ in standard form:

$$z = 5\left(-\frac{1}{\sqrt{2}} + \frac{1}{\sqrt{2}}i\right)$$

$$z = -\frac{5}{\sqrt{2}} + \frac{5}{\sqrt{2}}i, \text{ or } -\frac{5\sqrt{2}}{2} + \frac{5\sqrt{2}}{2}i$$

35. Write $z = \dfrac{3}{2}(\cos 150^\circ + i\sin 150^\circ)$ in standard form:

$$z = \frac{3}{2}\left(-\frac{\sqrt{3}}{2} + \frac{1}{2}i\right)$$

$$z = -\frac{3\sqrt{3}}{4} + \frac{3}{4}i$$

37. Write $z = 5\left[\cos 78^\circ 20' + i\sin 78^\circ 20'\right]$ in standard form: Remember to set your calculator mode to "degrees."

First, convent $78^\circ 20'$ to decimal form.

$78^\circ 20' = 78.\overline{3}$

Then, $\cos 78^\circ 20' \approx 0.2022$ and $\sin 78^\circ 20' \approx 0.9793$

$$z \approx 5(0.2022 + 0.9793i)$$

$$z \approx 1.01 + 4.9i$$

39. Let $z_1 z_2 = \left[4(\cos 60^\circ + i\sin 60^\circ)\right] \cdot \left[4(\cos 330^\circ + i\sin 330^\circ)\right]$.

Use the product formula to find $z_1 z_2$:

$$z_1 z_2 = 4 \cdot 4\left(\cos(60^\circ + 330^\circ) + i\sin(60^\circ + 330^\circ)\right)$$

$$= 16(\cos 390^\circ + i\sin 390^\circ)$$

$$= 16(\cos 30^\circ + i\sin 30^\circ)$$

$$= 16\left(\frac{\sqrt{3}}{2} + \frac{1}{2}i\right)$$

$$= 8\sqrt{3} + 8i$$

41. Let $z_1 z_2 = \left[\sqrt{2}\left(\cos\dfrac{5\pi}{4} + i\sin\dfrac{5\pi}{4}\right)\right] \cdot \left[3\sqrt{3}\left(\cos\dfrac{\pi}{6} + i\sin\dfrac{\pi}{6}\right)\right]$.

Use the product formula to find $z_1 z_2$:

$$z_1 z_2 = \sqrt{2} \cdot 3\sqrt{3}\left[\cos\left(\frac{5\pi}{4} + \frac{\pi}{6}\right) + i\sin\left(\frac{5\pi}{4} + \frac{\pi}{6}\right)\right]$$

$$= 3\sqrt{6}\left[\cos\frac{17\pi}{12} + i\sin\frac{17\pi}{12}\right]$$

$$\approx -1.9 - 7.1i$$

43. Let $z_1 z_2 = (-1+3i)(\sqrt{3}+i)$

Multipying, we have

$$z_1 z_2 = -\sqrt{3} + (3\sqrt{3}-1)i - 3$$
$$= (-\sqrt{3}-3) + (3\sqrt{3}-1)i$$

θ lies in quadrant 2, since $a<0$ and $b>0$.

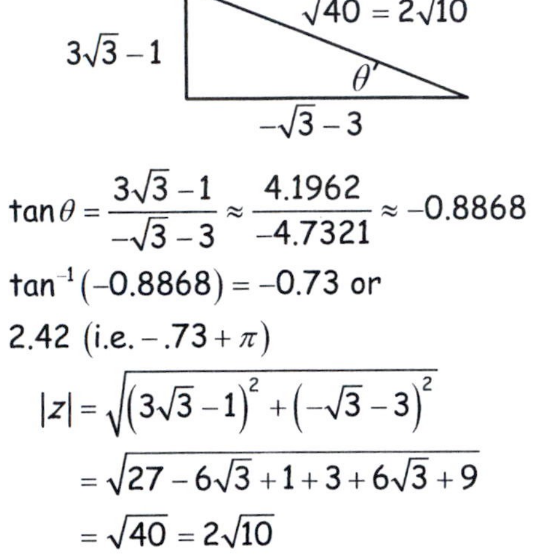

$$\tan\theta = \frac{3\sqrt{3}-1}{-\sqrt{3}-3} \approx \frac{4.1962}{-4.7321} \approx -0.8868$$

$\tan^{-1}(-0.8868) = -0.73$ or

2.42 (i.e. $-.73+\pi$)

$$|z| = \sqrt{(3\sqrt{3}-1)^2 + (-\sqrt{3}-3)^2}$$
$$= \sqrt{27 - 6\sqrt{3} + 1 + 3 + 6\sqrt{3} + 9}$$
$$= \sqrt{40} = 2\sqrt{10}$$

$$z_1 z_2 \approx 2\sqrt{10}(\cos 2.42 + i\sin 2.42)$$

45. Let $\dfrac{z_1}{z_2} = \dfrac{6(\cos 225^\circ + i\sin 225^\circ)}{3(\cos 45^\circ + i\sin 45^\circ)}$

Then, using the quotient formula:

$$\frac{z_1}{z_2} = \frac{6}{3}\left(\cos(225^\circ - 45^\circ) + i\sin(225^\circ - 45^\circ)\right)$$
$$\frac{z_1}{z_2} = 2(\cos 180^\circ + i\sin 180^\circ)$$
$$\frac{z_1}{z_2} = 2(-1+0) = -2$$

47. Let $\dfrac{z_1}{z_2} = \dfrac{10\left(\cos\frac{5\pi}{3} + i\sin\frac{5\pi}{3}\right)}{3\left(\cos\frac{7\pi}{6} + i\sin\frac{7\pi}{6}\right)}$

Then, using the quotient formula:

$$\frac{z_1}{z_2} = \frac{10}{3}\left(\cos\left(\frac{5\pi}{3} - \frac{7\pi}{6}\right) + i\sin\left(\frac{5\pi}{3} - \frac{7\pi}{6}\right)\right)$$
$$\frac{z_1}{z_2} = \frac{10}{3}\left(\cos\frac{\pi}{2} + i\sin\frac{\pi}{2}\right)$$
$$\frac{z_1}{z_2} = \frac{10}{3}(0 + i(1)) = \frac{10}{3}i$$

49. Let $\dfrac{z_1}{z_2} = \dfrac{-i}{1+i}$. Convert to trigonometric form. Then, use the quotient formula:

$$z_1 = (1)\left(\cos\frac{3\pi}{2} + i\sin\frac{3\pi}{2}\right)$$
$$z_2 = \sqrt{2}\left(\cos\frac{\pi}{4} + i\sin\frac{\pi}{4}\right)$$
$$\frac{z_1}{z_2} = \frac{1}{\sqrt{2}}\left(\cos\left(\frac{3\pi}{2} - \frac{\pi}{4}\right) + i\sin\left(\frac{3\pi}{2} - \frac{\pi}{4}\right)\right)$$
$$\frac{z_1}{z_2} = \frac{1}{\sqrt{2}}\left(\cos\frac{5\pi}{4} + i\sin\frac{5\pi}{4}\right)$$
$$\frac{z_1}{z_2} = \frac{1}{\sqrt{2}}\left(-\frac{1}{\sqrt{2}} - \frac{1}{\sqrt{2}}i\right) = -\frac{1}{2} - \frac{1}{2}i$$

51. Let $\dfrac{z_1}{z_2} = \dfrac{2e^{\frac{2\pi}{3}i}}{e^{\frac{\pi}{4}i}}$. Then,

$$\frac{z_1}{z_2} = 2e^{\left(\frac{2\pi}{3} - \frac{\pi}{4}\right)i}$$
$$= 2e^{\frac{5\pi}{12}i}$$
$$= 2\left(\cos\frac{5\pi}{12} + i\sin\frac{5\pi}{12}\right)$$
$$\approx 2(0.2588 + 0.9659i)$$
$$\approx 0.52 + 1.93i$$

53. $z_1 = 4\left(\cos\frac{5\pi}{6} + i\sin\frac{5\pi}{6}\right)$

$\approx (-3.46 + 2i)$ so plot $(-3.46, 2)$ on the complex plane

$z_2 = 2(\cos\pi + i\sin\pi)$

$\approx (-2 + 0i)$ so plot $(-2, 0)$ on the complex plane

$$z_1 z_2 = (4)(2)\left[\cos\left(\frac{5\pi}{6} + \pi\right) + i\sin\left(\frac{5\pi}{6} + \pi\right)\right]$$
$$= 8\left[\cos\frac{11\pi}{6} + i\sin\frac{11\pi}{6}\right]$$
$$= 8\left(\frac{\sqrt{3}}{2}\right) + 8\left(-\frac{1}{2}\right)i$$

$= 4\sqrt{3} - 4i \approx 6.93 - 4i$, so plot $(6.93,\ -4)$ on the complex plane.

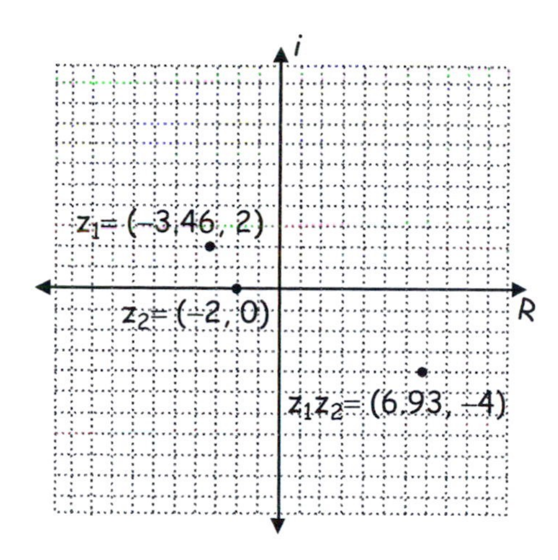

55. $z_1 = 2 - 5i$ and $z_2 = \sqrt{2} + 2i$
so plot $(2, -5)$ and $(1.41, 2)$ on the complex plane
Then, $z_1 z_2 = (2 - 5i)(\sqrt{2} + 2i)$

$$= 2\sqrt{2} + 10 - 5\sqrt{2}i + 4i$$
$$= (2\sqrt{2} + 10) - (5\sqrt{2} - 4)i$$
$$\approx 12.83 - 3.07i$$

so plot $(12.83, -3.07)$ on the complex plane
The result is shown below.

Plot:

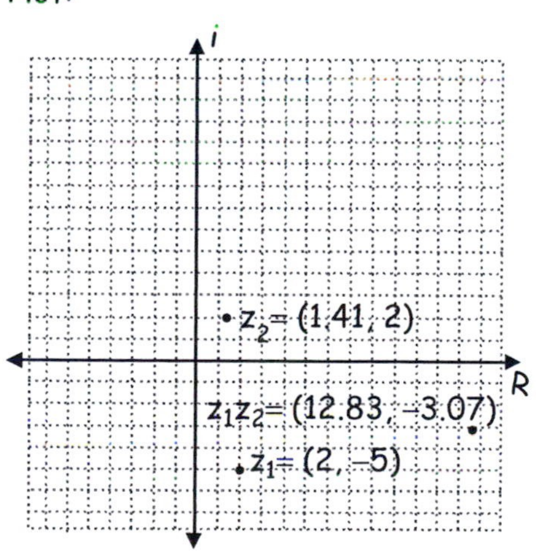

57. $z_1 = 2e^{\frac{\pi}{3}i}$ and $z_2 = 3e^{\frac{5\pi}{4}i}$

$$z_1 = 2\cos\frac{\pi}{3} + 2i\sin\frac{\pi}{3}$$
$$= 1 + \sqrt{3}i$$

So plot $(1, \sqrt{3})$ on the complex plane and

$$z_2 = 3\cos\frac{5\pi}{4} + 3i\sin\frac{5\pi}{4}$$
$$= \frac{-3\sqrt{2}}{2} - \frac{3\sqrt{2}}{2}i$$
$$\approx -2.12 - 2.12i$$

So plot $(-2.12, -2.12)$ on the complex plane

Then, $z_1 z_2 = 2e^{\frac{\pi}{3}i} \cdot 3e^{\frac{5\pi}{4}i}$

$$= 6e^{\left(\frac{\pi}{3} + \frac{5\pi}{4}\right)i}$$
$$= 6e^{\frac{19\pi}{12}i} = 6\cos\frac{19\pi}{12} + 6i\sin\frac{19\pi}{12}$$
$$\approx 1.55 - 5.80i$$

So plot $(1.55, -5.80)$ on the complex plane.

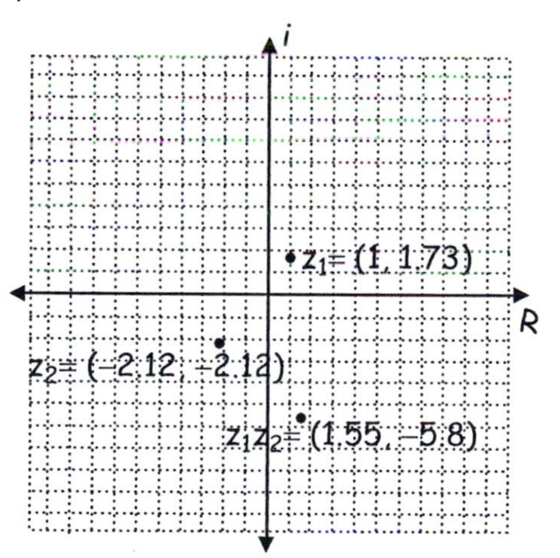

59. Calculate $\left(1-\sqrt{3}i\right)^5$ with DeMoivre's theorem: Let $z=1-\sqrt{3}i$. Then,

$$|z|=\sqrt{1^2+\left(\sqrt{3}\right)^2}=2$$

$$\tan\theta=\frac{-\sqrt{3}}{1}=-\sqrt{3}$$

(Note that θ lies in quadrant 4)

$$\theta=\frac{5\pi}{3}$$

$$z^5=2^5 e^{5\left(\frac{5\pi}{3}\right)i}$$

$$z^5=32e^{\frac{25\pi}{3}i}=32e^{\frac{\pi}{3}i}$$

61. Calculate $(5+3i)^{17}$ with DeMoivre's theorem: Let $z=5+3i$. Then,

$$|z|=\sqrt{5^2+3^2}=\sqrt{34}$$

$$\tan\theta=\frac{3}{5}$$

$$\theta\approx 0.54$$

$$z^{17}\approx\left(\sqrt{34}\right)^{17}e^{17(0.54)i}$$

$$\approx 1.04\times 10^{13}e^{9.19i}$$

$$\approx 1.04\times 10^{13}e^{2.9i}$$

63. Calculate $\left(\cos\frac{\pi}{4}+i\sin\frac{\pi}{4}\right)^8$ with DeMoivre's theorem:

Let $z=\cos\frac{\pi}{4}+i\sin\frac{\pi}{4}$. Then, $|z|=1$ and

$$z^8=\cos\frac{8\pi}{4}+i\sin\frac{8\pi}{4}$$

$$=\cos 2\pi+i\sin 2\pi,\text{ or}$$

$$z^8=e^{i(8)\left(\frac{\pi}{4}\right)}=e^{2\pi i}$$

65. Fourth roots of -1:

Let $z=-1=-1+0i$

Then, $\theta=\pi$ (Since, $\cos\theta=-1$).

$z=\cos\pi+i\sin\pi$, and $|z|=1$:

$$w_0=\left[\cos\left(\frac{\pi+2(0)\pi}{4}\right)+i\sin\left(\frac{\pi+2(0)\pi}{4}\right)\right]$$

$$=\cos\frac{\pi}{4}+i\sin\frac{\pi}{4}$$

$$=e^{\frac{\pi}{4}i},\text{ or }\frac{1}{\sqrt{2}}+\frac{1}{\sqrt{2}}i$$

which is approximately $(0.71, 0.71)$ in the complex plane

$$w_1=\left[\cos\left(\frac{\pi+2(1)\pi}{4}\right)+i\sin\left(\frac{\pi+2(1)\pi}{4}\right)\right]$$

$$=\cos\frac{3\pi}{4}+i\sin\frac{3\pi}{4}$$

$$=e^{\frac{3\pi}{4}i},\text{ or }-\frac{1}{\sqrt{2}}+\frac{1}{\sqrt{2}}i$$

which is approximately $(-0.71, 0.71)$ in the complex plane

$$w_2=\left[\cos\left(\frac{\pi+2(2)\pi}{4}\right)+i\sin\left(\frac{\pi+2(2)\pi}{4}\right)\right]$$

$$=\cos\frac{5\pi}{4}+i\sin\frac{5\pi}{4}$$

$$=e^{\frac{5\pi}{4}i},\text{ or }-\frac{1}{\sqrt{2}}-\frac{1}{\sqrt{2}}i$$

which is approximately $(-0.71, -0.71)$ in the complex plane

$$w_3=\left[\cos\left(\frac{\pi+2(3)\pi}{4}\right)+i\sin\left(\frac{\pi+2(3)\pi}{4}\right)\right]$$

$$=\cos\frac{7\pi}{4}+i\sin\frac{7\pi}{4}$$

$$=e^{\frac{7\pi}{4}i},\text{ or }-\frac{1}{\sqrt{2}}+\frac{1}{\sqrt{2}}i$$

which is approximately $(0.71, -0.71)$ in the complex plane

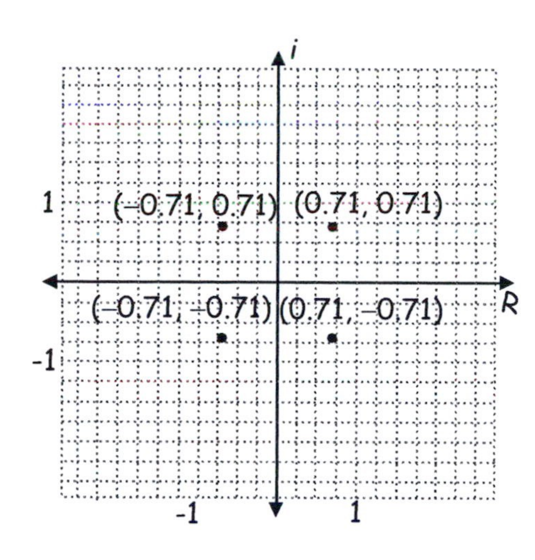

Plot:

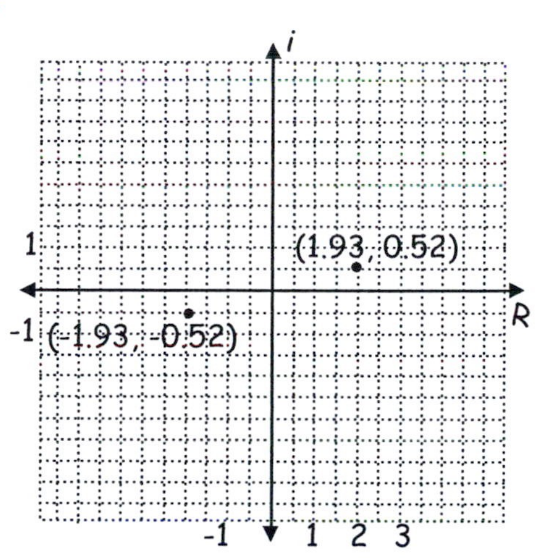

67. Square roots of $z = 2\sqrt{3} + 2i$:

$$|z| = \sqrt{\left(2\sqrt{3}\right)^2 + 2^2} = \sqrt{16} = 4$$

$$\tan\theta = \frac{2}{2\sqrt{3}} = \frac{1}{\sqrt{3}} \Rightarrow \theta = \frac{\pi}{6}$$

$$w_0 = 4^{\frac{1}{2}}\left[\cos\left(\frac{\frac{\pi}{6}}{2}\right) + i\sin\left(\frac{\frac{\pi}{6}}{2}\right)\right]$$

$$= 2\left(\cos\frac{\pi}{12} + i\sin\frac{\pi}{12}\right)$$

$$= 2e^{\frac{\pi}{12}i}$$

which is approximately $(1.93, 0.52)$ in the complex plane

$$w_1 = 4^{\frac{1}{2}}\left[\cos\left(\frac{\frac{\pi}{6} + 2\pi}{2}\right) + i\sin\left(\frac{\frac{\pi}{6} + 2\pi}{2}\right)\right]$$

$$= 2\left(\cos\frac{13\pi}{12} + i\sin\frac{13\pi}{12}\right)$$

$$= 2e^{\frac{13\pi}{12}i}$$

which is approximately $(-1.93, -0.52)$ in the complex plane

69. The fourth roots of $z = 256$:

$$|z| = \sqrt{256^2 + 0^2} = 256$$

$$\tan\theta = 0 \Rightarrow \theta = 0$$

$$w_0 = 256^{\frac{1}{4}}\left[\cos\frac{0}{4} + i\sin\frac{0}{4}\right]$$

$$= 4e^0 = 4$$

which is $(4, 0)$ in the complex plane

$$w_1 = 256^{\frac{1}{4}}e^{i\left(\frac{0+2\pi}{4}\right)} = 4e^{\frac{\pi}{2}i}$$

which is $(0, 4)$ in the complex plane

$$w_2 = 256^{\frac{1}{4}}e^{i\left(\frac{0+4\pi}{4}\right)} = 4e^{\pi i}$$

which is $(-4, 0)$ in the complex plane

$$w_3 = 256^{\frac{1}{4}}e^{i\left(\frac{0+6\pi}{4}\right)} = 4e^{\frac{3\pi}{2}i}$$

which is $(0, -4)$ in the complex plane

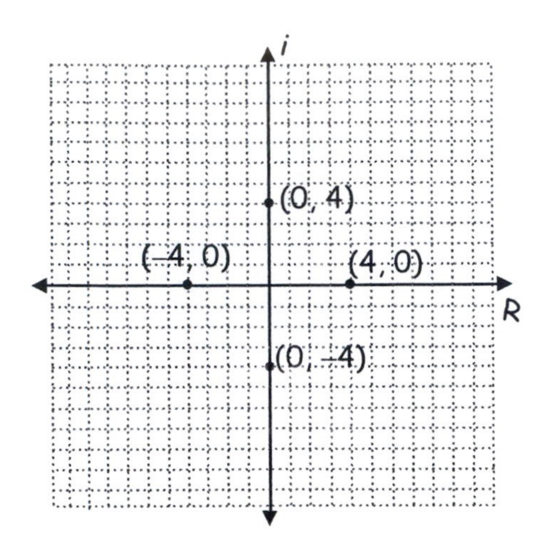

71. The square roots of

$z = 4(\cos 120^\circ + i\sin 120^\circ)$

Write as $z = 4e^{120^\circ i}$:

$w_0 = 4^{\frac{1}{2}}\left(e^{\frac{120^\circ}{2}i}\right) = 2e^{60^\circ i}$

which is approximately $(1, 1.732)$ in the complex plane

$w_1 = 4^{\frac{1}{2}}\left(e^{\frac{120^\circ+360^\circ}{2}i}\right) = 2e^{240^\circ i}$

which is approximately $(-1, -1.732)$ in the complex plane

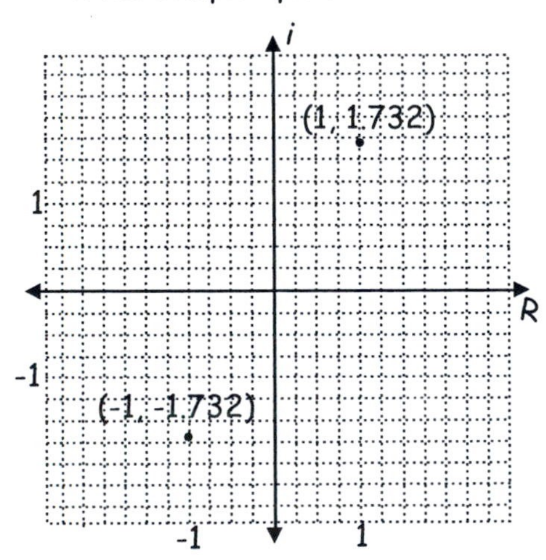

73. Solve: $z^2 - 4\sqrt{3} - 4i = 0$

Re-write as $z^2 = 4\sqrt{3} + 4i$.

Now, find the square roots of $4\sqrt{3} + 4i$.

$\tan\theta = \frac{4}{4\sqrt{3}} = \frac{1}{\sqrt{3}} \Rightarrow \theta = \frac{\pi}{6}$

$|z| = \sqrt{\left(4\sqrt{3}\right)^2 + 4^2} = \sqrt{64} = 8.$

Then, $4\sqrt{3} + 4i = 8e^{\frac{\pi}{6}i}$

The square roots of $4\sqrt{3} + 4i$ are

$$w_0 = 8^{\frac{1}{2}}e^{\left(\frac{\frac{\pi}{6}}{2}\right)i} = 2\sqrt{2}e^{\frac{\pi}{12}i}$$

$$w_1 = 8^{\frac{1}{2}}e^{\left(\frac{\frac{\pi}{6}+2\pi}{2}\right)i} = 2\sqrt{2}e^{\frac{13\pi}{12}i}$$

75. Solve $z^5 + 32 = 0$;

Re-write as $z^5 = -32$.

$|-32| = 32,\ \theta = \pi$

Find the fifth roots of -32:

$$w_0 = 32^{\frac{1}{5}}e^{\frac{\pi}{5}i} = 2e^{\frac{\pi}{5}i}$$

$$w_1 = 32^{\frac{1}{5}}e^{\left(\frac{\pi+2\pi}{5}\right)i} = 2e^{\frac{3\pi}{5}i}$$

$$w_2 = 32^{\frac{1}{5}}e^{\left(\frac{\pi+4\pi}{5}\right)i} = 2e^{\pi i}$$

$$w_3 = 32^{\frac{1}{5}}e^{\left(\frac{\pi+6\pi}{5}\right)i} = 2e^{\frac{7\pi}{5}i}$$

$$w_4 = 32^{\frac{1}{5}}e^{\left(\frac{\pi+8\pi}{5}\right)i} = 2e^{\frac{9\pi}{5}i}$$

77. Solve $z^2 + 25i = 0$;

Re-write as $z^2 = -25i$.

$|-25| = 25.$

$\theta = \frac{3\pi}{2}$

Find the square roots of $-25i$:

$-25i = 25e^{\frac{3\pi}{2}i}$

$$w_0 = 25^{\frac{1}{2}}e^{\left(\frac{\frac{3\pi}{2}}{2}\right)i} = 5e^{\frac{3\pi}{4}i}$$

$$w_1 = 25^{\frac{1}{2}}e^{\left(\frac{\frac{3\pi}{2}+2\pi}{2}\right)i} = 5e^{\frac{7\pi}{4}i},\ \text{or } 5e^{-\frac{\pi}{4}i}$$

End of Section 8.4

Chapter Eight: Section 8.5
Solutions to Odd-numbered Exercises

1. Given: **u** Plot: $-\mathbf{u}$, same magnitude as **u**, in the opposite direction.

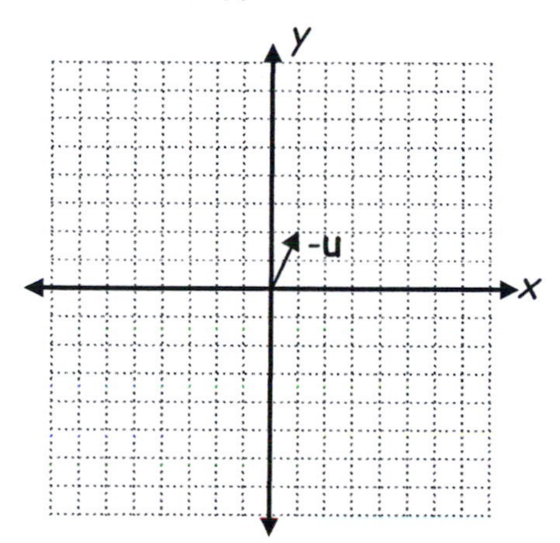

3. Given: **v** Plot: 3**v**, three time the magnitude of **v**, in the same direction.

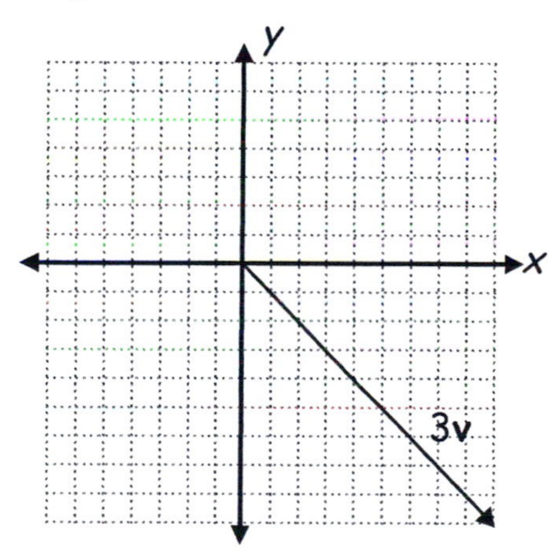

5. Given: **u, v** Plot: **2u - 2v**

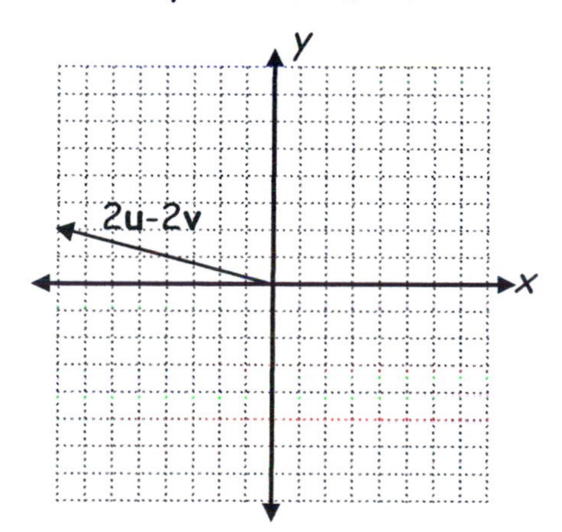

7. Since **v** has a horizontal displacement of 3 units in the positive direction and a vertical displacement of 3 units in the negative direction, $\mathbf{v} = \langle 3, -3 \rangle$

$\|\mathbf{v}\| = \sqrt{3^2 + (-3)^2} = \sqrt{18} = 3\sqrt{2}$

9. Since **v** ha a horizontal displacement of 5 units in the positive direction and a vertical displacement of 3 units in the positive direction, $\mathbf{v} = \langle 5, 3 \rangle$

$\|\mathbf{v}\| = \sqrt{5^2 + 3^2} = \sqrt{34}$

11. $\mathbf{v} = \langle 3 - (-2), 3 - 4 \rangle = \langle 5, -1 \rangle$

$\|\mathbf{v}\| = \sqrt{5^2 + (-1)^2} = \sqrt{26}$

13. $\mathbf{v} = \langle -2 - 5, 5 - (-2) \rangle = \langle -7, 7 \rangle$

$\|\mathbf{v}\| = \sqrt{(-7)^2 + 7^2} = \sqrt{98} = 7\sqrt{2}$

15. $\mathbf{v} = \langle -1 - 3, -2 - 4 \rangle = \langle -4, -6 \rangle$

$\|\mathbf{v}\| = \sqrt{(-4)^2 + (-6)^2} = \sqrt{52} = 2\sqrt{13}$

17. Given: $\mathbf{u} = \langle -2, 4 \rangle$, $\mathbf{v} = \langle 2, 0 \rangle$

a. $2\mathbf{u} + \mathbf{v} = 2\langle -2, 4 \rangle + \langle 2, 0 \rangle$
$= \langle -4, 8 \rangle + \langle 2, 0 \rangle$
$= \langle -2, 8 \rangle$

b. $-\mathbf{u} + 3\mathbf{v} = -\langle -2, 4 \rangle + 3\langle 2, 0 \rangle$
$= \langle 2, -4 \rangle + \langle 6, 0 \rangle$
$= \langle 8, -4 \rangle$

c. $-2\mathbf{v} = -2\langle 2, 0 \rangle$
$= \langle -4, 0 \rangle$

19. Given: $\mathbf{u}=\langle 2,0\rangle$, $\mathbf{v}=\langle -3,4\rangle$

a. $2\mathbf{u}+\mathbf{v}=2\langle 2,0\rangle+\langle -3,4\rangle$
$=\langle 4,0\rangle+\langle -3,4\rangle$
$=\langle 1,4\rangle$

b. $-\mathbf{u}+3\mathbf{v}=-\langle 2,0\rangle+3\langle -3,4\rangle$
$=\langle -2,0\rangle+\langle -9,12\rangle$
$=\langle -11,12\rangle$

c. $-2\mathbf{v}=-2\langle -3,4\rangle$
$=\langle 6,-8\rangle$

21. Given: $\mathbf{u}=\langle -1,-4\rangle$, $\mathbf{v}=\langle -3,-2\rangle$

a. $2\mathbf{u}+\mathbf{v}=2\langle -1,-4\rangle+\langle -3,-2\rangle$
$=\langle -2,-8\rangle+\langle -3,-2\rangle$
$=\langle -5,-10\rangle$

b. $-\mathbf{u}+3\mathbf{v}=-\langle -1,-4\rangle+3\langle -3,-2\rangle$
$=\langle 1,4\rangle+\langle -9,-6\rangle$
$=\langle -8,-2\rangle$

c. $-2\mathbf{v}=-2\langle -3,-2\rangle$
$=\langle 6,4\rangle$

23. Given (from the plot):
$\mathbf{u}=\langle 1,1\rangle$ and $\mathbf{v}=\langle 3,-3\rangle$
$-\mathbf{u}=-\langle 1,1\rangle=\langle -1,-1\rangle$
$2\mathbf{u}-\mathbf{v}=2\langle 1,1\rangle-\langle 3,-3\rangle$
$=\langle 2,2\rangle+\langle -3,3\rangle$
$=\langle -1,5\rangle$
$\mathbf{u}+\mathbf{v}=\langle 1,1\rangle+\langle 3,-3\rangle=\langle 4,-2\rangle$

$\|\mathbf{u}\|=\sqrt{1^2+1^2}=\sqrt{2}$
$\|\mathbf{v}\|=\sqrt{3^2+(-3)^2}=3\sqrt{2}$

25. Given (from the plot):
$\mathbf{u}=\langle -4,4\rangle$ and $\mathbf{v}=\langle 4,-4\rangle$
$-\mathbf{u}=-\langle -4,4\rangle=\langle 4,-4\rangle$
$2\mathbf{u}-\mathbf{v}=2\langle -4,4\rangle-\langle 4,-4\rangle$
$=\langle -8,8\rangle+\langle -4,4\rangle$
$=\langle -12,12\rangle$
$\mathbf{u}+\mathbf{v}=\langle -4,4\rangle+\langle 4,-4\rangle=\langle 0,0\rangle$
$\|\mathbf{u}\|=\sqrt{(-4)^2+4^2}=4\sqrt{2}$
$\|\mathbf{v}\|=\sqrt{4^2+(-4)^2}=4\sqrt{2}$

27. Given: $\mathbf{u}=\langle 6,-3\rangle$

a. $\|\mathbf{u}\|=\sqrt{6^2+(-3)^2}=3\sqrt{5}$
The scalar factor needed is
$\frac{1}{\|\mathbf{u}\|}=\frac{1}{3\sqrt{5}}$
The unit vector is
$\frac{1}{3\sqrt{5}}\langle 6,-3\rangle=\left\langle \frac{2}{\sqrt{5}},-\frac{1}{\sqrt{5}}\right\rangle$

b. $\mathbf{u}=\langle 6,-3\rangle=6\mathbf{i}-3\mathbf{j}$

29. Given: $\mathbf{u}=\langle -5,-1\rangle$

a. $\|\mathbf{u}\|=\sqrt{(-5)^2+(-1)^2}=\sqrt{26}$
The scalar factor needed is
$\frac{1}{\|\mathbf{u}\|}=\frac{1}{\sqrt{26}}$
The unit vector is
$\frac{1}{\sqrt{26}}\langle -5,-1\rangle=\left\langle -\frac{5}{\sqrt{26}},-\frac{1}{\sqrt{26}}\right\rangle$

b. $\mathbf{u}=\langle -5,-1\rangle=-5\mathbf{i}-\mathbf{j}$

31. Given: $\mathbf{u}=\langle 2,3\rangle$

a. $\|\mathbf{u}\|=\sqrt{2^2+3^2}=\sqrt{13}$
The scalar factor needed is
$\frac{1}{\|\mathbf{u}\|}=\frac{1}{\sqrt{13}}$
The unit vector is
$\frac{1}{\sqrt{13}}\langle 2,3\rangle=\left\langle \frac{2}{\sqrt{13}},\frac{3}{\sqrt{13}}\right\rangle$

b. $\mathbf{u}=\langle 2,3\rangle=2\mathbf{i}+3\mathbf{j}$

33. Given: $\mathbf{v} = 5(\cos 30°\mathbf{i} + \sin 30°\mathbf{j})$

Then, $\mathbf{v} = 5\left\langle \frac{\sqrt{3}}{2}, \frac{1}{2} \right\rangle$

$= \left\langle \frac{5\sqrt{3}}{2}, \frac{5}{2} \right\rangle$

$\|\mathbf{v}\| = \sqrt{\left(\frac{5\sqrt{3}}{2}\right)^2 + \left(\frac{5}{2}\right)^2}$

$= \sqrt{\frac{75}{4} + \frac{25}{4}}$

$= \frac{10}{2}$

$= 5$

The direction angle $\theta = 30°$.

35. Given: $\mathbf{v} = 4\mathbf{i} + 3\mathbf{j}$

Then, $\mathbf{v} = \langle 4, 3 \rangle$.

Magnitude: $\|\mathbf{v}\| = \sqrt{4^2 + 3^2} = 5$

Direction angle θ: $\tan\theta = \frac{3}{4} \Rightarrow$

$\theta = \tan^{-1}\frac{3}{4} \approx 0.644 \approx 36.9°$

37. Given: $\|\mathbf{v}\| = 6$ and $\theta = 30°$

The component form:

$\mathbf{v} = 6\langle \cos 30°, \sin 30° \rangle$

$= 6\left\langle \frac{\sqrt{3}}{2}, \frac{1}{2} \right\rangle$

$= \langle 3\sqrt{3}, 3 \rangle$

39. Given: $\|\mathbf{v}\| = 18$ and $\theta = 135°$

The component form:

$\mathbf{v} = 18\langle \cos 135°, \sin 135° \rangle$

$= 18\left\langle -\frac{\sqrt{2}}{2}, \frac{\sqrt{2}}{2} \right\rangle$

$= \langle -9\sqrt{2}, 9\sqrt{2} \rangle$

41. Given: $\|\mathbf{v}\| = 1$ and $\theta = 120°$

The component form:

$\mathbf{v} = (1)\langle \cos 120°, \sin 120° \rangle$

$= \left\langle -\frac{1}{2}, \frac{\sqrt{3}}{2} \right\rangle$

43. Given: $\|\mathbf{v}\| = 4$ and $\mathbf{v}$ in the direction of $2\mathbf{i} + 3\mathbf{j}$. Find the unit vector $\mathbf{u}$ in the direction of $2\mathbf{i} + 3\mathbf{j}$:

$\|2\mathbf{i} + 3\mathbf{j}\| = \sqrt{2^2 + 3^2} = \sqrt{13}$

Then, $\mathbf{v} = \frac{4}{\sqrt{13}}\langle 2, 3 \rangle = \left\langle \frac{8}{\sqrt{13}}, \frac{12}{\sqrt{13}} \right\rangle$

45. $\|\mathbf{v}\| = 4$; $\theta = 30°$

$\mathbf{v} = 4\langle \cos 30°, \sin 30° \rangle$

$= 4\left\langle \frac{\sqrt{3}}{2}, \frac{1}{2} \right\rangle$

$= \langle 2\sqrt{3}, 2 \rangle$

47. Let $\mathbf{u}$ = the sailboat's velocity

Let $\mathbf{v}$ = wind velocity

Then,

$\mathbf{u} = 45\langle \cos 149°, \sin 149° \rangle$

$\mathbf{v} = 15\langle \cos(-87°), \sin(-87°) \rangle$

Then,

$\mathbf{u} + \mathbf{v} \approx 45\langle -0.857, 0.515 \rangle +$

$15\langle 0.052, -0.999 \rangle$

$\approx \langle -38.573, 23.177 \rangle + \langle 0.785, -14.979 \rangle$

$\approx \langle -37.787, 8.197 \rangle$

$\|\mathbf{u} + \mathbf{v}\| \approx \sqrt{(-37.787)^2 + (8.197)^2}$

$\approx \sqrt{1427.86 + 67.19}$

≈ 38.67 mph

$\tan\theta \approx \frac{8.197}{-37.787} \approx -0.2169$

$\tan^{-1}(-0.2169) \approx -12.24$, or

$90° - 12.24° \approx 77.76°$

Or, equivalently, N 77.76° W

49. Let **v** = the force vector:

Then,

$$\|\mathbf{v}\| = \sqrt{150^2 + (-1235)^2}$$
$$= 1244.08 \text{ lbs.}$$
$$\left(@ \quad \theta = \tan^{-1}\left(\frac{-1235}{150}\right) \approx -83.1^\circ\right)$$

End of Section 8.5

Chapter Eight: Section 8.6
Solutions to Odd-numbered Exercises

1. $\langle 4, 3\rangle \cdot \langle 5, -1\rangle = 20 - 3 = 17$

3. $\langle 3, 5\rangle \cdot \langle 2, 0\rangle = 6 + 0 = 6$

5. $\langle 2, 2\rangle \cdot \langle 2, 2\rangle = 4 + 4 = 8$

7. $\langle -4, 3\rangle \cdot \langle 2, 3\rangle = -8 + 9 = 1$

9. $\langle 5, 1\rangle \cdot \langle -2, 3\rangle = -10 + 3 = -7$

11. $\langle 4, 4\rangle \cdot \langle 4, 4\rangle = 16 + 16 = 32$

13. $\|\langle -2, 3\rangle\| + 2 = \sqrt{(-2)^2 + 3^2} + 2$

$= \sqrt{13} + 2$

15. $\|\langle 6, 1\rangle\| = \sqrt{\mathbf{u} \cdot \mathbf{u}}$ for $\mathbf{u} = \langle 6, 1\rangle$

$= \sqrt{\langle 6, 1\rangle \cdot \langle 6, 1\rangle}$

$\sqrt{6^2 + 1^2}$

$= \sqrt{37}$

17. Given: $\mathbf{u} = 2\mathbf{i} + 7\mathbf{j} = \langle 2, 7\rangle$

$\|\mathbf{u}\| = \sqrt{\mathbf{u} \cdot \mathbf{u}} = \sqrt{2 \cdot 2 + 7 \cdot 7}$

$= \sqrt{53}$

19. Given: $\mathbf{u} = \langle -2, 3\rangle$, $\mathbf{v} = \langle 1, 0\rangle$

Find the angle between $\mathbf{u}$ and $\mathbf{v}$:

From $\mathbf{u} \cdot \mathbf{v} = \|\mathbf{u}\|\|\mathbf{v}\|\cos\theta$,

$$\cos\theta = \frac{\mathbf{u} \cdot \mathbf{v}}{\|\mathbf{u}\|\|\mathbf{v}\|}$$

$$\cos\theta = \frac{(-2)(1) + (3)(0)}{\sqrt{(-2)^2 + 3^2} \cdot \sqrt{1^2 + 0^2}} = \frac{-2}{\sqrt{13}}$$

$$\theta \approx \cos^{-1}\left(\frac{-2}{\sqrt{13}}\right) \approx 123.7°$$

21. Given: $\mathbf{u} = \langle 3,5\rangle, \mathbf{v} = \langle 4,4\rangle$

Find the angle between $\mathbf{u}$ and $\mathbf{v}$:

From $\mathbf{u} \cdot \mathbf{v} = \|\mathbf{u}\|\|\mathbf{v}\|\cos\theta$,

$$\cos\theta = \frac{\mathbf{u} \cdot \mathbf{v}}{\|\mathbf{u}\|\|\mathbf{v}\|}$$

$$\cos\theta = \frac{(3)(4) + (5)(4)}{\sqrt{3^2 + 5^2} \cdot \sqrt{4^2 + 4^2}} = \frac{4}{\sqrt{17}}$$

$$\theta \approx \cos^{-1}\left(\frac{4}{\sqrt{17}}\right) \approx 14.0°$$

23. Given: $\mathbf{u} = -\mathbf{i} + 2\mathbf{j}$, $\mathbf{v} = 3\mathbf{i} - 3\mathbf{j}$

Re-written: $\mathbf{u} = \langle -1, 2\rangle$, $\mathbf{v} = \langle 3, -3\rangle$

Find the angle between $\mathbf{u}$ and $\mathbf{v}$:

From $\mathbf{u} \cdot \mathbf{v} = \|\mathbf{u}\|\|\mathbf{v}\|\cos\theta$,

$$\cos\theta = \frac{\mathbf{u} \cdot \mathbf{v}}{\|\mathbf{u}\|\|\mathbf{v}\|}$$

$$\cos\theta = \frac{(-1)(3) + (2)(-3)}{\sqrt{(-1)^2 + 2^2} \cdot \sqrt{3^2 + (-3)^2}}$$

$$= \frac{-9}{\sqrt{5}\sqrt{18}} = \frac{-3}{\sqrt{10}}$$

$$\theta \approx \cos^{-1}\left(\frac{-3}{\sqrt{10}}\right) \approx 161.6°$$

25. Given: $\mathbf{u} = \cos\left(\frac{3\pi}{4}\right)\mathbf{i} + \sin\left(\frac{3\pi}{4}\right)\mathbf{j}$,

$\mathbf{v} = \cos\left(\frac{\pi}{2}\right)\mathbf{i} + \sin\left(\frac{\pi}{2}\right)\mathbf{j}$

Re-written: $\mathbf{u} = \left\langle -\frac{\sqrt{2}}{2}, \frac{\sqrt{2}}{2}\right\rangle$, $\mathbf{v} = \langle 0, 1\rangle$

Find the angle between $\mathbf{u}$ and $\mathbf{v}$:

From $\mathbf{u} \cdot \mathbf{v} = \|\mathbf{u}\|\|\mathbf{v}\|\cos\theta$,

$$\cos\theta = \frac{\mathbf{u} \cdot \mathbf{v}}{\|\mathbf{u}\|\|\mathbf{v}\|}$$

$$\cos\theta = \frac{\left(-\frac{\sqrt{2}}{2}\right)(0) + \left(\frac{\sqrt{2}}{2}\right)(1)}{\sqrt{\left(-\frac{\sqrt{2}}{2}\right)^2 + \left(\frac{\sqrt{2}}{2}\right)^2} \cdot \sqrt{1^2}}$$

$$= \frac{\frac{\sqrt{2}}{2}}{1} = \frac{\sqrt{2}}{2}$$

$$\theta = \cos^{-1}\left(\frac{\sqrt{2}}{2}\right) = \frac{\pi}{4}$$

27. Given the vertices of a triangle: $P(3, 3), Q(4, 2), R(-1, -6)$

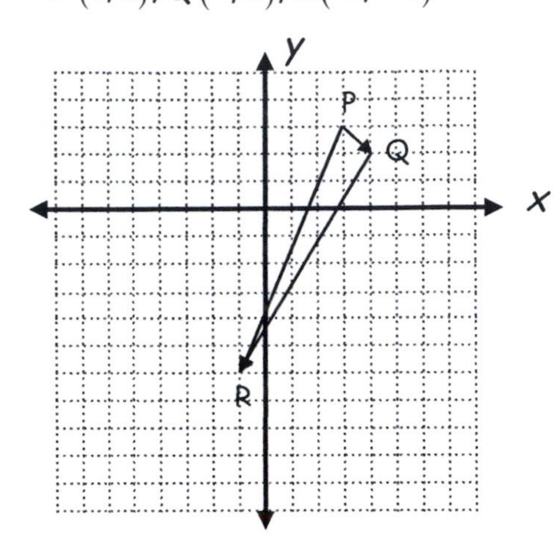

Let $\mathbf{u} = \overrightarrow{PQ} = \langle 4-3, 2-3 \rangle = \langle 1, -1 \rangle$
Let $\mathbf{v} = \overrightarrow{QR} = \langle -1-4, -6-2 \rangle = \langle -5, -8 \rangle$
Let $\mathbf{w} = \overrightarrow{PR} = \langle -1-3, -6-3 \rangle = \langle -4, -9 \rangle$
Note the directions of the vectors to ensure that the respective pairs of vectors have the same initial point for each angle.

$$\cos Q = \frac{-\mathbf{u} \cdot \mathbf{v}}{\|\mathbf{u}\| \|\mathbf{v}\|} = \frac{\langle -1, 1 \rangle \cdot \langle -5, -8 \rangle}{\sqrt{2}\sqrt{89}} = \frac{5-8}{\sqrt{178}} \approx -0.225$$

$$Q \approx \cos^{-1}(-0.225) \approx 103°$$

$$\cos P = \frac{\mathbf{u} \cdot \mathbf{w}}{\|\mathbf{u}\| \|\mathbf{w}\|} = \frac{\langle 1, -1 \rangle \cdot \langle -4, -9 \rangle}{\sqrt{2}\sqrt{97}} = \frac{-4+9}{\sqrt{194}} \approx 0.359$$

$$P \approx \cos^{-1}(0.359) \approx 69°$$

$$\cos R = \frac{(-\mathbf{v}) \cdot (-\mathbf{w})}{\|\mathbf{v}\| \|\mathbf{w}\|} = \frac{\langle 5, 8 \rangle \cdot \langle 4, 9 \rangle}{\sqrt{89}\sqrt{97}} = \frac{20+72}{\sqrt{8633}} \approx 0.990$$

$$R \approx \cos^{-1}(0.990) \approx 8°$$

29. Given the vertices of a triangle: $P(2, 4), Q(-2, -1), R(-4, 5)$

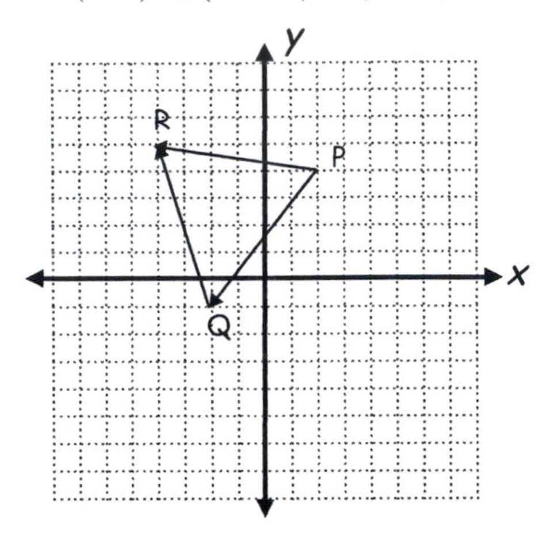

Let $\mathbf{u} = \overrightarrow{PQ} = \langle -2-2, -1-4 \rangle = \langle -4, -5 \rangle$
Let $\mathbf{v} = \overrightarrow{QR} = \langle -4+2, 5+1 \rangle = \langle -2, 6 \rangle$
Let $\mathbf{w} = \overrightarrow{PR} = \langle -4-2, 5-4 \rangle = \langle -6, 1 \rangle$
Note the directions of the vectors to ensure that the respective pairs of vectors have the same initial point for each angle.

$$\cos Q = \frac{-\mathbf{u} \cdot \mathbf{v}}{\|\mathbf{u}\| \|\mathbf{v}\|} = \frac{\langle 4, 5 \rangle \cdot \langle -2, 6 \rangle}{\sqrt{41}\sqrt{40}} = \frac{-8+30}{\sqrt{1640}} \approx 0.543$$

$$Q \approx \cos^{-1}(0.543) \approx 57.1°$$

$$\cos P = \frac{\mathbf{u} \cdot \mathbf{w}}{\|\mathbf{u}\| \|\mathbf{w}\|} = \frac{\langle -4, -5 \rangle \cdot \langle -6, 1 \rangle}{\sqrt{41}\sqrt{37}} = \frac{24-5}{\sqrt{1517}} \approx 0.488$$

$$P \approx \cos^{-1}(0.488) \approx 60.8°$$

$$\cos R = \frac{(-\mathbf{v}) \cdot (-\mathbf{w})}{\|\mathbf{v}\| \|\mathbf{w}\|} = \frac{\langle 2, -6 \rangle \cdot \langle 6, -1 \rangle}{\sqrt{40}\sqrt{37}} = \frac{12+6}{\sqrt{1480}} \approx 0.468$$

$$R \approx \cos^{-1}(0.468) \approx 62.1°$$

31. Given: $\|\mathbf{u}\| = 25$, $\|\mathbf{v}\| = 5$, $\theta = 120°$

Use $\cos\theta = \dfrac{\mathbf{u} \cdot \mathbf{v}}{\|\mathbf{u}\|\|\mathbf{v}\|}$:

$$\mathbf{u} \cdot \mathbf{v} = \|\mathbf{u}\|\|\mathbf{v}\|\cos\theta$$

$$\mathbf{u} \cdot \mathbf{v} = 25 \cdot 5 \cdot \cos 120° = 125(-0.5)$$

$$\approx -62.5$$

33. Given: $\|\mathbf{u}\| = 16$, $\|\mathbf{v}\| = 4$, $\theta = \dfrac{3\pi}{4}$

Use $\cos\theta = \dfrac{\mathbf{u} \cdot \mathbf{v}}{\|\mathbf{u}\|\|\mathbf{v}\|}$:

$$\mathbf{u} \cdot \mathbf{v} = \|\mathbf{u}\|\|\mathbf{v}\|\cos\theta$$

$$\mathbf{u} \cdot \mathbf{v} = 16 \cdot 4 \cdot \cos\frac{3\pi}{4} = -32\sqrt{2}$$

$$\approx 32(1.414)$$

$$\approx -45.25$$

35. There are an infinite number of vectors that are orthogonal to $\mathbf{u} = \langle 3, -3 \rangle$. If $\mathbf{v} = \langle a, b \rangle$ is orthogonal to $\mathbf{u}$, then $\mathbf{u} \cdot \mathbf{v} = 0$. This means that $\langle 3, -3 \rangle \cdot \langle a, b \rangle = 3a - 3b = 0$, which, in turn, means that $3(a - b) = 0 \Rightarrow a - b = 0 \Rightarrow a = b$. Therefore, any vector of the form $\langle a, a \rangle$ is orthogonal to $\mathbf{u}$.
Two examples are $\langle 1, 1 \rangle$ and $\langle 5, 5 \rangle$.

37. There are an infinite number of vectors that are orthogonal to $\mathbf{u} = \langle 2, -6 \rangle$. If $\mathbf{v} = \langle a, b \rangle$ is orthogonal to $\mathbf{u}$, then $\mathbf{u} \cdot \mathbf{v} = 0$. This means that $\langle 2, -6 \rangle \cdot \langle a, b \rangle = 2a - 6b = 0$, which, in turn, means that $2(a - 3b) = 0 \Rightarrow a - 3b = 0 \Rightarrow a = 3b$. Therefore, any vector of the form $\langle 3b, b \rangle$ is orthogonal to $\mathbf{u}$.
Two examples are $\langle 3, 1 \rangle$ and $\langle -6, -2 \rangle$.

39. Given: $\mathbf{u} = \langle 2, -3 \rangle$ and $\mathbf{v} = \langle 1, 6 \rangle$.
Test for orthogonality (i.e., $\mathbf{u} \cdot \mathbf{v} = 0$), and/or for whether the vectors are parallel (i.e., the directional angle of $\mathbf{v}$, say $\theta_{\mathbf{v}}$, is equal to the directional angle, say $\theta_{\mathbf{u}}$, of $\mathbf{u}$). The latter can be tested by determining whether $\tan\theta_{\mathbf{v}} = \tan\theta_{\mathbf{u}}$.

$$\mathbf{u} \cdot \mathbf{v} = \langle 2, -3 \rangle \cdot \langle 1, 6 \rangle = 2(1) + (-3)(6)$$

$$= 2 - 18 = -16 \neq 0$$

Then, $\mathbf{u}$ and $\mathbf{v}$ are not orthogonal.

Test whether the vectors are parallel:

$\tan\theta_{\mathbf{u}} = \dfrac{-3}{2}$ and $\tan\theta_{\mathbf{v}} = \dfrac{6}{1}$

Therefore $\theta_{\mathbf{u}} \neq \theta_{\mathbf{v}}$.
The vectors are not parallel.

41. Given: $\mathbf{u} = 2\mathbf{i} - 2\mathbf{j} = \langle 2, -2 \rangle$ and
$\mathbf{v} = -\mathbf{i} - \mathbf{j} = \langle -1, -1 \rangle$

Test $\mathbf{u} \cdot \mathbf{v} = 0$:

$$\langle 2, -2 \rangle \cdot \langle -1, -1 \rangle = 2(-1) + (-2)(-1)$$

$$= -2 + 2$$

$$= 0$$

Therefore $\mathbf{u}$ and $\mathbf{v}$ are orthogonal.

43. Given: $\mathbf{u} = \langle 1, 3 \rangle$, $\mathbf{v} = \langle 4, 2 \rangle$
Find $\text{proj}_{\mathbf{v}}\mathbf{u}$:

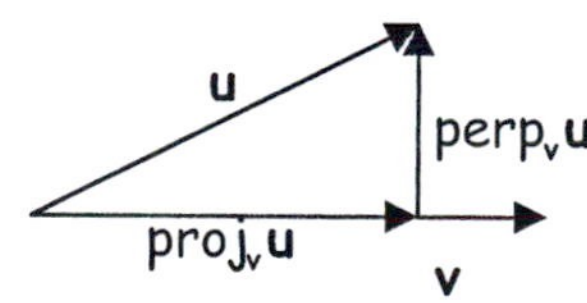

$$\text{proj}_{\mathbf{v}}\mathbf{u} = \left(\frac{\mathbf{u} \cdot \mathbf{v}}{\|\mathbf{v}\|^2}\right)\mathbf{v}$$

$$= \left(\frac{\langle 1, 3 \rangle \cdot \langle 4, 2 \rangle}{4^2 + 2^2}\right)\langle 4, 2 \rangle$$

$$= \left(\frac{4 + 6}{20}\right)\langle 4, 2 \rangle$$

$$= \langle 2, 1 \rangle$$

$$\text{perp}_{\mathbf{v}}\mathbf{u} = \langle 1, 3 \rangle - \langle 2, 1 \rangle$$

$$= \langle -1, 2 \rangle$$

$$(\mathbf{u} = \text{proj}_{\mathbf{v}}\mathbf{u} + \text{perp}_{\mathbf{v}}\mathbf{u})$$

45. Given: $\mathbf{u}=\langle 3,-5\rangle$, $\mathbf{v}=\langle 6,2\rangle$

Find $\text{proj}_{\mathbf{v}}\mathbf{u}$:

u
perp$_{\mathbf{v}}$u
proj$_{\mathbf{v}}$u
v

$$\text{proj}_{\mathbf{v}}\mathbf{u}=\left(\frac{\mathbf{u}\cdot\mathbf{v}}{\|\mathbf{v}\|^2}\right)\mathbf{v}$$
$$=\left(\frac{\langle 3,-5\rangle\cdot\langle 6,2\rangle}{6^2+2^2}\right)\langle 6,2\rangle$$
$$=\left(\frac{18-10}{40}\right)\langle 6,2\rangle$$
$$=\frac{1}{5}\langle 6,2\rangle$$
$$=\left\langle\frac{6}{5},\frac{2}{5}\right\rangle$$
$$\text{perp}_{\mathbf{v}}\mathbf{u}=\langle 3,-5\rangle-\left\langle\frac{6}{5},\frac{2}{5}\right\rangle$$
$$=\left\langle\frac{9}{5},-\frac{27}{5}\right\rangle$$

$(\mathbf{u}=\text{proj}_{\mathbf{v}}\mathbf{u}+\text{perp}_{\mathbf{v}}\mathbf{u})$

47. Given: $\mathbf{u}=\langle -3,-3\rangle$, $\mathbf{v}=\langle -4,-1\rangle$

Find $\text{proj}_{\mathbf{v}}\mathbf{u}$:

u
perp$_{\mathbf{v}}$u
proj$_{\mathbf{v}}$u
v

$$\text{proj}_{\mathbf{v}}\mathbf{u}=\left(\frac{\mathbf{u}\cdot\mathbf{v}}{\|\mathbf{v}\|^2}\right)\mathbf{v}$$
$$=\left(\frac{\langle -3,-3\rangle\cdot\langle -4,-1\rangle}{(-4)^2+(-1)^2}\right)\langle -4,-1\rangle$$
$$=\left(\frac{12+3}{17}\right)\langle -4,-1\rangle$$
$$=\left\langle-\frac{60}{17},-\frac{15}{17}\right\rangle$$
$$\text{perp}_{\mathbf{v}}\mathbf{u}=\langle -3,-3\rangle-\left\langle-\frac{60}{17},-\frac{15}{17}\right\rangle$$
$$=\left\langle\frac{9}{17},-\frac{36}{17}\right\rangle$$

$(\mathbf{u}=\text{proj}_{\mathbf{v}}\mathbf{u}+\text{perp}_{\mathbf{v}}\mathbf{u})$

49. Use $\mathbf{W}=\mathbf{F}\cdot\mathbf{D}$. Given: $J=(1,4)$, $K=(5,6)$,

$$\mathbf{v}=\mathbf{F}=\langle 2,3\rangle$$
$$\mathbf{D}=\langle 5-1,6-4\rangle=\langle 4,2\rangle$$
$$\mathbf{W}=\langle 2,3\rangle\cdot\langle 4,2\rangle=8+6=14$$

51. Use $\mathbf{W}=\mathbf{F}\cdot\mathbf{D}$. Given: $J=(3,0)$,

$$K=(-4,-2),\ \mathbf{v}=-\mathbf{i}+2\mathbf{j}$$
$$\mathbf{D}=\langle -4-3,-2-0\rangle=\langle -7,-2\rangle$$
$$\mathbf{F}=\mathbf{v}=\langle -1,2\rangle$$
$$\mathbf{W}=\langle -1,2\rangle\cdot\langle -7,-2\rangle=7-4=3$$

53. Let $\mathbf{u}=\langle 0,-25000\rangle$.

Use $\tan 8^\circ\approx 0.1405$ to write the vector $\mathbf{v}$: $\mathbf{v}=\langle -1,-0.1405\rangle$

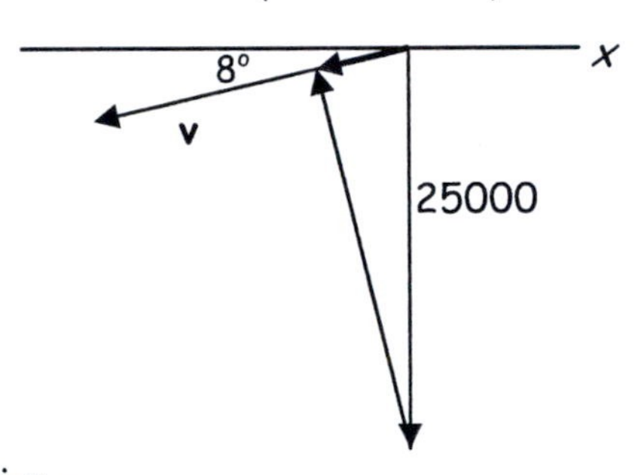

$$\text{proj}_{\mathbf{v}}\mathbf{u}=$$
$$\left(\frac{\langle 0,-25000\rangle\cdot\langle -1,-0.1405\rangle}{(-1)^2+(-0.1405)^2}\right)\langle -1,-0.1405\rangle$$
$$\text{proj}_{\mathbf{v}}\mathbf{u}\approx\left(\frac{-3513.5}{1+0.0197}\right)\langle -1,-0.1405\rangle$$
$$\approx 3445.5\langle -1,-0.1405\rangle$$
$$\approx\langle -3445.5,-484.2\rangle$$
$$\|\text{proj}_{\mathbf{v}}\mathbf{u}\|\approx\sqrt{(-3445.5)^2+(-483.97)^2}$$
$$\approx 3479.3\ \text{lb.}$$

55. Let $\mathbf{u} = \langle 0, -155 \rangle$. Use $\tan 45° = 1$

Then, $\mathbf{v} = \langle -1, -1 \rangle$

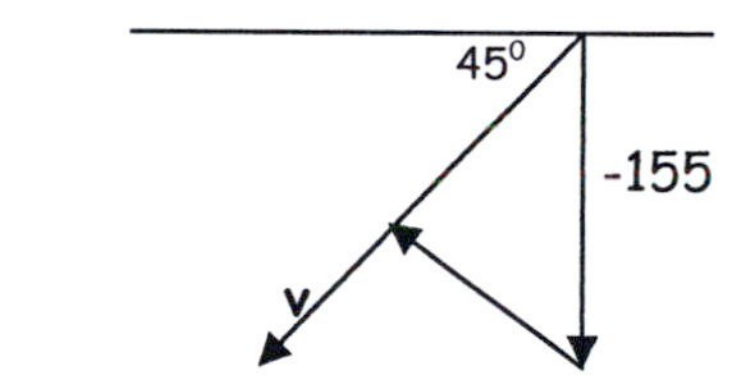

$$\text{proj}_{\mathbf{v}}\mathbf{u} = \left(\frac{\langle 0, -155 \rangle \cdot \langle -1, -1 \rangle}{(-1)^2 + (-1)^2} \right) \langle -1, -1 \rangle$$

$$= \left(\frac{155}{2} \right) \langle -1, -1 \rangle$$

$$= \langle -77.5, -77.5 \rangle$$

$$\|\text{proj}_{\mathbf{v}}\mathbf{u}\| \approx \sqrt{(-77.5)^2 + (-77.5)^2}$$

$$\approx 109.6 \text{ lbs.}$$

57. Let $\mathbf{F} = 3000 \langle \cos 15°, \sin 15° \rangle$, and

$\mathbf{D} = \langle 200, 0 \rangle$

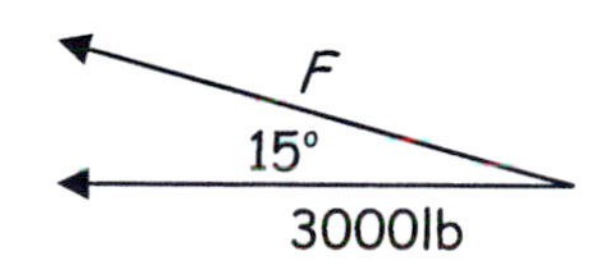

$$\mathbf{W} = \mathbf{F} \cdot \mathbf{D}$$

$$= 3000 \langle \cos 15°, \sin 15° \rangle \cdot \langle 200, 0 \rangle$$

$$\approx 3000 \langle 0.966, 0.259 \rangle \cdot \langle 200, 0 \rangle$$

$$\approx \langle 2898, 777 \rangle \cdot \langle 200, 0 \rangle$$

$$\approx 579{,}555.5 \text{ ft-lbs.}$$

End of Section 8.6

Chapter Eight Test
Solutions to Odd-Numbered Exercises

1.

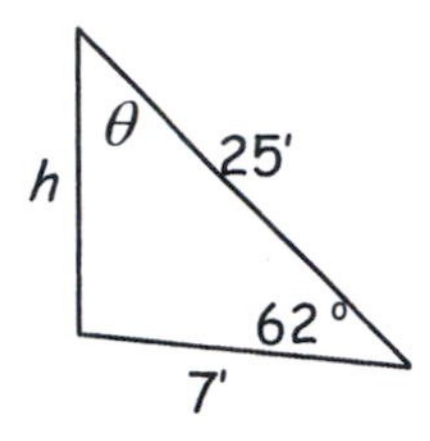

Using the Law of Cosines,

$n^2 = 7^2 + 25^2 - 2(7)(25)\cos 62°$

$n^2 \approx 509.6850$ feet

$n \approx 22.5762$ feet.

Then, using the Law of Sines,

$$\frac{\sin\theta}{7} = \frac{\sin 62°}{22.5762}$$

$$\sin\theta = \frac{7\sin 62°}{22.5762} \approx 0.2738$$

$$\theta \approx 15.8886$$

3. Given: $a = 15, c = 13, C = 57°$

Use Law of Sines:

$$\frac{\sin A}{a} = \frac{\sin C}{c}$$

$$\sin A = \frac{15\sin 57°}{13}$$

$$\sin A \approx \frac{12.58}{13} \approx 0.9677 \Rightarrow A \approx 75.4°$$

$$B = 180° - (75.4° + 57°) \approx 47.6°$$

$$\frac{b}{\sin 47.6°} \approx \frac{13}{\sin 57°}$$

$$b \approx \frac{13\sin 47.6°}{\sin 57°}$$

$$b \approx 11.45$$

5. Given: $b = 8, c = 13, C = 78°$

Use Law of Sines:

$$\frac{\sin B}{b} = \frac{\sin C}{c}$$

$$\sin B \approx \frac{8\sin 78°}{13} \approx 0.602$$

$$B \approx \sin^{-1}(0.602) \approx 37.01°$$

$$A \approx 180° - (78° + 37.01°)$$

$$\approx 64.99°$$

$$\frac{a}{\sin A} = \frac{c}{\sin C}$$

$$a = \frac{13\sin 64.99°}{\sin 78°}$$

$$a \approx 12.04$$

7. Given: $a = 6, c = 14, B = 94°7'$

$(B \approx 94.12°)$

Use Law of Cosines:

$$b^2 = a^2 + c^2 - 2ac\cos B$$

$$= 6^2 + 14^2 - 2(6)(14)\cos 94°7'$$

$$\approx 244.06$$

$$b \approx 15.62$$

Use Law of Sines:

$$\frac{\sin B}{b} = \frac{\sin A}{a}$$

$$\sin A \approx \frac{6\sin 94°7'}{15.62} \approx 0.3831$$

$$A \approx \sin^{-1}(0.3831) \approx 22.52°$$

$$C \approx 180° - (22°31' + 94°7')$$

$$\approx 63°22'$$

9. Given: $a = 10.8, b = 13.4, c = 6$

Use Law of Cosines:

$$\cos A = \frac{a^2 - b^2 - c^2}{-2bc}$$

$$= \frac{10.8^2 - (13.4)^2 - 6^2}{-2(13.4)(6)}$$

$$= \frac{-98.92}{-160.8} \approx 0.6152$$

$$A = \cos^{-1}(0.6152) \approx 52.04°$$

Use Law of Sines:

$$\frac{\sin B}{b} = \frac{\sin A}{a}$$

$$\sin B \approx \frac{13.4 \sin 52.04^\circ}{10.8} \approx 0.9782$$

$B \approx \sin^{-1}(0.9782) \Rightarrow B \approx 78.01^\circ$ or

$B \approx 101.99^\circ$ (now, determine which)

$b > a$ and $b > c \Rightarrow B > 90^\circ$

Therefore $B \approx 101.99^\circ$

$$C \approx 180^\circ - (101.99^\circ + 52.04^\circ)$$
$$\approx 25.98^\circ$$

11. Polar to Cartesian coordinates:

$$\left[7, \frac{7\pi}{6}\right] \rightarrow \left[7\cos\frac{7\pi}{6}, 7\sin\frac{7\pi}{6}\right]$$
$$\rightarrow (-6.06, -3.5)$$

13. Cartesian to Polar coordinates:

$r^2 = 10^2 + 12^2 = 244 \Rightarrow r = 15.62$

$$\tan\theta = \frac{y}{x} = \frac{12}{10} = \frac{6}{5}$$

θ lies in quadrant 1

$$\tan\theta = \frac{6}{5}$$

$$\theta = \tan^{-1}\left(\frac{6}{5}\right) \approx 0.88$$

Then, $(10, 12) \rightarrow (15.62, 0.88)$

15. Rectangular to Polar form:

Given: $x^2 + y^2 = 9ax$

Then, with $x = r\cos\theta$ and $y = r\sin\theta$

$r^2\cos^2\theta + r^2\sin^2\theta - 9ar\cos\theta = 0$

Then, $r^2 - 9ar\cos\theta = 0$

$$r = 9a\cos\theta$$

17. Polar to rectangular form:

Given: $r = \dfrac{16}{4\cos\theta + 4\sin\theta}$

Use $r\cos\theta = x$, or $\cos\theta = \dfrac{x}{r}$

Divide both sides by r:

$$\frac{r}{r} = \frac{16}{4r\cos\theta + 4r\sin\theta}$$

$$1 = \frac{4}{x + y}$$

$$x + y = 4$$

19. Sketch $r^2 = 25\cos 2\theta$.

The pattern represents a Lemniscate:

Symmetric with respect to polar axis:

$r^2 = 5^2\cos 2\theta$

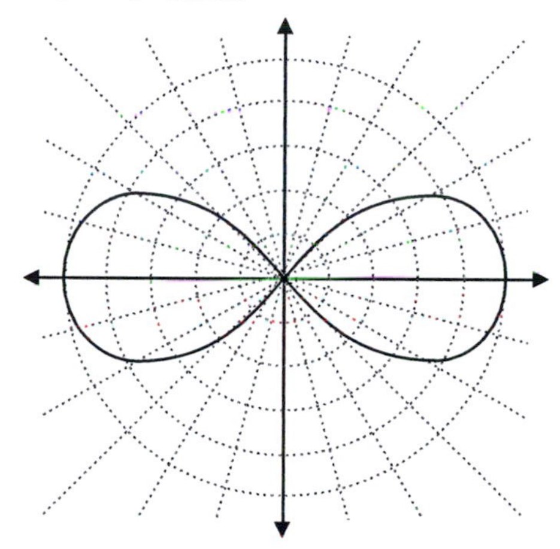

21. Sketch with $x = t + 5$ and $y = |t - 2|$

Then $t = x - 5$. Substituting,

$y = |x - 5 - 2| = |x - 7|$

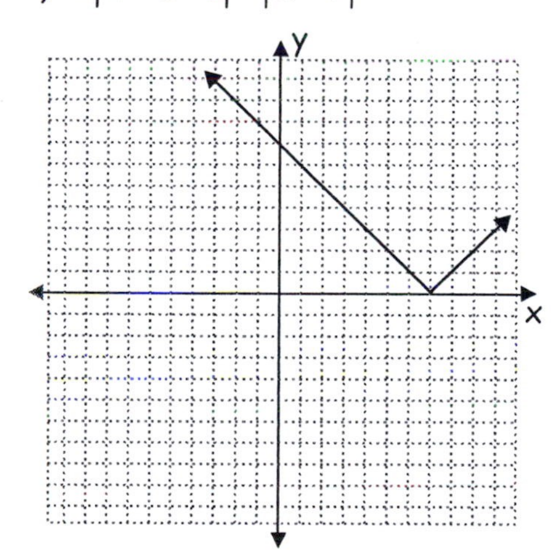

23. Sketch with $x = 4\sin\theta$ and $y = \cos\theta + 1$

The latter equation gives $\cos\theta = y - 1$ and corresponds to the figure:

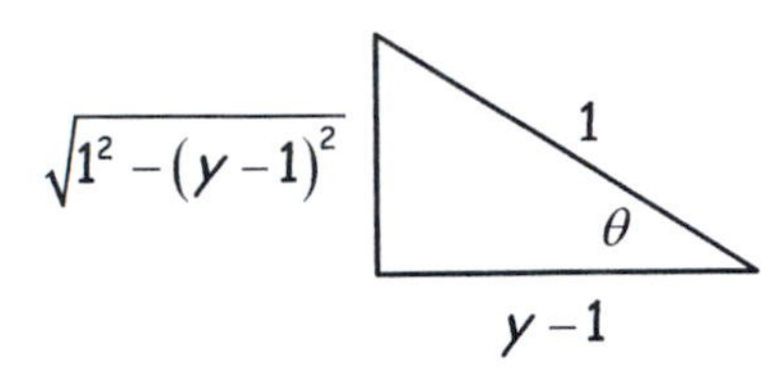

Then $\sin\theta = \sqrt{1^2 - (y-1)^2}$

$= \sqrt{-y^2 + 2y}$

so $x = 4\sqrt{-y^2 + 2y}$

or $\frac{x^2}{16} + (y-1)^2 = 1$

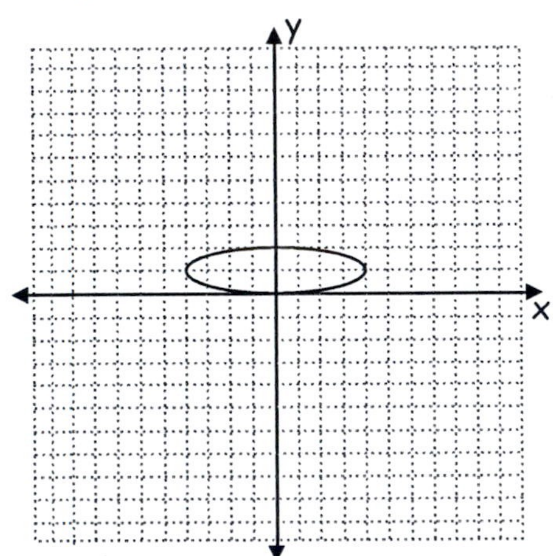

25. Given: $6x = 2 - y$

Let $t = x$

Then, $6t = 2 - y$, or $y = 2 - 6t$.

27. Parametric equation for the circle with Center $(1, 1)$ and radius 1:

$(x-1)^2 + (y-1)^2 = 1$

Answers will vary.

If we let $x = 1 + \cos$, then substitute in the original equation to find an expression for y.

$(1 + \cos\theta - 1)^2 + (y-1)^2 = 1$

$(y-1)^2 = 1 - \cos^2\theta$

$(y-1)^2 = \sin^2\theta$

$y - 1 = -\sin\theta$

$y = 1 - \sin\theta$

Solution: $x = 1 + \cos\theta$, $y = 1 - \sin\theta$

29. Given: $z = -3 + 3i$

$|z| = \sqrt{(-3)^2 + 3^2} = 3\sqrt{2}$ and $z = (-3, 3)$

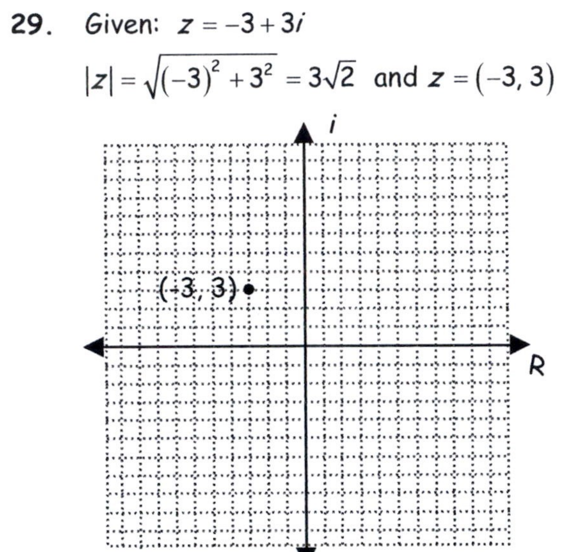

31. Given: $z_1 = 4 + 2i$, $z_2 = -5 + i$

$z_1 + z_2 = -1 + 3i$

$z_1 z_2 = (4 + 2i)(-5 + i)$

$= -20 - 10i + 4i - 2$

$= -22 - 6i$

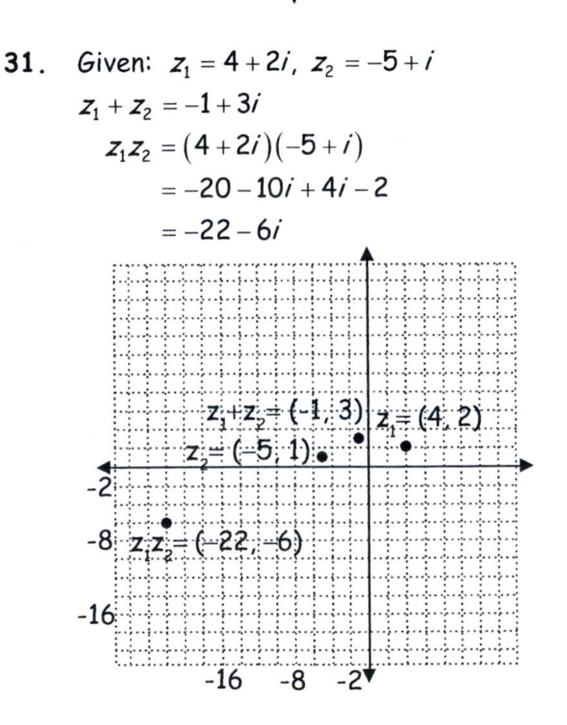

33. Graph region of the complex plane described by $\{z = a + bi | a > 2, b > 3\}$:

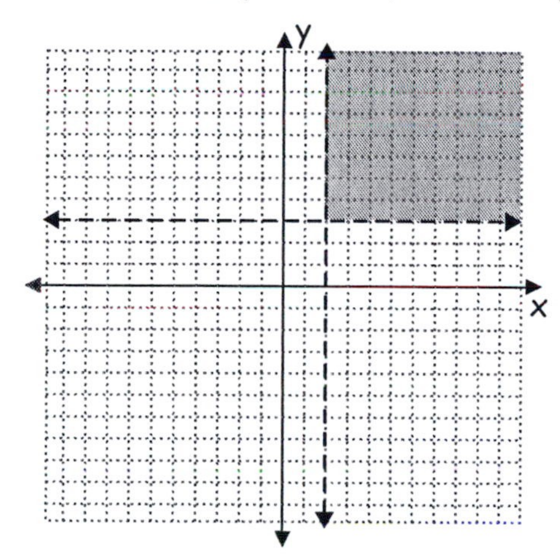

35. Write $z = 1 + 4i$ in trigonometric form:

$z = |z|\cos\theta + |z| i \sin\theta$, where

$|z| = \sqrt{1^2 + 4^2} = \sqrt{17}$

and $\tan\theta = \frac{4}{1} = 4$. Then,

$\theta \approx \tan^{-1} 4 \approx 1.33$

$z \approx \sqrt{17}\,(\cos(1.33) + i\sin(1.33))$

37. Write $z = 3(\cos 60° + i\sin 60°)$ in standard form:

$\cos 60° = .5$ and $\sin 60° = \frac{\sqrt{3}}{2}$

$z = \frac{3}{2} + \frac{3i\sqrt{3}}{2}$

39. Simplify and express $\frac{z_1}{z_2}$ in trigonometric form and standard form:

Given:

$z_1 = 5(\cos 240° + i\sin 240°)$ and

$z_2 = \cos 120° + i\sin 120°$

Use the quotient formula:

$\frac{z_1}{z_2} =$

$\frac{5}{1}(\cos(240° - 120°) + i\sin(240° - 120°))$

$\frac{z_1}{z_2} = 5(\cos 120° + i\sin 120°)$

[trigonometric form]

$\frac{z_1}{z_2} = -\frac{5}{2} + \frac{5i\sqrt{3}}{2} \approx -2.5 + 4.33i$

[standard form]

41. $\left(12e^{35° i}\right)\left(2e^{280° i}\right) = 24e^{(35° + 280°)i}$

$= 24e^{315° i}$

$= 24(\cos 315° + i\sin 315°)$

[trigonometric form]

$= 24\left(\frac{\sqrt{2}}{2} - \frac{\sqrt{2}}{2}i\right)$

$= 12\sqrt{2} - 12\sqrt{2}i$

[standard form]

43. Use DeMoivre's Theorem with

$z = \left[3(\cos 240° + i\sin 240°)\right]^{11}$

$z^{11} = |3|^{11}\left(\cos\left[11(240°)\right] + i\sin\left[11(240°)\right]\right)$

$= 177{,}147(\cos 2640° + i\sin 2640°)$

$= 177{,}147(\cos 120° + i\sin 120°)$

45. The cube roots of

$z = 125\left[\cos\frac{7\pi}{4} + i\sin\frac{7\pi}{4}\right]$:

Note that $|z| = 125$.

$$w_0 = (125)^{\frac{1}{3}}\left[\cos\left(\frac{\frac{7\pi}{4} + 0}{3}\right) + i\sin\left(\frac{\frac{7\pi}{4} + 0}{3}\right)\right]$$

$$= 5\left[\cos\frac{7\pi}{12} + i\sin\frac{7\pi}{12}\right]$$

$$= 5e^{\frac{7\pi}{12}i}$$

$$w_1=(125)^{\frac{1}{3}}\left[\cos\left(\frac{\frac{7\pi}{4}+2\pi}{3}\right)+i\sin\left(\frac{\frac{7\pi}{4}+2\pi}{3}\right)\right]$$
$$=5\left[\cos\frac{15\pi}{12}+i\sin\frac{15\pi}{12}\right]$$
$$=5e^{\frac{15\pi}{12}i}\text{, or }5e^{\frac{5\pi}{4}i}$$

$$w_2=(125)^{\frac{1}{3}}\left[\cos\left(\frac{\frac{7\pi}{4}+4\pi}{3}\right)+i\sin\left(\frac{\frac{7\pi}{4}+4\pi}{3}\right)\right]$$
$$=5\left[\cos\frac{23\pi}{12}+i\sin\frac{23\pi}{12}\right]$$
$$=5e^{\frac{23\pi}{12}i}$$

47. Solve: $z^3+4\sqrt{2}-4i\sqrt{2}=0$.

Re-written, the equation is

$z^3=-4\sqrt{2}+4\sqrt{2}i$

The problem requires finding the third roots of $w=-4\sqrt{2}+4\sqrt{2}i$:

Note: $|w|=\sqrt{\left(-4\sqrt{2}\right)^2+\left(4\sqrt{2}\right)^2}=8$

To write in trigonometric form, note that θ lies in quadrant 2.

$$\tan\theta=\frac{4\sqrt{2}}{-4\sqrt{2}}=-1\Rightarrow\theta=\frac{3\pi}{4}$$

$$w=8\left[\cos\frac{3\pi}{4}+i\sin\frac{3\pi}{4}\right]$$

$$w_0=8^{\frac{1}{3}}\left[\cos\left(\frac{\frac{3\pi}{4}+0}{3}\right)+i\sin\left(\frac{\frac{3\pi}{4}+0}{3}\right)\right]$$
$$=2\left[\cos\frac{\pi}{4}+i\sin\frac{\pi}{4}\right]$$
$$=2e^{\frac{\pi}{4}i}$$

$$w_1=8^{\frac{1}{3}}\left[\cos\left(\frac{\frac{3\pi}{4}+2\pi}{3}\right)+i\sin\left(\frac{\frac{3\pi}{4}+2\pi}{3}\right)\right]$$
$$=2\left[\cos\frac{11\pi}{12}+i\sin\frac{11\pi}{12}\right]$$
$$=2e^{\frac{11\pi}{12}i}$$

$$w_2=8^{\frac{1}{3}}\left[\cos\left(\frac{\frac{3\pi}{4}+4\pi}{3}\right)+i\sin\left(\frac{\frac{3\pi}{4}+4\pi}{3}\right)\right]$$
$$=2\left[\cos\frac{19\pi}{12}+i\sin\frac{19\pi}{12}\right]$$
$$=2e^{\frac{19\pi}{12}i}$$

49. Given: initial point $(6, 5)$ and terminal point $(-4, -1)$:

$\mathbf{v}=\langle -4-6, -1-5\rangle=\langle -10, -6\rangle$

$\|\mathbf{v}\|=\sqrt{(-10)^2+(-6)^2}=\sqrt{136}=2\sqrt{34}$

51. Given: $\mathbf{u}=\langle 1,-1\rangle$, $\mathbf{v}=\langle 4,-3\rangle$

a. $2\mathbf{u}+\mathbf{v}=2\langle 1,-1\rangle+\langle 4,-3\rangle$
$=\langle 2,-2\rangle+\langle 4,-3\rangle$
$=\langle 6,-5\rangle$

b. $-\mathbf{u}+3\mathbf{v}=-\langle 1,-1\rangle+3\langle 4,-3\rangle$
$=\langle -1,1\rangle+\langle 12,-9\rangle$
$=\langle 11,-8\rangle$

c. $-2\mathbf{v}=-2\langle 4,-3\rangle$
$=\langle -8,6\rangle$

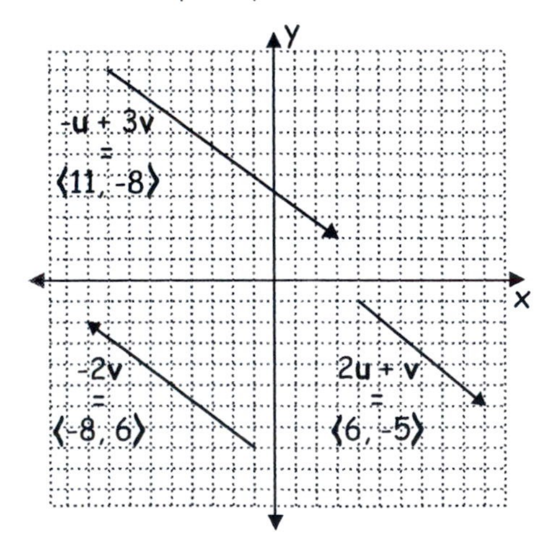

53. Use the initial and terminal points of **u** and **v**. The initial and terminal points for **u** are $(-4, -4)$ and $(1, -3)$, respectively. The initial and terminal points for **v** are $(-1, 1)$ and $(2, 2)$, respectively.

$$\mathbf{u} = \langle 1-(-4), -3-(-4)\rangle = \langle 5, 1\rangle$$
$$\mathbf{v} = \langle 2-(-1), 2-1\rangle = \langle 3, 1\rangle$$
$$-\mathbf{u} = -\langle 5, 1\rangle = \langle -5, -1\rangle$$
$$2\mathbf{u}-\mathbf{v} = 2\langle 5, 1\rangle - \langle 3, 1\rangle = \langle 10, 2\rangle - \langle 3, 1\rangle = \langle 7, 1\rangle$$
$$\mathbf{u}+\mathbf{v} = \langle 5, 1\rangle + \langle 3, 1\rangle = \langle 8, 2\rangle$$
$$\|\mathbf{u}\| = \sqrt{5^2+1^2} = \sqrt{26}$$
$$\|\mathbf{v}\| = \sqrt{3^2+1^2} = \sqrt{10}$$

55. Given: $\mathbf{u} = \langle 6, 3\rangle$

Calculate the scalar $a = \frac{1}{\|\mathbf{u}\|}$

$$a = \frac{1}{\sqrt{6^2+3^2}} = \frac{1}{3\sqrt{5}}$$

a. $a\,\mathbf{u} = \frac{1}{3\sqrt{5}}\langle 6, 3\rangle = \left\langle \frac{2}{\sqrt{5}}, \frac{1}{\sqrt{5}}\right\rangle$

b. $\mathbf{u} = 6\langle 1, 0\rangle + 3\langle 0, 1\rangle = 6\mathbf{i}+3\mathbf{j}$

57. Given: $\mathbf{v} = 5\mathbf{i}-\mathbf{j} = \langle 5, -1\rangle$

$$\|\mathbf{v}\| = \sqrt{5^2+(-1)^2} = \sqrt{26}$$
$$\tan\theta = \frac{-1}{5} \Rightarrow \theta = \tan^{-1}(-0.2) \approx -11.31^\circ$$

59. Given: $\|\mathbf{v}\| = 6$, **v** in the direction of $3\mathbf{i}-4\mathbf{j}$

Then, $\tan\theta = \frac{-4}{3}$, and

$$\theta = \tan^{-1}\left(\frac{-4}{3}\right) \approx -53.13^\circ$$

$$\mathbf{v} \approx 6\langle \cos(-53.13^\circ), \sin(-53.13^\circ)\rangle \approx 6\langle 0.60, -0.80\rangle \approx \langle 3.6, -4.8\rangle$$

61. Let $\mathbf{u}$ = sailboat vector
Let $\mathbf{v}$ = the wind vector

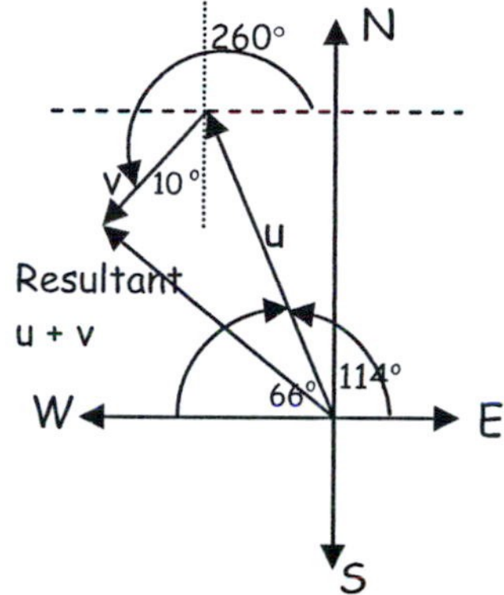

We want to find the resultant vector $\mathbf{u}+\mathbf{v}$. First, put the given angles of the vectors into standard form: The direction of the sailboat vector, **u**, given as W 66° N is 114° in standard form. The wind vector, **v**, given as S 10° W is 260°. Then,

$$\mathbf{u} = 55\langle \cos 114^\circ, \sin 114^\circ\rangle$$
$$\mathbf{v} = 20\langle \cos 260^\circ, \sin 260^\circ\rangle$$

The resultant vector is $\mathbf{u}+\mathbf{v}$:

$$55\langle \cos 114^\circ, \sin 114^\circ\rangle + 20\langle \cos 260^\circ, \sin 260^\circ\rangle =$$
$$55\langle -0.407, 0.914\rangle + 20\langle -0.174, -0.985\rangle =$$
$$\langle -22.57, 50.25\rangle + \langle -3.47, -19.7\rangle =$$
$$\langle -25.84, 30.55\rangle$$

The directional angle of the resultant vector lies in quadrant 2 with

$$\tan\theta \approx \frac{30.55}{-25.84} \approx -1.18$$
$$\theta \approx \tan^{-1}(-1.18) \approx 130.23^\circ$$

This standard angle equates to W 49.77° N

$$\|\mathbf{v}+\mathbf{u}\| = \sqrt{(-25.84)^2+(30.55)^2} \approx 40.01 \text{ mph}$$

63. Given: $\mathbf{u}=\langle 1,-4\rangle$, $\mathbf{v}=\langle 2,5\rangle$

$$\begin{aligned}(\mathbf{u}\cdot\mathbf{v})3\mathbf{v}&=\langle 1,-4\rangle\cdot\langle 2,5\rangle 3\langle 2,5\rangle\\&=(2-20)3\langle 2,5\rangle\\&=-18\langle 6,15\rangle\\&=\langle -108,-270\rangle\end{aligned}$$

65. Given: $\mathbf{u}=-\mathbf{i}-3\mathbf{j}$, or $\langle -1,-3\rangle$

$$\begin{aligned}\|\mathbf{u}\|&=\sqrt{\mathbf{u}\cdot\mathbf{u}}\\&=\sqrt{(-1)(-1)+(-3)(-3)}\\&=\sqrt{1+9}\\&=\sqrt{10}\end{aligned}$$

67. Given: $\mathbf{u}=\cos\frac{\pi}{4}\mathbf{i}+\sin\frac{\pi}{4}\mathbf{j}$, or

$\mathbf{u}=\left\langle\frac{\sqrt2}{2},\frac{\sqrt2}{2}\right\rangle$, and $\mathbf{v}=\cos\frac{2\pi}{3}\mathbf{i}+\sin\frac{2\pi}{3}\mathbf{j}$,

or $\mathbf{v}=\left\langle-\frac{1}{2},\frac{\sqrt3}{2}\right\rangle$

Find the angle between the two vectors:
From $\mathbf{u}\cdot\mathbf{v}=\|\mathbf{u}\|\|\mathbf{v}\|\cos\theta$ we have

$$\begin{aligned}\cos\theta&=\frac{\mathbf{u}\cdot\mathbf{v}}{\|\mathbf{u}\|\|\mathbf{v}\|}\\&=\frac{\frac{-\sqrt2}{4}+\frac{\sqrt6}{4}}{\sqrt{\left(\frac{\sqrt2}{2}\right)^2+\left(\frac{\sqrt2}{2}\right)^2}\sqrt{\left(-\frac12\right)^2+\left(\frac{\sqrt3}{2}\right)^2}}\\&=\frac{-\sqrt2+\sqrt6}{4}\approx 0.2588\end{aligned}$$

Then,

$$\theta\approx\cos^{-1}(0.2588)=75^\circ\text{ or }\frac{5\pi}{2}$$

69. Given: $\|\mathbf{u}\|=8$, $\|\mathbf{v}\|=9$, $\theta=\frac{2\pi}{3}$

Find $\mathbf{u}\cdot\mathbf{v}$:

$$\begin{aligned}\mathbf{u}\cdot\mathbf{v}&=\|\mathbf{u}\|\|\mathbf{v}\|\cos\theta\\&=8\cdot9\cdot\cos\frac{2\pi}{3}\\&=72\left(-\frac12\right)=-36\end{aligned}$$

71. Find the projection of $\mathbf{u}=\langle 4,-1\rangle$ onto $\mathbf{v}=\langle 2,2\rangle$. First, find $\|\mathbf{v}\|$:

$$\|\mathbf{v}\|=\sqrt{2^2+2^2}=2\sqrt2$$

$$\begin{aligned}\text{proj}_{\mathbf{v}}\mathbf{u}&=\left(\frac{\mathbf{u}\cdot\mathbf{v}}{\|\mathbf{v}\|^2}\right)\mathbf{v}\\&=\left(\frac{6}{8}\right)\langle 2,2\rangle\\&=\left\langle\frac32,\frac32\right\rangle\end{aligned}$$

$$\begin{aligned}\text{perp}_{\mathbf{v}}\mathbf{u}&=\mathbf{u}-\text{proj}_{\mathbf{v}}\mathbf{u}\\&=\langle 4,-1\rangle-\left\langle\frac32,\frac32\right\rangle\\&=\left\langle\frac52,-\frac52\right\rangle\end{aligned}$$

73. Given: $J=(-5,3)$, $K=(0,4)$ and $\mathbf{v}=\langle 5,6\rangle$. Use $\mathbf{W}=\mathbf{F}\cdot\mathbf{D}$:

$$\mathbf{F}=\mathbf{v}=\langle 5,6\rangle$$
$$\mathbf{D}=\langle 0-(-5),4-3\rangle=\langle 5,1\rangle$$
$$\mathbf{W}=\langle 5,6\rangle\cdot\langle 5,1\rangle=25+6=31$$

75. Let $\mathbf{F}=2650\langle\cos10^\circ,\sin10^\circ\rangle$

$\mathbf{D}=\langle 160,0\rangle$, $\|\mathbf{D}\|=160$

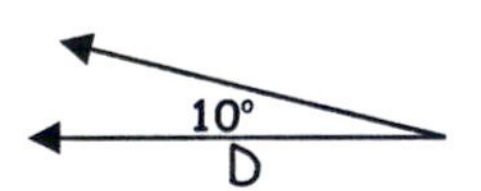

$$\begin{aligned}\mathbf{W}&=\mathbf{FD}\\&=2650\langle\cos10^\circ,\sin10^\circ\rangle\cdot\langle 160,0\rangle\\&\approx2650\langle 0.9848,0.1736\rangle\cdot\langle 160,0\rangle\\&\approx\langle 2609.74,460.17\rangle\cdot\langle 160,0\rangle\\&\approx417{,}558.5\ \text{ft-lbs}\end{aligned}$$

End of Chapter 8 Test

Chapter Nine: Section 9.1
Solutions to Odd-numbered Exercises

1. Using the general form of the ellipse and re-writing the given equation, you can determine the values of a, b and c to find the coordinates of the foci and vertices:
$\frac{(x-h)^2}{b^2}+\frac{(y-k)^2}{a^2}=1$ where Center $(h,k)=(5,2)$, $a^2=25$ and $b^2=4$.
Then, $a=5$ and $b=2$.
$c^2=a^2-b^2 \Rightarrow c^2=21 \Rightarrow c=\sqrt{21}$
The foci are $\left(5, 2+\sqrt{21}\right)$ and $\left(5, 2-\sqrt{21}\right)$.
The vertices are $(5, k+a)$ and $(5, k-a)$
$\Rightarrow$ the vertices are $(5,7)$ and $(5,-3)$

3. Re-write in standard form of an ellipse (Divide by 9):
$\frac{(x+2)^2}{9}+\frac{(y+5)^2}{3}=1$, or
$\frac{(x+2)^2}{(3)^2}+\frac{(y+5)^2}{\left(\sqrt{3}\right)^2}=1$
Then, $a=3$ and $b=\sqrt{3}$.
$c^2=9-3=6 \Rightarrow c=\sqrt{6}$
The ellipse is horizontally oriented with Center $(-2,-5)$, vertices $(1,-5)$and $(-5,-5)$ and foci $\left(-2+\sqrt{6},-5\right)$ and $\left(-2-\sqrt{6},-5\right)$.

5. First, complete the squares in x and y:
$\left(x^2+6x+9\right)+$
$2\left(y^2-4y+4\right)+13-9-8=0$
$(x+3)^2+2(y-2)^2=4$
Re-write in standard form (divide by 4):
$\frac{(x+3)^2}{4}+\frac{(y-2)^2}{2}=1$
Then, $a=2$, $b=\sqrt{2}$ and $c^2=4-2=2$
$\Rightarrow c=\sqrt{2}$
The ellipse is horizontally oriented with Center $(-3,2)$, Vertices $(-5,2)$ and $(-1,2)$ and Foci $\left(-3+\sqrt{2},\ 2\right)$ and $\left(-3-\sqrt{2},\ 2\right)$

7. Complete the square in x and y:
$4\left(x^2+10x+25\right)+\left(y^2-2y+1\right)+$
$85-100-1=0$
Re-write in standard form:
$4(x+5)^2+(y-1)^2=16$ (Divide by 16)
$\frac{(x+5)^2}{4}+\frac{(y-1)^2}{16}=1$
Then, $a^2=16$, $b^2=4$, $c^2=a^2-b^2=12$
$\Rightarrow c=2\sqrt{3}$, $a=4$, $b=2$
The ellipse is oriented vertically with Center $(-5,1)$,Vertices $(-5,5)$ and $(-5,-3)$, and Foci $\left(-5,1-2\sqrt{3}\right)$ and $\left(-5,1+2\sqrt{3}\right)$.

9. Complete the square in x and y:
$\left(x^2+8x+16\right)+3\left(y^2-4y+4\right)+$
$1-16-12=0$
Re-write in standard form:
$(x+4)^2+3(y-2)^2=27$ (Divide by 27)
$\frac{(x+4)^2}{27}+\frac{(y-2)^2}{9}=1$
Then, $a^2=27$, $b^2=9$, $c^2=a^2-b^2=18$
$\Rightarrow c=3\sqrt{2}$, $a=3\sqrt{3}$, $b=3$
The ellipse is oriented horizontally with Center $(-4,2)$, Vertices $\left(-4+3\sqrt{3},2\right)$ and $\left(-4-3\sqrt{3},2\right)$, and Foci $\left(-4-3\sqrt{2},2\right)$ and $\left(-4+3\sqrt{2},2\right)$.

11. Complete the square in x and y:
$(x^2 - 4x + 4) + 5(y^2) - 1 - 4 = 0$
Re-write in standard form:
$(x-2)^2 + 5y^2 = 5$ (Divide by 5)
$$\frac{(x-2)^2}{5} + y^2 = 1$$
Then, $a^2 = 5, b^2 = 1, c^2 = a^2 - b^2 = 4$
$\Rightarrow c = 2, a = \sqrt{5}, b = 1$
The ellipse is oriented horizontally with Center $(2, 0)$, Vertices $(2+\sqrt{5}, 0)$ and $(2-\sqrt{5}, 0)$, and Foci $(0, 0)$ and $(4, 0)$.

13. Matches c with Center $(3, 2)$, Vertices $(3, -3)$ and $(3, 7)$.

15. Matches b with Center $(0, 3)$, Vertices $(0, 5)$ and $(0, 1)$.

17. Matches e as a circle with Center $(2, -2)$, radius 3.

19. Matches g with Center $(1, 0)$, Vertices $(3, 0)$ and $(-1, 0)$.

21. Using $c^2 = a^2 - b^2$, $a = 3, b = 1$
Major axis is horizontal.
$c^2 = 3^2 - 1^2 = 8;\ c = 2\sqrt{2}$
Center: $(h, k) = (3, -1)$
Foci: $(3-2\sqrt{2}, -1)$ and $(3+2\sqrt{2}, -1)$

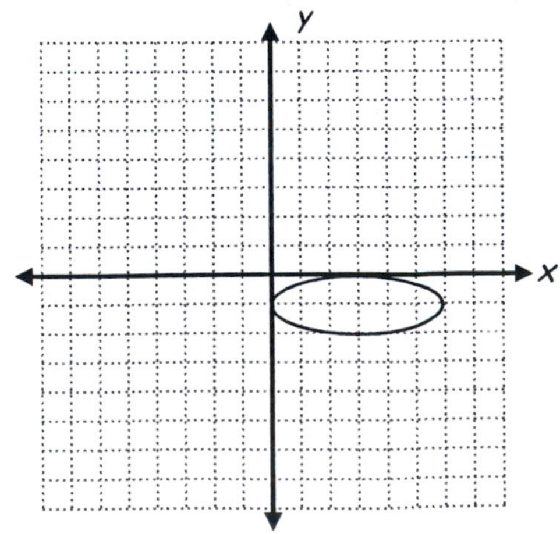

23. Using $c^2 = a^2 - b^2$, $a = 3, b = 2$
$c^2 = 3^2 - 2^2 = 5;\ c = \sqrt{5}$
Major axis is horizontal.
Center: $(h, k) = (3, 4)$

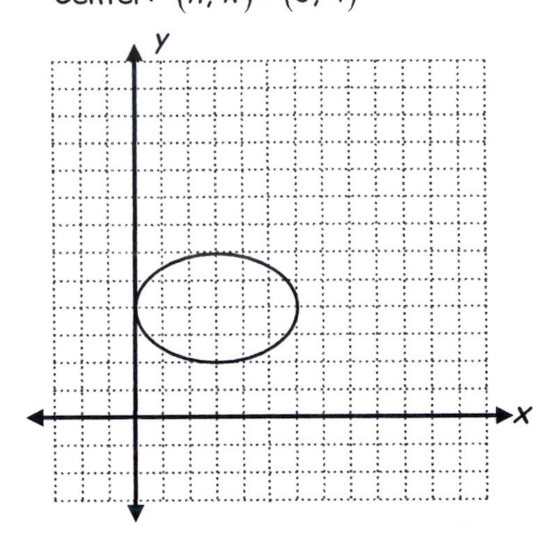

25. Using $c^2 = a^2 - b^2$, $a = 2, b = 1$
$c^2 = 2^2 - 1^2 = 3;\ c = \sqrt{3}$
Major axis is vertical.
Center: $(h, k) = (1, 4)$
Foci: $(1, 4-\sqrt{3})$ and $(1, 4+\sqrt{3})$

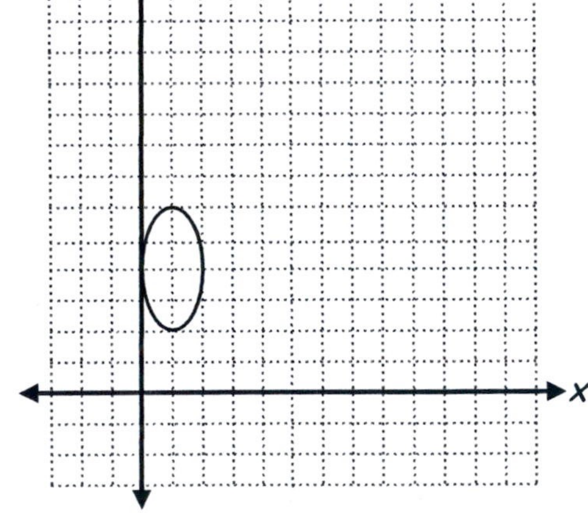

27. Using $c^2 = a^2 - b^2$, $a = 5$, $b = 2$

$c^2 = 5^2 - 2^2 = 21;\ c = \sqrt{21}$

Major axis is horizontal.

Center: $(h, k) = (-1, -5)$

Foci: $\left(-1-\sqrt{21},\ -5\right)$ and $\left(-1+\sqrt{21},\ -5\right)$

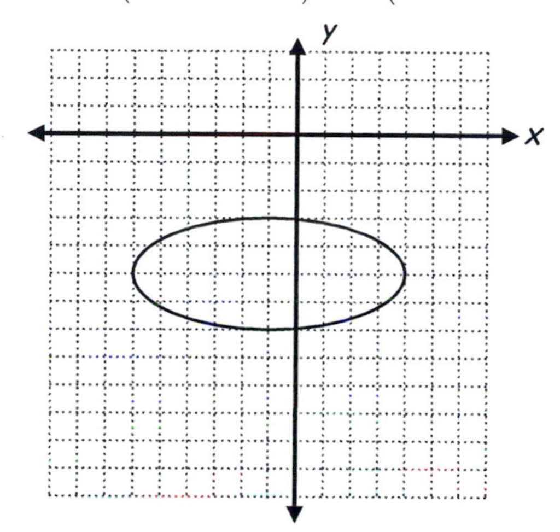

Using $c^2 = a^2 - b^2$, $a = 4$, $b = 3$

$c^2 = 4^2 - 3^2 = 7;\ c = \sqrt{7}$

Major axis is horizontal.

Center: $(h, k) = (-1, 2)$

Foci: $\left(-1-\sqrt{7}, 2\right)$ and $\left(-1+\sqrt{7}, 2\right)$

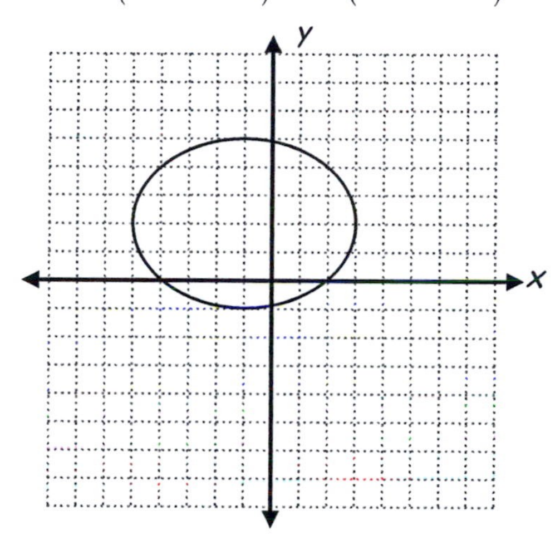

29. Using $c^2 = a^2 - b^2$, $a = 4$, $b = 3$

$c^2 = 4^2 - 3^2 = 7;\ c = \sqrt{7}$

Major axis is horizontal.

Center: $(h, k) = (-2, -1)$

Foci: $\left(-2-\sqrt{7}, -1\right)$ and $\left(-2+\sqrt{7}, -1\right)$

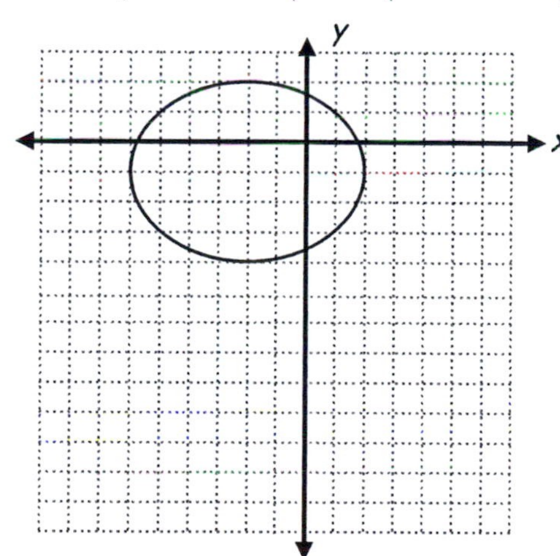

31. Complete the square:

$$9x^2 + 16y^2 + 18x - 64y = 71$$

$$9\left(x^2 + 2x + 1\right) + 16\left(y^2 - 4y + 4\right) = 71 + 73$$

$$\frac{(x+1)^2}{16} + \frac{(y-2)^2}{9} = 1$$

33. Complete the square:

$$16x^2 + y^2 + 160x - 6y = -393$$

$$16\left(x^2 + 10x + 25\right) + \left(y^2 - 6y + 9\right) = -393 + 409$$

$$\frac{(x+5)^2}{1} + \frac{(y-3)^2}{16} = 1$$

Using $c^2 = a^2 - b^2$, $a = 4$, $b = 1$

$c^2 = 4^2 - 1^2 = 15;\ c = \sqrt{15}$

Major axis is vertical.

Center: $(h, k) = (-5, 3)$

Foci: $\left(-5, 3-\sqrt{15}\right)$ and $\left(-5, 3+\sqrt{15}\right)$

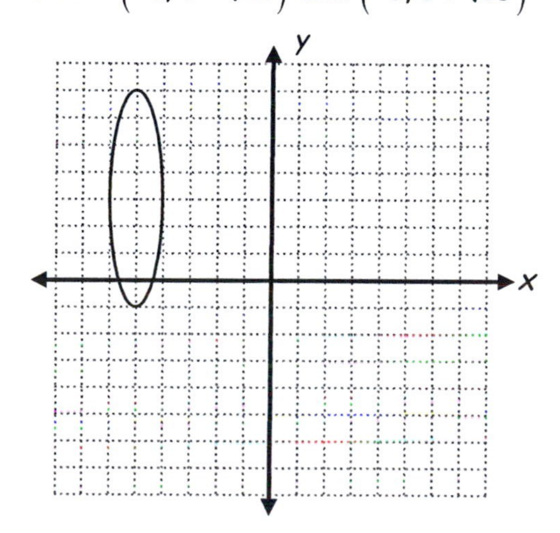

35. Complete the square:

$$4x^2+9y^2+40x+90y+289=0$$
$$4\left(x^2+10x+25\right)+9\left(y^2+10y+25\right)=$$
$$-289+325$$
$$\frac{(x+5)^2}{9}+\frac{(y+5)^2}{4}=1$$

Using $c^2=a^2-b^2$, $a=3$, $b=2$

$c^2=3^2-2^2=5;\ c=\sqrt{5}$

Major axis is horizontal.

Center: $(h,k)=(-5,-5)$

Foci: $\left(-5-\sqrt{5},-5\right)$ and $\left(-5+\sqrt{5},-5\right)$

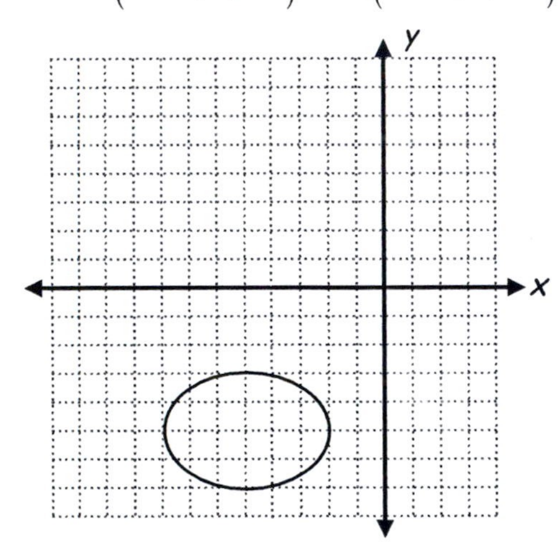

37. Complete the square:

$$4x^2+y^2+4y=0$$
$$4x^2+\left(y^2+4y+4\right)=4$$
$$\frac{(x-0)^2}{1}+\frac{(y+2)^2}{4}=1$$

Using $c^2=a^2-b^2$, $a=2$, $b=1$

$c^2=2^2-1^2=3;\ c=\sqrt{3}$

Major axis is vertical.

Center: $(h,k)=(0,-2)$

Foci: $\left(0,-2-\sqrt{3}\right)$ and $\left(0,-2+\sqrt{3}\right)$

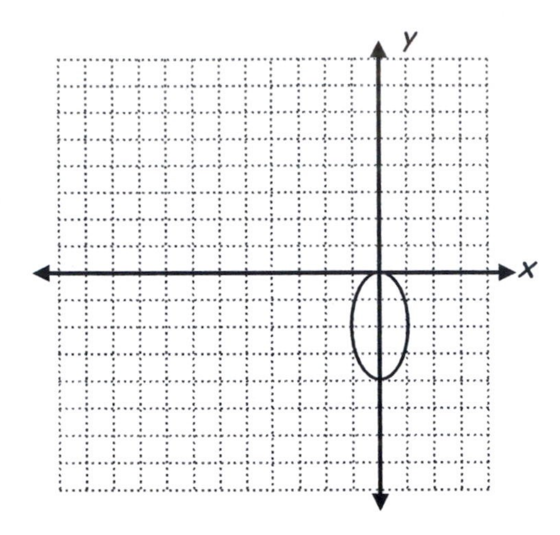

39. Given the conditions, write the equation of the ellipse:

Center: $(h,k)=(0,0)$

Major axis is vertical.

Using $b^2=a^2-c^2$, $a=5$, $c=3$

$b^2=5^2-3^2=16;\ b=4$

$$\frac{(x-0)^2}{b^2}+\frac{(y-0)^2}{a^2}=1$$
$$\frac{x^2}{16}+\frac{y^2}{25}=1$$

41. Given the conditions, write the equation of the ellipse:

Vertices: $(1,4)$ and $(1,-2)$ indicate that the major axis is vertical.

The center is the midpoint of the vertices, so center is $(1,1)$.

Given that the foci are $2\sqrt{2}$ units from the center, the foci are $\left(1,1+2\sqrt{2}\right)$ and $\left(1,1-2\sqrt{2}\right)$.

$c=2\sqrt{2}$ and $a=3$.

Using $b^2=a^2-c^2$, $b^2=1$ and $b=1$.

The equation of the ellipse is

$$\frac{(x-1)^2}{1}+\frac{(y-1)^2}{9}=1$$

43. Given the conditions, write the equation of the ellipse:
Foci at $(0, 0)$ and $(6, 0)$ means that the major axis is horizontal and that $c = 3$ and that center $(h, k) = (3, 0)$.
Eccentricity $e = \frac{1}{2}$ means that
$e = \frac{c}{a} = \frac{1}{2} \Rightarrow a = 2c$. Therefore, $a = 6$.
From $b^2 = a^2 - c^2$, $b^2 = 36 - 9$; $b = 3\sqrt{3}$.
The equation is $\frac{(x-3)^2}{36} + \frac{y^2}{27} = 1$

45. Given the conditions, write the equation of the ellipse:
Vertices: $(-2, -1)$ and $(-2, -5)$ indicate that the major axis is vertical.
and that $a = 2$ and the center $(h, k) = (-2, -3)$. That the minor axis has length 2 means that $b = 1$. Then, the equation can be written as

$$(x+2)^2 + \frac{(y+3)^2}{4} = 1$$

47. Given the conditions, write the equation of the ellipse:
Vertices: $(1, 3)$ and $(9, 3)$ indicate that the major axis is horizontal, that $a = 4$ and that the center $(h, k) = (5, 3)$.
That one focus is at $(6, 3)$ means that the other focus is at $(4, 3)$ and that $c = 1$.
Using $b^2 = a^2 - c^2$, $b^2 = 16 - 1$.
Then, $b = \sqrt{15}$.
The equation is $\frac{(x-5)^2}{16} + \frac{(y-3)^2}{15} = 1$

49. From the given graph: Center $(h, k) = (2, -2)$, $a = 3$ and $b = 2$. The major axis is vertical, so the equation is

$$\frac{(x-2)^2}{4} + \frac{(y+2)^2}{9} = 1$$

51. From the given graph: Center $(h, k) = (1, 0)$, $a = 4$ and $b = 3$. The major axis is vertical, so the equation is

$$\frac{(x-1)^2}{9} + \frac{y^2}{16} = 1$$

53. Given: $\frac{x^2}{100} + \frac{y^2}{144} = 1$
Orientation is vertical with
$a^2 = 144 \Rightarrow a = 12$
$b^2 = 100 \Rightarrow b = 10$
$c^2 = a^2 - b^2 = 44 \Rightarrow c = \sqrt{44} = 2\sqrt{11}$
Eccentricity $= \frac{c}{a} = \frac{2\sqrt{11}}{12} = \frac{\sqrt{11}}{6}$
Length of major axis: $2a = 24$
Length of minor axis: $2b = 20$

55. Given: $x^2 + 9y^2 = 36$, or $\frac{x^2}{36} + \frac{y^2}{4} = 1$
Orientation is horizontal with
$a^2 = 36 \Rightarrow a = 6$
$b^2 = 4 \Rightarrow b = 2$
$c^2 = a^2 - b^2 = 32 \Rightarrow c = \sqrt{32} = 4\sqrt{2}$
Eccentricity $= \frac{c}{a} = \frac{4\sqrt{2}}{6} = \frac{2\sqrt{2}}{3}$
Length of major axis: $2a = 12$
Length of minor axis: $2b = 4$

57. Given: $4x^2 + 16y^2 = 16$, or $\frac{x^2}{4} + y^2 = 1$

Orientation is horizontal with

$a^2 = 4 \Rightarrow a = 2$

$b^2 = 1 \Rightarrow b = 1$

$c^2 = a^2 - b^2 = 3 \Rightarrow c = \sqrt{3}$

Eccentricity $= \frac{c}{a} = \frac{\sqrt{3}}{2}$

Length of major axis: $2a = 4$

Length of minor axis: $2b = 2$

59. Given: $20x^2 + 10y^2 = 40$, or $\frac{x^2}{2} + \frac{y^2}{4} = 1$

Orientation is vertical with

$a^2 = 4 \Rightarrow a = 2$

$b^2 = 2 \Rightarrow b = \sqrt{2}$

$c^2 = a^2 - b^2 = 4 - 2 = 2 \Rightarrow c = \sqrt{2}$

Eccentricity $= \frac{c}{a} = \frac{\sqrt{2}}{2}$

Length of major axis: $2a = 4$

Length of minor axis: $2b = 2\sqrt{2}$

61. Given: $x^2 = 49 - 7y^2$, or $\frac{x^2}{49} + \frac{y^2}{7} = 1$

Orientation is horizontal with

$a^2 = 49 \Rightarrow a = 7$

$b^2 = 7 \Rightarrow b = \sqrt{7}$

$c^2 = a^2 - b^2 = 49 - 7 = 42 \Rightarrow c = \sqrt{42}$

Eccentricity $= \frac{c}{a} = \frac{\sqrt{42}}{7}$

Length of major axis: $2a = 14$

Length of minor axis: $2b = 2\sqrt{7}$

63. Eccentricity of Pluto:

The figure indicates the relevant lengths along the major axis:

Sun

7.37×10^9 km

c c

4.43×10^9 km

a

$$2c = 7.37 \times 10^9 - 4.43 \times 10^9$$

$$c = \frac{(7.37 - 4.43) \times 10^9}{2} \text{km}$$

$$c = \frac{2.94 \times 10^9}{2} \text{km} = 1.47 \times 10^9 \text{ km}$$

$$a = c + 4.43 \times 10^9 \text{ km}$$

$$a = (1.47 + 4.43) \times 10^9 \text{ km}$$

$$a = 5.9 \times 10^9 \text{ km}$$

$$\text{Eccentricity} = \frac{c}{a} = \frac{1.47 \times 10^9 \text{ km}}{5.9 \times 10^9 \text{ km}} \approx 0.249$$

65. Given: Eccentricity $= 0.017$

Find the length of the minor axis, b, shown in the figure:

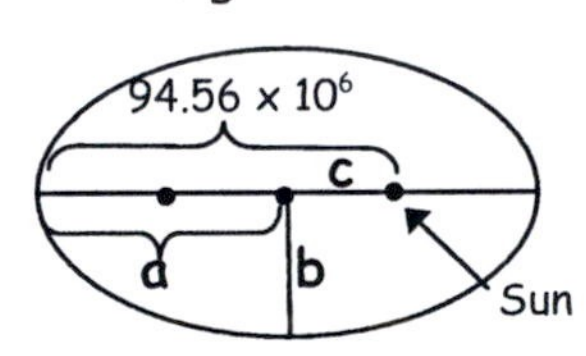

$a + c = 94.56 \times 10^6$

$\frac{c}{a} = 0.017 \Rightarrow c = 0.017a$

Then,

$$a + 0.017a = 94.56 \times 10^6$$

$$a(1 + 0.017) = 94.56 \times 10^6$$

$$a = \frac{94.56 \times 10^6}{1.017}$$

$$a \approx 92.979 \times 10^6$$

Then,

$$c \approx 94.56 \times 10^6 - 92.979 \times 10^6$$

$$c \approx 1.581 \times 10^6$$

So,

$$b^2 = a^2 - c^2$$

$$b^2 \approx (8645.0944 - 2.4996) \times 10^{12}$$

$$b \approx \sqrt{8645.5948 \times 10^{12}}$$

$$b \approx 92.97 \times 10^6 \text{ miles}$$

The length of the minor axis is

$2b \approx 185.93 \times 10^6$ miles

End of Section 9.1

Chapter Nine: Section 9.2
Solutions to Odd-numbered Exercises

1. Given $(x+1)^2 = 4(y-3)$
   Set $4p = 4$. Solve: $p = 1$
   Vertex: $(-1, 3)$
   Since $p > 0$ the parabola opens upward.
   Find x-intercepts (i.e., where $y = 0$):
   $(x+1)^2 = -12 \Rightarrow x+1 = \sqrt{-12}$.
   Therefore there are no x-intercepts.
   Find y-intercepts (set $x = 0$ and solve for y):
   $4y - 12 = 1$
   $y = \frac{13}{4}$
   Focus: $(-1, 3+1) = (-1, 4)$;
   Directrix: line $y = 3 - 1 = 2$

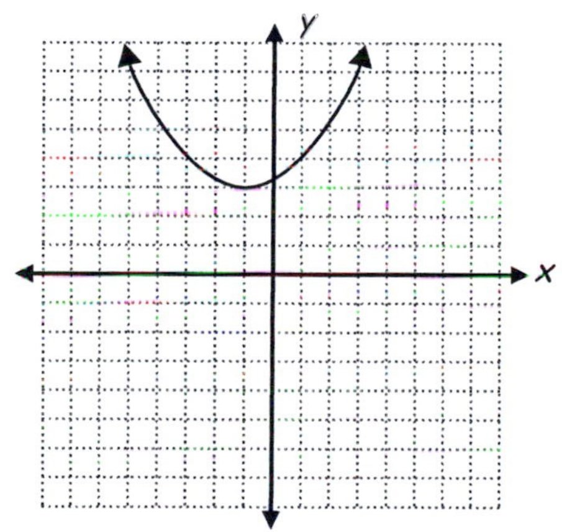

3. Given: $y^2 - 4y = 8x + 4$
   Complete the square:
   $y^2 - 4y + 4 = 8x + 4 + 4$
   $(y-2)^2 = 8(x+1)$
   Set $4p = 8$. Solve: $p = 2$
   Vertex: $(-1, 2)$
   Since $p > 0$, the parabola opens to the right. For the x-intercept, set $y = 0$ and solve for x:
   $8x + 8 = 4$
   $x = -\frac{1}{2}$

   For the y-intercepts, set $x = 0$ and solve for y:
   $(y-2)^2 = 8$
   $y - 2 = \pm 2\sqrt{2}$
   $y = 2 \pm 2\sqrt{2}$
   Focus: $(-1+2, 2) = (1, 2)$;
   Directrix: line $x = -1 - 2 = -3$

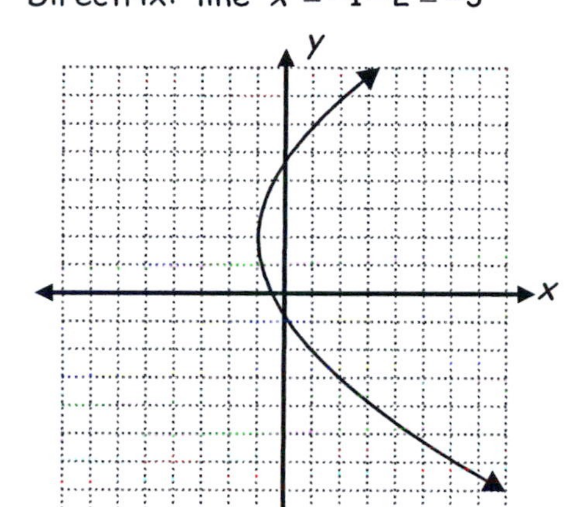

5. Given: $x^2 - 8y = 6x - 1$
   Complete the square:
   $x^2 - 6x + 9 = 8y - 1 + 9$
   $(x-3)^2 = 8(y+1)$
   Set $4p = 8$. Solve: $p = 2$
   Vertex: $(3, -1)$
   The parabola opens upward.
   Find the x-intercepts by letting $y = 0$:
   $(x-3)^2 = 8$
   $x - 3 = \pm\sqrt{8}$
   $x = 3 \pm 2\sqrt{2}$
   Focus: $(3, -1+2) = (3, 1)$;
   Directrix: line $y = -1 - 2 = -3$

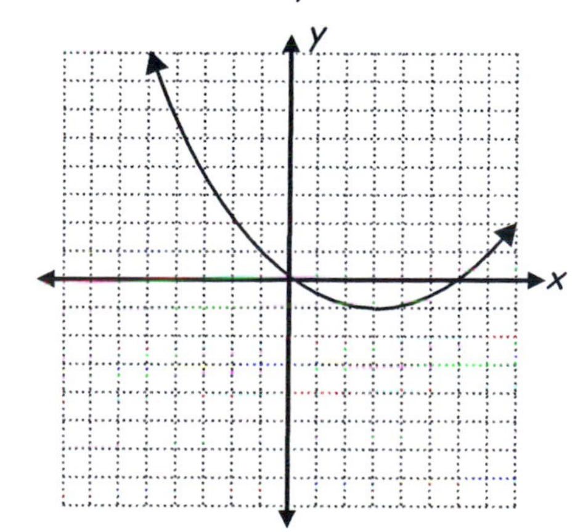

7. Given: $(y-1)^2 = 8(x+3)$

Set $4p = 8$. Solve: $p = 2$.

Vertex: $(-3, 1)$;

Graph opens to the right.

Find the y-intercepts by setting $x = 0$ and solving for y:

$$(y-1)^2 = 24$$
$$y - 1 = \pm\sqrt{24}$$
$$y = 1 \pm 2\sqrt{6}$$

Find the x-intercept by setting $y = 0$:

$$8x + 24 = 1$$
$$x = -\frac{23}{8}$$

Focus: $(-1, 1)$; Directrix: line $x = -5$

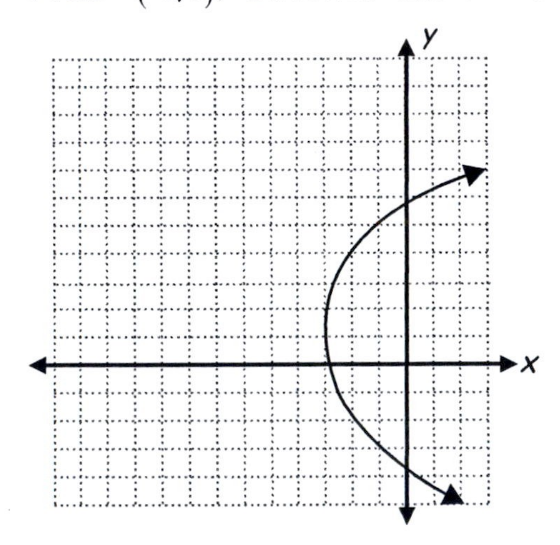

9. Given $(y+1)^2 = -12(x+1)$

Set $4p = -12$. Solve: $p = -3$

Vertex: $(-1, -1)$

Graph opens to the left.

Find the x-intercepts by letting $y = 0$:

$$-12(x+1) = 1$$
$$-12x = 13$$
$$x = -\frac{13}{12}$$

There are no y-intercepts.

Focus: $(-1 + (-3), -1) = (-4, -1)$;

Directrix: line $x = -1 - (-3) = 2$

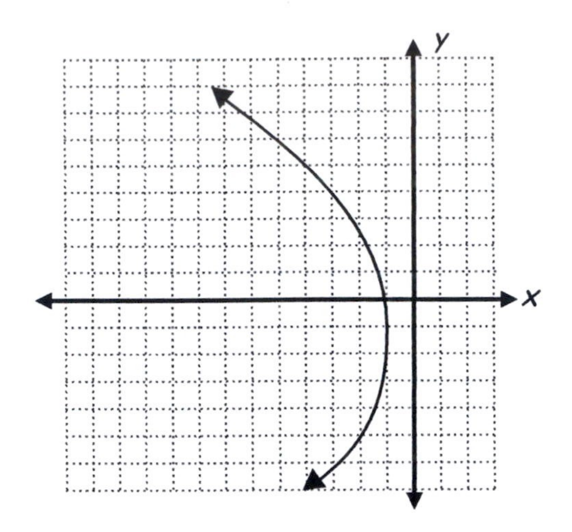

11. Given: $y^2 + 6y - 2x + 13 = 0$

Complete the square:

$$y^2 + 6y + 9 = 2x - 13 + 9$$
$$(y+3)^2 = 2(x-2)$$

Set $4p = 2$. Solve: $p = \frac{1}{2}$

Vertex: $(2, -3)$

The parabola opens to the right.

For the x-intercept, set $y = 0$ and solve for x:

$$2x - 4 = 9$$
$$2x = 13$$
$$x = \frac{13}{2}$$

There are no y-intercepts.

Focus: $\left(2 + \frac{1}{2}, -3\right) = \left(\frac{5}{2}, -3\right)$;

Directrix: line $x = 2 - \frac{1}{2} = \frac{3}{2}$

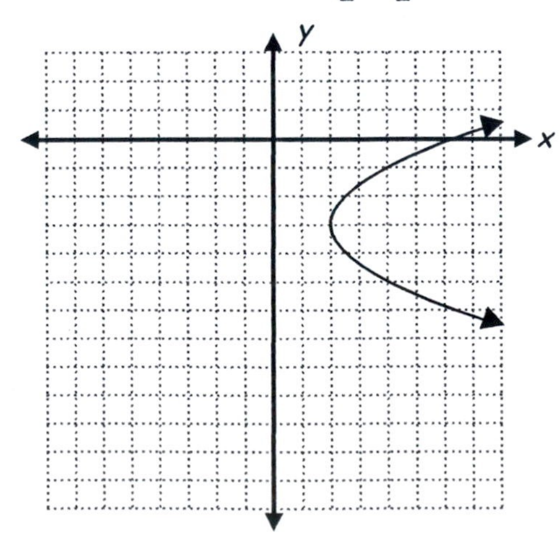

13. Given $y^2 = 6x$

Set $4p = 6$. Solve: $p = \frac{3}{2}$

Vertex: $(0, 0)$

Graph opens to the right.

Find the x-intercepts by letting $y = 0$:

$6x = 0 \Rightarrow x = 0 \Rightarrow y$-intercept at $(0, 0)$

Focus: $\left(\frac{3}{2}, 0\right)$; Directrix: line $x = -\frac{3}{2}$

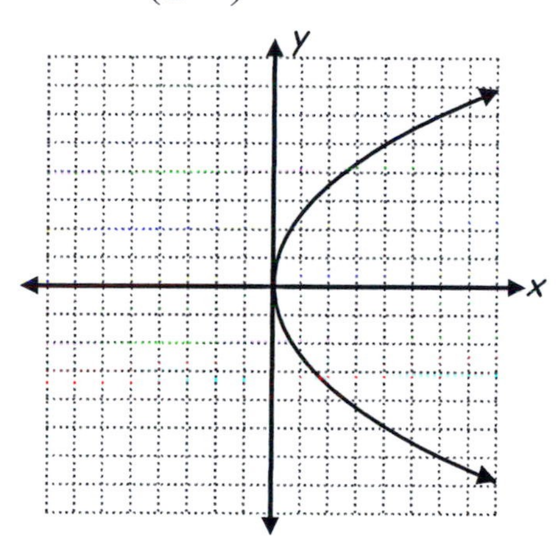

15. Given $x^2 = 7y$

Set $4p = 7$. Solve: $p = \frac{7}{4}$

Vertex: $(0, 0)$

The graph opens upward.

Find x-intercepts (i.e., where $y = 0$):

x-intercept at $(0,0)$

Focus: $\left(0, \frac{7}{4}\right)$; Directrix: line $y = -\frac{7}{4}$

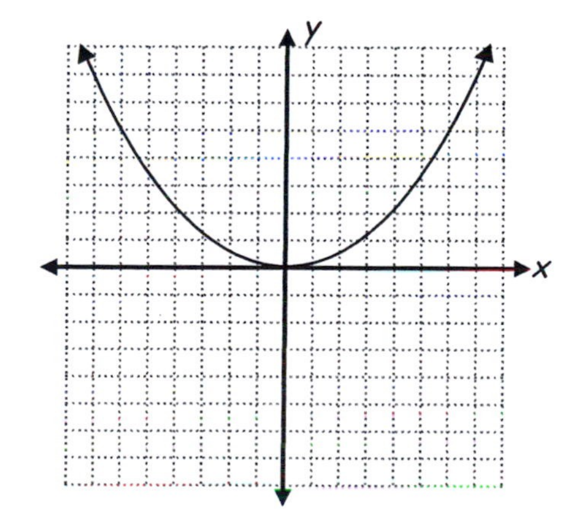

17. Given: $y = -12x^2$, or $x^2 = -\frac{1}{12}y$

Set $4p = -\frac{1}{12}$. Solve: $p = -\frac{1}{48}$

Vertex: $(0, 0)$

The graph opens downward.

Find x-intercepts (i.e., where $y = 0$):

x-intercept at $(0, 0)$

Focus: $\left(0, -\frac{1}{48}\right)$; Directrix: line $y = \frac{1}{48}$

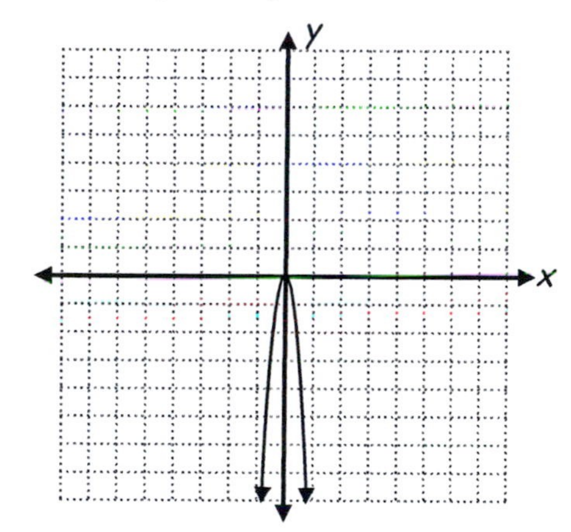

19. Given: $x = \frac{1}{6}y^2$, or $y^2 = 6x$

Set $4p = 6$. Solve: $p = \frac{3}{2}$

Vertex: $(0, 0)$

Graph opens to the right.

Focus: $\left(\frac{3}{2}, 0\right)$; Directrix: line $x = -\frac{3}{2}$

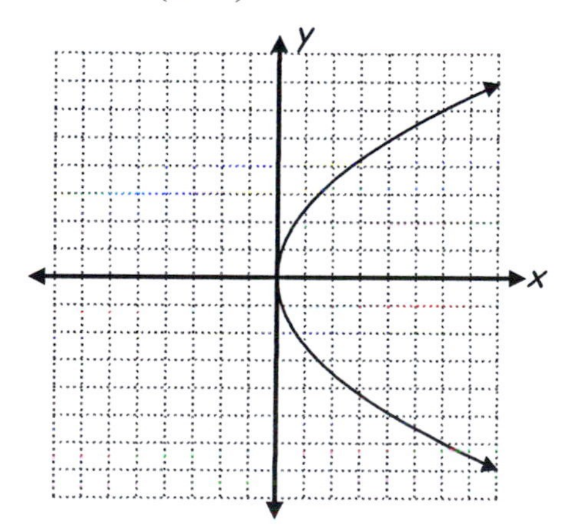

21. Given: $y^2 + 16x = 0$, or $y^2 = -16x$

Set $4p = -16$. Solve: $p = -4$

Vertex: $(0, 0)$

Graph opens to the left.

Focus: $(-4, 0)$; Directrix: line $x = 4$

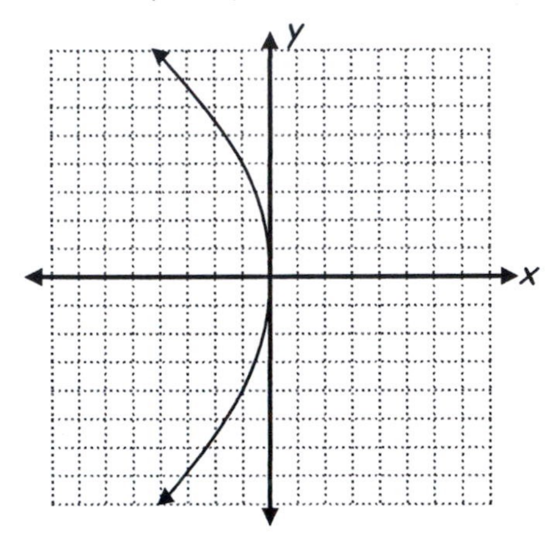

23. Given: $4y + 2x^2 = 4$, or $x^2 = -2(y-1)$

Set $4p = -2$. Solve: $p = -\frac{1}{2}$;

Vertex: $(0, 1)$

The graph opens downward.

Find x-intercepts (i.e., where $y = 0$):

$x^2 = 2 \Rightarrow x = \pm\sqrt{2}$

Focus: $\left(0, \frac{1}{2}\right)$; Directrix: line $y = \frac{3}{2}$

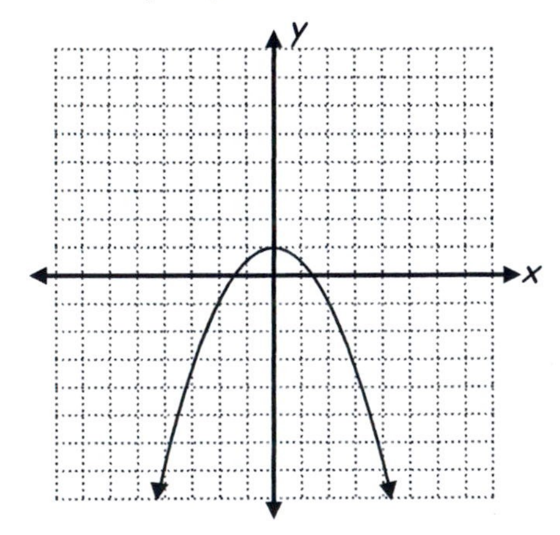

25. Given: Focus at $(-2, 1)$ and the Directrix line $x = 0$.

Vertex is $(-1, 1)$.

Graph opens to the left.

$p = -1 \Rightarrow 4p = -4$

$(y-1)^2 = -4(x+1)$

27. Given: Vertex at $(3, -1)$ and Focus at $(3, 1)$.

$-1 + p = 1 \Rightarrow p = 2 \Rightarrow 4p = 8$

Graph opens upward.

$(x-3)^2 = 8(y+1)$

29. Given: Vertex at $(3, -2)$ and Directrix line $x = -3$; $p = 6$, parabola opens to the right. $(y+2)^2 = 24(x-3)$

31. Given: Focus at $\left(-3, -\frac{3}{2}\right)$ and Directrix line $y = -\frac{1}{2}$.

Vertex: $(-3, -1)$

$4p = 4\left(-\frac{1}{2}\right) = -2$

Graph opens downward.

$(x+3)^2 = -2(y+1)$

33. Given: Vertex at $(-4, 3)$ and Focus at $\left(-\frac{3}{2}, 3\right)$.

$p = \frac{5}{2} \Rightarrow 4p = 10$

the graph opens to the right.

$(y-3)^2 = 10(x+4)$

35. From the given graph:

Focus is at $(0, -1)$ and the Vertex is at $(2, -1)$.

$p = -2 \Rightarrow 4p = -8$.

$(y+1)^2 = -8(x-2)$

37. The parabola $(x+2)^2 = 3(y-1)$ opens upward with a Vertex of $(-2, 1)$. This description matches the graph of g.

39. The parabola $y^2 = 4(x+1)$ opens to the right with a Vertex of $(-1, 0)$. This description matches the graph of b.

41. The parabola $(x-1)^2 = -(y-2)$ opens downward with a Vertex of $(1, 2)$. This description matches the graph of e.

43. The parabola $(x-2)^2 = 4y$ opens upward with a Vertex of $(2, 0)$. This description matches the graph of d.

45. The parabola $x^2 = -4(y+1)$ opens downward with a Vertex of $(0, -1)$ and passes through the points $(-2, -2)$ and $(2, -2)$.

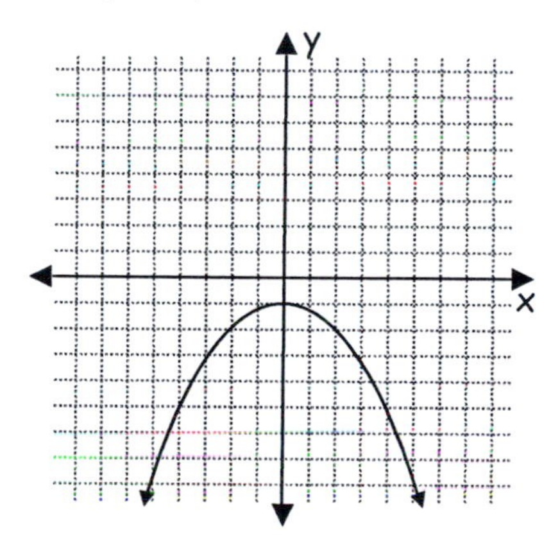

47. The parabola $(x-5)^2 = 2y$ opens upward with a Vertex of $(5, 0)$ and passes through the points $\left(4, \frac{1}{2}\right)$ and $\left(6, \frac{1}{2}\right)$.

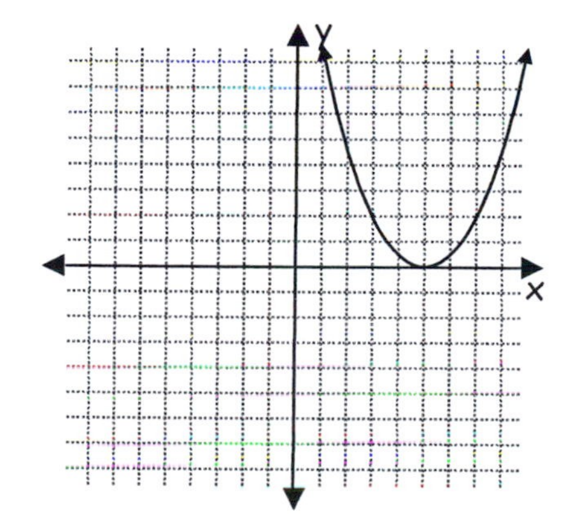

49. Complete the square in terms of y:

$(y^2 + 2y + 1) - 1 - 2x - 5 = 0$

Rewrite in standard parabolic form:

$(y+1)^2 = 2(x+3)$

The parabola $(y+1)^2 = 2(x+3)$ opens to the right with a Vertex of $(-3, -1)$ and passes through the points $\left(-\frac{5}{2}, -2\right)$ and $\left(-\frac{5}{2}, 0\right)$.

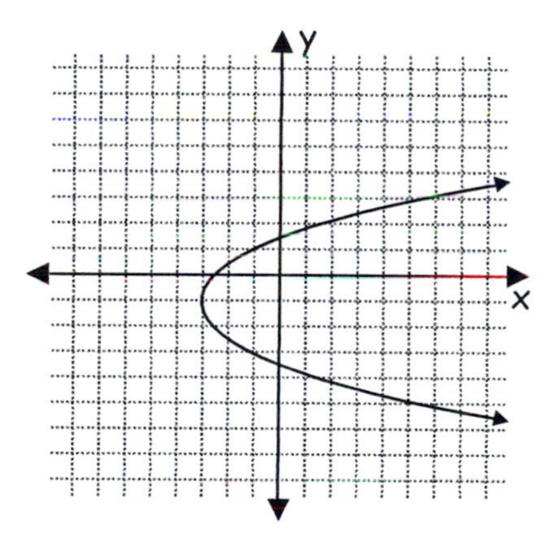

51. The parabola $(y-1)^2 = 6(x+2)$ opens to the right with a Vertex of $(-2, 1)$ and passes through the points $\left(-\frac{1}{2}, -2\right)$ and $\left(-\frac{1}{2}, 4\right)$.

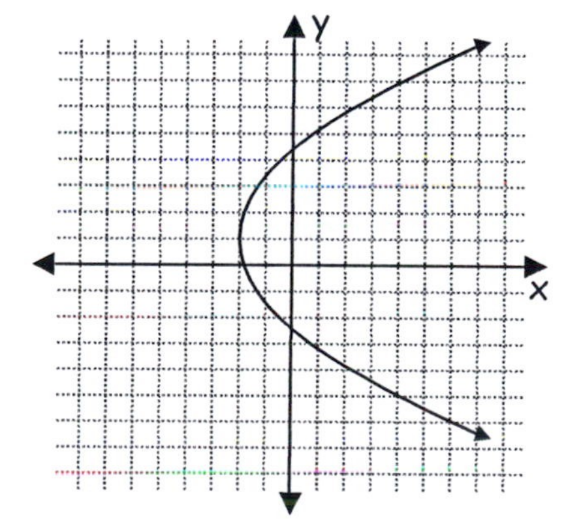

53. The parabola $(y+3)^2 = 5(x-1)$ opens to the right with a Vertex of $(1, -3)$ and passes through the points $\left(\frac{9}{4}, -\frac{1}{2}\right)$ and $\left(\frac{9}{4}, -\frac{11}{2}\right)$.

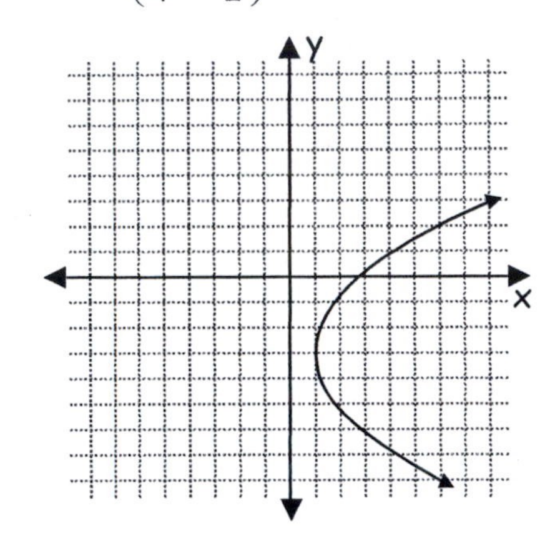

55. Placing the vertex at $(0, 0)$ would allow us to describe the "cross-section" of the parabolic dish with $x^2 = 4py$. Then, to locate the focus, find p that satisfies one of the know points on the given parabola, namely, $(-3, 1)$ or $(3, 1)$: $3^2 = 4p(1) \Rightarrow p = \frac{9}{4}$

This is the value we need. We would place the receiver at the focus of the parabolic dish $\left(0, \frac{9}{4}\right)$, or $\frac{9}{4}$ ft. from the vertex.

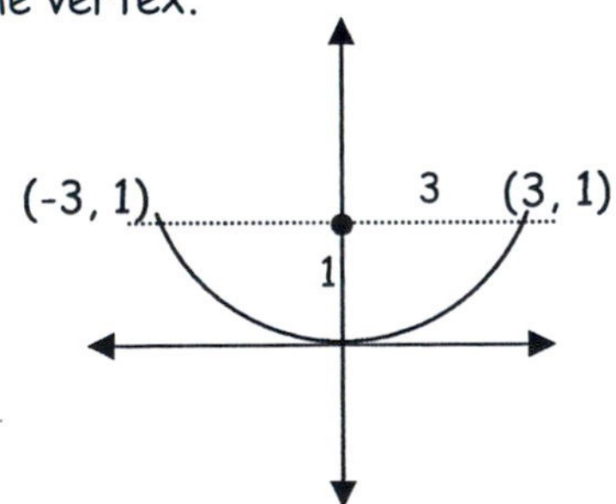

End of Section 9.2

Chapter Nine: Section 9.3
Solutions to Odd-numbered Exercises

1. Given: $\frac{(x+3)^2}{4}-\frac{(y+1)^2}{9}=1$

   The foci are aligned horizontally.

   Center: $(-3,-1)$

   $a=2, b=3, c=\sqrt{13}$

   Asymptotes: $y+1=\pm\frac{3}{2}(x+3)$.

   Foci: $\left(-3-\sqrt{13},-1\right)$ and $\left(-3+\sqrt{13},-1\right)$

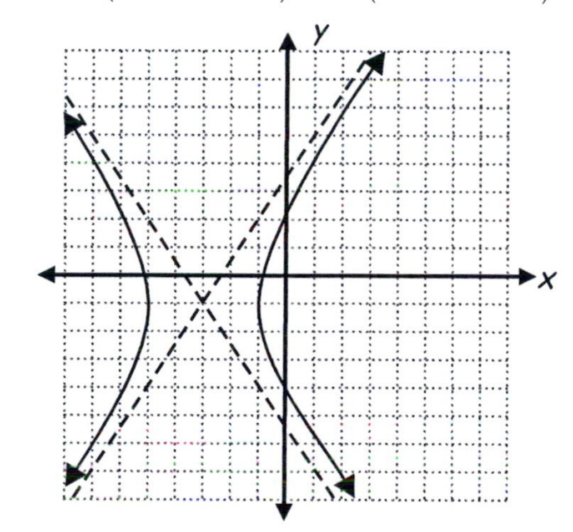

3. Given: $4y^2-x^2-24y+2x=-19$

   Complete the square:

   $$4\left(y^2-6y+9\right)-\left(x^2-2x+1\right)=-19+35$$

   $$\frac{(y-3)^2}{4}-\frac{(x-1)^2}{16}=1$$

   $a=2$ and $b=4 \Rightarrow c=2\sqrt{5}$

   Center is at $(1,3)$.

   Asymptotes: $y-3=\pm\frac{1}{2}(x-2)$.

   Foci are $\left(1,3-2\sqrt{5}\right)$ and $\left(1,3+2\sqrt{5}\right)$.

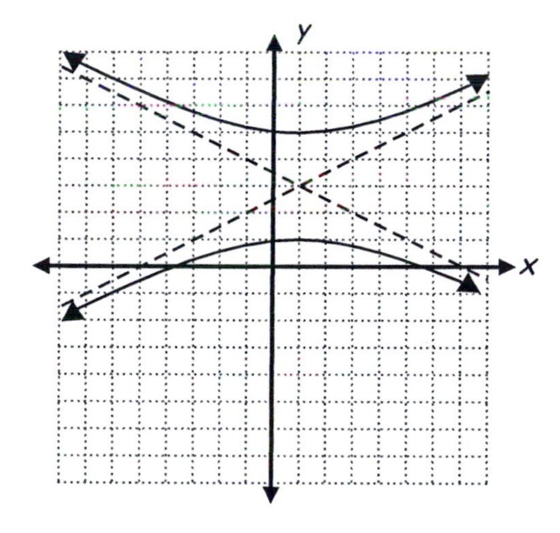

5. Given: $9x^2-25y^2=18x-50y+241$

   Complete the square:

   $$9\left(x^2-2x+1\right)-25\left(y^2-2y+1\right)=241-16$$

   $$\frac{(x-1)^2}{25}-\frac{(y-1)^2}{9}=1$$

   $a=5$ and $b=3 \Rightarrow c=\sqrt{34}$

   Center is at $(1,1)$.

   Asymptotes: $y-1=\pm\frac{3}{5}(x-1)$.

   Foci at $\left(1-\sqrt{34},1\right)$ and $\left(1+\sqrt{34},1\right)$.

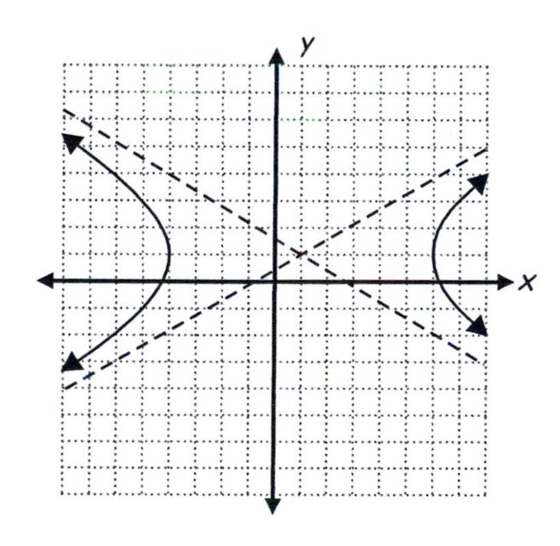

7. Given: $\dfrac{x^2}{16}-\dfrac{(y-2)^2}{4}=1$

The foci are aligned horizontally.

Center: $(0,2)$

$a=4,\ b=2,\ c=2\sqrt{5}$

Asymptotes: $y-2=\pm\dfrac{1}{2}x$

Foci: $\left(-2\sqrt{5},2\right)$ and $\left(2\sqrt{5},2\right)$

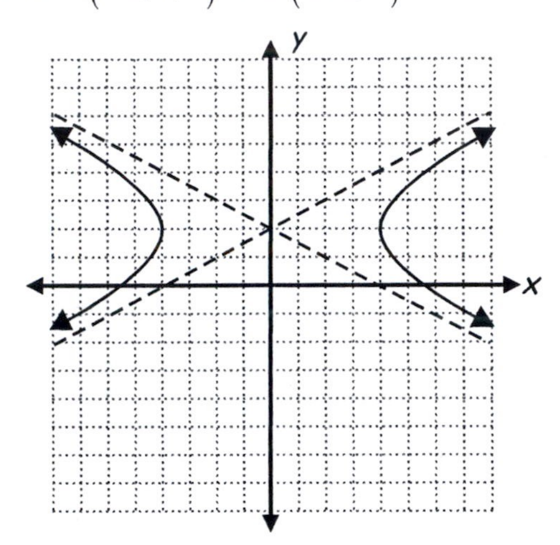

9. Given: $9y^2-25x^2-36y-100x=289$

Complete the square:

$9\left(y^2-4y+4\right)-25\left(x^2+4x+4\right)=225$

$$\frac{(y-2)^2}{25}-\frac{(x+2)^2}{9}=1$$

$a=5$ and $b=3\ \Rightarrow c=\sqrt{34}$

Center is at $(-2,2)$.

Asymptotes: $y-2=\pm\dfrac{5}{3}(x+2)$.

Foci are $\left(-2,2-\sqrt{34}\right)$ and $\left(-2,2+\sqrt{34}\right)$.

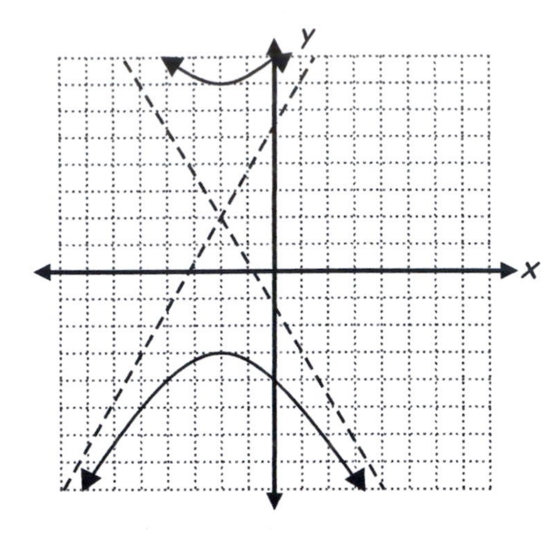

11. Given: $9x^2-16y^2-36x+32y-124=0$

Complete the square:

$9\left(x^2-4x+4\right)-16\left(y^2-2y+1\right)=144$

$$\frac{(x-2)^2}{16}-\frac{(y-1)^2}{9}=1$$

Center is at $(2,1)$.

$a=4$ and $b=3\ \Rightarrow c=5$

Asymptotes: $y-\ 1=\pm\dfrac{3}{4}(x-2)$.

Foci at $(-3,1)$ and $(7,1)$.

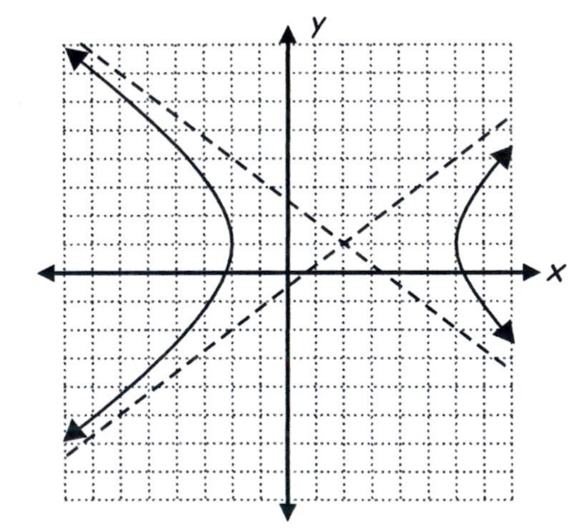

13. Given: $\dfrac{(x+3)^2}{4}-\dfrac{(y-2)^2}{9}=1$

The foci are aligned horizontally.

$a^2=4\Rightarrow a=2$

$b^2=9\Rightarrow b=3$

$c^2=13\Rightarrow c=\sqrt{13}$

Center: $(-3,2)$

Foci: $\left(-3-\sqrt{13},2\right)$ and $\left(-3+\sqrt{13},2\right)$

Vertices: $(-3+2,2)=(-1,2)$ and $(-3-2,2)=(-5,2)$

15. Given: $3(x-1)^2-(y+4)^2=9$

Rewrite in standard form:

$\dfrac{(x-1)^2}{3}-\dfrac{(y+4)^2}{9}=1$

The foci are aligned horizontally.

$a^2=3\Rightarrow a=\sqrt{3}$

$b^2=9\Rightarrow b=3$

$c^2=12\Rightarrow c=2\sqrt{3}$

Center: $(1,-4)$

Foci: $\left(1-2\sqrt{3},-4\right)$ and $\left(1+2\sqrt{3},-4\right)$

Vertices: $\left(1+\sqrt{3},-4\right)$ and $\left(1-\sqrt{3},-4\right)$

17. Given: $(x+2)^2-5(y-1)^2=25$

Rewrite in standard form:

$\dfrac{(x+2)^2}{25}-\dfrac{(y-1)^2}{5}=1$

The foci are aligned horizontally.

$a^2=25\Rightarrow a=5$

$b^2=5\Rightarrow b=\sqrt{5}$

$c^2=30\Rightarrow c=\sqrt{30}$

Center: $(-2,1)$

Foci: $\left(-2-\sqrt{30},1\right)$ and $\left(-2+\sqrt{30},1\right)$

Vertices: $(-2-5,1)=(-7,1)$ and $(-2+5,1)=(3,1)$

19. Given: $2x^2+12x-y^2-2y+9=0$

Complete the squares in x and y:

$2\left(x^2+6x+9\right)-\left(y^2+2y+1\right)=-9+18-1.$

$2(x+3)^2-(y+1)^2=8$

Rewrite in standard form:

$\dfrac{(x+3)^2}{4}-\dfrac{(y+1)^2}{8}=1$

The foci are aligned horizontally.

$a^2=4\Rightarrow a=2$

$b^2=8\Rightarrow b=2\sqrt{2}$

$c^2=12\Rightarrow c=2\sqrt{3}$

Center: $(-3,-1)$

Foci: $\left(-3-2\sqrt{3},-1\right)$ and $\left(-3+2\sqrt{3},-1\right)$

Vertices: $(-3+2,-1)=(-1,-1)$ and $(-3-2,-1)=(-5,-1)$

21. Given: $x^2-4y^2-2x=0$

Complete the squares in x and y and rewrite in standard form:

$\left(x^2-2x+1\right)-4\left(y^2\right)=1$

$(x-1)^2-4y^2=1$

$\dfrac{(x-1)^2}{1}-\dfrac{y^2}{\frac{1}{4}}=1$

The foci are aligned horizontally.

$a^2=1\Rightarrow a=1$

$b^2=\dfrac{1}{4}\Rightarrow b=\dfrac{1}{2}$

$c^2=\dfrac{5}{4}\Rightarrow c=\dfrac{\sqrt{5}}{2}$

Center: $(1,0)$

Foci: $\left(1-\dfrac{\sqrt{5}}{2},0\right)$ and $\left(1+\dfrac{\sqrt{5}}{2},0\right)$

Vertices: $(1+1,0)=(2,0)$ and $(1-1,0)=(0,0)$

23. Given: $4x^2 - y^2 - 64x + 10y + 167 = 0$
Complete the squares in x and y and rewrite in standard form:

$$4(x^2 - 16x + 64) - (y^2 - 10y + 25) = -167 + 256 - 25$$

$$4(x-8)^2 - (y-5)^2 = 64$$

$$\frac{(x-8)^2}{16} - \frac{(y-5)^2}{64} = 1$$

The foci are aligned horizontally.

$$a^2 = 16 \Rightarrow a = 4$$

$$b^2 = 64 \Rightarrow b = 8$$

$$c^2 = 80 \Rightarrow c = 4\sqrt{5}$$

Center: $(8, 5)$

Foci: $(8 - 4\sqrt{5}, 5)$ and $(8 + 4\sqrt{5}, 5)$

Vertices: $(8-4, 5) = (4, 5)$ and $(8+4, 5) = (12, 5)$

25. Given: Foci at $(-3, 0)$ and $(3, 0)$ with vertices at $(-2, 0)$ and $(2, 0)$. The foci are horizontally aligned (on the x-axis) with center at $(0, 0)$.
$a = 2$, $c = 3$, and $b^2 = c^2 - a^2 = 5$.
So, the equation of the hyperbola is

$$\frac{x^2}{4} - \frac{y^2}{5} = 1$$

27. Given: Asymptotes of $y = \pm 2x$ and vertices at $(0, -1)$ and $(0, 1)$.
The vertices are aligned vertically with center at $(0, 0)$.

$$\frac{a}{b} = \pm 2 \Rightarrow b = \frac{a}{2} \Rightarrow b^2 = \frac{a^2}{4}.$$

Then, with $a = 1$, $b^2 = \frac{1}{4}$.
So, the equation of the hyperbola is

$$\frac{y^2}{1} - \frac{x^2}{\frac{1}{4}} = 1$$

29. Given: Foci at $(2, 4)$ and $(-2, 4)$, and asymptotes of $y = \pm 3x + 4$ (i.e., $y - 4 = \pm 3(x - 0)$):
The foci are aligned horizontally.

$$\frac{b}{a} = 3 \Rightarrow b^2 = 9a^2.$$

Center is at $(0, 4)$ and $c = 2$.
Using $a^2 = c^2 - b^2$,

$$4 = a^2 + b^2$$

$$4 = a^2 + 9a^2$$

$$4 = 10a^2 \Rightarrow a^2 = \frac{2}{5} \Rightarrow b^2 = \frac{18}{5}$$

So, the equation of the hyperbola is

$$\frac{x^2}{\frac{2}{5}} - \frac{(y-4)^2}{\frac{18}{5}} = 1$$

31. Given: Foci at $(2, 5)$ and $(10, 5)$, and vertices at $(3, 5)$ and $(9, 5)$.
The foci are aligned horizontally with center at $(6, 5)$.
$c = 4$ and $a = 3 \Rightarrow b^2 = 16 - 9 = 7$
So, the equation of the hyperbola is

$$\frac{(x-6)^2}{9} - \frac{(y-5)^2}{7} = 1$$

33. Given: Asymptotes of $y = \pm(2x + 8) + 3$ and vertices at $(-6, 3)$ and $(-2, 3)$.
The asymptotes are $y - 3 = \pm 2(x + 4)$.
The center is at $(-4, 3)$ and the vertices are aligned horizontally.

$$\frac{b}{a} = 2 \Rightarrow b = 2a \text{ and then } a = 2 \Rightarrow b = 4.$$

So, the equation of the hyperbola is

$$\frac{(x+4)^2}{4} - \frac{(y-3)^2}{16} = 1$$

35. From the given graph, the foci are aligned horizontally. The vertices are at $(-2, -1)$ and $(4, -1)$ and center at $(1, -1)$.

$a = 3$ and $b = 2$.

$$\frac{(x-1)^2}{9} - \frac{(y+1)^2}{4} = 1$$

37. From the given graph, the foci are aligned vertically. The vertices are at $(3, 0)$ and $(3, -8)$ and center at $(3, -4)$.

$a = 4$ and $b = 5$.

$$\frac{(y+4)^2}{16} - \frac{(x-3)^2}{25} = 1$$

39. Given: $\dfrac{x^2}{9} - \dfrac{y^2}{1} = 1$

The foci are aligned horizontally.

Center $= (0, 0)$.

$a^2 = 9 \Rightarrow a = 3$

$b^2 = 1 \Rightarrow b = 1$

$c^2 = 10 \Rightarrow c = \sqrt{10}$

The description matches graph a.

41. Given: $x^2 - \dfrac{(y-3)^2}{4} = 1$

The foci are aligned horizontally.

Center $= (0, 3)$, $a = 1$, $b = 2$

The description matches graph b.

43. Given: $(y+2)^2 - \dfrac{(x-2)^2}{4} = 1$

The foci are aligned vertically.

Center $= (2, -2)$, $a = 1$, $b = 2$

The description matches graph g.

45. Given: $\dfrac{y^2}{4} - (x-1)^2 = 1$

The foci are aligned vertically.

Center $= (1, 0)$, $a = 2$, $b = 1$

The description matches graph h.

47. Use a graphing calculator. First, rewrite the equation in standard form.

$$\frac{x^2}{15} - \frac{y^2}{\frac{15}{6}} = 1 \text{ or } \frac{x^2}{15} - \frac{y^2}{\frac{5}{2}} = 1$$

With the Apps-Conic-Hyperbola functions on the TI-83 Plus, use the following values, which are derived from the equation:

$a = \sqrt{15}$, $b = \sqrt{\dfrac{5}{2}}$, $h = 0$ and $k = 0$

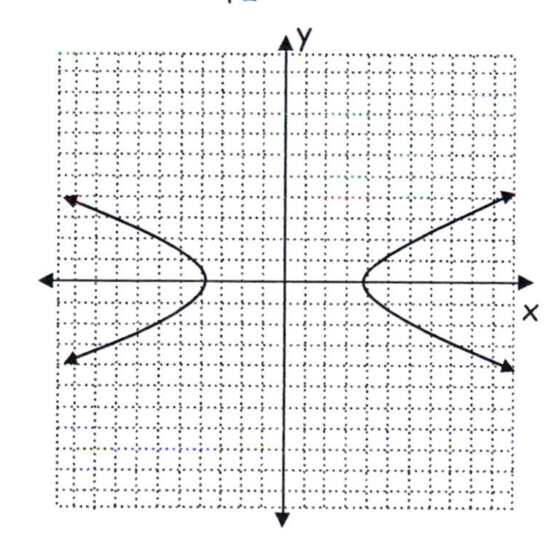

49. Use a graphing calculator. First, rewrite the equation in standard form.

$$\frac{y^2}{12}-\frac{x^2}{2}=1$$

With the Apps-Conic-Hyperbola functions on the TI-83 Plus, use the following values, which are derived from the equation:

$a=2\sqrt{3}$, $b=\sqrt{2}$, $h=0$ and $k=0$

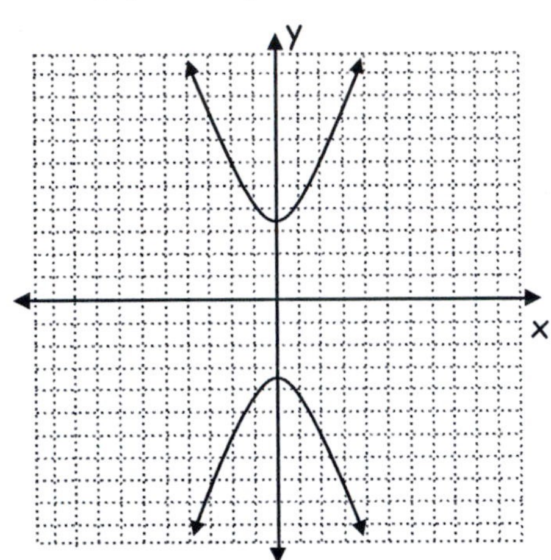

51. Use a graphing calculator. First, rewrite the equation in standard form.

$$\frac{(y+2)^2}{20}-\frac{x^2}{4}=1$$

With the Apps-Conic-Hyperbola functions on the TI-83 Plus, use the following values, which are derived from the equation:

$a=2\sqrt{5}$, $b=2$, $h=0$ and $k=-2$

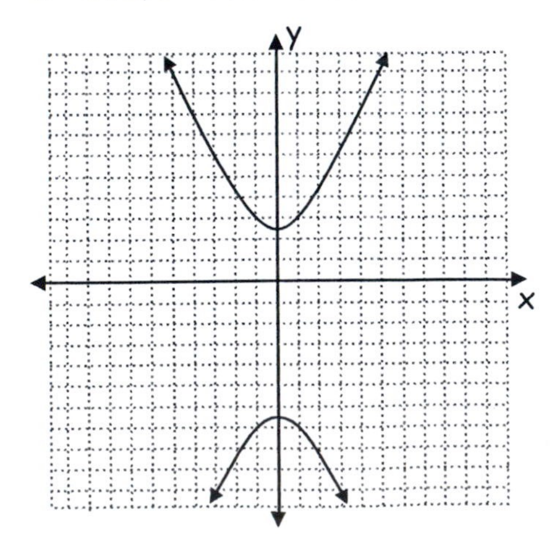

53. Given the graph and description of the comet, $c=94{,}000{,}000$ and $a=60{,}000{,}000$. Using $c^2=a^2+b^2$, we have that $b^2=c^2-a^2$. Substituting, we have that $b^2\approx 5.236\times 10^{15}$. Note that this is approximately the same as $\left(7.2\times 10^7\right)^2$.

The equation of the path of the comet can be written as

$$\frac{x^2}{3.6\times 10^{15}}-\frac{y^2}{5.236\times 10^{15}}=1, \text{ or}$$

equivalently, in standard form as

$$\frac{x^2}{\left(6\times 10^7\right)^2}-\frac{y^2}{\left(7.2\times 10^7\right)^2}=1$$

End of Section 9.3

Chapter Nine: Section 9.4
Solutions to Odd-numbered Exercises

1. Given: $(8, 6)$, $\theta = 30°$

$$x' = 8\cos 30° + 6\sin 30°$$
$$= 8\left(\frac{\sqrt{3}}{2}\right) + 6\left(\frac{1}{2}\right)$$
$$= 4\sqrt{3} + 3$$
$$y' = -8\sin 30° + 6\cos 30°$$
$$= -8\left(\frac{1}{2}\right) + 6\left(\frac{\sqrt{3}}{2}\right)$$
$$= -4 + 3\sqrt{3}$$
$$(x', y') = \left(4\sqrt{3} + 3, -4 + 3\sqrt{3}\right)$$

3. Given: $\left(-\frac{1}{2}, -\frac{1}{8}\right)$, $\theta = \frac{\pi}{4}$

$$x' = -\frac{1}{2}\cos\frac{\pi}{4} + \left(-\frac{1}{8}\right)\sin\frac{\pi}{4}$$
$$= -\frac{1}{2}\left(\frac{\sqrt{2}}{2}\right) - \frac{1}{8}\left(\frac{\sqrt{2}}{2}\right)$$
$$= -\frac{\sqrt{2}}{4} - \frac{\sqrt{2}}{16}$$
$$= -\frac{5\sqrt{2}}{16}$$
$$y' = -\left(-\frac{1}{2}\right)\sin\frac{\pi}{4} + \left(-\frac{1}{8}\right)\cos\frac{\pi}{4}$$
$$= \frac{1}{2}\left(\frac{\sqrt{2}}{2}\right) - \frac{1}{8}\left(\frac{\sqrt{2}}{2}\right)$$
$$= \frac{\sqrt{2}}{4} - \frac{\sqrt{2}}{16}$$
$$= \frac{3\sqrt{2}}{16}$$
$$(x', y') = \left(-\frac{5\sqrt{2}}{16}, \frac{3\sqrt{2}}{16}\right)$$

5. Given: $(13, -4)$, $\theta = 78°$

$$x' = 13\cos 78° + (-4)\sin 78°$$
$$\approx 13(0.2079) + (-4)(0.9781)$$
$$\approx -1.2097$$
$$y' = -13\sin 78° + (-4)\cos 78°$$
$$\approx -13(0.9781) + (-4)(0.2079)$$
$$\approx -13.5476$$
$$(x', y') = (-1.2097, -13.5476)$$

7. Given: $\left(3.65, \frac{3}{8}\right)$, $\theta = \frac{\pi}{6}$

$$x' = 3.65\cos\frac{\pi}{6} + \frac{3}{8}\sin\frac{\pi}{6}$$
$$= 3.65\left(\frac{\sqrt{3}}{2}\right) + \frac{3}{8}\left(\frac{1}{2}\right)$$
$$\approx 3.1610 + \frac{3}{16}$$
$$\approx 3.3485$$
$$y' = -3.65\sin\frac{\pi}{6} + \frac{3}{8}\cos\frac{\pi}{6}$$
$$= -3.65\left(\frac{1}{2}\right) + \frac{3}{8}\left(\frac{\sqrt{3}}{2}\right)$$
$$\approx -1.5002$$
$$(x', y') \approx (3.3485, -1.5002)$$

9. Given: $2x^2 - 3xy + 2y^2 - 2x = 0$

$A = 2$, $B = -3$, $C = 2$

$$B^2 - 4AC = (-3)^2 - 4(2)(2)$$
$$= 9 - 16$$
$$= -7$$

$B^2 - 4AC < 0 \Rightarrow$

the equation represents an ellipse.

11. Given: $3x^2 + 8xy + 4y^2 - 7 = 0$

$A = 3$, $B = 8$, $C = 4$

$$B^2 - 4AC = 8^2 - 4(3)(4)$$
$$= 64 - 48$$
$$= 16$$

$B^2 - 4AC > 0 \Rightarrow$

the equation represents a hyperbola.

13. Given: $-2x^2-8xy+2y^2+2y+5=0$

$A=-2,\ B=-8,\ C=2$

$$\begin{aligned} B^2-4AC &= (-8)^2-4(-2)(2)\\ &= 64+16\\ &= 80 \end{aligned}$$

$B^2-4AC>0 \Rightarrow$

the equation represents a hyperbola.

15. Given: $x^2-xy+4y^2+2x-3y+1=0$

$A=1,\ B=-1,\ C=4$

$$\begin{aligned} B^2-4AC &= (-1)^2-4(1)(4)\\ &= 1-16\\ &= -15 \end{aligned}$$

$B^2-4AC<0 \Rightarrow$

the equation represents an ellipse.

17. Given: $xy=2$ $\left(\text{or } \frac{1}{2}xy=1\right)$

$A=0,\ B=\frac{1}{2},\ C=0$

$$\begin{aligned} B^2-4AC &= \left(\frac{1}{2}\right)^2-4(0)(0)\\ &= \frac{1}{4} \end{aligned}$$

$B^2-4AC>0 \Rightarrow$

the equation represents a hyperbola.

Find the angle of rotation, θ:

$$\cot 2\theta = \frac{0-0}{\frac{1}{2}} = 0 \Rightarrow 2\theta = \frac{\pi}{2} \Rightarrow \theta = \frac{\pi}{4}$$

$$\begin{aligned} x &= x'\cos\frac{\pi}{4}-y'\sin\frac{\pi}{4}\\ &= x'\left(\frac{\sqrt{2}}{2}\right)-y'\left(\frac{\sqrt{2}}{2}\right) \end{aligned}$$

$$\begin{aligned} y &= x'\sin\frac{\pi}{4}+y'\cos\frac{\pi}{4}\\ &= x'\left(\frac{\sqrt{2}}{2}\right)+y'\left(\frac{\sqrt{2}}{2}\right) \end{aligned}$$

$xy=$

$$\left(\left(\frac{\sqrt{2}}{2}\right)x'-\left(\frac{\sqrt{2}}{2}\right)y'\right)\left(\left(\frac{\sqrt{2}}{2}\right)x'+\left(\frac{\sqrt{2}}{2}\right)y'\right)=2$$

Simplifying:

$$\left(\frac{\sqrt{2}}{2}\right)^2(x'-y')(x'+y')=2$$

$$\frac{1}{2}\left((x')^2-(y')^2\right)=2$$

$$\frac{(x')^2}{4}-\frac{(y')^2}{4}=1$$

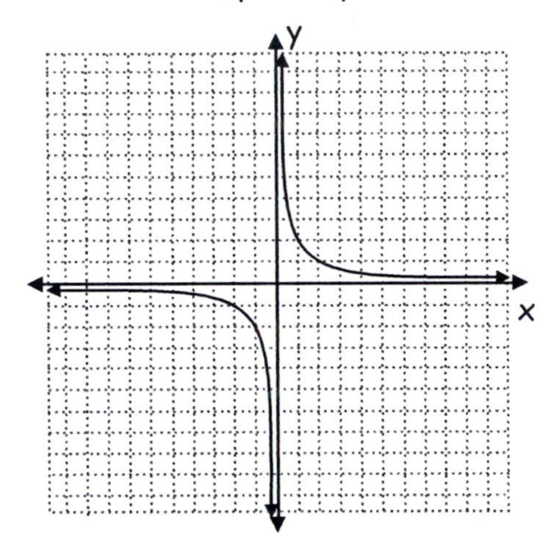

19. Given: $x^2+2xy+y^2-x+y=0$

$A=1,\ B=2,\ C=1$

$$\begin{aligned} B^2-4AC &= 2^2-4(1)(1)\\ &= 4-4=0 \end{aligned}$$

$B^2-4AC=0 \Rightarrow$

the equation represents a parabola.

Find the angle of rotation, θ:

$$\cot 2\theta = \frac{A-C}{B} = \frac{1-1}{2} = 0 \Rightarrow$$

$$2\theta = \frac{\pi}{2} \Rightarrow \theta = \frac{\pi}{4}$$

$$\begin{aligned} x &= x'\cos\frac{\pi}{4}-y'\sin\frac{\pi}{4}\\ &= x'\left(\frac{\sqrt{2}}{2}\right)-y'\left(\frac{\sqrt{2}}{2}\right) \end{aligned}$$

$$\begin{aligned} y &= x'\sin\frac{\pi}{4}+y'\cos\frac{\pi}{4}\\ &= x'\left(\frac{\sqrt{2}}{2}\right)+y'\left(\frac{\sqrt{2}}{2}\right) \end{aligned}$$

Substituting in the given equation:

$$\left[x'\left(\frac{\sqrt{2}}{2}\right)-y'\left(\frac{\sqrt{2}}{2}\right)\right]^2+$$
$$2\left[x'\left(\frac{\sqrt{2}}{2}\right)-y'\left(\frac{\sqrt{2}}{2}\right)\right]\left[x'\left(\frac{\sqrt{2}}{2}\right)+y'\left(\frac{\sqrt{2}}{2}\right)\right]+$$
$$\left[x'\left(\frac{\sqrt{2}}{2}\right)+y'\left(\frac{\sqrt{2}}{2}\right)\right]^2-\left[x'\left(\frac{\sqrt{2}}{2}\right)-y'\left(\frac{\sqrt{2}}{2}\right)\right]+$$
$$\left[x'\left(\frac{\sqrt{2}}{2}\right)+y'\left(\frac{\sqrt{2}}{2}\right)\right]=0$$

The equation simplifies to

$$(x')^2=-\frac{\sqrt{2}}{2}y',\ \text{or}\ y'=-\sqrt{2}x'^2$$

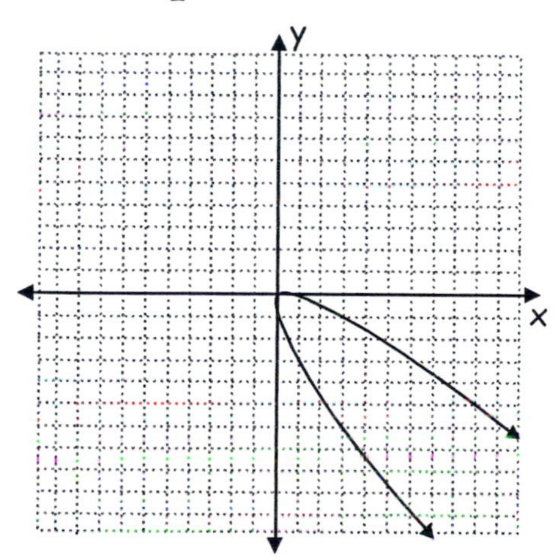

21. Given:

$$22x^2+6\sqrt{3}xy+16y^2-49=276$$
$$A=22,\ B=6\sqrt{3},\ C=16$$
$$B^2-4AC=\left(6\sqrt{3}\right)^2-4(22)(16)$$
$$=-1300<0$$
$B^2-4AC<0\Rightarrow$

the equation represents an ellipse.

Find the angle of rotation, θ:

$$\cot 2\theta=\frac{22-16}{6\sqrt{3}}=\frac{1}{\sqrt{3}}\Rightarrow$$
$$2\theta=\frac{\pi}{3}\Rightarrow\theta=\frac{\pi}{6}$$
$$x=x'\cos\frac{\pi}{6}-y'\sin\frac{\pi}{6}$$
$$=x'\left(\frac{\sqrt{3}}{2}\right)-y'\left(\frac{1}{2}\right)$$
$$y=x'\sin\frac{\pi}{6}+y'\cos\frac{\pi}{6}$$
$$=x'\left(\frac{1}{2}\right)+y'\left(\frac{\sqrt{3}}{2}\right)$$

Substituting in the given equation:

$$22\left[x'\left(\frac{\sqrt{3}}{2}\right)-y'\left(\frac{1}{2}\right)\right]^2+$$
$$6\sqrt{3}\left[x'\left(\frac{\sqrt{3}}{2}\right)-y'\left(\frac{1}{2}\right)\right]\left[x'\left(\frac{1}{2}\right)+y'\left(\frac{\sqrt{3}}{2}\right)\right]+$$
$$16\left[x'\left(\frac{1}{2}\right)+y'\left(\frac{\sqrt{3}}{2}\right)\right]^2=325$$

The equation simplifies to

$$25(x')^2+\frac{25}{2}(y')^2=325,\ \text{or}$$
$$\frac{(x')^2}{13}+\frac{(y')^2}{25}=1$$

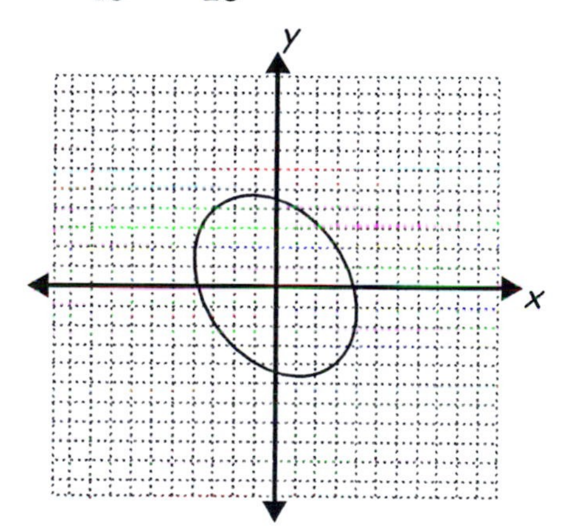

23. Given:

$$34x^2+8\sqrt{3}xy+42y^2=1380$$
$$A=34,\ B=8\sqrt{3},\ C=42$$
$$B^2-4AC=\left(8\sqrt{3}\right)^2-4(34)(42)$$
$$=-5520<0$$
$B^2-4AC<0\Rightarrow$

the equation represents an ellipse.

Find the angle of rotation, θ:

$$\cot 2\theta=\frac{A-C}{B}=\frac{34-42}{8\sqrt{3}}=-\frac{1}{\sqrt{3}}=0\Rightarrow$$
$$2\theta=\frac{2\pi}{3}\ (\text{note: } 0<2\theta<\pi)\ \Rightarrow\theta=\frac{\pi}{3}$$

$$x = x'\cos\frac{\pi}{3} - y'\sin\frac{\pi}{3}$$
$$= x'\left(\frac{1}{2}\right) - y'\left(\frac{\sqrt{3}}{2}\right)$$
$$y = x'\sin\frac{\pi}{3} + y'\cos\frac{\pi}{3}$$
$$= x'\left(\frac{\sqrt{3}}{2}\right) + y'\left(\frac{1}{2}\right)$$

Substituting in the given equation:

$$34\left[x'\left(\frac{1}{2}\right) - y'\left(\frac{\sqrt{3}}{2}\right)\right]^2 +$$
$$8\sqrt{3}\left[x'\left(\frac{1}{2}\right) - y'\left(\frac{\sqrt{3}}{2}\right)\right]\left[x'\left(\frac{\sqrt{3}}{2}\right) + y'\left(\frac{1}{2}\right)\right] +$$
$$42\left[x'\left(\frac{\sqrt{3}}{2}\right) + y'\left(\frac{1}{2}\right)\right]^2 = 1380$$

The equation simplifies to

$$\frac{17}{2}\left(x'^2 - 2\sqrt{3}x'y' + 3y'^2\right) +$$
$$2\sqrt{3}\left(\sqrt{3}x'^2 - 2x'y' - \sqrt{3}y'^2\right) +$$
$$\frac{21}{2}\left(3(x')^2 + 2\sqrt{3}x'y' + (y')^2\right) = 1380$$

This equation simplifies to

$$46(x')^2 + 30(y')^2 = 1380, \text{ or}$$
$$\frac{(x')^2}{30} + \frac{(y')^2}{46} = 1$$

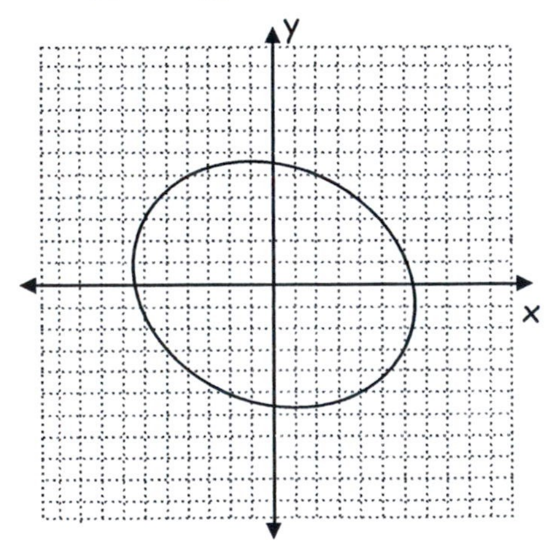

25. Use *Mathematica* or another program to graph $x^2 + 6xy + y^2 = 18$

$A = 1, B = 6, C = 1$

$B^2 - 4AC = 36 - 4 = 32 > 0$

Therefore the equation represents a hyperbola.

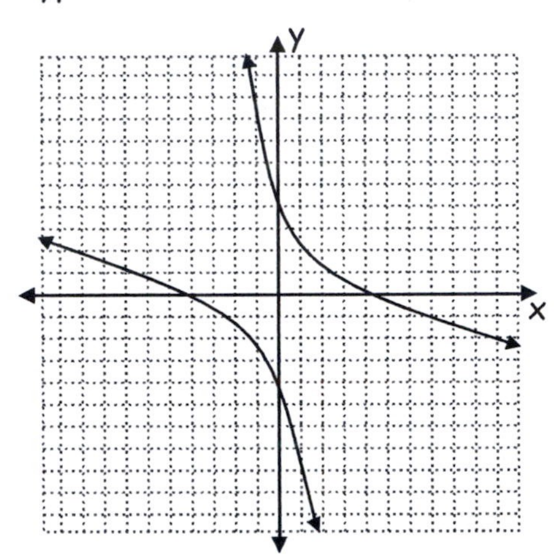

27. Use *Mathematica* or another program to graph $9x^2 + 14xy - 9y^2 = 15$

$A = 9, B = 14, C = -9$

$B^2 - 4AC = 196 - 4(9)(-9) = 520 > 0$

Therefore the equation represents a hyperbola.

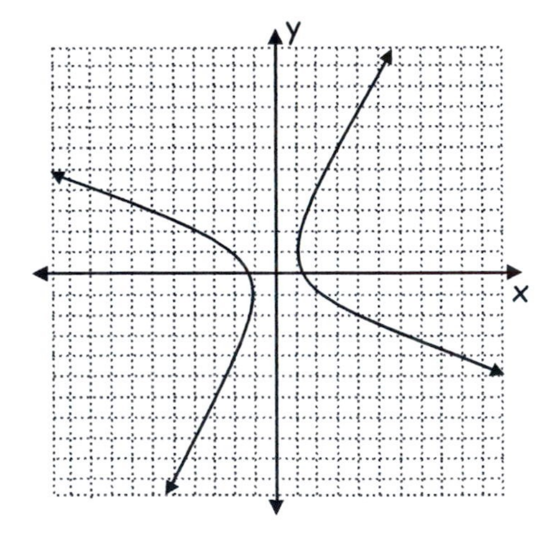

29. Use *Mathematica* or another program to graph

$40x^2+20xy+10y^2+\left(2\sqrt{2}-6\right)x-\left(4\sqrt{2}+8\right)y=90$

$A=40,\ B=20,\ C=10$

$B^2-4AC=20^2-4(40)(10)=-1200<0$

Therefore the equation represents an ellipse.

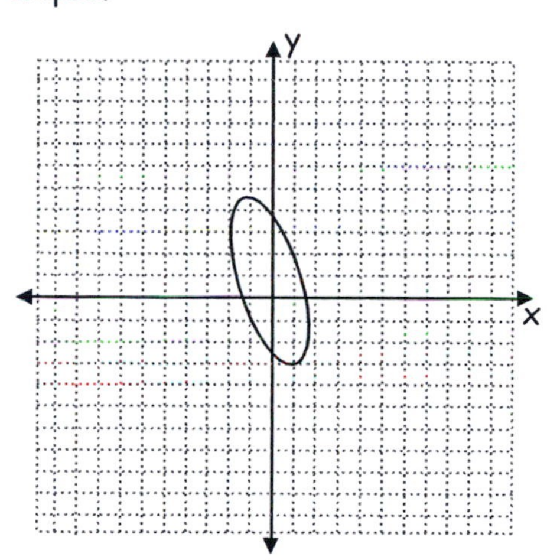

31. Use *Mathematica* or another program to graph $48x^2+15xy+7y^2=28$

$A=48,\ B=15,\ C=7$

$B^2-4AC=225-4(48)(7)=-1119<0$

Therefore the equation represents an ellipse.

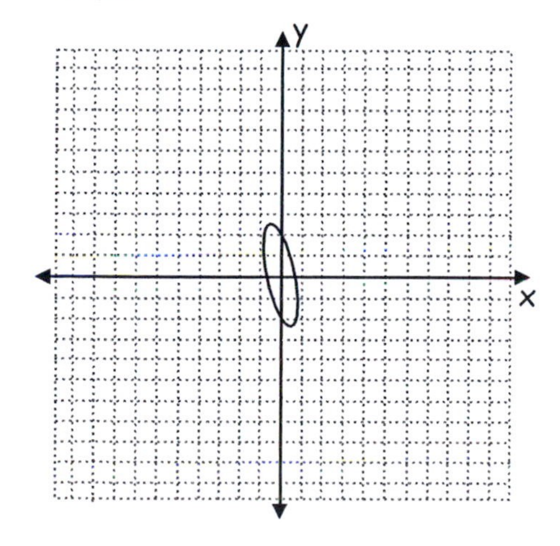

33. Given: $3x^2+2xy+y^2-10=0$

$A=3,\ B=2,\ C=1$

$B^2-4AC=4-4(3)(1)=-8<0$

Therefore the equation represents an ellipse.

The angle of rotation, θ:

$\cot 2\theta=\dfrac{A-C}{B}=\dfrac{3-1}{2}=1$

Then, $2\theta=\dfrac{\pi}{4}\Rightarrow\theta=\dfrac{\pi}{8}$

This description (ellipse) matches graph c.

35. Given: $xy-1=0$

$A=0,\ B=1,\ C=0$

$B^2-4AC=1-4(0)(0)=1>0$

Therefore the equation represents a hyperbola.

The angle of rotation, θ:

$\cot 2\theta=\dfrac{A-C}{B}=\dfrac{0-0}{1}=0$

Then, $2\theta=\dfrac{\pi}{2}\Rightarrow\theta=\dfrac{\pi}{4}$

$$x=x'\cos\frac{\pi}{4}-y'\sin\frac{\pi}{4}$$

$$=x'\left(\frac{\sqrt{2}}{2}\right)-y'\left(\frac{\sqrt{2}}{2}\right)$$

$$y=x'\sin\frac{\pi}{4}+y'\cos\frac{\pi}{4}$$

$$=x'\left(\frac{\sqrt{2}}{2}\right)+y'\left(\frac{\sqrt{2}}{2}\right)$$

$xy=$

$$\left(\left(\frac{\sqrt{2}}{2}\right)x'-\left(\frac{\sqrt{2}}{2}\right)y'\right)\left(\left(\frac{\sqrt{2}}{2}\right)x'+\left(\frac{\sqrt{2}}{2}\right)y'\right)=1$$

Simplifying:

$$\left(\frac{\sqrt{2}}{2}\right)^2(x'-y')(x'+y')=1$$

$$\frac{1}{2}\left((x')^2-(y')^2\right)=1$$

$$\frac{(x')^2}{2}-\frac{(y')^2}{2}=1$$

$a=\sqrt{2}$ and $b=\sqrt{2}$, center $(0,0)$.

The description matches graph a.

37. Given: $x^2 - y^2 - 16 = 0$, or

$$\frac{x^2}{16} - \frac{y^2}{16} = 1$$

This equation represents a hyperbola. There is no rotation necessary (i.e., $B = 0$). $a = b = 4$. The foci are aligned horizontally. The description matches graph g.

39. Given: $4xy - 9 = 0$

$A = 0, B = 4, C = 0$

$B^2 - 4AC = 4 - 4(0)(0) = 4 > 0$

Therefore the equation represents a hyperbola.

The angle of rotation, θ:

$$\cot 2\theta = \frac{A-C}{B} = \frac{0-0}{1} = 0$$

Then, $2\theta = \frac{\pi}{2} \Rightarrow \theta = \frac{\pi}{4}$

$$x = x'\cos\frac{\pi}{4} - y'\sin\frac{\pi}{4} = x'\left(\frac{\sqrt{2}}{2}\right) - y'\left(\frac{\sqrt{2}}{2}\right)$$

$$y = x'\sin\frac{\pi}{4} + y'\cos\frac{\pi}{4} = x'\left(\frac{\sqrt{2}}{2}\right) + y'\left(\frac{\sqrt{2}}{2}\right)$$

$4xy =$

$$4\left(\left(\frac{\sqrt{2}}{2}\right)x' - \left(\frac{\sqrt{2}}{2}\right)y'\right)\left(\left(\frac{\sqrt{2}}{2}\right)x' + \left(\frac{\sqrt{2}}{2}\right)y'\right) = 9$$

Simplifying:

$4xy =$

$$4\left[\left(\frac{\sqrt{2}}{2}\right)^2 (x'-y')(x'+y')\right] = 9$$

$$2\left((x')^2 - (y')^2\right) = 9$$

$$\frac{(x')^2}{9/2} - \frac{(y')^2}{9/2} = 1$$

$a = \frac{3}{2}\sqrt{2}$, and $b = \frac{3}{2}\sqrt{2}$, center $(0, 0)$

The description matches graph e.

41. The angle of rotation must have the effect of rendering the coefficient of the $x'y'$ term to be 0. If the process fails to do this, then there has been an error in the substitution or simplification.

43. Given: $7x^2 - 6\sqrt{3}xy + 13y^2 - 16 = 0$

a. Rotate the axes to be parallel to the axis of the conic and show $F = F'$ (i.e., $F' = -16$).

$$\cot 2\theta = \frac{A-C}{B} = \frac{7-13}{-6\sqrt{3}} = \frac{1}{\sqrt{3}} \Rightarrow 2\theta = \frac{\pi}{3}.$$

Then, $\theta = \frac{\pi}{6}$.

$$x = x'\cos\frac{\pi}{6} - y'\sin\frac{\pi}{6} = x'\left(\frac{\sqrt{3}}{2}\right) - y'\left(\frac{1}{2}\right)$$

$$y = x'\sin\frac{\pi}{6} + y'\cos\frac{\pi}{6} = x'\left(\frac{1}{2}\right) + y'\left(\frac{\sqrt{3}}{2}\right)$$

Substituting in the given equation:

$$7\left[x'\left(\frac{\sqrt{3}}{2}\right) - y'\left(\frac{1}{2}\right)\right]^2 - 6\sqrt{3}\left[x'\left(\frac{\sqrt{3}}{2}\right) - y'\left(\frac{1}{2}\right)\right]\left[x'\left(\frac{1}{2}\right) + y'\left(\frac{\sqrt{3}}{2}\right)\right] + 13\left[x'\left(\frac{1}{2}\right) + y'\left(\frac{\sqrt{3}}{2}\right)\right]^2 - 16 = 0$$

Without having simplifying the equation, it is clear that $F' = -16$.

b. From the expanded form in part a above, the coefficients A' and C' of $(x')^2$ and $(y')^2$, respectively, can be determined:

$$A' = 7\left(\frac{\sqrt{3}}{2}\right)^2 + \left(-6\sqrt{3}\right)\left(\frac{\sqrt{3}}{2}\right)\left(\frac{1}{2}\right) + 13\left(\frac{1}{2}\right)^2$$
$$= 7\left(\frac{3}{4}\right) - 18\left(\frac{1}{4}\right) + 13\left(\frac{1}{4}\right)$$
$$= \frac{21}{4} - \frac{18}{4} + \frac{13}{4}$$
$$= \frac{16}{4} = 4$$

$$C' = 7\left(\frac{1}{2}\right)^2 - \left(6\sqrt{3}\right)\left(-\frac{1}{2}\right)\left(\frac{\sqrt{3}}{2}\right) + 13\left(\frac{\sqrt{3}}{2}\right)^2$$
$$= \frac{7}{4} + \frac{18}{4} + \frac{39}{4}$$
$$= \frac{64}{4} = 16$$

Then, $A' + C' = 4 + 16 = 20 = A + C$

c. Simplification of the equation in part a yields

$$4(x')^2 + 16(y')^2 - 16 = 0$$

Then, $A' = 4$, $B' = 0$, $C' = 16$

$$(B')^2 - 4(A')(C') = -4(4)(16) = -256$$

Also, from the original equation

$$B^2 - 4AC = \left(6\sqrt{3}\right)^2 - 4(7)(13)$$
$$= 108 - 364$$
$$= -256$$

Therefore,

$$B^2 - 4AC = (B')^2 - 4(A')(C')$$

End of Section 9.4

Chapter Nine: Section 9.5
Solutions to Odd-numbered Exercises

1. Given: $r = \dfrac{3}{4 - \cos\theta}$

Change to standard polar form:

$r = \dfrac{ed}{1 - e\cos\theta}$ by dividing numerator and denominator by 4:

$$r = \frac{\frac{3}{4}}{1 - \frac{1}{4}\cos\theta}$$

Then, $e = \frac{1}{4}$, and this indicates that the equation represents an ellipse.

"$\cos\theta$" $\Rightarrow$ the directrix is vertical and that the major axis is horizontal.

Using $ed = \frac{3}{4} = \frac{1}{4}d$, then $d = 3$.

The equation of the directrix: $x = -3$

(Note the negative sign on $\cos\theta$.)

Locate the vertices: For $\cos\theta$, use $\theta = 0$ to obtain $r = 1$, and $\theta = \pi$ to obtain $r = \frac{3}{5}$.

In rectangular coordinates, the right vertex is $(1, 0)$ and the left vertex is $\left(-\frac{3}{5}, 0\right)$.

Then, $2a = 1 + \frac{3}{5} = \frac{8}{5} \Rightarrow a = \frac{4}{5}$.

From $e = \frac{c}{a}$, $c = ea = \frac{1}{4} \cdot \frac{4}{5} = \frac{1}{5}$

$b^2 = \left|a^2 - c^2\right| = \frac{16}{25} - \frac{1}{25} = \frac{15}{25}$

Then, $b = \frac{\sqrt{15}}{5} \Rightarrow$ length of minor axis is $2b = \frac{2\sqrt{15}}{5}$

The equation matches graph c.

3. Given: $r = \dfrac{3}{3 + 4\sin\theta}$

Change to standard polar form:

$r = \dfrac{ed}{1 + e\sin\theta}$ by dividing numerator and denominator by 3:

$$r = \frac{1}{1 + \frac{4}{3}\sin\theta}$$

Then, $e = \frac{4}{3}$, and this indicates that the equation represents a hyperbola.

"$\sin\theta$" $\Rightarrow$ the directrix is horizontal and that the foci are vertically aligned.

Using $ed = 1 = \frac{4}{3}d$, then $d = \frac{3}{4}$.

The equation of the directrix: $y = \frac{3}{4}$

Locate the vertices: For $\sin\theta$, use $\theta = \frac{\pi}{2}$ to obtain $r = \frac{3}{7}$, and $\theta = \frac{3\pi}{2}$ to obtain $r = -3$.

In rectangular coordinates, the lower vertex is $\left(0, \frac{3}{7}\right)$ and the upper vertex is $(0, 3)$.

Then, $2a = 3 - \frac{3}{7} = \frac{18}{7} \Rightarrow a = \frac{9}{7}$.

From $e = \frac{c}{a}$, $c = ea = \frac{4}{3} \cdot \frac{9}{7} = \frac{12}{7}$

$b^2 = \left|a^2 - c^2\right| = \left|\frac{81}{49} - \frac{144}{49}\right| = \frac{63}{49}$

Then, $b = \frac{\sqrt{63}}{7} \Rightarrow$ length of minor axis is $2b = \frac{2\sqrt{63}}{7}$

The equation matches graph f.

5. Given: $r=\dfrac{6}{1+3\sin\theta}$

The given form is in standard polar form:

$$r=\frac{ed}{1+e\sin\theta}$$

Then, $e=3$, and this indicates that the equation represents a hyperbola.

"$\sin\theta$" $\Rightarrow$ the directrix is horizontal and that the foci are aligned vertically.

Using $ed=3d=6$, then $d=2$.

The equation of the directrix: $y=2$

Locate the vertices: For $\sin\theta$, use $\theta=\dfrac{\pi}{2}$ to obtain $r=\dfrac{3}{2}$, and $\theta=\dfrac{3\pi}{2}$ to obtain $r=-3$.

In rectangular coordinates, the lower vertex is $\left(0,\dfrac{3}{2}\right)$ and the upper vertex is $(0,3)$.

Then, $2a=3-\dfrac{3}{2}=\dfrac{3}{2}\Rightarrow a=\dfrac{3}{4}$.

From $e=\dfrac{c}{a}$, $c=ea=3\cdot\dfrac{3}{4}=\dfrac{9}{4}$

$b^2=|a^2-c^2|=\left|\dfrac{9}{16}-\dfrac{81}{16}\right|=\dfrac{72}{16}$

Then, $b=\dfrac{3\sqrt{2}}{2}\Rightarrow$ length of minor axis is $2b=3\sqrt{2}$

The equation matches graph b.

7. Given: $r=\dfrac{7}{1+6\sin\theta}$

From the basic equation $r=\dfrac{ed}{1+e\sin\theta}$,

we observe that $e=6>1$.

Then, the equation represents a hyperbola.

$ed=6d=7\Rightarrow d=\dfrac{7}{6}$

The (+) sign on the $\sin\theta$ term $\Rightarrow$

the equation of the directrix is $y=\dfrac{7}{6}$.

9. Given: $r=\dfrac{3}{4-\cos\theta}$

Transform to the the basic form by dividing the numerator and denominator

by 4: $r=\dfrac{\frac{3}{4}}{1-\frac{1}{4}\cos\theta}=\dfrac{ed}{1+e\cos\theta}$,

we observe that $e=\dfrac{1}{4}<1$.

Then, the equation represents an ellipse.

$ed=\dfrac{1}{4}d=\dfrac{3}{4}\Rightarrow d=3$

The (-) sign on the $\cos\theta$ term $\Rightarrow$

the equation of the directrix is $x=-3$.

11. Given: $r=\dfrac{1}{1+3\cos\theta}$

From the basic equation $r=\dfrac{ed}{1+e\sin\theta}$,

we observe that $e=3>1$.

Then, the equation represents a hyperbola.

$ed=3d=1\Rightarrow d=\dfrac{1}{3}$

The (+) sign on the $\cos\theta$ term $\Rightarrow$

the equation of the directrix is $x=\dfrac{1}{3}$.

13. Given: $r=\dfrac{5}{2+\cos\theta}$

Transform to the the basic form by dividing the numerator and denominator

by 2: $r=\dfrac{\frac{5}{2}}{1+\frac{1}{2}\cos\theta}=\dfrac{ed}{1+e\cos\theta}$,

we observe that $e=\dfrac{1}{2}<1$.

Then, the equation represents an ellipse.

$ed = \frac{1}{2}d = \frac{5}{2} \Rightarrow d = 5$

The (+) sign on the $\cos\theta$ term $\Rightarrow$ the equation of the directrix is $x = 5$.

15. Given: $r = \dfrac{6}{3 - 5\cos\theta}$

Transform to the the basic form by dividing the numerator and denominator

by 3: $r = \dfrac{2}{1 - \frac{5}{3}\cos\theta} = \dfrac{ed}{1 - e\cos\theta}$,

we observe that $e = \frac{5}{3} > 1$.

Then, the equation represents a hyperbola.

$ed = \frac{5}{3}d = 2 \Rightarrow d = \frac{6}{5}$

The (-) sign on the $\cos\theta$ term $\Rightarrow$ the equation of the directrix is $x = -\frac{6}{5}$.

17. Given: $r = \dfrac{3}{2 + 2\sin\theta}$

Transform to the the basic form by dividing the numerator and denominator

by 2: $r = \dfrac{\frac{3}{2}}{1 + \sin\theta} = \dfrac{ed}{1 + e\cos\theta}$,

we observe that $e = 1$.

Then, the equation represents a parabola.

$ed = 1d = \frac{3}{2} \Rightarrow d = \frac{3}{2}$

The (+) sign on the $\sin\theta$ term $\Rightarrow$ the equation of the directrix is $y = \frac{3}{2}$.

19. Given: $r = \dfrac{4}{6 - 7\cos\theta}$

Transform to the the basic form by dividing the numerator and denominator

by 6: $r = \dfrac{\frac{2}{3}}{1 - \frac{7}{6}\cos\theta} = \dfrac{ed}{1 - e\cos\theta}$,

we observe that $e = \frac{7}{6} > 1$.

Then, the equation represents a hyperbola.

$ed = \frac{7}{6}d = \frac{2}{3} \Rightarrow d = \frac{4}{7}$

The (-) sign on the $\cos\theta$ term $\Rightarrow$ the equation of the directrix is $x = -\frac{4}{7}$.

21. Given: Parabola with $e = 1$, and Directrix $x = -2$ $(d = 2)$

Then, $ed = (1)(2) = 2$, and

$r = \dfrac{2}{1 - \cos\theta}$

23. Given: Hyperbola with $e = 4$, and Directrix $y = -\frac{3}{4}$ $\left(d = \frac{3}{4}\right)$

Then, $ed = (4)\left(\frac{3}{4}\right) = 3$, and

$r = \dfrac{3}{1 - 4\sin\theta}$

25. Given: Ellipse with $e = \frac{1}{4}$, and Directrix $x = 12$ $(d = 12)$

Then, $ed = \left(\frac{1}{4}\right)(12) = 3$, and

$r = \dfrac{3}{1 + \frac{1}{4}\cos\theta}$

27. Sketch $r=\dfrac{5}{1+3\cos\theta}$

In this form, the equation provides direct information about the conic:
$e=3\Rightarrow$ the graph is a hyperbola

$ed=3d=5\Rightarrow d=\dfrac{5}{3}$

The equation of the directrix is $x=\dfrac{5}{3}$,

The foci are aligned horizontally.

The vertices occur at $\theta=0$ and $\theta=\pi$.

At $\theta=0$, $r=\dfrac{5}{1+3(1)}=\dfrac{5}{4}$, and

at $\theta=\pi$, $r=\dfrac{5}{1-3}=-\dfrac{5}{2}$

The vertices are $\left(-\dfrac{5}{2},\pi\right)$ and $\left(\dfrac{5}{4},0\right)$.

Find a, b and c:

$$2a=\left(\frac{5}{2}-\left(-\frac{5}{4}\right)\right)=\frac{5}{4}$$

$$a=\frac{5}{8}$$

From $e=\dfrac{c}{a}$, $c=3\left(\dfrac{5}{8}\right)=\dfrac{15}{8}$

$$b^2=\left|c^2-a^2\right|=\left(\frac{15}{8}\right)^2-\left(\frac{5}{8}\right)^2=\frac{25}{8}\Rightarrow$$

$$b=\frac{5\sqrt{2}}{4}\approx 1.77$$

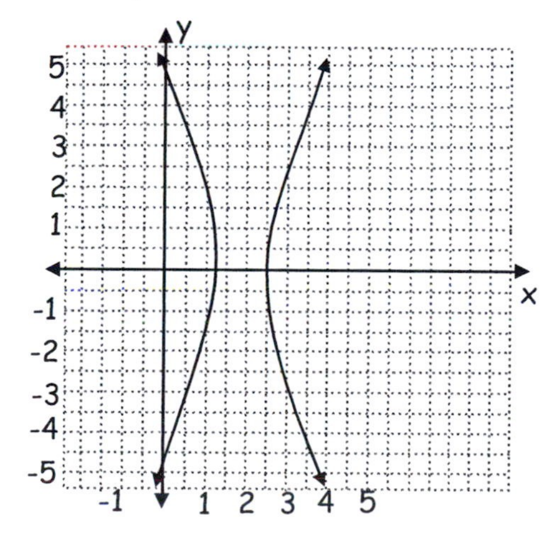

29. Sketch $r=\dfrac{4}{1-2\sin\theta}$

In this form, the equation provides direct information about the conic:
$e=2\Rightarrow$ the graph is a hyperbola

$ed=2d=4\Rightarrow d=\dfrac{4}{2}=2$

The equation of the directrix is $y=-2$,

The foci are vertically aligned.

The vertices occur at $\theta=\dfrac{\pi}{2}$ and $\theta=\dfrac{3\pi}{2}$.

At $\theta=\dfrac{\pi}{2}$, $r=\dfrac{4}{1-2(1)}=-4$, and

at $\theta=\dfrac{3\pi}{2}$, $r=\dfrac{4}{1-2(-1)}=\dfrac{4}{3}$

The vertices are $\left(-4,\dfrac{\pi}{2}\right)$ and $\left(\dfrac{4}{3},\dfrac{3\pi}{2}\right)$.

Note that the rectangular coordinates for the vertices are $(0,-4)$ and $\left(0,-\dfrac{4}{3}\right)$

and center at $\left(0,-\dfrac{8}{3}\right)$.

Find a, b and c:

$$2a=\left(-\frac{4}{3}-(-4)\right)=\frac{8}{3}\Rightarrow a=\frac{4}{3}$$

From $e=\dfrac{c}{a}$, $c=2\left(\dfrac{4}{3}\right)=\dfrac{8}{3}$

$$b^2=\left|c^2-a^2\right|=\left(\frac{8}{3}\right)^2-\left(\frac{4}{3}\right)^2=\frac{16}{3}\Rightarrow$$

$$b=\frac{4\sqrt{3}}{3}\approx 2.31$$

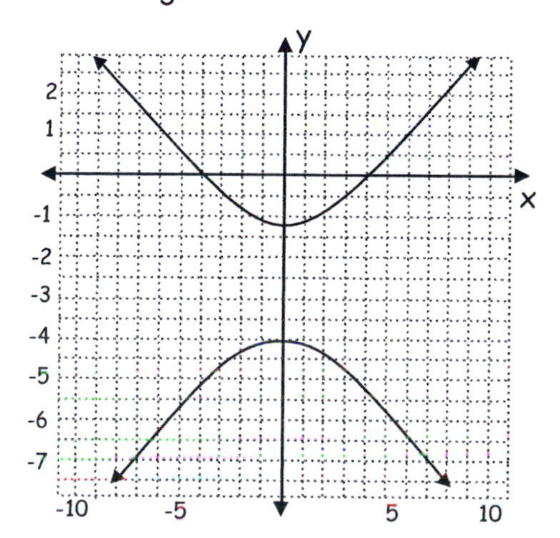

31. Sketch $r = \dfrac{9}{3-2\cos\theta}$

Transform to the the basic form by dividing the numerator and denominator

by 3: $r = \dfrac{3}{1-\frac{2}{3}\cos\theta}$,

we observe that $e = \frac{2}{3} < 1$.

Then, the equation represents an ellipse.

$ed = 3 = \frac{2}{3}d \Rightarrow d = \frac{9}{2}$

The equation of the directrix is $x = -\frac{9}{2}$,

The major axis is horizontal.

At $\theta = 0$, $r = \dfrac{3}{1-\frac{2}{3}(1)} = 9$, and

at $\theta = \pi$, $r = \dfrac{3}{1-\frac{2}{3}(-1)} = \dfrac{9}{5}$

The vertices are $(9, 0)$ and $\left(\frac{9}{5}, \pi\right)$.

To determine the rectangular coordinates for the vertices:

$x = 9\cos(0) = 9$; $y = 9\sin(0) = 0$ yields $(9, 0)$.

$x = \frac{9}{5}\cos\pi = -\frac{9}{5}$; $y = \frac{9}{5}\sin\pi = 0$ yields $\left(-\frac{9}{5}, 0\right)$.

The center is $\left(\frac{18}{5}, 0\right)$.

Find a, b and c:

$2a = \left(9 - \frac{9}{5}\right) = \frac{54}{5} \Rightarrow a = \frac{27}{5}$

From $e = \frac{c}{a}$, $c = ea = \frac{2}{3}\left(\frac{27}{5}\right) = \frac{18}{5}$

$b^2 = \left|c^2 - a^2\right| = \left|\left(\frac{18}{5}\right)^2 - \left(\frac{27}{5}\right)^2\right| = \frac{81}{5} \Rightarrow$

$b = \frac{9\sqrt{5}}{5}$

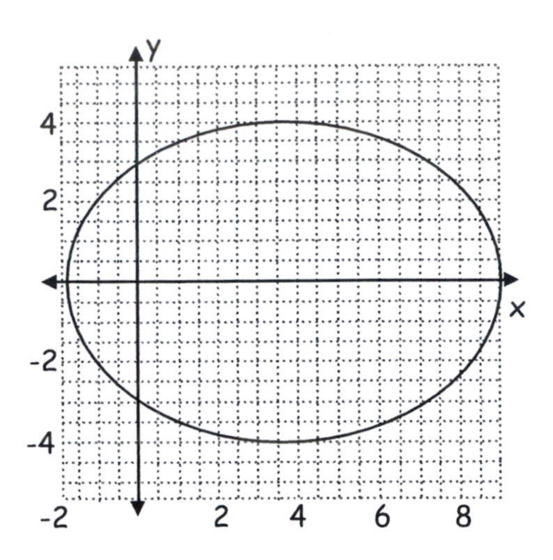

33. Sketch $r = \dfrac{4}{1+2\cos\theta}$

By inspection of the equation we observe that $e = 2 > 1$.

Then, the equation represents a hyperbola with foci aligned horizontally.

$ed = 4 = 2d \Rightarrow d = 2$

The equation of the directrix is $x = 2$,

At $\theta = 0$, $r = \dfrac{4}{1+2(1)} = \dfrac{4}{3}$, and

at $\theta = \pi$, $r = \dfrac{4}{1+2(-1)} = -4$

The vertices are $\left(\frac{4}{3}, 0\right)$ and $(-4, \pi)$.

To determine the rectangular coordinates for the vertices:

$x = \frac{4}{3}\cos(0) = \frac{4}{3}$; $y = \frac{4}{3}\sin(0) = 0$

yields $\left(\frac{4}{3}, 0\right)$.

$x = -4\cos\pi = 4$; $y = -4\sin\pi = 0$ yields $(4, 0)$.

The center is $\left(\frac{8}{3}, 0\right)$.

Find a, b and c:

$2a = \left(4 - \frac{4}{3}\right) = \frac{8}{3} \Rightarrow a = \frac{4}{3}$

From $e = \frac{c}{a}$, $c = ea = 2\left(\frac{4}{3}\right) = \frac{8}{3}$

$b^2 = |c^2 - a^2| = \left(\frac{8}{3}\right)^2 - \left(\frac{4}{3}\right)^2 = \frac{16}{3} \Rightarrow$

$b = \frac{4\sqrt{3}}{3}$

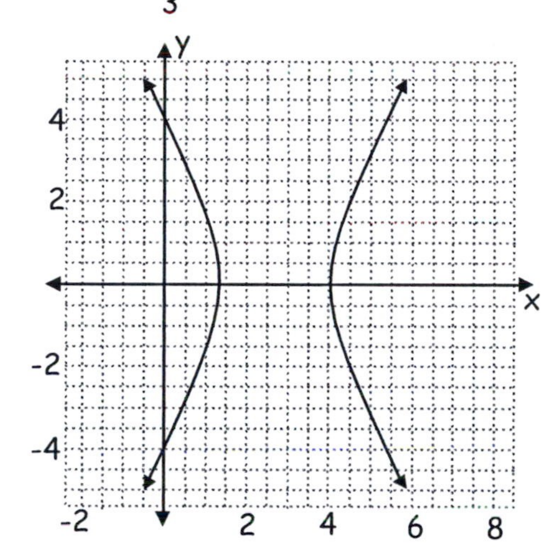

35. Sketch $r = \dfrac{2}{1+\cos\left(\theta - \frac{\pi}{4}\right)}$

By inspection of the equation we observe that $e = 1$.

Then, the equation represents a parabola with a rotation of $\frac{\pi}{4}$ counterclockwise relative to the xy-axis.

$ed = 2 \Rightarrow (1)d = 2 \Rightarrow d = 2$

At $\theta = \frac{\pi}{4}$, $r = \frac{2}{1+1} = 1$.

The vertex is $\left(1, \frac{\pi}{4}\right)$.

The graph is a rotation of $r = \dfrac{2}{1+\cos\theta}$.

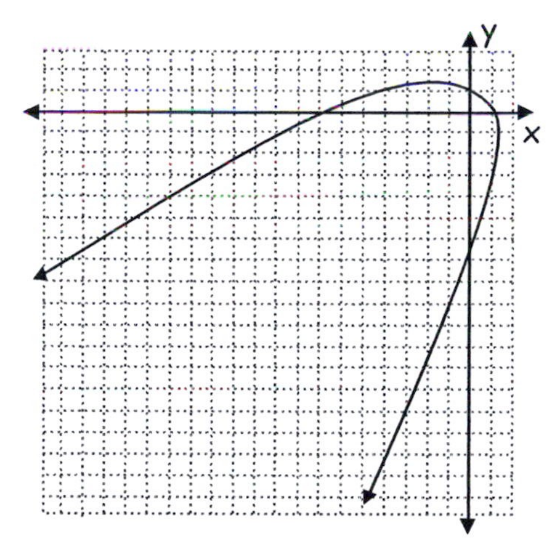

37. Use a graphing calculator for

$r = \dfrac{-3}{4 - 9\cos\theta}$.

Remember to set Mode to Polar.

Also, it is helpful to adjust the scale, min, and max for x and y. The graph is a hyperbola.

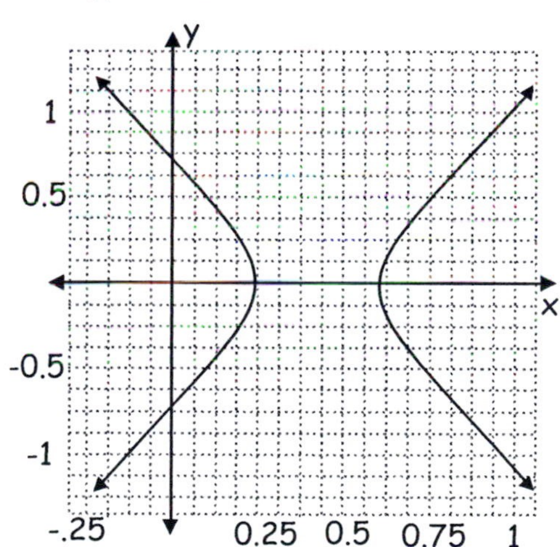

39. Use a graphing calculator for

$r = \dfrac{-11}{3 - \cos\theta}$.

Remember to set Mode to Polar.

Also, it is helpful to adjust the scale, min, and max for x and y. The graph is an ellipse.

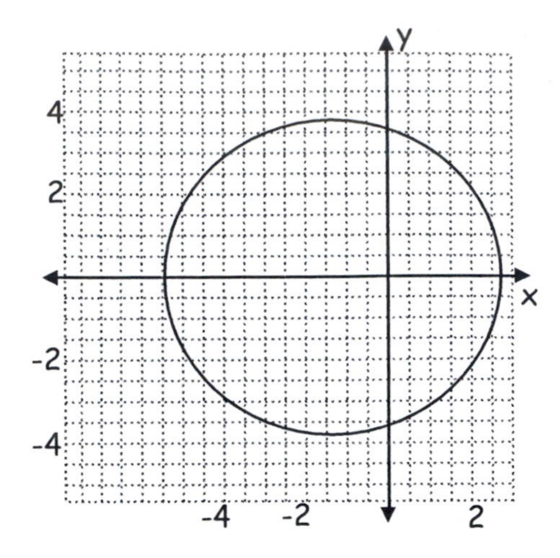

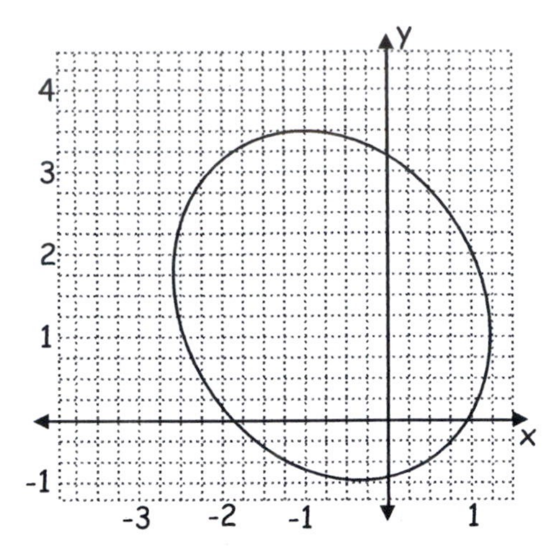

41. Use a graphing calculator for

$$r = \frac{3}{7+3\cos\theta}.$$

Remember to set Mode to Polar.
Also, it is helpful to adjust the scale, min, and max for x and y. The graph is an ellipse.

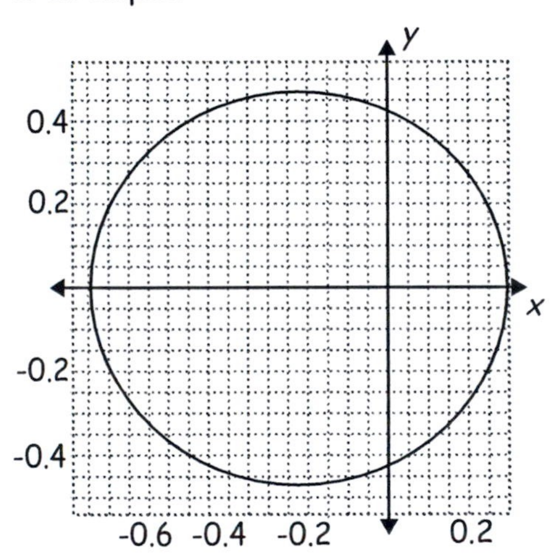

43. Use a graphing calculator for

$$r = \frac{-7}{5+3\sin\left(\theta - \frac{\pi}{6}\right)}.$$

Remember to set Mode to Polar
Also, it is helpful to adjust the scale, min, and max for x and y.
The graph is an ellipse with a rotation of $\frac{\pi}{6}$.

45. Use a graphing calculator for

$$r = \frac{4}{-3-2\cos\left(\theta + \frac{\pi}{3}\right)} = \frac{-\frac{4}{3}}{1+\frac{2}{3}\cos\left(\theta + \frac{\pi}{3}\right)}$$

$e = \frac{2}{3} \Rightarrow$ the graph is an ellipse.

Remember to set Mode to Polar.
Also, it is helpful to adjust the scale, min, and max for x and y.
The graph is an ellipse with a rotation of $\frac{\pi}{6}$.

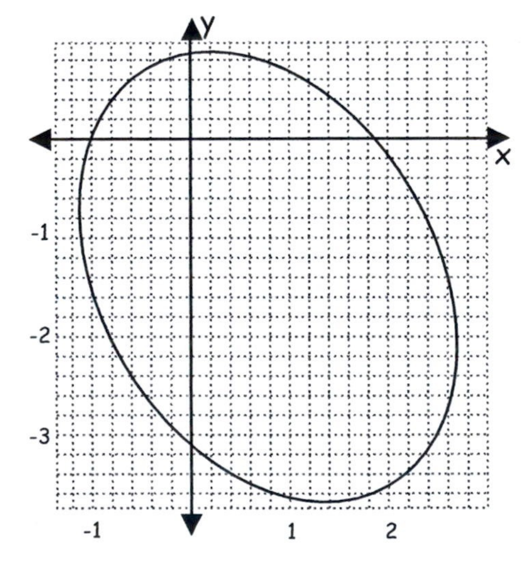

End of Section 9.5

Chapter Nine Test
Solutions to Odd-Numbered Exercises

1. Given: $\dfrac{(x+1)^2}{9}+\dfrac{(y-2)^2}{16}=1$

Center at $(-1, 2)$; $a=4$, $b=3 \Rightarrow$

$c^2=16-9=7 \Rightarrow c=\sqrt{7}$

Foci at $\left(-1, 2-\sqrt{7}\right)$ and $\left(-1, 2+\sqrt{7}\right)$

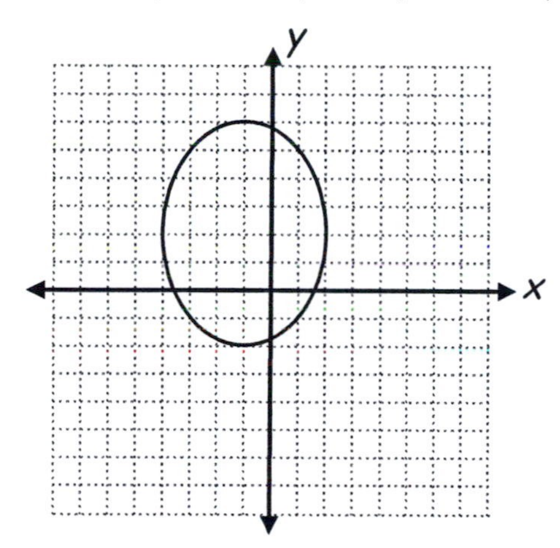

3. Given: Ellipse with center at $(-1, 4)$, the major axis is vertical and of length 8. Foci are $\sqrt{7}$ from the center.

$a=4$ and $c=\sqrt{7} \Rightarrow b^2=16-7=9$

The equation is $\dfrac{(x+1)^2}{9}+\dfrac{(y-4)^2}{16}=1.$

5. Given: Ellipse with vertices at $\left(\dfrac{7}{2}, -1\right)$ and $\left(\dfrac{1}{2}, -1\right)$, and with $e=0$. The ellipse is a circle with center at $(2, -1)$.

$a=b=\dfrac{3}{2}$ (the radius). The equation in standard form for an ellipse is

$$\frac{4(x-2)^2}{9}+\frac{4(y+1)^2}{9}=1$$

7. Given: $(y+1)^2=-12(x+3)$.

A parabola with vertex at $(-3, -1)$.

$4p=-12 \Rightarrow p=-3$.

The parabola opens to the left.

Directrix: line $x=0$

Focus: $(-6, -1)$

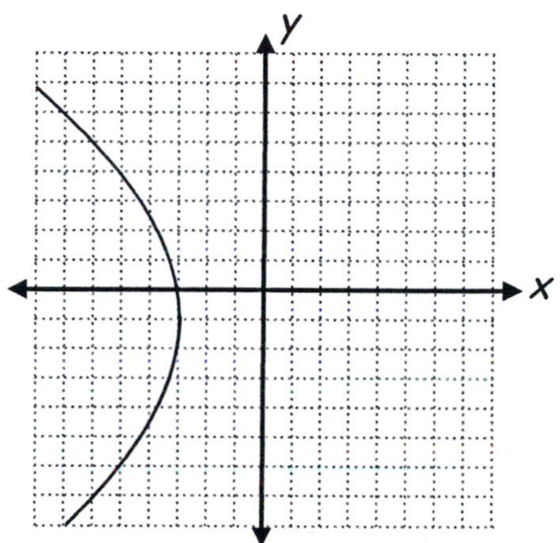

9. Given: Vertex at $(-2,3)$

Directrix is line $y=2$.

$p=1$

The equation is $(x+2)^2=4(y-3)$

11. Given: Focus at $(3, -1)$

Directrix is line $x=2$.

$p=\dfrac{1}{2}$ and the vertex is at $\left(\dfrac{5}{2}, -1\right)$

The equation is $(y+1)^2=2\left(x-\dfrac{5}{2}\right)$.

13. Given: $\dfrac{(y+2)^2}{9}-\dfrac{(x-2)^2}{16}=1$

The foci are aligned vertically with center at $(2, -2)$.

$a=3$ and $b=4 \Rightarrow c^2=9+16 \Rightarrow c=5$.

Foci are at $(2, 3)$ and $(2, -7)$.

Vertices are at $(2, 1)$ and $(2, -5)$.

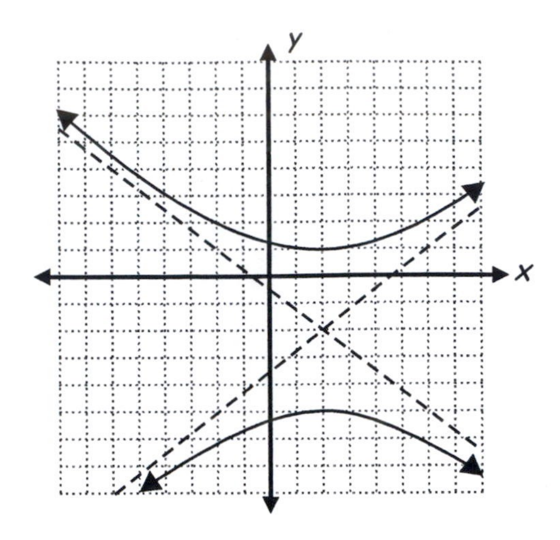

15. Given: Vertices are at $(4,-1)$ and $(-2,-1)$, and foci at $(5,-1)$ and $(-3,-1)$.
The center is at $(1,-1)$.
$a=3$ and $c=4 \Rightarrow b^2=16-9=7$.
The equation is $\frac{(x-1)^2}{9}-\frac{(y+1)^2}{7}=1$

17. Given: Foci at $(-1,-2)$ and $(-1,8)$ with asymptotes of $y=\pm\left(\frac{3}{4}x+\frac{3}{4}\right)+3$.
The asymptotes may be re-written as $y-3=\frac{3}{4}(x+1)$.
The center is at $(-1,3)$; $\frac{a}{b}=\frac{3}{4}$, $c=5$.
$a=\frac{3b}{4}$. Using $c^2=a^2+b^2$, substitute in the equation for a and solve for b:

$$25=\left(\frac{3b}{4}\right)^2+b^2$$

$$25=\frac{9}{16}b^2+b^2 \Rightarrow b^2=16 \Rightarrow a=3$$

The equation is $\frac{(y-3)^2}{9}-\frac{(x+1)^2}{16}=1$.

19. Given: $(x,y)=(-8,7)$. Find x' and y' rotated $\theta=\frac{\pi}{4}$.

$$x'=-8\cos\frac{\pi}{4}+7\sin\frac{\pi}{4}$$
$$=-8\left(\frac{\sqrt{2}}{2}\right)+7\left(\frac{\sqrt{2}}{2}\right)$$
$$=-\frac{\sqrt{2}}{2}$$
$$y'=-(-8)\sin\frac{\pi}{4}+7\cos\frac{\pi}{4}$$
$$=8\left(\frac{\sqrt{2}}{2}\right)+7\left(\frac{\sqrt{2}}{2}\right)$$
$$=\frac{15\sqrt{2}}{2}$$
$$(x',y')=\left(-\frac{\sqrt{2}}{2},\frac{15\sqrt{2}}{2}\right)$$

21. Given: $(x,y)=(4.6,-8.9)$. Find x' and y' rotated $\theta=53°$.

$$x'=4.6\cos53°+(-8.9)\sin53°$$
$$\approx 4.6(0.6018)-8.9(0.7986)$$
$$\approx -4.3395$$
$$y'=-4.6\sin53°+(-8.9)\cos53°$$
$$\approx -4.6(0.7986)-8.9(0.6018)$$
$$\approx -9.0299$$
$$(x',y')=(-4.3395,-9.0299)$$

23. Given: $(x,y)=(5,-32.1)$. Find x' and y' rotated $\theta=2.7°$.

$$x'=5\cos2.7°+(-32.1)\sin2.7°$$
$$\approx 3.4823$$
$$y'=-5\sin2.7°+(-32.1)\cos2.7°$$
$$\approx -32.2999$$
$$(x',y')=(3.4823,-32.2999)$$

25. Given: $xy - 6 = 0$

$A = 0, B = 1, C = 0$

$B^2 - 4AC = 1 > 0 \Rightarrow$ the graph is a hyperbola.

$$\cot 2\theta = \frac{A-C}{B} = 0 \Rightarrow 2\theta = \frac{\pi}{2} \Rightarrow \theta = \frac{\pi}{4}$$

$$x = x'\cos\frac{\pi}{4} - y'\sin\frac{\pi}{4}$$

$$= \frac{\sqrt{2}}{2}(x' - y')$$

$$y = x'\sin\frac{\pi}{4} + y'\cos\frac{\pi}{4}$$

$$= \frac{\sqrt{2}}{2}(x' + y')$$

Substituting:

$$\left[\frac{\sqrt{2}}{2}(x' - y')\right]\left[\frac{\sqrt{2}}{2}(x' + y')\right] - 6 = 0$$

$$\frac{1}{2}(x')^2 - \frac{1}{2}(y')^2 = 6$$

$$\frac{(x')^2}{12} - \frac{(y')^2}{12} = 1$$

Then, on the $x'y'-$ axis, the center is $(0, 0)$, $a = 2\sqrt{3}$ and $b = 2\sqrt{3}$.

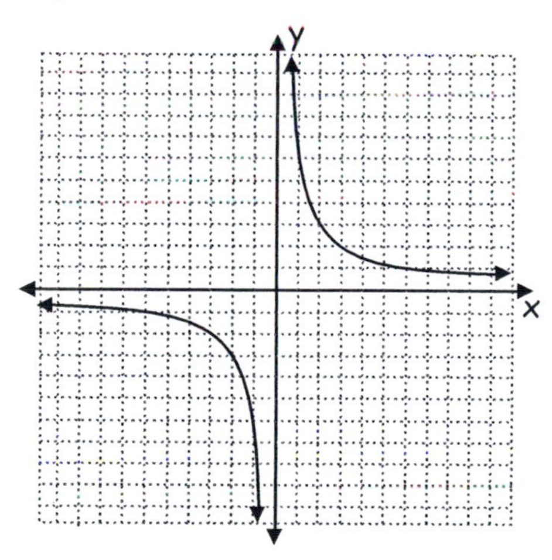

27. Given:

$$10x^2 + 2\sqrt{3}xy + 12y^2 - y = 0$$

$A = 10, B = 2\sqrt{3}, C = 12$

$B^2 - AC = 12 - 4(10)(12) = -468 < 0.$

Then, the equation represents an ellipse.

$$\cot 2\theta = \frac{A-C}{B} = \frac{10-12}{2\sqrt{3}} = -\frac{1}{\sqrt{3}}$$

Then, $2\theta = \frac{2\pi}{3} \Rightarrow \theta = \frac{\pi}{3}$

$$x = x'\cos\frac{\pi}{3} - y'\sin\frac{\pi}{3}$$

$$= \frac{1}{2}x' - \frac{\sqrt{3}}{2}y'$$

$$y = x'\sin\frac{\pi}{3} + y'\cos\frac{\pi}{3}$$

$$= \frac{\sqrt{3}}{2}x' + \frac{1}{2}y'$$

Substituting in the original equation:

$$10\left[\frac{1}{2}x' - \frac{\sqrt{3}}{2}y'\right]^2 +$$

$$2\sqrt{3}\left[\frac{1}{2}x' - \frac{\sqrt{3}}{2}y'\right]\left[\frac{\sqrt{3}}{2}x' + \frac{1}{2}y'\right] +$$

$$12\left[\frac{\sqrt{3}}{2}x' + \frac{1}{2}y'\right]^2 - \left[\frac{\sqrt{3}}{2}x' + \frac{1}{2}y'\right] = 0$$

This simplifies to

$$\left[10\left(\frac{1}{4}\right) + \frac{3}{2} + 12\left(\frac{3}{4}\right)\right](x')^2 +$$

$$\left[10\left(\frac{3}{4}\right) - 2\sqrt{3}\left(\frac{\sqrt{3}}{4}\right) + 12\left(\frac{1}{4}\right)\right](y')^2 +$$

$$\left[10\left(\frac{-2\sqrt{3}}{4}\right) + 2\sqrt{3}\left(-\frac{3}{4} + \frac{1}{4}\right) + 12\left(\frac{2\sqrt{3}}{4}\right)\right]x'y' +$$

$$-\frac{\sqrt{3}}{2}x' - \frac{1}{2}y' = 0$$

This further simplifies to

$$\left[\frac{5}{2} + \frac{3}{2} + 9\right](x')^2 + \left[\frac{15}{2} - \frac{3}{2} + 3\right](y')^2 +$$

$$\left[-5\sqrt{3} - \sqrt{3} + 6\sqrt{3}\right]x'y' +$$

$$-\frac{\sqrt{3}}{2}x' - \frac{1}{2}y' = 0$$

Further simplification leads to the need to complete the square:

$$13(x')^2 - \frac{\sqrt{3}}{2}x' + 9(y')^2 - \frac{1}{2}y' = 0$$

Completing the square:

$$13\left[(x')^2 - \frac{\sqrt{3}}{26}x' + \frac{3}{52^2}\right] +$$

$$9\left[(y')^2 - \frac{1}{18}y' + \frac{1}{36^2}\right] =$$

$$13\left((x') - \frac{\sqrt{3}}{52}\right)^2 + 9\left((y') - \frac{1}{36}\right)^2 = \frac{5}{234}$$

Divide both sides by $\frac{5}{234}$ to put in standard form for an ellipse:

$$\frac{\left((x') - \frac{\sqrt{3}}{52}\right)^2}{\frac{5}{3042}} + \frac{\left((y') - \frac{1}{36}\right)^2}{\frac{5}{2106}} = 1$$

Note the small scale of the graph:

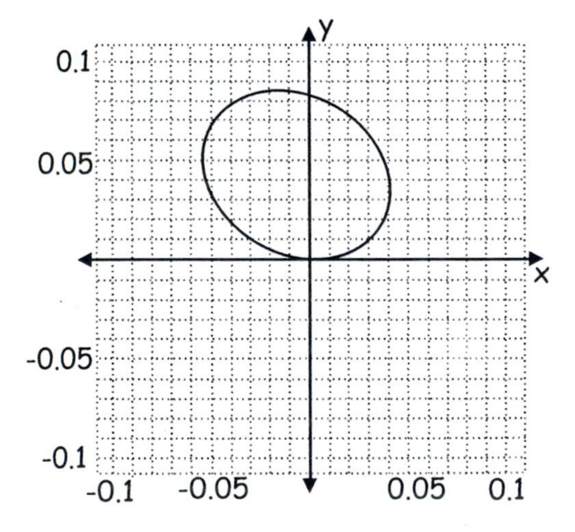

29. Given: $x^2 + 2xy + y^2 + x - y = 0$

$A = 1, B = 2, C = 1$

$B^2 - 4AC = 4 - 4 = 0 \Rightarrow$ the graph is a parabola.

$$\cot 2\theta = \frac{A - C}{B} = 0 \Rightarrow 2\theta = \frac{\pi}{2} \Rightarrow \theta = \frac{\pi}{4}$$

$$x = x'\cos\frac{\pi}{4} - y'\sin\frac{\pi}{4}$$

$$= \frac{\sqrt{2}}{2}(x' - y')$$

$$y = x'\sin\frac{\pi}{4} + y'\cos\frac{\pi}{4}$$

$$= \frac{\sqrt{2}}{2}(x' + y')$$

Substitute in $x^2 + 2xy + y^2 + x - y = 0$:

$$\left[\frac{\sqrt{2}}{2}(x' - y')\right]^2 + 2\left[\frac{\sqrt{2}}{2}(x' - y')\right]\left[\frac{\sqrt{2}}{2}(x' + y')\right] +$$

$$\left[\frac{\sqrt{2}}{2}(x' + y')\right]^2 + \left[\frac{\sqrt{2}}{2}(x' - y')\right] - \left[\frac{\sqrt{2}}{2}(x' + y')\right] = 0$$

$$\left[\frac{2}{4} + 2\left(\frac{2}{4}\right) + \frac{2}{4}\right](x')^2 + \left[\frac{2}{4} - 2\left(\frac{2}{4}\right) + \frac{2}{4}\right](y')^2 +$$

$$\left[-2\left(\frac{2}{4}\right) + 2\left(\frac{2}{4}\right) - 2\left(\frac{2}{4}\right) + 2\left(\frac{2}{4}\right)\right]x'y' -$$

$$2\left(\frac{\sqrt{2}}{2}\right)y' = 0$$

The equation is simplified and put in standard form for a parabola:

$$2(x')^2 - \sqrt{2}y' = 0$$

$$y' = \frac{2(x')^2}{\sqrt{2}}$$

$$y' = \sqrt{2}(x')^2$$

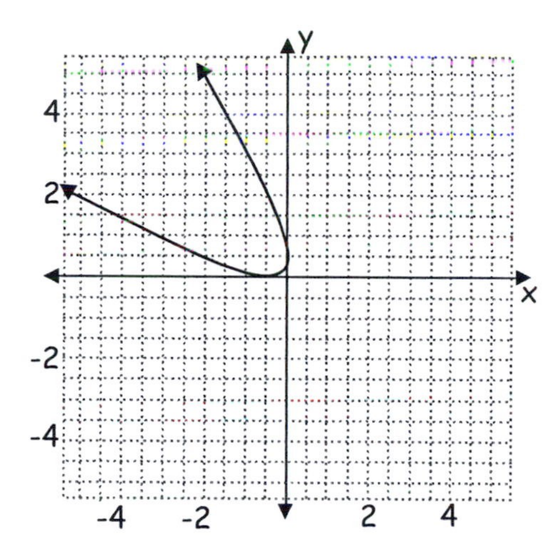

31. Given: $r = \dfrac{8}{1+2\sin\theta}$

From the basic equation $r = \dfrac{ed}{1+e\sin\theta}$,

we observe that $e = 2 > 1$.

Then, the equation represents a hyperbola.

$ed = 2d = 8 \Rightarrow d = 4$

The (+) sign on the $\sin\theta$ term $\Rightarrow$ the equation of the directrix is the line $y = 4$.

33. Given: $r = \dfrac{3}{7+6\sin\theta}$

Rewrite the equation to the basic form by dividing the numerator and

denominator by 7: $r = \dfrac{\frac{3}{7}}{1+\frac{6}{7}\sin\theta}$

From the basic equation $r = \dfrac{ed}{1+e\sin\theta}$,

we observe that $e = \dfrac{6}{7} < 1$.

Then, the equation represents an ellipse.

$ed = \dfrac{6}{7}d = \dfrac{3}{7} \Rightarrow d = \dfrac{1}{2}$

The (+) sign on the $\sin\theta$ term $\Rightarrow$ the equation of the directrix is the line $y = \dfrac{1}{2}$.

35. Given: $r = \dfrac{7}{4+4\sin\theta}$

Rewrite the equation to the basic form by dividing the numerator and

denominator by 4: $r = \dfrac{\frac{7}{4}}{1+\sin\theta}$

From the basic equation $r = \dfrac{ed}{1+e\sin\theta}$,

we observe that $e = 1$.

Then, the equation represents a parabola.

$ed = 1d = \dfrac{7}{4} \Rightarrow d = \dfrac{7}{4}$

The (+) sign on the $\sin\theta$ term $\Rightarrow$ the equation of the directrix is the line $y = \dfrac{7}{4}$.

37. Given: $r = \dfrac{7}{5+2\cos\theta}$

Rewrite the equation to the basic form by dividing the numerator and

denominator by 5: $r = \dfrac{\frac{7}{5}}{1+\frac{2}{5}\cos\theta}$

From the basic equation $r = \dfrac{ed}{1+e\cos\theta}$,

we observe that $e = \dfrac{2}{5} < 1$.

Then, the equation represents an ellipse.

$ed = \dfrac{2}{5}d = \dfrac{7}{5} \Rightarrow d = \dfrac{7}{2}$

The (+) sign on the $\cos\theta$ term $\Rightarrow$ the equation of the directrix is the line $x = \dfrac{7}{2}$.

39. Given: $e = 4$, Directrix: $y = 3\ (d = 3)$

$$r = \frac{12}{1 + 4\sin\theta}$$

41. Given: $e = \frac{1}{4}$, Directrix: $x = 16\ (d = 16)$

$$r = \frac{\frac{1}{4}(16)}{1 + \frac{1}{4}\cos\theta} = \frac{4}{1 + \frac{1}{4}\cos\theta}$$

43. Given: $e = 9$, Directrix: $x = \frac{1}{3}\left(d = \frac{1}{3}\right)$

$$r = \frac{9\left(\frac{1}{3}\right)}{1 + 9\cos\theta} = \frac{3}{1 + 9\cos\theta}$$

End of Chapter 9 Test

Chapter Ten: Section 10.1
Solutions to Odd-numbered Exercises

1. $\begin{cases} 2x - y = -12 \\ 3x + y = -13 \end{cases} \longrightarrow \begin{cases} y = 2x + 12 \\ 3x + (2x + 12) = -13 \end{cases}$

$5x + 12 = -13$

$5x = -25$

$x = -5$

$y = 2(-5) + 12$

$y = 2$

Solution: $(-5, 2)$

3. $\begin{cases} 3y = 9 \\ x + 2y = 11 \end{cases} \longrightarrow \begin{cases} y = 3 \\ x + 2(3) = 11 \end{cases}$

$x = 5$

Solution: $(5, 3)$

5. $\begin{cases} 2x + y = -2 \\ -4x - 2y = 5 \end{cases} \longrightarrow \begin{cases} y = -2x - 2 \\ -4x - 2(-2x - 2) = 5 \end{cases}$

$-4x + 4x + 4 = 5$

$4 \neq 5$

Solution Set: $\varnothing$

7. $\begin{cases} 2x - y = -3 \\ -4x + 2y = 6 \end{cases}$

$y = 2x + 3$

$-4x + 2(2x + 3) = 6$

$-4x + 4x + 6 = 6$

$0 = 0$

Solution Set: $\{(x, 2x + 3)\}$, or $\left\{\left(\frac{y-3}{2}, y\right)\right\}$

9. $\begin{cases} 2x + 5y = 33 \\ 3x = -3 \end{cases}$

$x = -1$

$2(-1) + 5y = 33$

$-2 + 5y = 33$

$y = 7$

Solution: $(-1, 7)$

11. $\begin{cases} -2x + y = 5 \\ 9x - 2y = 5 \end{cases} \longrightarrow \begin{cases} y = 2x + 5 \\ 9x - 2(2x + 5) = 5 \end{cases}$

$9x - 4x - 10 = 5$

$5x = 15$

$x = 3$

$y = 2(3) + 5$

$y = 11$

Solution: $(3, 11)$

13. $\begin{cases} 4x - y = -1 \\ -8x + 2y = 2 \end{cases}$

$y = 4x + 1$

$-8x + 2(4x + 1) = 2$

$-8x + 8x + 2 = 2$

$2 = 2$

Solution Set: $\{(x, 4x + 1)\}$, or $\left\{\left(\frac{y-1}{4}, y\right)\right\}$

15. $\begin{cases} 9x - y = -1 \\ 3x + 2y = 44 \end{cases}$

$y = 9x + 1$

$3x + 2(9x + 1) = 44$

$3x + 18x + 2 = 44$

$21x = 42$

$x = 2$

$y = 9(2) + 1$

$y = 19$

Solution: $(2, 19)$

17. Solve: $\begin{cases} x-2y=3 \\ 3x+y=30 \end{cases}$

From equation 1: $x=2y+3$

Substitute in equation 2:

$$\begin{aligned} 3(2y+3)+y&=30 \\ 7y&=21 \\ y&=3 \\ x&=2(3)+3=9 \end{aligned}$$

Solution: $(9,3)$

19. Solve: $\begin{cases} 4x-4y=12 \\ x+y=7 \end{cases}$

From equation 2: $x=7-y$

Substitute in equation 1:

$$\begin{aligned} 4(7-y)-4y&=12 \\ -8y&=-16 \\ y&=2 \\ x&=7-2=5 \end{aligned}$$

Solution: $(5,2)$

21. Solve: $\begin{cases} 2x+y=4 \\ 3x-y=6 \end{cases}$

From equation 2: $y=3x-6$

Substitute in equation 1:

$$\begin{aligned} 2x+3x-6&=4 \\ 5x&=10 \\ x&=2 \\ y&=3(2)-6=0 \end{aligned}$$

Solution: $(2,0)$

23. Solve: $\begin{cases} 3x+4y=5 \\ 2x+2y=3 \end{cases}$

From equation 2: $x=-y+\dfrac{3}{2}$

Substitute in equation 1:

$$\begin{aligned} 3\left(-y+\frac{3}{2}\right)+4y&=5 \\ y&=5-\frac{9}{2} \\ y&=\frac{1}{2} \\ x&=-\frac{1}{2}+\frac{3}{2}=1 \end{aligned}$$

Solution: $\left(1,\dfrac{1}{2}\right)$

25. Solve: $\begin{cases} 7x-5y=11 \\ x+5y=13 \end{cases}$

From equation 2: $x=13-5y$

Substitute in equation 1:

$$\begin{aligned} 7(13-5y)-5y&=11 \\ 91-35y-5y&=11 \\ -40y&=-80 \\ y&=2 \\ x&=13-5(2)=3 \end{aligned}$$

Solution: $(3,2)$

27. Solve: $\begin{cases} 4x-2y=6 \\ -3x+2y=2 \end{cases}$

From equation 1: $y=2x-3$

Substitute in equation 2:

$$\begin{aligned} -3x+2(2x-3)&=2 \\ x&=8 \\ y&=2(8)-3=13 \end{aligned}$$

Solution: $(8,13)$

29. Solve: $\begin{cases} 7x+2y=5 \\ 14x+6y=8 \end{cases}$

From equation 2: $y=\dfrac{4}{3}-\dfrac{7}{3}x$

Substitute in equation 1:

$$\begin{aligned} 7x+2\left(\frac{4}{3}-\frac{7}{3}x\right)&=5 \\ 7x+\frac{8}{3}-\frac{14}{3}x&=5 \\ \frac{7}{3}x&=\frac{7}{3} \\ x&=1 \\ y&=\frac{4}{3}-\frac{7}{3}(1)=-1 \end{aligned}$$

Solution: $(1,-1)$

31. Solve: $\begin{cases} 5x+13y=38 \\ -20x+36y=-64 \end{cases}$

From equation 2: $x=\frac{9}{5}y+\frac{16}{5}$

Substitute in equation 1:

$$5\left(\frac{9}{5}y+\frac{16}{5}\right)+13y=38$$
$$9y+16+13y=38$$
$$22y=22$$
$$y=1$$
$$x=\frac{9}{5}(1)+\frac{16}{5}=5$$

Solution: $(5,1)$

33. Solve: $\begin{cases} 28x+29y=-72 \\ 13x+25y=-218 \end{cases}$

From equation 1: $x=\frac{-29y-72}{28}$

Substitute in equation 2:

$$13\left(\frac{-29y-72}{28}\right)+25y=-218$$
$$\left(\frac{-377y-936}{28}\right)+25y=-218$$
$$\left(-\frac{377}{28}+\frac{700}{28}\right)y=\frac{-6104+936}{28}$$
$$\frac{323}{28}y=-\frac{5168}{28}$$
$$y=-16$$
$$x=\frac{-29(-16)-72}{28}$$
$$x=14$$

Solution: $(14,-16)$

35. Solve: $\begin{cases} 6x+4y+3z=-11 \\ 5x+y+2z=-13 \\ x-4y+5z=7 \end{cases}$

From equation 3: $x=4y-5z+7$

Substitute in equations 1 and 2:

$$6(4y-5z+7)+4y+3z=-11$$
$$5(4y-5z+7)+y+2z=-13$$

Simplifying:

$$28y-27z=-53$$
$$21y-23z=-48$$

From the second equation above:

$$y=\frac{23z-48}{21}$$

Substituting in equation 1:

$$28\left(\frac{23z-48}{21}\right)-27z=-53$$
$$\frac{4(23z-48)}{3}-27z=-53$$
$$92z-192-81z=-159$$
$$11z=33$$
$$z=3$$

Back substituting:

$$y=\frac{69-48}{21}=1$$
$$x=4(1)-5(3)+7=-4$$

Solution: $(-4,1,3)$

37. Solve: $\begin{cases} 2x-5y+7z=-16 \\ 8x-4y+5z=0 \\ x+2y-10z=10 \end{cases}$

From equation 3: $x=10z-2y+10$

Substitute in equations 1 and 2:

$$2(10z-2y+10)-5y+7z=-16$$
$$8(10z-2y+10)-4y+5z=0$$

Simplifying:

$27z-9y=-36$, or $3z-y=-4$

$85z-20y=-80$, or $17z-4y=-16$

From the first equation above:

$$y=3z+4$$

Substituting in equation 2:

$$17z-4(3z+4)=-16$$
$$5z=0$$
$$z=0$$

Back substituting:

$$y=3(0)+4=4$$
$$x=10(0)-2(4)+10=2$$

Solution: $(2,4,0)$

39. Solve: $\begin{cases} 8x - y + 13z = -13 \\ 6x + 5y + 4z = 19 \\ 13x + 19y - 11z = 36 \end{cases}$

From equation 1: $y = 8x + 13z + 13$

Substitute in equations 2 and 3:

$$6x + 5(8x + 13z + 13) + 4z = 19$$
$$13x + 19(8x + 13z + 13) - 11z = 36$$

Simplifying:

$$46x + 69z = -46$$
$$165x + 236z = -211$$

From the first equation above:

$$x = -1 - \frac{3}{2}z$$

Substituting in equation 2:

$$165\left(-1 - \frac{3}{2}z\right) + 236z = -211$$
$$-165 - \frac{495}{2}z + 236z = -211$$
$$-\frac{23}{2}z = -46$$
$$z = 4$$

Back substituting:

$$x = -1 - \frac{3(4)}{2} = -7$$
$$y = 8(-7) + 13(4) + 13 = 9$$

Solution: $(-7, 9, 4)$

41. $\begin{cases} -2x + 3y = 13 \\ 4x + 2y = -18 \end{cases} \xrightarrow{2E_1}$

$\begin{cases} -4x + 6y = 26 \\ 4x + 2y = -18 \end{cases}$ Add:

$$8y = 8$$
$$y = 1$$
$$-2x + 3(1) = 13$$
$$-2x = 10$$
$$x = -5$$

Solution: $(-5, 1)$

43. $\begin{cases} x + 2y = 17 \\ 3x + 4y = 39 \end{cases} \xrightarrow{-2E_1}$

$\begin{cases} -2x - 4y = -34 \\ 3x + 4y = 39 \end{cases}$ Add:

$$x = 5$$
$$5 + 2y = 17$$
$$2y = 12$$
$$y = 6$$

Solution: $(5, 6)$

45. $\begin{cases} -2x - 2y = 4 \\ 3x + 3y = -6 \end{cases} \xrightarrow[2E_2]{3E_1}$

$\begin{cases} -6x - 6y = 12 \\ 6x + 6y = -12 \end{cases}$ Add:

$$0 = 0$$

Solution Set: $\{(x, -x-2)\}$, or $\{(-y-2, y)\}$

47. $\begin{cases} \frac{x}{5} - y = -\frac{11}{5} \\ \frac{x}{4} + y = 4 \end{cases} \xrightarrow[-4E_2]{5E_1}$

$\begin{cases} x - 5y = -11 \\ -x - 4y = -16 \end{cases}$ Add:

$$-9y = -27$$
$$y = 3$$
$$x - 5(3) = -11$$
$$x = 4$$

Solution: $(4, 3)$

49. $\begin{cases} 4x + y = 11 \\ 3x - 2y = 0 \end{cases} \xrightarrow{2E_1}$

$\begin{cases} 8x + 2y = 22 \\ 3x - 2y = 0 \end{cases}$ Add:

$$11x = 22$$
$$x = 2$$
$$3(2) - 2y = 0$$
$$y = 3$$

Solution: $(2, 3)$

51. $\begin{cases} -x-5y=-6 \\ \frac{3}{5}x+3y=1 \end{cases} \xrightarrow[5E_2]{3E_1}$

$\begin{cases} -3x-15y=-18 \\ 3x+15y=5 \end{cases}$ Add:

$0 \neq 13$

No solution

53. $\begin{cases} -2x+4y=6 \\ 3x-y=-4 \end{cases} \xrightarrow{4E_2}$

$\begin{cases} -2x+4y=6 \\ 12x-4y=-16 \end{cases}$ Add:

$10x=-10$

$x=-1$

$-2(-1)+4y=6$

$4y=4$

$y=1$

Solution: $(-1,1)$

55. Solve: $\begin{cases} 5x-3y=-4 \\ -5x+6y=13 \end{cases}$

by elimination: Add:

$3y=9$

$y=3$

Back substituting:

$5x-3(3)=-4$

$5x=5$

$x=1$

Solution: $(1,3)$

57. Solve: $\begin{cases} 7x-3y=11 \\ 9x+6y=24 \end{cases}$

by elimination: First, rewrite, then add:

$\begin{cases} 14x-6y=22 \\ 9x+6y=24 \end{cases}$

$23x=46$

$x=2$

Back substituting:

$7(2)-3y=11$

$-3y=-3$

$y=1$

Solution: $(2,1)$

59. Solve: $\begin{cases} -7x+4y=-1 \\ 14x-7y=7 \end{cases}$

by elimination: First, rewrite, then add:

$\begin{cases} -14x+8y=-2 \\ 14x-7y=7 \end{cases}$

$y=5$

Back substituting:

$-7x+4(5)=-1$

$-7x=-21$

$x=3$

Solution: $(3,5)$

61. Solve: $\begin{cases} 2x+5y=-11 \\ 3x+4y=-13 \end{cases}$

by elimination: First, rewrite, then add:

$\begin{cases} -6x-15y=33 \\ 6x+8y=-26 \end{cases}$

$-7y=7$

$y=-1$

Back substituting:

$2x+5(-1)=-11$

$2x=-6$

$x=-3$

Solution: $(-3,-1)$

63. Solve: $\begin{cases} 3x+17y=9 \\ 7x-23y=21 \end{cases}$

by elimination: First, rewrite equation 1 by multiplying both sides by 7, and equation 2 by -3. Then add:

$$\begin{cases} 21x+119y=63 \\ -21x+69y=-63 \end{cases}$$

$$188y=0$$

$$y=0$$

Back substituting:

$$3x+17(0)=9$$

$$x=3$$

Solution: $(3, 0)$

65. Solve: $\begin{cases} 13x+15y=-14 \\ 19x+11y=56 \end{cases}$

by elimination: First, rewrite equation 1 by multiplying both sides by 19, and equation 2 by -13. Then add:

$$\begin{cases} 247x+285y=-266 \\ -247x-143y=-728 \end{cases}$$

$$142y=-994$$

$$y=-7$$

Back substituting:

$$13x+15(-7)=-14$$

$$13x=91$$

$$x=7$$

Solution: $(7, -7)$

67. Solve: $\begin{cases} 2x-4y+3z=11 \\ 3x-8y+6z=17 \\ x-4y=-3 \end{cases}$

by elimination: First, rewrite equation 1 by multiplying both sides by -2. Add the first two equations to eliminate y and z:

$$\begin{cases} -4x+8y-6z=-22 \\ 3x-8y+6z=17 \end{cases}$$

$$-x=-5$$

$$x=5$$

From equation 3:

$$y=\frac{x+3}{4}=\frac{5+3}{4}=2$$

Substituting in equation 1:

$$2(5)-4(2)+3z=11$$

$$3z=9$$

$$z=3$$

Solution: $(5, 2, 3)$

69. Solve: $\begin{cases} 4x-3y+5z=13 \\ 7x+9y-2z=19 \\ 8x-5y=-2 \end{cases}$

by elimination: First, rewrite equation 1 by multiplying both sides by 2, and equation 2 by 5. Then add to eliminate z:

$$\begin{cases} 8x-6y+10z=26 \\ 35x+45y-10z=95 \end{cases}$$

$$\begin{cases} 43x+39y=121 \\ 8x-5y=-2 \end{cases}$$

Rewrite equations 1 and 2 above by multiplying the first by 5 and the second by 39. Add to eliminate y:

$$\begin{cases} 215x+195y=605 \\ 312x-195y=-78 \end{cases}$$

$$527x=527$$

$$x=1$$

Back substituting in equation 3:

$$8(1)-5y=-2$$

$$y=2$$

Back substituting in equation 1:

$$4(1)-3(2)+5z=13$$

$$z=3$$

Solution: $(1, 2, 3)$

71. Solve: $\begin{cases} 7x-5y=19 \\ 3x-10y+7z=37 \\ 3x+4y+8z=26 \end{cases}$

by elimination: First, rewrite equation 2 by multiplying both sides by 8, and equation 3 by -7. Then add to eliminate z:

$$\begin{cases} 24x-80y+56z=296 \\ -21x-28y-56z=-182 \end{cases}$$

$$\begin{cases} 3x-108y=114 \\ 7x-5y=19 \end{cases}$$

Rewrite equations 1 and 2 above by multiplying the first by 7 and the second by -3. Add to eliminate x:

$$\begin{cases} 21x-756y=798 \\ -21x+15y=-57 \end{cases}$$

$$-741y=741$$

$$y=-1$$

Back substituting in first equation 1:

$$7x-5(-1)=19$$

$$x=2$$

Back substituting in equation 2:

$$3(2)-10(-1)+7z=37$$

$$7z=21$$

$$z=3$$

Solution: $(2,-1,3)$

73. $\begin{cases} x-y+4z=-4 \\ 4x+y-2z=-1 \\ -y+2z=-3 \end{cases} \xrightarrow[2E_3]{2E_2}$

$\begin{cases} x-y+4z=-4 \\ 8x+2y-4z=-2 \\ -2y+4z=-6 \end{cases} \xrightarrow[E_3+E_2]{E_1+E_2}$

$\begin{cases} 9x+y=-6 \\ 8x=-8 \end{cases}$

$x=-1$

$9(-1)+y=-6$

$y=3$

$-1-3+4z=-4$

$4z=0$

$z=0$

Solution: $(-1,3,0)$

75. $\begin{cases} x+y=4 \\ y+3z=-1 \\ 2x-2y+5z=-5 \end{cases}$

$x=4-y$

$\begin{cases} y+3z=-1 \\ 2(4-y)-2y+5z=-5 \end{cases}$ Simplify

$\begin{cases} y+3z=-1 \\ -4y+5z=-13 \end{cases} \xrightarrow{4E_1}$

$\begin{cases} 4y+12z=-4 \\ -4y+5z=-13 \end{cases}$ Add

$17z=-17$

$z=-1$

$y=-3(-1)-1=2$

$x=4-2=2$

Solution: $(2,2,-1)$

77. $\begin{cases} 3x-y+z=2 \\ -6x+2y-2z=-4 \\ -3x+y-z=-2 \end{cases}$

Multiplying equation 1 by -2 and equation 3 by 2, we see the three equations are identical.

So the system is dependant.

$E_1: x=\dfrac{y-z+2}{3}$

Solution: $\left(\dfrac{y-z+2}{3}, y, z\right)$

79. $\begin{cases} 3x-y+z=2 \\ -6x+2y-2z=1 \\ 5x+2y-3z=2 \end{cases} \xrightarrow{2E_1+E_2}$

$0\neq 5$

Inconsistent system

No Solution

81. $\begin{cases} 3x+8z=3 \\ x+3z=1 \end{cases}$

$$x=1-3z$$
$$3(1-3z)+8z=3$$
$$3-9z+8z=3$$
$$z=0$$
$$x=1-3(0)$$
$$x=1$$

Solution Set: $\{(1, y, 0) \mid y \in \mathbb{R}\}$

83. $\begin{cases} 2x-7y-4z=7 \\ -x+4y+2z=-3 \\ 3y-4z=-1 \end{cases} \xrightarrow{2E_2}$

$\begin{cases} 2x-7y-4z=7 \\ -2x+8y+4z=-6 \\ 3y-4z=-1 \end{cases} \xrightarrow{E_1+E_2}$

$\begin{cases} y=1 \\ -2x+8y+4z=-6 \\ 3y-4z=-1 \end{cases}$ Substitute

$$3(1)-4z=-1$$
$$z=1$$
$$2x-7(1)-4(1)=7$$
$$x=9$$

Solution: $(9, 1, 1)$

85. $\begin{cases} 2x+3y+4z=1 \\ 3x-4y+5z=-5 \\ 4x+5y+6z=5 \end{cases} \xrightarrow{-2E_1}$

$\begin{cases} -4x-6y-8z=-2 \\ 3x-4y+5z=-5 \\ 4x+5y+6z=5 \end{cases} \xrightarrow{E_1+E_3}$

$\begin{cases} -y-2z=3 \\ 3x-4y+5z=-5 \\ 4x+5y+6z=5 \end{cases} \xrightarrow{-4E_2+3E_3}$

$\begin{cases} -y-2z=3 \\ 31y-2z=35 \\ 4x+5y+6z=5 \end{cases}$

$$y=-2z-3$$
$$31(-2z-3)-2z=35$$
$$-62z-93-2z=35$$
$$-64z=128$$
$$z=-2$$
$$y=-2(-2)-3$$
$$y=1$$
$$2x+3(1)+4(-2)=1$$
$$x=3$$

Solution: $(3, 1, -2)$

87. $\begin{cases} x+2y+3z=29 \\ 2x-y-z=-2 \\ 3x+2y-6z=-8 \end{cases} \xrightarrow[2E_2+E_3]{2E_2+E_1}$

$\begin{cases} 5x+z=25 \\ 2x-y-z=-2 \\ 7x-8z=-12 \end{cases} \xrightarrow{8E_1+E_3}$

$$47x=188$$
$$x=4$$
$$5(4)+z=25$$
$$z=5$$
$$2(4)-y-5=-2$$
$$y=5$$

Solution: $(4, 5, 5)$

89. Solve: $\begin{cases} 2x+5y=6 \\ 3y+8z=-6 \\ x+4y=-5 \end{cases}$

Solve for x in equation 3: $x=-4y-5$

Substitute in equation 1:

$$2(-4y-5)+5y=6$$
$$-8y-10+5y=6$$
$$-3y=16$$
$$y=-\frac{16}{3}$$

Then, $x=-4\left(-\frac{16}{3}\right)-5=\frac{49}{3}$

From the equation 2:

$$z=\frac{-3\left(\frac{-16}{3}\right)-6}{8}=\frac{10}{8}=\frac{5}{4}$$

Solution: $\left(\frac{49}{3}, -\frac{16}{3}, \frac{5}{4}\right)$

91. Solve: $\begin{cases} 9x+4y-8z=-4 \\ -6x+3y-9z=-9 \\ 8y-3z=18 \end{cases}$

Solve for z in equation 3:

$$z=\frac{8y-18}{3}$$

Substitute in equations 1 and 2:

$$\begin{cases} 9x+4y-\dfrac{64y}{3}+\dfrac{144}{3}=-4 \\ -6x+3y-\dfrac{72y}{3}+\dfrac{162}{3}=-9 \end{cases}$$

$$\begin{cases} 9x-\dfrac{52y}{3}=-\dfrac{156}{3} \\ -6x-\dfrac{63y}{3}=-\dfrac{189}{3} \end{cases}$$

Simplify:

$$\begin{cases} 9x-\dfrac{52y}{3}=-52 \\ -6x-21y=-63 \end{cases}$$

Multiply the first equation by 3 and divide the second equation by -3:

$$\begin{cases} 27x-52y=-156 \\ 2x+7y=21 \end{cases}$$

Solve for x in equation 2:

$$x=\frac{21-7y}{2}$$

Substitute in equation 1:

$$27\left(\frac{21-7y}{2}\right)-52y=-156$$

$$\frac{567-189y}{2}-52y=-156$$

$$567-189y-104y=-312$$

$$293y=879$$

$$y=3$$

Substituting in the solution for z:

$$z=\frac{8(3)-18}{3}=2$$

Solving for x:

$$9x+4(3)-8(2)=-4$$

$$9x=0$$

$$x=0$$

Solution: $(0, 3, 2)$

93. Solve: $\begin{cases} 3x-7y+z=4 \\ 5x+2y-9z=2 \\ -2x+7y-z=1 \end{cases}$

Solve for z in equation 3:

$z=-2x+7y-1$

Substitute in equations 1 and 2:

$$\begin{cases} 3x-7y-2x+7y-1=4 \\ 5x+2y+18x-63y+9=2 \end{cases}$$

Simplify:

$$\begin{cases} x-1=4 \\ 23x-61y=-7 \end{cases}$$

From equation 1, $x=5$.

Substitute in equation 2:

$$23(5)-61y=-7$$

$$-61y=-122$$

$$y=2$$

Back substitute to solve for z:

$z=-2(5)+7(2)-1=3$

Solution: $(5, 2, 3)$

95. Solve: $\begin{cases} 3y+8z=-11 \\ 5x-7y-6z=-10 \\ 15x-7y-2z=4 \end{cases}$

Multiply equation 2 by -3 and add to equation 3 to eliminate x:

$$\begin{cases} -15x+21y+18z=30 \\ 15x-7y-2z=4 \end{cases}$$

$$14y+16z=34$$

Simplify by dividing the equation by 2 and multiply equation 1 by -1. Then, add:

$$\begin{cases} 7y+8z=17 \\ -3y-8z=11 \end{cases}$$

$$4y=28, \quad y=7$$

Use equation 1 to solve for z:

$$3(7)+8z=-11$$

$$z=-4$$

Use equation 2 to solve for x:

$$5x-7(7)-6(-4)=-10$$

$$5x=15$$

$$x=3$$

Solution: $(3, 7, -4)$

97. Solve: $\begin{cases} 21x - 7y + 51z = 141 \xrightarrow[7E_2]{9E_1} \\ 13x + 9y - 5z = -19 \\ 19x - 8y + 23z = 30 \end{cases}$

$\begin{cases} 189x - 63y + 459z = 1269 \\ 91x + 63y - 35z = -133 \end{cases}$

Add equations 1 and 2 to eliminate y:

(A) $280x + 424z = 1136$

In similar fashion, we can construct another equation in x and z from another manipulation of two equations in the original set:

$\begin{cases} 21x - 7y + 51z = 141 \xrightarrow[9E_3]{8E_2} \\ 13x + 9y - 5z = -19 \\ 19x - 8y + 23z = 30 \end{cases}$

$\begin{cases} 104x + 72y - 40z = -152 \\ 171x - 72y + 207z = 270 \end{cases}$

Add equations 1 and 2 to eliminate y:

(B) $275x + 167z = 118$

Simplify equation A by dividing both sides by 8. Then, use equations A and B to eliminate x (multiply equation A by 275 and equation B by 35):

$$\begin{aligned} 9625x + 14575z &= 39050 \\ -9625x - 5845z &= -4130 \\ 8730z &= 34920 \\ z &= 4 \end{aligned}$$

Solve for x in $280x + 424(4) = 1136$

$x = -2$

Solve for y in $21(-2) - 7y + 51(4) = 141$

$y = 3$

Solution: $(-2, 3, 4)$

99. Solve: $\begin{cases} -23x + 17y - 7z = -51 \xrightarrow{-4E_1} \\ -13x + 25y - 11z = 45 \\ 51x - 21y - 28z = -58 \end{cases}$

$\begin{cases} 92x - 68y + 28z = 204 \\ 51x - 21y - 28z = -58 \end{cases}$

Add equations 1 and 2 to eliminate z:

(A) $143x - 89y = 146$

In similar fashion, we can construct another equation in x and z from another manipulation of two equations in the original set:

$\begin{cases} -23x + 17y - 7z = -51 \xrightarrow[7E_2]{-11E_1} \\ -13x + 25y - 11z = 45 \end{cases}$

$\begin{cases} 253x - 187y + 77z = 561 \\ -91x + 175y - 77z = 315 \end{cases}$

Add equations 1 and 2 (and divide by 2) to eliminate z:

(B) $81x - 6y = 438$

Eliminate y by multiplying equation A by 6 and equation B by -89. Then, add:

$$\begin{aligned} 858x - 534y &= 876 \\ -7209x + 534y &= -38982 \\ -6351x &= -38106 \\ x &= 6 \end{aligned}$$

Solve for y in equation B:

$$\begin{aligned} 81(6) - 6y &= 438 \\ y &= 8 \end{aligned}$$

Solve for z by substituting in equation 1:

$$\begin{aligned} -23(6) + 17(8) - 7z &= -51 \\ z &= 7 \end{aligned}$$

Solution: $(6, 8, 7)$

101. The system of equations can be constructed on the basis

$$\begin{cases} y\text{-intercept is } 5 \\ \text{a line contains } (1,3) \\ \text{a line contains } (2,0) \end{cases} \longrightarrow$$

$$\begin{cases} c=5 \\ a+b+c=3 \\ 4a+2b+c=0 \end{cases} \xrightarrow[-1E_1+E_3]{-1E_1+E_2}$$

$$\begin{cases} c=5 \\ a+b=-2 \\ 4a+2b=-5 \end{cases} \xrightarrow{-2E_2+E_3}$$

$$\begin{cases} c=5 \\ a+b=-2 \\ 2a=-1 \end{cases}$$

$$a=-\frac{1}{2},\ b=-\frac{3}{2},\ c=5$$

103. Let $x=$ amount @ 8%
$y=$ amount @ 15%

$$\begin{cases} x-y=2000 \\ 0.08x+0.15y=1310 \end{cases} \xrightarrow{100E_2}$$

$$\begin{cases} x-y=2000 \\ 8x+15y=131{,}000 \end{cases} \xrightarrow{15E_1+E_2}$$

$$\begin{cases} x-y=2000 \\ 23x=161{,}000 \end{cases}$$

$x=\$7000$ (amount at 8%)
$y=\$5000$ (amount at 15%)

105. Let $x=$ no. of ounces of 12% alc. sol.
$y=$ no. of ounces of 30% alc. sol.

$$\begin{cases} x+y=60 \\ 0.12x+0.30y=0.18(60) \end{cases} \xrightarrow{100E_2}$$

$$\begin{cases} x+y=60 \\ 12x+30y=18(60) \end{cases} \xrightarrow{-12E_1+E_2}$$

$$\begin{cases} x+y=60 \\ 18y=360 \end{cases}$$

$y=20$ ounces of 30% alcohol solution
$x=40$ ounces of 12% alcohol solution

107. Let $B=$ Bob's age now
$M=$ Marla's age now

$$\begin{cases} \text{3 years ago Bob was twice as old as Marla} \\ \text{15 years ago Bob was 3 times as old as Marla} \end{cases} \longrightarrow$$

$$\begin{cases} B-3=2(M-3) \\ B-15=3(M-15) \end{cases} \xrightarrow{-1E_1+E_2}$$

$$\begin{cases} B-2M=-3 \\ -M=-27 \end{cases}$$

$M=27$
$B-2(27)=-3$
$B=51$

Bob is 51 years old.

109. Let $x=$ amount invested in stocks
Let $y=$ amount invested in bonds

$$\begin{cases} x+y=1000 \\ x=4y \end{cases}$$

Substituting equation 2 in equation 1, solve for y:

$4y+y=1000$
$5y=1000$
$y=\$200$

Then, $x=1000-y=\$800$ (in stocks).

111. Let $x=$ number of night-time tickets
Let $y=$ number of day-time tickets

$$\begin{cases} 7x+5y=740 \\ x+y=120 \end{cases}$$

Solve for x in equation 2 and substitute in equation 1. Then, solve for y:

$7(120-y)+5y=740$
$840-7y+5y=740$
$-2y=-100$
$y=50$ day-time tickets

113. Let x = amount invested in CD (5%)

Let y = amount invested in bonds (4%)

Let z = amount invested in stocks (13.5%)

$$\begin{cases} x+y+z=10{,}000 \\ z=2x \\ 0.05x+0.04y+0.135z=1000 \end{cases}$$

Simplify equation 3:

$50x+40y+135z=1{,}000{,}000$

Substitute $z=2x$ in equation 1 and in equation 3:

$$\begin{cases} x+y+2x=10{,}000 \\ 50x+40y+270x=1{,}000{,}000 \end{cases}$$

Simplifying:

$$\begin{cases} 3x+y=10{,}000, \text{ or } y=10{,}000-3x \\ 320x+40y=1{,}000{,}000 \end{cases}$$

Substituting in equation 2, solve for x:

$$320x+40(10{,}000-3x)=1{,}000{,}000$$
$$200x=600{,}000$$
$$x=\$3{,}000$$
$$z=\$6{,}000$$
$$y=\$1{,}000$$

Solution: \$3,000 in CDs
\$6,000 in stocks
\$1,000 in bonds

End of Section 10.1

Chapter Ten: Section 10.2
Solutions to Odd-numbered Exercises

1. Given: $A = \begin{bmatrix} 4 & -1 \\ 0 & 3 \\ 9 & -5 \end{bmatrix}$

a. The order is 3 x 2 since there are 3 rows and 2 columns.

b. $a_{12} = -1$ since -1 is in the 1st row, 2nd column.

c. There is no 3rd column.

3. Given: $C = \begin{bmatrix} 1 & 0 \\ 5 & -3 \\ 2 & 9 \\ \pi & e \\ 10 & -7 \end{bmatrix}$

a. The order is 5 x 2 since there are 5 rows and 2 columns.

b. There is no 3rd column.

c. $c_{51} = 10$ since 10 is in the 5th row, 1st column.

5. Given: $E = \begin{bmatrix} -443 & 951 & 165 & 274 \\ 286 & -653 & 812 & -330 \\ 909 & 377 & 429 & -298 \end{bmatrix}$

a. The order is 3 x 4 since there are 3 rows and 4 columns.

b. There is no 4th row.

c. $e_{21} = 286$ since 286 is in the 2nd row, 1st column.

7. Given: $B = \begin{bmatrix} 8 & 1 \\ 3 & 0 \\ 6 & 7 \end{bmatrix}$

a. order of B: 3 x 2 since there are 3 rows and 2 columns.

b. $b_{12} = 1$ since 1 is in the 1st row, 2nd column.

c. b_{13} in not applicable for B, since there isn't a 3rd column.

9. Given: $D = \begin{bmatrix} 4 & 9 & 7 & 1 & 8 \\ 5 & 3 & 0 & 2 & 6 \end{bmatrix}$

a. order of D: 2 x 5 since there are 2 rows and 5 columns.

b. $d_{21} = 5$ since 5 is in the 2nd row, 1st column.

c. $d_{24} = 2$ since 2 is in the 2nd row, 4th column.

11. $\begin{cases} \dfrac{2-3x}{2} = y \\ 3z + 2(x+y) = 0 \\ 2x - y = 2(x - 3z) \end{cases} \longrightarrow$

$\begin{cases} -\dfrac{3}{2}x - y = -1 \\ 2x + 2y + 3z = 0 \\ -y + 6z = 0 \end{cases} \longrightarrow$

$$\left[\begin{array}{ccc|c} -\frac{3}{2} & -1 & 0 & -1 \\ 2 & 2 & 3 & 0 \\ 0 & -1 & 6 & 0 \end{array}\right]$$

13. $\begin{cases} \dfrac{12x-1}{5} + \dfrac{y}{2} = \dfrac{3z}{2} \\ y - (x + 3z) = -(1 - y) \\ 2x - 2 - z - 2y = 7x \end{cases} \longrightarrow$

$\begin{cases} \dfrac{12}{5}x + \dfrac{1}{2}y - \dfrac{3}{2}z = \dfrac{1}{5} \\ x + 3z = 1 \\ 5x + 2y + z = -2 \end{cases} \longrightarrow$

$$\left[\begin{array}{ccc|c} \frac{12}{5} & \frac{1}{2} & -\frac{3}{2} & \frac{1}{5} \\ 1 & 0 & 3 & 1 \\ 5 & 2 & 1 & -2 \end{array}\right]$$

15. First, put each equation in standard form:

$$\begin{cases} \frac{3}{2}x+2y-3z=6 \\ 3x-6y+27z=0 \\ 2x+6y+z=3 \end{cases}$$

$$\left[\begin{array}{ccc|c} \frac{3}{2} & 2 & -3 & 6 \\ 3 & -6 & 27 & 0 \\ 2 & 6 & 1 & 3 \end{array}\right]$$

17. First, put each equation in standard form:

$$\begin{cases} -3x+y-2z=-4 \\ \frac{1}{2}x-4y-z=1 \\ -3y+3z=1 \end{cases}$$

$$\left[\begin{array}{ccc|c} -3 & 1 & -2 & -4 \\ \frac{1}{2} & -4 & -1 & 1 \\ 0 & -3 & 3 & 1 \end{array}\right]$$

19. $\begin{cases} \frac{1}{2}x-14y-\frac{1}{4}z=-8 \\ \frac{1}{5}x-\frac{7}{6}y+\frac{1}{4}z=-3 \\ 5x-5y+\frac{8}{3}z=-5 \end{cases}$

$$\left[\begin{array}{ccc|c} \frac{1}{2} & -14 & -\frac{1}{4} & -8 \\ \frac{1}{5} & -\frac{7}{6} & \frac{1}{4} & -3 \\ 5 & -5 & \frac{8}{3} & -5 \end{array}\right]$$

21. $\left[\begin{array}{cc|c} 2 & -5 & 3 \\ -4 & 3 & -1 \end{array}\right] \xrightarrow{2R_1+R_2} \left[\begin{array}{cc|c} 2 & -5 & 3 \\ 0 & -7 & 5 \end{array}\right]$

23. $\left[\begin{array}{cc|c} 9 & -2 & 7 \\ 1 & 3 & -2 \end{array}\right] \xrightarrow{R_1 \leftrightarrow R_2} \left[\begin{array}{cc|c} 1 & 3 & -2 \\ 9 & -2 & 7 \end{array}\right]$

25. $\left[\begin{array}{cc|c} 8 & -2 & -4 \\ 3 & -1 & 7 \end{array}\right] \xrightarrow{-2R_2} \left[\begin{array}{cc|c} 8 & -2 & -4 \\ -6 & 2 & -14 \end{array}\right]$

27. $\left[\begin{array}{cc|c} 4 & 12 & -6 \\ 7 & 3 & 9 \end{array}\right] \xrightarrow[\frac{1}{2}R_1+R_2]{} \left[\begin{array}{cc|c} 4 & 12 & -6 \\ 9 & 9 & 6 \end{array}\right]$

29. $\left[\begin{array}{cc|c} 8 & -2 & 10 \\ 9 & -3 & 0 \end{array}\right] \xrightarrow[-\frac{2}{3}R_2]{\frac{1}{2}R_1} \left[\begin{array}{cc|c} 4 & -1 & 5 \\ -6 & 2 & 0 \end{array}\right]$

31. $\left[\begin{array}{ccc|c} 6 & -2 & 5 & 14 \\ -7 & 19 & 2 & 3 \\ -9 & 11 & -4 & 7 \end{array}\right] \xrightarrow[0.5R_3]{3R_1}$

$$\left[\begin{array}{ccc|c} 18 & -6 & 15 & 42 \\ -7 & 19 & 2 & 3 \\ -4.5 & 5.5 & -2 & 3.5 \end{array}\right]$$

33. $\left[\begin{array}{ccc|c} 8 & 11 & 18 & 2 \\ 14 & 33 & -3 & -5 \\ -9 & 21 & 12 & 9 \end{array}\right] \xrightarrow[-2R_3+R_2]{\frac{1}{3}R_3+R_1}$

$$\left[\begin{array}{ccc|c} 5 & 18 & 22 & 5 \\ 32 & -9 & -27 & -23 \\ -9 & 21 & 12 & 9 \end{array}\right]$$

35. $\left[\begin{array}{ccc|c} 2 & 3 & -3 & 5 \\ 1 & 1 & 3 & 4 \\ 3 & 3 & 9 & 12 \end{array}\right] \xrightarrow[-3R_2+R_3]{-2R_2+R_1}$

$$\left[\begin{array}{ccc|c} 0 & 1 & -9 & -3 \\ 1 & 1 & 3 & 4 \\ 0 & 0 & 0 & 0 \end{array}\right]$$

37. $\left[\begin{array}{cc|c} -5 & 20 & -15 \\ 2 & -12 & 5 \end{array}\right] \xrightarrow[\frac{1}{2}R_2]{\frac{1}{5}R_1} \left[\begin{array}{cc|c} -1 & 4 & -3 \\ 1 & -6 & \frac{5}{2} \end{array}\right]$

39. $\left[\begin{array}{ccc|c} 1 & 5 & -9 & 11 \\ 1 & 4 & -1 & 4 \\ 4 & 3 & 5 & 45 \end{array}\right] \xrightarrow[-4R_1+R_3]{-R_1+R_2}$

$$\left[\begin{array}{ccc|c} 1 & 5 & -9 & 11 \\ 0 & -1 & 8 & -7 \\ 0 & -17 & 41 & 1 \end{array}\right]$$

41. $\begin{cases} 2x-5y=11 \\ 3x+2y=7 \end{cases}$

$$\left[\begin{array}{cc|c} 2 & -5 & 11 \\ 3 & 2 & 7 \end{array}\right] \xrightarrow{\frac{1}{2}R_1}$$

$$\left[\begin{array}{cc|c} 1 & -\frac{5}{2} & \frac{11}{2} \\ 3 & 2 & 7 \end{array}\right] \xrightarrow{-3R_1+R_2} \left[\begin{array}{cc|c} 1 & -\frac{5}{2} & \frac{11}{2} \\ 0 & \frac{19}{2} & -\frac{19}{2} \end{array}\right]$$

This matrix translates to $\frac{19}{2}y=-\frac{19}{2}$,

or $y=-1$.

Then, substituting we get

$x-\frac{5}{2}(-1)=\frac{11}{2}$, or $x=3$

The solution is $(3,-1)$.

43. Solve: $\begin{cases} x-4y=-11 \\ 7x-y=4 \end{cases}$

$$\left[\begin{array}{cc|c} 1 & -4 & -11 \\ 7 & -1 & 4 \end{array}\right] \xrightarrow{-7R_1+R_2}$$

$$\left[\begin{array}{cc|c} 1 & -4 & -11 \\ 0 & 27 & 81 \end{array}\right] \xrightarrow{\frac{1}{27}R_2} \left[\begin{array}{cc|c} 1 & -4 & -11 \\ 0 & 1 & 3 \end{array}\right]$$

This matrix translates to $x-4y=-11$,

or $x=4y-11$, and $y=3$

Substituting: $x=4(3)-11=1$

Solution: $(1,3)$

45. Solve: $\begin{cases} 2x+6y=4 \\ -4x-7y=7 \end{cases}$

$$\left[\begin{array}{cc|c} 2 & 6 & 4 \\ -4 & -7 & 7 \end{array}\right] \xrightarrow[2R_1+R_2]{\frac{1}{2}R_1}$$

$$\left[\begin{array}{cc|c} 1 & 3 & 2 \\ 0 & 5 & 15 \end{array}\right] \xrightarrow{\frac{1}{5}R_2} \left[\begin{array}{cc|c} 1 & 3 & 2 \\ 0 & 1 & 3 \end{array}\right]$$

This matrix translates to $x+3y=2$,

or $x=-3y+2$, and $y=3$

Substituting: $x=-3(3)+2=-7$

Solution: $(-7,3)$

47. Solve: $\begin{cases} 2x+y=-2 \\ -4x-2y=5 \end{cases}$

$$\left[\begin{array}{cc|c} 2 & 1 & -2 \\ -4 & -2 & 5 \end{array}\right] \xrightarrow{\frac{1}{2}R_1}$$

$$\left[\begin{array}{cc|c} 1 & \frac{1}{2} & -1 \\ -4 & -2 & 5 \end{array}\right] \xrightarrow{4R_1+R_2} \left[\begin{array}{cc|c} 1 & \frac{1}{2} & -1 \\ 0 & 0 & 1 \end{array}\right]$$

This matrix translates to the system

$\begin{cases} x+\frac{1}{2}y=-1 \\ 0=1 \end{cases}$ which is an inconsistency

because $0\neq 1$. The solution set: $\varnothing$

49. Solve: $\begin{cases} 2x-3y=0 \\ 5x+y=17 \end{cases}$

$$\left[\begin{array}{cc|c} 2 & -3 & 0 \\ 5 & 1 & 17 \end{array}\right] \xrightarrow[-\frac{5}{2}R_1+R_2]{\frac{1}{2}R_1}$$

$$\left[\begin{array}{cc|c} 1 & -\frac{3}{2} & 0 \\ 0 & \frac{17}{2} & 17 \end{array}\right] \xrightarrow{\frac{2}{17}R_2} \left[\begin{array}{cc|c} 1 & -\frac{3}{2} & 0 \\ 0 & 1 & 2 \end{array}\right]$$

This matrix translates to $x-\frac{3}{2}y=0$,

or $x=\frac{3}{2}y$, and $y=2$

Substituting: $x=\frac{3}{2}(2)=3$

Solution: $(3,2)$

51. Solve: $\begin{cases} 3x+6y=-12 \\ 2x+4y=-8 \end{cases}$

$$\left[\begin{array}{cc|c} 3 & 6 & -12 \\ 2 & 4 & -8 \end{array}\right] \xrightarrow[\frac{1}{2}R_2]{\frac{1}{3}R_1}$$

$$\left[\begin{array}{cc|c} 1 & 2 & -4 \\ 1 & 2 & -4 \end{array}\right] \xrightarrow{-R_1+R_2} \left[\begin{array}{cc|c} 1 & 2 & -4 \\ 0 & 0 & 0 \end{array}\right]$$

This matrix translates to $x+2y=-4$,

or $x=-2y-4$.

The solution set is $\{(-2y-4, y) \mid y\in\mathbb{R}\}$.

53. Solve: $\begin{cases} \frac{2}{3}x + 2y = 1 \\ x + 3y = 0 \end{cases}$

$$\left[\begin{array}{cc|c} \frac{2}{3} & 2 & 1 \\ 1 & 3 & 0 \end{array}\right] \xrightarrow[3R_1]{R_1 \leftrightarrow R_2}$$

$$\left[\begin{array}{cc|c} 1 & 3 & 0 \\ 2 & 6 & 3 \end{array}\right] \xrightarrow{-2R_1+R_2} \left[\begin{array}{cc|c} 1 & 3 & 0 \\ 0 & 0 & 3 \end{array}\right]$$

This matrix translates to the system $\begin{cases} x + 3y = 0 \\ 0 = 3 \end{cases}$ which is an inconsistency because $0 \neq 3$. The solution set: $\varnothing$

55. Solve: $\begin{cases} 3x - 9y - 7z = -9 \\ 5x + 11y - z = 17 \\ -4x - 8y + 7z = 5 \end{cases}$

$$\left[\begin{array}{ccc|c} 3 & -9 & -7 & -9 \\ 5 & 11 & -1 & 17 \\ -4 & -8 & 7 & 5 \end{array}\right] \xrightarrow[R_1 \longleftrightarrow R_3]{R_2+R_3}$$

$$\left[\begin{array}{ccc|c} 1 & 3 & 6 & 22 \\ 5 & 11 & -1 & 17 \\ 3 & -9 & -7 & -9 \end{array}\right] \xrightarrow[-3R_1+R_3]{-5R_1+R_2}$$

$$\left[\begin{array}{ccc|c} 1 & 3 & 6 & 22 \\ 0 & -4 & -31 & -93 \\ 0 & -18 & -25 & -75 \end{array}\right] \xrightarrow[-\frac{1}{4}R_2]{}$$

$$\left[\begin{array}{ccc|c} 1 & 3 & 6 & 22 \\ 0 & 1 & \frac{31}{4} & \frac{93}{4} \\ 0 & -18 & -25 & -75 \end{array}\right] \xrightarrow{18R_2-R_3}$$

$$\left[\begin{array}{ccc|c} 1 & 3 & 6 & 22 \\ 0 & 1 & \frac{31}{4} & \frac{93}{4} \\ 0 & 0 & \frac{329}{2} & \frac{987}{2} \end{array}\right] \xrightarrow{\frac{2}{329}R_3}$$

$$\left[\begin{array}{ccc|c} 1 & 3 & 6 & 22 \\ 0 & 1 & \frac{31}{4} & \frac{93}{4} \\ 0 & 0 & 1 & 3 \end{array}\right]$$

Then, $z = 3$.
Back substituting:

$$y + \frac{31}{4}(3) = \frac{93}{4}$$
$$y = 0$$
$$x + 3(0) + 6(3) = 22$$
$$x = 4$$

Solution: $(4, 0, 3)$

57. Solve: $\begin{cases} 17x + 13y + 8z = 46 \\ -12x + 3y + 28z = -19 \\ 14x + 5y - 15z = -15 \end{cases}$

$$\left[\begin{array}{ccc|c} 17 & 13 & 8 & 46 \\ -12 & 3 & 28 & -19 \\ 14 & 5 & -15 & -15 \end{array}\right] \xrightarrow[R_2+R_3]{R_2+R_1}$$

$$\left[\begin{array}{ccc|c} 5 & 16 & 36 & 27 \\ -12 & 3 & 28 & -19 \\ 2 & 8 & 13 & -34 \end{array}\right] \xrightarrow[6R_3+R_2]{\frac{1}{2}R_3 \leftrightarrow R_1}$$

$$\left[\begin{array}{ccc|c} 1 & 4 & \frac{13}{2} & -17 \\ 0 & 51 & 106 & -223 \\ 5 & 16 & 36 & 27 \end{array}\right] \xrightarrow[-5R_1+R_3]{}$$

$$\left[\begin{array}{ccc|c} 1 & 4 & \frac{13}{2} & -17 \\ 0 & 51 & 106 & -223 \\ 0 & -4 & \frac{7}{2} & 112 \end{array}\right] \xrightarrow{4R_2+51R_3}$$

$$\left[\begin{array}{ccc|c} 1 & 4 & \frac{13}{2} & -17 \\ 0 & 51 & 106 & -223 \\ 0 & 0 & \frac{1205}{2} & 4820 \end{array}\right] \xrightarrow[\frac{2}{1205}R_3]{\frac{1}{51}R_2}$$

$$\left[\begin{array}{ccc|c} 1 & 4 & \frac{13}{2} & -17 \\ 0 & 1 & \frac{106}{51} & -\frac{223}{51} \\ 0 & 0 & 1 & 8 \end{array}\right]$$

Then, $z = 8$.
Back substituting:

$$y+\frac{106}{51}(8)=-\frac{223}{51}$$
$$y=-21$$
$$x+4(-21)+\frac{13}{2}(8)=-17$$
$$x=15$$

Solution: $(15,-21,8)$

59. Solve $\begin{cases} \frac{2}{3}x+y=-3 \\ 3x+\frac{5}{2}y=-\frac{7}{2} \end{cases}$ with

Gauss-Jordan elimination:

$$\left[\begin{array}{cc|c} \frac{2}{3} & 1 & -3 \\ 3 & \frac{5}{2} & -\frac{7}{2} \end{array}\right] \xrightarrow{\frac{3}{2}R_1}$$

$$\left[\begin{array}{cc|c} 1 & \frac{3}{2} & -\frac{9}{2} \\ 3 & \frac{5}{2} & -\frac{7}{2} \end{array}\right] \xrightarrow{-3R_1+R_2}$$

$$\left[\begin{array}{cc|c} 1 & \frac{3}{2} & -\frac{9}{2} \\ 0 & -2 & 10 \end{array}\right] \xrightarrow{-\frac{1}{2}R_2}$$

$$\left[\begin{array}{cc|c} 1 & \frac{3}{2} & -\frac{9}{2} \\ 0 & 1 & -5 \end{array}\right] \xrightarrow{-\frac{3}{2}R_2+R_1} \left[\begin{array}{cc|c} 1 & 0 & 3 \\ 0 & 1 & -5 \end{array}\right]$$

This matrix translates to $x=3$ and $y=-5$

Solution: $(3,-5)$

61. Solve $\begin{cases} 6x+2y=-4 \\ -9x-3y=6 \end{cases}$ with

Gauss-Jordan elimination:

$$\left[\begin{array}{cc|c} 6 & 2 & -4 \\ -9 & -3 & 6 \end{array}\right] \xrightarrow[-\frac{1}{3}R_2]{\frac{1}{2}R_1}$$

$$\left[\begin{array}{cc|c} 3 & 1 & -2 \\ 3 & 1 & -2 \end{array}\right] \xrightarrow[\frac{1}{3}R_1]{-R_1+R_2} \left[\begin{array}{cc|c} 1 & \frac{1}{3} & -\frac{2}{3} \\ 0 & 0 & 0 \end{array}\right]$$

This matrix translates to

$x=-\frac{1}{3}y-\frac{2}{3}$ or $y=-3x-2$

Solution Set: $\{(x,-3x-2)\mid x\in\mathbb{R}\}$

63. Solve $\begin{cases} 3x+8y=-4 \\ x+2y=-2 \end{cases}$ with

Gauss-Jordan elimination:

$$\left[\begin{array}{cc|c} 3 & 8 & -4 \\ 1 & 2 & -2 \end{array}\right] \xrightarrow[R_1\leftrightarrow R_2]{-3R_2+R_1}$$

$$\left[\begin{array}{cc|c} 1 & 2 & -2 \\ 0 & 2 & 2 \end{array}\right] \xrightarrow{\frac{1}{2}R_2}$$

$$\left[\begin{array}{cc|c} 1 & 2 & -2 \\ 0 & 1 & 1 \end{array}\right] \xrightarrow{-2R_2+R_1} \left[\begin{array}{cc|c} 1 & 0 & -4 \\ 0 & 1 & 1 \end{array}\right]$$

This matrix translates to

$x=-4$, or $y=1$

Solution Set: $(-4,1)$

65. Solve $\begin{cases} 9x-11y=10 \\ -4x+3y=-12 \end{cases}$ with

Gauss-Jordan elimination:

$$\left[\begin{array}{cc|c} 9 & -11 & 10 \\ -4 & 3 & -12 \end{array}\right] \xrightarrow{\frac{1}{9}R_1}$$

$$\left[\begin{array}{cc|c} 1 & -\frac{11}{9} & \frac{10}{9} \\ -4 & 3 & -12 \end{array}\right] \xrightarrow{4R_1+R_2}$$

$$\left[\begin{array}{cc|c} 1 & -\frac{11}{9} & \frac{10}{9} \\ 0 & -\frac{17}{9} & -\frac{68}{9} \end{array}\right] \xrightarrow{-\frac{9}{17}R_2}$$

$$\left[\begin{array}{cc|c} 1 & -\frac{11}{9} & \frac{10}{9} \\ 0 & 1 & 4 \end{array}\right] \xrightarrow{\frac{11}{9}R_2+R_1} \left[\begin{array}{cc|c} 1 & 0 & 6 \\ 0 & 1 & 4 \end{array}\right]$$

This matrix translates to $x=6$ and $y=4$

Solution: $(6,4)$

67. Solve $\begin{cases} 3x-8y=7 \\ 18x-35y=-23 \end{cases}$ with

Gauss-Jordan elimination:

$$\left[\begin{array}{cc|c} 3 & -8 & 7 \\ 18 & -35 & -23 \end{array}\right] \xrightarrow[\frac{1}{3}R_1]{-6R_1+R_2} \left[\begin{array}{cc|c} 1 & -\frac{8}{3} & \frac{7}{3} \\ 0 & 13 & -65 \end{array}\right] \xrightarrow{\frac{1}{13}R_2} \left[\begin{array}{cc|c} 1 & -\frac{8}{3} & \frac{7}{3} \\ 0 & 1 & -5 \end{array}\right] \xrightarrow{\frac{8}{3}R_2+R_1} \left[\begin{array}{cc|c} 1 & 0 & -11 \\ 0 & 1 & -5 \end{array}\right]$$

This matrix translates to

$x=-11$ and $y=-5$

Solution Set: $(-11,-5)$

69. Solve $\begin{cases} -5x+9y+3z=1 \\ 3x+2y-6z=9 \\ x+4y-z=16 \end{cases}$ with

Gauss-Jordan elimination:

$$\left[\begin{array}{ccc|c} -5 & 9 & 3 & 1 \\ 3 & 2 & -6 & 9 \\ 1 & 4 & -1 & 16 \end{array}\right] \xrightarrow{R_3 \leftrightarrow R_1} \left[\begin{array}{ccc|c} 1 & 4 & -1 & 16 \\ 3 & 2 & -6 & 9 \\ -5 & 9 & 3 & 1 \end{array}\right] \xrightarrow[5R_1+R_3]{-3R_1+R_2} \left[\begin{array}{ccc|c} 1 & 4 & -1 & 16 \\ 0 & -10 & -3 & -39 \\ 0 & 29 & -2 & 81 \end{array}\right] \xrightarrow[-\frac{1}{10}R_2]{}$$

$$\left[\begin{array}{ccc|c} 1 & 4 & -1 & 16 \\ 0 & 1 & \frac{3}{10} & \frac{39}{10} \\ 0 & 29 & -2 & 81 \end{array}\right] \xrightarrow{-29R_2+R_3} \left[\begin{array}{ccc|c} 1 & 4 & -1 & 16 \\ 0 & 1 & \frac{3}{10} & \frac{39}{10} \\ 0 & 0 & -\frac{107}{10} & -\frac{321}{10} \end{array}\right] \xrightarrow{-\frac{10}{107}R_3} \left[\begin{array}{ccc|c} 1 & 4 & -1 & 16 \\ 0 & 1 & \frac{3}{10} & \frac{39}{10} \\ 0 & 0 & 1 & 3 \end{array}\right] \xrightarrow{-\frac{3}{10}R_3+R_2}$$

$$\left[\begin{array}{ccc|c} 1 & 4 & -1 & 16 \\ 0 & 1 & 0 & 3 \\ 0 & 0 & 1 & 3 \end{array}\right] \xrightarrow[R_3+R_1]{-4R_2+R_1} \left[\begin{array}{ccc|c} 1 & 0 & 0 & 7 \\ 0 & 1 & 0 & 3 \\ 0 & 0 & 1 & 3 \end{array}\right] \longrightarrow \begin{cases} x=7 \\ y=3 \\ z=3 \end{cases} \text{ or } (7,3,3)$$

71. Solve $\begin{cases} x+y=4 \\ y+3z=-1 \\ 2x-2y+5z=-5 \end{cases}$ with

Gauss-Jordan elimination:

$$\left[\begin{array}{ccc|c} 1 & 1 & 0 & 4 \\ 0 & 1 & 3 & -1 \\ 2 & -2 & 5 & -5 \end{array}\right] \xrightarrow[-R_2+R_1]{-2R_1+R_3} \left[\begin{array}{ccc|c} 1 & 0 & -3 & 5 \\ 0 & 1 & 3 & -1 \\ 0 & -4 & 5 & -13 \end{array}\right] \xrightarrow{4R_2+R_3} \left[\begin{array}{ccc|c} 1 & 0 & -3 & 5 \\ 0 & 1 & 3 & -1 \\ 0 & 0 & 17 & -17 \end{array}\right] \xrightarrow{\frac{1}{17}R_3}$$

$$\left[\begin{array}{ccc|c} 1 & 0 & -3 & 5 \\ 0 & 1 & 3 & -1 \\ 0 & 0 & 1 & -1 \end{array}\right] \xrightarrow{-3R_3+R_2} \left[\begin{array}{ccc|c} 1 & 0 & -3 & 5 \\ 0 & 1 & 0 & 2 \\ 0 & 0 & 1 & -1 \end{array}\right] \xrightarrow{3R_3+R_1} \left[\begin{array}{ccc|c} 1 & 0 & 0 & 2 \\ 0 & 1 & 0 & 2 \\ 0 & 0 & 1 & -1 \end{array}\right] \longrightarrow \begin{cases} x=2 \\ y=2 \\ z=-1 \end{cases} \text{ or } (2,2,-1)$$

73. Solve $\begin{cases} 3x+8z=3 \\ -3x-7z=-3 \\ x+3z=1 \end{cases}$ with

Gauss-Jordan elimination:

$$\left[\begin{array}{ccc|c} 3 & 0 & 8 & 3 \\ -3 & 0 & -7 & -3 \\ 1 & 0 & 3 & 1 \end{array}\right] \xrightarrow{R_1 \leftrightarrow R_3} \left[\begin{array}{ccc|c} 1 & 0 & 3 & 1 \\ -3 & 0 & -7 & -3 \\ 3 & 0 & 8 & 3 \end{array}\right] \xrightarrow{R_3+R_2}$$

$$\left[\begin{array}{ccc|c} 1 & 0 & 3 & 1 \\ 0 & 0 & 1 & 0 \\ 3 & 0 & 8 & 3 \end{array}\right] \xrightarrow{-3R_1+R_3} \left[\begin{array}{ccc|c} 1 & 0 & 3 & 1 \\ 0 & 0 & 1 & 0 \\ 0 & 0 & -1 & 0 \end{array}\right] \xrightarrow[-3R_2+R_1]{R_2+R_3}$$

$$\left[\begin{array}{ccc|c} 1 & 0 & 0 & 1 \\ 0 & 0 & 1 & 0 \\ 0 & 0 & 0 & 0 \end{array}\right] \longrightarrow \begin{cases} x=1 \\ y=y \text{ for } y \in \mathbb{R} \\ z=0 \end{cases}$$

or $\{(1,\ y,\ 0) \mid y \in \mathbb{R}\}$

75. Solve $\begin{cases} x+2y=-1 \\ y+3z=7 \\ 2x+5z=21 \end{cases}$ with

Gauss-Jordan elimination:

$$\left[\begin{array}{ccc|c} 1 & 2 & 0 & -1 \\ 0 & 1 & 3 & 7 \\ 2 & 0 & 5 & 21 \end{array}\right] \xrightarrow[-2R_2+R_1]{-2R_1+R_3} \left[\begin{array}{ccc|c} 1 & 0 & -6 & -15 \\ 0 & 1 & 3 & 7 \\ 0 & -4 & 5 & 23 \end{array}\right] \xrightarrow{4R_2+R_3}$$

$$\left[\begin{array}{ccc|c} 1 & 0 & -6 & -15 \\ 0 & 1 & 3 & 7 \\ 0 & 0 & 17 & 51 \end{array}\right] \xrightarrow{\frac{1}{17}R_3} \left[\begin{array}{ccc|c} 1 & 0 & -6 & -15 \\ 0 & 1 & 3 & 7 \\ 0 & 0 & 1 & 3 \end{array}\right] \xrightarrow[6R_3+R_1]{-3R_3+R_2}$$

$$\left[\begin{array}{ccc|c} 1 & 0 & 0 & 3 \\ 0 & 1 & 0 & -2 \\ 0 & 0 & 1 & 3 \end{array}\right] \longrightarrow \begin{cases} x=3 \\ y=-2 \\ z=3 \end{cases}$$

or $(3, -2, 3)$

77. Solve $\begin{cases} 7x-8y+2z=-2 \\ 5x-3y-z=-3 \\ 8x+y-3z=7 \end{cases}$ with

Gauss-Jordan elimination:

$$\left[\begin{array}{ccc|c} 7 & -8 & 2 & -2 \\ 5 & -3 & -1 & -3 \\ 8 & 1 & -3 & 7 \end{array}\right] \xrightarrow{R_3-R_1} \left[\begin{array}{ccc|c} 1 & 9 & -5 & 9 \\ 5 & -3 & -1 & -3 \\ 8 & 1 & -3 & 7 \end{array}\right] \xrightarrow[-8R_1+R_3]{-5R_1+R_2}$$

$$\left[\begin{array}{ccc|c} 1 & 9 & -5 & 9 \\ 0 & -48 & 24 & -48 \\ 0 & -71 & 37 & -65 \end{array}\right] \xrightarrow{\frac{-1}{48}R_2} \left[\begin{array}{ccc|c} 1 & 9 & -5 & 9 \\ 0 & 1 & -\frac{1}{2} & 1 \\ 0 & -71 & 37 & -65 \end{array}\right] \xrightarrow[71R_2+R_3]{-9R_2+R_1}$$

$$\left[\begin{array}{ccc|c} 1 & 0 & -\frac{1}{2} & 0 \\ 0 & 1 & -\frac{1}{2} & 1 \\ 0 & 0 & \frac{3}{2} & 6 \end{array}\right] \xrightarrow{\frac{2}{3}R_3} \left[\begin{array}{ccc|c} 1 & 0 & -\frac{1}{2} & 0 \\ 0 & 1 & -\frac{1}{2} & 1 \\ 0 & 0 & 1 & 4 \end{array}\right] \xrightarrow[\frac{1}{2}R_3+R_2]{\frac{1}{2}R_3+R_1}$$

$$\left[\begin{array}{ccc|c} 1 & 0 & 0 & 2 \\ 0 & 1 & 0 & 3 \\ 0 & 0 & 1 & 4 \end{array}\right] \longrightarrow \begin{cases} x=2 \\ y=3 \\ z=4 \end{cases} \text{ or } (2, 3, 4)$$

79. Solve $\begin{cases} 8x+5y+3z=-2 \\ 12x-y-18z=1 \\ 7x+6y+10z=19 \end{cases}$ with

Gauss-Jordan elimination:

$$\left[\begin{array}{ccc|c} 8 & 5 & 3 & -2 \\ 12 & -1 & -18 & 1 \\ 7 & 6 & 10 & 19 \end{array}\right] \xrightarrow{-R_3+R_1} \left[\begin{array}{ccc|c} 1 & -1 & -7 & -21 \\ 12 & -1 & -18 & 1 \\ 7 & 6 & 10 & 19 \end{array}\right] \xrightarrow[-7R_1+R_3]{-12R_1+R_2}$$

$$\left[\begin{array}{ccc|c} 1 & -1 & -7 & -21 \\ 0 & 11 & 66 & 253 \\ 0 & 13 & 59 & 166 \end{array}\right] \xrightarrow[\frac{1}{11}R_2]{} \left[\begin{array}{ccc|c} 1 & -1 & -7 & -21 \\ 0 & 1 & 6 & 23 \\ 0 & 13 & 59 & 166 \end{array}\right] \xrightarrow[R_2+R_1]{-13R_2+R_3}$$

$$\left[\begin{array}{ccc|c} 1 & 0 & -1 & 2 \\ 0 & 1 & 6 & 23 \\ 0 & 0 & -19 & -133 \end{array}\right] \xrightarrow{-\frac{1}{19}R_3} \left[\begin{array}{ccc|c} 1 & 0 & -1 & 2 \\ 0 & 1 & 6 & 23 \\ 0 & 0 & 1 & 7 \end{array}\right] \xrightarrow[-6R_3+R_2]{R_3+R_1}$$

$$\left[\begin{array}{ccc|c} 1 & 0 & 0 & 9 \\ 0 & 1 & 0 & -19 \\ 0 & 0 & 1 & 7 \end{array}\right] \longrightarrow \begin{cases} x=9 \\ y=-19 \\ z=7 \end{cases}$$

or $(9,-19,7)$

81. $\begin{cases} w-x+2z=9 \\ 2w+3y=-1 \\ -2w-5y-z=0 \\ x+2y=-4 \end{cases}$

$$\left[\begin{array}{cccc|c} 1 & -1 & 0 & 2 & 9 \\ 2 & 0 & 3 & 0 & -1 \\ -2 & 0 & -5 & -1 & 0 \\ 0 & 1 & 2 & 0 & -4 \end{array}\right] \xrightarrow{R_2 \longleftrightarrow R_4} \left[\begin{array}{cccc|c} 1 & -1 & 0 & 2 & 9 \\ 0 & 1 & 2 & 0 & -4 \\ -2 & 0 & -5 & -1 & 0 \\ 2 & 0 & 3 & 0 & -1 \end{array}\right] \xrightarrow[R_3+R_4]{2R_1+R_3}$$

$$\left[\begin{array}{cccc|c} 1 & -1 & 0 & 2 & 9 \\ 0 & 1 & 2 & 0 & -4 \\ 0 & -2 & -5 & 3 & 18 \\ 0 & 0 & -2 & -1 & -1 \end{array}\right] \xrightarrow[-R_4]{2R_2+R_3} \left[\begin{array}{cccc|c} 1 & -1 & 0 & 2 & 9 \\ 0 & 1 & 2 & 0 & -4 \\ 0 & 0 & -1 & 3 & 10 \\ 0 & 0 & 2 & 1 & 1 \end{array}\right] \xrightarrow{-R_3}$$

$$\left[\begin{array}{cccc|c} 1 & -1 & 0 & 2 & 9 \\ 0 & 1 & 2 & 0 & -4 \\ 0 & 0 & 1 & -3 & -10 \\ 0 & 0 & 2 & 1 & 1 \end{array}\right] \xrightarrow{-2R_3+R_4} \left[\begin{array}{cccc|c} 1 & -1 & 0 & 2 & 9 \\ 0 & 1 & 2 & 0 & -4 \\ 0 & 0 & 1 & -3 & -10 \\ 0 & 0 & 0 & 7 & 21 \end{array}\right] \xrightarrow[\frac{1}{7}R_4]{R_2+R_1}$$

$$\left[\begin{array}{cccc|c} 1 & 0 & 2 & 2 & 5 \\ 0 & 1 & 2 & 0 & -4 \\ 0 & 0 & 1 & -3 & -10 \\ 0 & 0 & 0 & 1 & 3 \end{array}\right] \xrightarrow[-2R_3+R_2]{-2R_3+R_1} \left[\begin{array}{cccc|c} 1 & 0 & 0 & 8 & 25 \\ 0 & 1 & 0 & 6 & 16 \\ 0 & 0 & 1 & -3 & -10 \\ 0 & 0 & 0 & 1 & 3 \end{array}\right] \xrightarrow[-6R_4+R_2]{3R_4+R_3}$$

$$\left[\begin{array}{cccc|c} 1 & 0 & 0 & 8 & 25 \\ 0 & 1 & 0 & 0 & -2 \\ 0 & 0 & 1 & 0 & -1 \\ 0 & 0 & 0 & 1 & 3 \end{array}\right] \xrightarrow{-8R_4+R_1} \left[\begin{array}{cccc|c} 1 & 0 & 0 & 0 & 1 \\ 0 & 1 & 0 & 0 & -2 \\ 0 & 0 & 1 & 0 & -1 \\ 0 & 0 & 0 & 1 & 3 \end{array}\right]$$

Solution: $(1,-2,-1,3)$

End of Section 10.2

Chapter Ten: Section 10.3
Solutions to Odd-numbered Exercises

1. $\begin{vmatrix} 4 & -3 \\ 1 & 2 \end{vmatrix} = 4(2)-(1)(-3)=11$

3. $\begin{vmatrix} 0 & 3 \\ -5 & 2 \end{vmatrix} = 0(2)-(-5)(3)=15$

5. $\begin{vmatrix} a & x \\ x & b \end{vmatrix} = ab - x^2$

7. $\begin{vmatrix} -2 & 2 \\ -2 & -2 \end{vmatrix} = -2(-2)-(-2)(2)=8$

9. $\begin{vmatrix} -1 & 2 \\ 3 & 4 \end{vmatrix} = (-1)(4)-(3)(2)=-10$

11. $\begin{vmatrix} -2 & 9 \\ 5 & -3 \end{vmatrix} = (-2)(-3)-(5)(9)=-39$

13. Solve for x:

$$\begin{vmatrix} x-2 & 2 \\ 2 & x+1 \end{vmatrix} = 0$$

$$(x-2)(x+1)-2(2)=0$$
$$x^2-x-2-4=0$$
$$x^2-x-6=0$$
$$(x-3)(x+2)=0$$

Set each factor equal to zero and solve:

$x=3,\ x=-2$

Solution Set: $\{-2, 3\}$

15. Solve for x:

$$\begin{vmatrix} x+1 & 8 \\ 1 & x+3 \end{vmatrix} = 0$$

$$(x+1)(x+3)-8(1)=0$$
$$x^2+4x+3-8=0$$
$$x^2+4x-5=0$$
$$(x-1)(x+5)=0$$

Set each factor equal to zero and solve:

$x=1,\ x=-5$

Solution Set: $\{-5, 1\}$

17. Solve for x:

$$\begin{vmatrix} x+6 & 2 \\ -1 & x+3 \end{vmatrix} = 0$$

$$(x+6)(x+3)-(-1)(2)=0$$
$$x^2+9x+18+2=0$$
$$x^2+9x+20=0$$
$$(x+4)(x+5)=0$$

Set each factor equal to zero and solve:

$x=-5,\ x=-4$

Solution Set: $\{-5, -4\}$

19. Solve for x:

$$\begin{vmatrix} x+5 & 3 \\ 3 & x-3 \end{vmatrix} = 0$$

$$(x+5)(x-3)-(3)(3)=0$$
$$x^2+2x-15-9=0$$
$$x^2+2x-24=0$$
$$(x+6)(x-4)=0$$

Set each factor equal to zero and solve:

$x=-6,\ x=4$

Solution Set: $\{-6, 4\}$

21. Solve for x:

$$\begin{vmatrix} x-3 & 2 \\ 1 & x-4 \end{vmatrix} = 0$$

$$(x-3)(x-4)-(1)(2)=0$$
$$x^2-7x+12-2=0$$
$$x^2-7x+10=0$$
$$(x-5)(x-2)=0$$

Set each factor equal to zero and solve:

$x=5,\ x=2$

Solution Set: $\{2, 5\}$

For items 23 - 29, use

$$A=\begin{bmatrix}2 & -1 & 5\\ 0 & 1 & 3\\ 1 & 0 & -2\end{bmatrix}$$

23. Cofactor of a_{12}:

$$(-1)^{1+2}\begin{vmatrix}0 & 3\\ 1 & -2\end{vmatrix}=(-1)\left[(0)(-2)-(1)(3)\right]$$
$$=-(-3)$$
$$=3$$

25. Cofactor of a_{22}:

$$(-1)^{2+2}\begin{vmatrix}2 & 5\\ 1 & -2\end{vmatrix}=\left[(2)(-2)-(1)(5)\right]$$
$$=-4-5$$
$$=-9$$

27. Cofactor of a_{33}:

$$(-1)^{3+3}\begin{vmatrix}2 & -1\\ 0 & 1\end{vmatrix}=\left[(2)(1)-(0)(-1)\right]$$
$$=2$$

29. Cofactor of a_{21}:

$$(-1)^{2+1}\begin{vmatrix}-1 & 5\\ 0 & -2\end{vmatrix}=(-1)\left[(-1)(-2)-(0)(5)\right]$$
$$=-(2)$$
$$=-2$$

31. Expand $\begin{vmatrix}4 & 5 & 3\\ -1 & 2 & 7\\ 11 & 6 & 2\end{vmatrix}$ along row 3:

$$(-1)^{3+1}11\begin{vmatrix}5 & 3\\ 2 & 7\end{vmatrix}+(-1)^{3+2}6\begin{vmatrix}4 & 3\\ -1 & 7\end{vmatrix}+(-1)^{3+3}2\begin{vmatrix}4 & 5\\ -1 & 2\end{vmatrix}=$$
$$11(35-6)-6(28+3)+2(8+5)=$$
$$11(29)-6(31)+2(13)=159$$

33. Expand $\begin{vmatrix}5 & 8 & 5\\ 0 & -6 & 3\\ 2 & 4 & -1\end{vmatrix}$ along row 1:

$$(-1)^{1+1}5\begin{vmatrix}-6 & 3\\ 4 & -1\end{vmatrix}+(-1)^{1+2}8\begin{vmatrix}0 & 3\\ 2 & 1\end{vmatrix}+(-1)^{1+3}5\begin{vmatrix}0 & -6\\ 2 & 4\end{vmatrix}=$$
$$5(6-12)-8(0-6)+5(0+12)=$$
$$-30+48+60=78$$

35. Expand $\begin{vmatrix}13 & 0 & -7\\ 4 & 2 & 3\\ 1 & 4 & 0\end{vmatrix}$ along row 2:

$$(-1)^{2+1}4\begin{vmatrix}0 & -7\\ 4 & 0\end{vmatrix}+(-1)^{2+2}2\begin{vmatrix}13 & -7\\ 1 & 0\end{vmatrix}+(-1)^{2+3}3\begin{vmatrix}13 & 0\\ 1 & 4\end{vmatrix}=$$
$$-4(0+28)+2(0+7)-3(52-0)=$$
$$-112+14-156=-254$$

37. Expand $=\begin{vmatrix}8 & 0 & -7 & 5\\ 4 & -2 & 3 & 3\\ -1 & 1 & 0 & 2\\ 2 & 0 & 6 & 0\end{vmatrix}$ along row 4:

$$=(-1)^{4+1}2\begin{vmatrix}0 & -7 & 5\\ -2 & 3 & 3\\ 1 & 0 & 2\end{vmatrix}+(-1)^{4+2}0\begin{vmatrix}8 & -7 & 5\\ 4 & 3 & 3\\ -1 & 0 & 2\end{vmatrix}$$
$$+(-1)^{4+3}(-6)\begin{vmatrix}8 & 0 & 5\\ 4 & -2 & 3\\ -1 & 1 & 2\end{vmatrix}+(-1)^{4+4}0\begin{vmatrix}8 & 0 & -7\\ 4 & -2 & 3\\ -1 & 1 & 0\end{vmatrix}$$

$2R_3+R_2$; $-2R_2+R_1,\ 4R_3+R_2$

$$=-2\begin{vmatrix}0 & -7 & 5\\ 0 & 3 & 7\\ 1 & 0 & 2\end{vmatrix}-6\begin{vmatrix}0 & 4 & -1\\ 0 & 2 & 11\\ -1 & 1 & 2\end{vmatrix}$$
$$=-2(-1)^{3+1}(1)\begin{vmatrix}-7 & 5\\ 3 & 7\end{vmatrix}-6(-1)^{3+1}(-1)\begin{vmatrix}4 & -1\\ 2 & 11\end{vmatrix}$$
$$=-2(-49-15)+6(44+2)$$
$$=-2(-64)+6(46)$$
$$=404$$

39. Evaluate:

$$|A| = \begin{vmatrix} 2 & 0 & 1 \\ -5 & 1 & 0 \\ 3 & -1 & 1 \end{vmatrix} \overset{-R_1+R_3}{=} \begin{vmatrix} 2 & 0 & 1 \\ -5 & 1 & 0 \\ 1 & -1 & 0 \end{vmatrix}$$

$$= (-1)^{1+3}(1)\begin{vmatrix} -5 & 1 \\ 1 & -1 \end{vmatrix}$$

$$= (-5)(-1) - 1(1)$$

$$= 4$$

41. Evaluate: $|A| = \begin{vmatrix} 12 & 3 & 6 \\ 2 & 2 & -4 \\ 0 & 2 & 0 \end{vmatrix}$

$$|A| = 2(2)(3)\begin{vmatrix} 4 & 1 & 2 \\ 1 & 1 & -2 \\ 0 & 1 & 0 \end{vmatrix}$$

$$= 12(-1)^{3+2}\begin{vmatrix} 4 & 2 \\ 1 & -2 \end{vmatrix}$$

$$= (-12)(-8-2)$$

$$= 120$$

43. Evaluate:

$$|A| = \begin{vmatrix} x & x & x & x \\ 0 & x & x & x \\ 0 & 0 & x & x \\ 0 & 0 & 0 & x \end{vmatrix} = x^4 \begin{vmatrix} 1 & 1 & 1 & 1 \\ 0 & 1 & 1 & 1 \\ 0 & 0 & 1 & 1 \\ 0 & 0 & 0 & 1 \end{vmatrix}$$

$$= x^4(-1)^{4+4}\begin{vmatrix} 1 & 1 & 1 \\ 0 & 1 & 1 \\ 0 & 0 & 1 \end{vmatrix}$$

$$= x^4(-1)^{3+3}\begin{vmatrix} 1 & 1 \\ 0 & 1 \end{vmatrix}$$

$$= x^4(1-0)$$

$$= x^4$$

45. Evaluate:

$$|A| = \begin{vmatrix} 0 & 2 & 0 & 0 \\ -2 & -4 & 5 & 9 \\ 1 & 3 & -1 & 1 \\ 0 & 7 & 0 & 2 \end{vmatrix}$$

$$= (-1)^{1+2}(2)\begin{vmatrix} -2 & 5 & 9 \\ 1 & -1 & 1 \\ 0 & 0 & 2 \end{vmatrix}$$

$$= -2(-1)^{3+3}(2)\begin{vmatrix} -2 & 5 \\ 1 & -1 \end{vmatrix}$$

$$= -4(2-5)$$

$$= 12$$

47. Evaluate:

$$|A| = \begin{vmatrix} x & x & 0 & 0 \\ yz & x^3 & z & x^4 \\ z & xy & x & 0 \\ x^2 & 0 & 0 & 0 \end{vmatrix}$$

$$= x^2(x)x^4\begin{vmatrix} 1 & 1 & 0 & 0 \\ yz & x^3 & z & 1 \\ z & xy & x & 0 \\ 1 & 0 & 0 & 0 \end{vmatrix}$$

$$= (-1)^{2+4}x^7(1)\begin{vmatrix} 1 & 1 & 0 \\ z & xy & x \\ 1 & 0 & 0 \end{vmatrix}$$

$$= x^7\left[(-1)^{3+1}(1)\begin{vmatrix} 1 & 0 \\ xy & x \end{vmatrix}\right]$$

$$= x^7(x-0)$$

$$= x^8$$

49. Use the ⟨2nd⟩⟨*MATRX*⟩ and ⟨MATH⟩ features of a graphing calculator to find the value of the determinant:

$$|A| = \begin{vmatrix} 0.1 & 0.3 & 0.1 \\ 0.2 & -0.2 & -0.1 \\ -0.1 & -0.4 & 0.5 \end{vmatrix} = -0.051$$

51. Use the $\langle 2\text{nd}\rangle\langle MATRX\rangle$ and $\langle \text{MATH}\rangle$ features of a graphing calculator to find the value of the determinant:

$$|A| = \begin{vmatrix} 3.1 & 0.6 & -1.1 \\ 1.2 & 5.2 & -7.3 \\ -0.1 & -4.1 & 6.5 \end{vmatrix} = 12.595$$

53. Use the $\langle 2\text{nd}\rangle\langle MATRX\rangle$ and $\langle \text{MATH}\rangle$ features of a graphing calculator to find the value of the determinant:

$$|A| = \begin{vmatrix} 25 & 32 & 17 \\ -13 & 14 & -24 \\ 16 & 26 & 36 \end{vmatrix} = 21{,}334$$

55. Use Cramer's Rule to solve the given system of equations:

$$\begin{cases} 5x + 7y = 9 \\ 2x + 3y = -7 \end{cases}$$

$$D = \begin{vmatrix} 5 & 7 \\ 2 & 3 \end{vmatrix} = 15 - 14 = 1$$

$$D_x = \begin{vmatrix} 9 & 7 \\ -7 & 3 \end{vmatrix} = 27 - (-49) = 76$$

$$D_y = \begin{vmatrix} 5 & 9 \\ 2 & -7 \end{vmatrix} = -35 - (18) = -53$$

$$x = \frac{D_x}{D} = \frac{76}{1} = 76$$

$$y = \frac{D_y}{D} = \frac{-53}{1} = -53$$

Solution: $(76, -53)$

57. Use Cramer's Rule to solve the given system of equations:

$$\begin{cases} -2x - 2y = 4 \\ 3x + 3y = -6 \end{cases}$$

Note: This system is equivalent to

$$\begin{cases} x = -y - 2 \\ y = -x - 2 \end{cases}$$

The solution set can be constructed from this form:

$$\{(-y-2, y) \mid y \in \mathbb{R}\}$$

By Cramer's Rule this is demonstrated by showing the determinants are zero:

$$D = \begin{vmatrix} -2 & -2 \\ 3 & 3 \end{vmatrix} = -6 - (-6) = 0$$

$$D_x = \begin{vmatrix} 4 & -2 \\ -6 & 3 \end{vmatrix} = 12 - 12 = 0$$

$$D_y = \begin{vmatrix} -2 & 4 \\ 3 & -6 \end{vmatrix} = 12 - 12 = 0$$

Therefore, there are an infinite number of ordered pairs that satisfy the given system of equations.

59. Use Cramer's Rule to solve the given system of equations:

$$\begin{cases} \frac{2}{3}x + 2y = 1 \\ x + 3y = 0 \end{cases}$$

First, simplify by multiplying the first equation by 3:

$$\begin{cases} 2x + 6y = 3 \\ x + 3y = 0 \end{cases}$$

$$D = \begin{vmatrix} 2 & 6 \\ 1 & 3 \end{vmatrix} = 6 - 6 = 0$$

$$D_x = \begin{vmatrix} 3 & 6 \\ 0 & 3 \end{vmatrix} = 9 - 0 = 9 \text{ (i.e., } \neq 0)$$

Therefore, there are no solutions.

61. Use Cramer's Rule to solve the given system of equations:

$$\begin{cases} 5x - 4y = -49 \\ 24x - 19y = 179 \end{cases}$$

$$D = \begin{vmatrix} 5 & -4 \\ 24 & -19 \end{vmatrix} = 5(-19) - (-4)(24) = 1$$

$$D_x = \begin{vmatrix} -49 & -4 \\ 179 & -19 \end{vmatrix} = (-49)(-19) - (-4)(179) = 1647$$

$$D_y = \begin{vmatrix} 5 & -49 \\ 24 & 179 \end{vmatrix} = 5(179) - (-49)(24) = 2071$$

$$x = \frac{D_x}{D} = \frac{1647}{1} = 1647$$

$$y = \frac{D_y}{D} = \frac{2071}{1} = 2071$$

Solution: $(1647, 2071)$

63. Use Cramer's Rule to solve the given system of equations:

$$\begin{cases} -5x + 10y = 3 \\ \frac{7}{2}x - 7y = 20 \end{cases}$$

First, simplify by multiplying the second equation by 2:

$$\begin{cases} -5x + 10y = 3 \\ 7x - 14y = 40 \end{cases}$$

$$D = \begin{vmatrix} -5 & 10 \\ 7 & -14 \end{vmatrix} = (-5)(-14) - (10)(7) = 0$$

$$D_x = \begin{vmatrix} 3 & 10 \\ 40 & -14 \end{vmatrix} = -42 - 400 = -442$$

(i.e., $\neq 0$)

Therefore, there are no solutions.

65. Use Cramer's Rule to solve the given system of equations:

$$\begin{cases} 2x - y = 0 \\ 5x - 3y - 3z = 5 \\ 2x + 6z = -10 \end{cases}$$

$$D = \begin{vmatrix} 2 & -1 & 0 \\ 5 & -3 & -3 \\ 2 & 0 & 6 \end{vmatrix} = 2\begin{vmatrix} 2 & -1 & 0 \\ 5 & -3 & -3 \\ 1 & 0 & 3 \end{vmatrix}$$

$$\overset{R_3 + R_2}{=} \begin{vmatrix} 2 & -1 & 0 \\ 6 & -3 & 0 \\ 1 & 0 & 3 \end{vmatrix} = 3\begin{vmatrix} 2 & -1 & 0 \\ 2 & -1 & 0 \\ 1 & 0 & 3 \end{vmatrix}$$

$$= 3\left[(-1)^{3+3}(3)\begin{vmatrix} 2 & -1 \\ 2 & -1 \end{vmatrix}\right] = 0$$

$$D_x = \begin{vmatrix} 0 & -1 & 0 \\ 5 & -3 & -3 \\ -10 & 0 & 6 \end{vmatrix} = (-1)^{1+2}(-1)\begin{vmatrix} 5 & -3 \\ -10 & 6 \end{vmatrix}$$

$$= 30 - 30 = 0$$

$$D_y = \begin{vmatrix} 2 & 0 & 0 \\ 5 & 5 & -3 \\ 2 & -10 & 6 \end{vmatrix} = (-1)^{1+1}(2)\begin{vmatrix} 5 & -3 \\ -10 & 6 \end{vmatrix}$$

$$= 2(30 - 30) = 0$$

$$D_z = \begin{vmatrix} 2 & -1 & 0 \\ 5 & -3 & 5 \\ 2 & 0 & -10 \end{vmatrix}$$

$$= (-1)^{1+1}2\begin{vmatrix} -3 & 5 \\ 0 & -10 \end{vmatrix} + (-1)^{1+2}(-1)\begin{vmatrix} 5 & 5 \\ 2 & -10 \end{vmatrix}$$

$$= 2(30 - 0) + (-50 - 10)$$

$$= 0$$

Therefore, the system is dependent.

$x = -3z - 5$ and $y = -6z - 10$

Solution Set: $\{(-3z - 5, -6z - 10, z) \mid z \in \mathbb{R}\}$

67. Solve: $\begin{cases} 3w - x + 5y + 3z = 2 \\ -4w - 10y - 2z = 10 \\ w - x + 2z = 7 \\ 4w - 2x + 5y + 5z = 9 \end{cases}$

by Cramer's Rule.

Rewrite by using row manipulation to simplify the matrix:

$$A = \left[\begin{array}{cccc|c} 3 & -1 & 5 & 3 & 2 \\ -4 & 0 & -10 & -2 & 10 \\ 1 & -1 & 0 & 2 & 7 \\ 4 & -2 & 5 & 5 & 9 \end{array}\right] \xrightarrow[-2R_3+R_4]{-R_3+R_1}$$

$$\left[\begin{array}{cccc|c} 2 & 0 & 5 & 1 & -5 \\ -4 & 0 & -10 & -2 & 10 \\ 2 & 0 & 5 & 1 & -5 \\ 4 & -2 & 5 & 5 & 9 \end{array}\right]$$

$$D = \begin{vmatrix} 2 & 0 & 5 & 1 \\ -4 & 0 & -10 & -2 \\ 2 & 0 & 5 & 1 \\ 4 & -2 & 5 & 5 \end{vmatrix}$$

$$= (-1)^{4+2}(-2)\begin{vmatrix} 2 & 5 & 1 \\ -4 & -10 & -2 \\ 2 & 5 & 1 \end{vmatrix}$$

$$\overset{R_3-R_1}{=} -2\begin{vmatrix} 0 & 0 & 0 \\ -4 & -10 & -2 \\ 2 & 5 & 1 \end{vmatrix} = 0$$

$$D_w = \begin{vmatrix} -5 & 0 & 5 & 1 \\ 10 & 0 & -10 & -2 \\ -5 & 0 & 5 & 1 \\ 9 & -2 & 5 & 5 \end{vmatrix}$$

$$= (-1)^{4+2}(-2)\begin{vmatrix} -5 & 5 & 1 \\ 10 & -10 & -2 \\ -5 & 5 & 1 \end{vmatrix}$$

$$= -2(5)(5)(2)\begin{vmatrix} -1 & 1 & 1 \\ 1 & -1 & -1 \\ -1 & 1 & 1 \end{vmatrix}$$

$$\underset{R_2+R_3}{\overset{R_2+R_1}{=}} -100\begin{vmatrix} 0 & 0 & 0 \\ 1 & -1 & -1 \\ 0 & 0 & 0 \end{vmatrix} = 0$$

Similarly, each of D_x, D_y and D_z can be shown to equal zero.

Therefore, there is an infinite set of solutions. To describe the solution set, we can manipulate the original matrix A with row operations and reduce the rows to enable us to"read" the solution:

$$A = \left[\begin{array}{cccc|c} 3 & -1 & 5 & 3 & 2 \\ -4 & 0 & -10 & -2 & 10 \\ 1 & -1 & 0 & 2 & 7 \\ 4 & -2 & 5 & 5 & 9 \end{array}\right] \xrightarrow[R_4+R_2]{R_1 \leftrightarrow R_3}$$

$$\left[\begin{array}{cccc|c} 1 & -1 & 0 & 2 & 7 \\ 0 & -2 & -5 & 3 & 19 \\ 3 & -1 & 5 & 3 & 2 \\ 4 & -2 & 5 & 5 & 9 \end{array}\right] \xrightarrow[-4R_1+R_4]{-3R_1+R_3}$$

$$\left[\begin{array}{cccc|c} 1 & -1 & 0 & 2 & 7 \\ 0 & -2 & -5 & 3 & 19 \\ 0 & 2 & 5 & -3 & -19 \\ 0 & 2 & 5 & -3 & -19 \end{array}\right] \xrightarrow[-R_3+R_4]{R_3+R_2}$$

$$\left[\begin{array}{cccc|c} 1 & -1 & 0 & 2 & 7 \\ 0 & 0 & 0 & 0 & 0 \\ 0 & 2 & 5 & -3 & -19 \\ 0 & 0 & 0 & 0 & 0 \end{array}\right]$$

The reductions of rows 2 and 4 to all zeros also indicate, as found with Cramer's Rule, that there are an infinite number of solutions. Expressed in terms of x and z:

$w = x - 2z + 7$ and $y = \dfrac{-2x + 3z - 19}{5}$

The solution set is

$$\left\{\left(x - 2z + 7, x, \frac{-2x + 3z - 19}{5}, z\right); x, z \in \mathbb{R}\right\}$$

Or, equivalently,

$w = \dfrac{-5y - z - 5}{2}$ and $x = \dfrac{-5y + 3z - 19}{2}$

$$\left\{\left(\frac{-5y - z - 5}{2}, \frac{-5y + 3z - 19}{2}, y, z\right); y, z \in \mathbb{R}\right\}$$

69. Solve: $\begin{cases} 3w-2x+y-5z=-1 \\ w+x-y+4z=2 \\ 4w-x-z=1 \\ 5w-x=9 \end{cases}$

by Cramer's Rule.

Rewrite by using row manipulation to simplify the matrix:

$$A=\left[\begin{array}{cccc|c} 3 & -2 & 1 & -5 & -1 \\ 1 & 1 & -1 & 4 & 2 \\ 4 & -1 & 0 & -1 & 1 \\ 5 & -1 & 0 & 0 & 9 \end{array}\right]\xrightarrow{R_1+R_2}$$

$$\left[\begin{array}{cccc|c} 3 & -2 & 1 & -5 & -1 \\ 4 & -1 & 0 & -1 & 1 \\ 4 & -1 & 0 & -1 & 1 \\ 5 & -1 & 0 & 0 & 9 \end{array}\right]$$

$$D=\begin{vmatrix} 3 & -2 & 1 & -5 \\ 4 & -1 & 0 & -1 \\ 4 & -1 & 0 & -1 \\ 5 & -1 & 0 & 0 \end{vmatrix}=(-1)^{1+3}(1)\begin{vmatrix} 4 & -1 & -1 \\ 4 & -1 & -1 \\ 5 & -1 & 0 \end{vmatrix}$$

$$=(-1)^{1+3}(-1)\begin{vmatrix} 4 & -1 \\ 5 & -1 \end{vmatrix}+(-1)^{2+3}(-1)\begin{vmatrix} 4 & -1 \\ 5 & -1 \end{vmatrix}$$

$$=-(4(-1)-(-1)(5))+(-1)^6(4(-1)-(-1)(5))$$

$$=-1+1$$

$$=0$$

$$D_w=\begin{vmatrix} -1 & -2 & 1 & -5 \\ 1 & -1 & 0 & -1 \\ 1 & -1 & 0 & -1 \\ 9 & -1 & 0 & 0 \end{vmatrix}=(-1)^{1+3}(1)\begin{vmatrix} 1 & -1 & -1 \\ 1 & -1 & -1 \\ 9 & -1 & 0 \end{vmatrix}$$

$$=(-1)^{1+3}(-1)\begin{vmatrix} 1 & -1 \\ 9 & -1 \end{vmatrix}+(-1)^{2+3}(-1)\begin{vmatrix} 1 & -1 \\ 9 & -1 \end{vmatrix}$$

$$=(-1)(-1+9)+(1)(-1+9)$$

$$=0$$

Similarly, each of D_x, D_y and D_z can be shown to equal zero.

Therefore, there is an infinite set of solutions. To describe the solution set, we can manipulate the original matrix A with row operations and reduce the rows to enable us to "read" the solution:

$$A=\left[\begin{array}{cccc|c} 3 & -2 & 1 & -5 & -1 \\ 1 & 1 & -1 & 4 & 2 \\ 4 & -1 & 0 & -1 & 1 \\ 5 & -1 & 0 & 0 & 9 \end{array}\right]\xrightarrow{R_1\leftrightarrow R_2}$$

$$\left[\begin{array}{cccc|c} 1 & 1 & -1 & 4 & 2 \\ 3 & -2 & 1 & -5 & -1 \\ 4 & -1 & 0 & -1 & 1 \\ 5 & -1 & 0 & 0 & 9 \end{array}\right]\xrightarrow[\substack{-4R_1+R_3 \\ -5R_1+R_4}]{-3R_1+R_2}$$

$$\left[\begin{array}{cccc|c} 1 & 1 & -1 & 4 & 2 \\ 0 & -5 & 4 & -17 & -7 \\ 0 & -5 & 4 & -17 & -7 \\ 0 & -6 & 5 & -20 & -1 \end{array}\right]\xrightarrow{-R_2+R_3}$$

$$\left[\begin{array}{cccc|c} 1 & 1 & -1 & 4 & 2 \\ 0 & -5 & 4 & -17 & -7 \\ 0 & 0 & 0 & 0 & 0 \\ 0 & -6 & 5 & -20 & -1 \end{array}\right]\xrightarrow[R_3\leftrightarrow R_4]{-\frac{1}{5}R_2}$$

$$\left[\begin{array}{cccc|c} 1 & 1 & -1 & 4 & 2 \\ 0 & 1 & -\frac{4}{5} & \frac{17}{5} & \frac{7}{5} \\ 0 & -6 & 5 & -20 & -1 \\ 0 & 0 & 0 & 0 & 0 \end{array}\right]\xrightarrow[6R_2+R_3]{-R_2+R_1}$$

$$\left[\begin{array}{cccc|c} 1 & 0 & -\frac{1}{5} & \frac{3}{5} & \frac{3}{5} \\ 0 & 1 & -\frac{4}{5} & \frac{17}{5} & \frac{7}{5} \\ 0 & 0 & \frac{1}{5} & \frac{2}{5} & \frac{37}{5} \\ 0 & 0 & 0 & 0 & 0 \end{array}\right]\xrightarrow{5R_3}$$

$$\left[\begin{array}{cccc|c} 1 & 0 & -\frac{1}{5} & \frac{3}{5} & \frac{3}{5} \\ 0 & 1 & -\frac{4}{5} & \frac{17}{5} & \frac{7}{5} \\ 0 & 0 & 1 & 2 & 37 \\ 0 & 0 & 0 & 0 & 0 \end{array}\right] \longrightarrow$$

With back substitution, we have

$$\begin{cases} w = -z+8 \\ x = -5z+31 \\ y = -2z+37 \\ z \in \mathbb{R} \end{cases}, \text{ or equivalently,}$$

The solution set is

$\{(-z+8, -5z+31, -2z+37, z); z \in \mathbb{R}\}$

71. Solve: $\begin{cases} w - x + y - z = 2 \\ 2w - x + 3y = -5 \\ x - 2z = 7 \\ 3w + 4x = -13 \end{cases}$

by Cramer's Rule.

The augmented matrix:

$$A = \left[\begin{array}{cccc|c} 1 & -1 & 1 & -1 & 2 \\ 2 & -1 & 3 & 0 & -5 \\ 0 & 1 & 0 & -2 & 7 \\ 3 & 4 & 0 & 0 & -13 \end{array}\right]$$

$$D = \begin{vmatrix} 1 & -1 & 1 & -1 \\ 2 & -1 & 3 & 0 \\ 0 & 1 & 0 & -2 \\ 3 & 4 & 0 & 0 \end{vmatrix} \overset{-3R_1+R_2}{=} \begin{vmatrix} 1 & -1 & 1 & -1 \\ -1 & 2 & 0 & 3 \\ 0 & 1 & 0 & -2 \\ 3 & 4 & 0 & 0 \end{vmatrix}$$

$$= (-1)^{1+3}(1)\begin{vmatrix} -1 & 2 & 3 \\ 0 & 1 & -2 \\ 3 & 4 & 0 \end{vmatrix}$$

$$= (-1)^{3+1} 3\begin{vmatrix} 2 & 3 \\ 1 & -2 \end{vmatrix} + (-1)^{3+2} 4\begin{vmatrix} -1 & 3 \\ 0 & -2 \end{vmatrix}$$

$$= 3(-4-3) - 4(2) = -21 - 8$$

$$= -29$$

$$D_w = \begin{vmatrix} 2 & -1 & 1 & -1 \\ -5 & -1 & 3 & 0 \\ 7 & 1 & 0 & -2 \\ -13 & 4 & 0 & 0 \end{vmatrix}$$

$$= (-3)\begin{vmatrix} 2 & -1 & -1 \\ 7 & 1 & -2 \\ -13 & 4 & 0 \end{vmatrix} + (1)\begin{vmatrix} -5 & -1 & 0 \\ 7 & 1 & -2 \\ -13 & 4 & 0 \end{vmatrix}$$

$$\overset{-2R_1+R_2}{=} (-3)\begin{vmatrix} 2 & -1 & -1 \\ 3 & 3 & 0 \\ -13 & 4 & 0 \end{vmatrix} + (-1)^{2+3}(-2)\begin{vmatrix} -5 & -1 \\ -13 & 4 \end{vmatrix}$$

$$= (-3)(-1)^{1+3}(-1)\begin{vmatrix} 3 & 3 \\ -13 & 4 \end{vmatrix} + 2(-20-13)$$

$$= 3(12+39) - 66 = 153 - 66$$

$$= 87$$

$$w = \frac{D_w}{D} = \frac{87}{-29} = -3$$

$$D_x = \begin{vmatrix} 1 & 2 & 1 & -1 \\ 2 & -5 & 3 & 0 \\ 0 & 7 & 0 & -2 \\ 3 & -13 & 0 & 0 \end{vmatrix}$$

$$= (1)\begin{vmatrix} 2 & -5 & 0 \\ 0 & 7 & -2 \\ 3 & -13 & 0 \end{vmatrix} - 3\begin{vmatrix} 1 & 2 & -1 \\ 0 & 7 & -2 \\ 3 & -13 & 0 \end{vmatrix}$$

$$= (2)\begin{vmatrix} 2 & -5 \\ 3 & -13 \end{vmatrix} - 3\left[(1)\begin{vmatrix} 7 & -2 \\ -13 & 0 \end{vmatrix} + (3)\begin{vmatrix} 2 & -1 \\ 7 & -2 \end{vmatrix}\right]$$

$$= 2(-26+15) - 3[-26+3(-4+7)]$$

$$= -22 - 3[-26+9]$$

$$= 51 - 22$$

$$= 29$$

$$x = \frac{D_x}{D} = \frac{29}{-29} = -1$$

$$D_y = \begin{vmatrix} 1 & -1 & 2 & -1 \\ 2 & -1 & -5 & 0 \\ 0 & 1 & 7 & -2 \\ 3 & 4 & -13 & 0 \end{vmatrix} \overset{-2R_1+R_3}{=} \begin{vmatrix} 1 & -1 & 2 & -1 \\ 2 & -1 & -5 & 0 \\ -2 & 3 & 3 & 0 \\ 3 & 4 & -13 & 0 \end{vmatrix}$$

$$= (-1)^{1+4}(-1)\begin{vmatrix} 2 & -1 & -5 \\ -2 & 3 & 3 \\ 3 & 4 & -13 \end{vmatrix}$$

$$= (2)\begin{vmatrix} 3 & 3 \\ 4 & -13 \end{vmatrix} + (1)\begin{vmatrix} -2 & 3 \\ 3 & -13 \end{vmatrix} - 5\begin{vmatrix} -2 & 3 \\ 3 & 4 \end{vmatrix}$$

$$= 2(-39-12) + (26-9) - 5(-8-9)$$

$$= 0$$

$$y = \frac{D_y}{D} = \frac{0}{-29} = 0$$

$$D_z = \begin{vmatrix} 1 & -1 & 1 & 2 \\ 2 & -1 & 3 & -5 \\ 0 & 1 & 0 & 7 \\ 3 & 4 & 0 & -13 \end{vmatrix} \overset{-3R_1+R_2}{=} \begin{vmatrix} 1 & -1 & 1 & 2 \\ -1 & 2 & 0 & -11 \\ 0 & 1 & 0 & 7 \\ 3 & 4 & 0 & -13 \end{vmatrix}$$

$$= (1)\begin{vmatrix} -1 & 2 & -11 \\ 0 & 1 & 7 \\ 3 & 4 & -13 \end{vmatrix}$$

$$= (-1)\begin{vmatrix} 1 & 7 \\ 4 & -13 \end{vmatrix} + (3)\begin{vmatrix} 2 & -11 \\ 1 & 7 \end{vmatrix}$$

$$= -(-13-28) + 3(14+11)$$

$$= 116$$

$$z = \frac{D_z}{D} = \frac{116}{-29} = -4$$

The solution is $(-3, -1, 0, -4)$.

73. Use a graphing calculator to solve the given system $\begin{cases} 2x+4y+z=1 \\ x-2y-3z=2 \\ x+y-z=-1 \end{cases}$

Use $\langle$2nd$\rangle\langle$MATRX$\rangle$ feature to define each matrix and then use the $\langle$2nd$\rangle\langle$MATRX$\rangle\langle$MATH$\rangle$ feature to obtain the determinant values for D, D_x, D_y and D_z.

$$D = \begin{vmatrix} 2 & 4 & 1 \\ 1 & -2 & -3 \\ 1 & 1 & -1 \end{vmatrix} = 5$$

$$D_x = \begin{vmatrix} 1 & 4 & 1 \\ 2 & -2 & -3 \\ -1 & 1 & -1 \end{vmatrix} = 25; \text{ Then,}$$

$$x = \frac{D_x}{D} = \frac{25}{5} = 5$$

$$D_y = \begin{vmatrix} 2 & 1 & 1 \\ 1 & 2 & -3 \\ 1 & -1 & -1 \end{vmatrix} = -15; \text{ Then,}$$

$$y = \frac{D_y}{D} = \frac{-15}{5} = -3$$

$$D_z = \begin{vmatrix} 2 & 4 & 1 \\ 1 & -2 & 2 \\ 1 & 1 & -1 \end{vmatrix} = 15; \text{ Then,}$$

$$z = \frac{D_z}{D} = \frac{15}{5} = 3$$

The solution is $(5, -3, 3)$.

End of Section 10.3

Chapter Ten: Section 10.4
Solutions to Odd-numbered Exercises

1. $$3A-B=3\begin{bmatrix}3 & -2\\ 1 & 0\\ 0 & 5\end{bmatrix}-\begin{bmatrix}4 & -5\\ 3 & 0\\ -2 & 2\end{bmatrix}$$
$$=\begin{bmatrix}9 & -6\\ 3 & 0\\ 0 & 15\end{bmatrix}-\begin{bmatrix}4 & -5\\ 3 & 0\\ -2 & 2\end{bmatrix}$$
$$=\begin{bmatrix}5 & -1\\ 0 & 0\\ 2 & 13\end{bmatrix}$$

3. $$3C=3\begin{bmatrix}2 & -1\\ 6 & 10\\ -3 & 7\end{bmatrix}$$
$$=\begin{bmatrix}6 & -3\\ 18 & 30\\ -9 & 21\end{bmatrix}$$

5. This addition is not possible because the matrices to be added are not of the same order. One is of order 2 x 3 and the other is 3 x 2.

7. $$2A+2B=2\begin{bmatrix}3 & -2\\ 1 & 0\\ 0 & 5\end{bmatrix}+2\begin{bmatrix}4 & -5\\ 3 & 0\\ -2 & 2\end{bmatrix}$$
$$=\begin{bmatrix}6 & -4\\ 2 & 0\\ 0 & 10\end{bmatrix}+\begin{bmatrix}8 & -10\\ 6 & 0\\ -4 & 4\end{bmatrix}$$
$$=\begin{bmatrix}14 & -14\\ 8 & 0\\ -4 & 14\end{bmatrix}$$

9. $$C-3A=\begin{bmatrix}2 & -1\\ 6 & 10\\ -3 & 7\end{bmatrix}-3\begin{bmatrix}3 & -2\\ 1 & 0\\ 0 & 5\end{bmatrix}$$
$$=\begin{bmatrix}2 & -1\\ 6 & 10\\ -3 & 7\end{bmatrix}-\begin{bmatrix}9 & -6\\ 3 & 0\\ 0 & 15\end{bmatrix}$$
$$=\begin{bmatrix}-7 & 5\\ 3 & 10\\ -3 & -8\end{bmatrix}$$

11. $4A-3D$ is not possible. The orders of A and D are not the same.

13. $$\begin{bmatrix}a & 2b & c\end{bmatrix}+3\begin{bmatrix}a & 2 & -c\end{bmatrix}=\begin{bmatrix}8 & 2 & 2\end{bmatrix}$$
$$\begin{bmatrix}a & 2b & c\end{bmatrix}+\begin{bmatrix}3a & 6 & -3c\end{bmatrix}=\begin{bmatrix}8 & 2 & 2\end{bmatrix}$$
$$\begin{bmatrix}4a & 2b+6 & -2c\end{bmatrix}=\begin{bmatrix}8 & 2 & 2\end{bmatrix}$$
$$4a=8\Rightarrow a=2$$
$$2b+6=2\Rightarrow b=-2$$
$$-2c=2\Rightarrow c=-1$$

15. $\begin{bmatrix}3x\\ 2y\end{bmatrix}+\begin{bmatrix}x\\ -y\\ z\end{bmatrix}=\begin{bmatrix}4\\ 0\\ 2\end{bmatrix}$ Not possible. The addition of two matrices requires that the addends have the same order.

17. $$\begin{bmatrix}x\\ 3x\end{bmatrix}-\begin{bmatrix}y\\ 2y\end{bmatrix}=\begin{bmatrix}5\\ 20\end{bmatrix}$$
$$\begin{bmatrix}x-y\\ 3x-2y\end{bmatrix}=\begin{bmatrix}5\\ 20\end{bmatrix}$$
$$\begin{cases}x-y=5\\ 3x-2y=20\end{cases}$$
$$x=y+5$$
$$3(y+5)-2y=20$$
$$3y+15-2y=20$$
$$y=5$$
$$x=10$$

19. $2\begin{bmatrix} x \\ 2y \end{bmatrix} - 3\begin{bmatrix} 5y \\ -3x \end{bmatrix} = \begin{bmatrix} -9 \\ 31 \end{bmatrix}$

$\begin{bmatrix} 2x \\ 4y \end{bmatrix} - \begin{bmatrix} 15y \\ -9x \end{bmatrix} = \begin{bmatrix} -9 \\ 31 \end{bmatrix}$

$\begin{cases} 2x - 15y = -9 \\ 9x + 4y = 31 \end{cases}$

$D = \begin{vmatrix} 2 & -15 \\ 9 & 4 \end{vmatrix} = 8 + 135 = 143$

$D_x = \begin{vmatrix} -9 & -15 \\ 31 & 4 \end{vmatrix} = -36 + 465 = 429$

$x = \frac{D_x}{D} = \frac{429}{143} = 3$

$D_y = \begin{vmatrix} 2 & -9 \\ 9 & 31 \end{vmatrix} = 62 + 81 = 143$

$y = \frac{D_y}{D} = \frac{143}{143} = 1$

21. $2\begin{bmatrix} 2x^2 & x \\ 7x & 4 \end{bmatrix} - \begin{bmatrix} 5x \\ x-2 \end{bmatrix} = \begin{bmatrix} 2x & 0 \\ 6 & x^2 \end{bmatrix}$ is not possible. Subtraction requires the orders of the matrices to be the same.

23. $\begin{bmatrix} 3 & -2 & 1 \end{bmatrix}\begin{bmatrix} 5 & -1 \\ 0 & 3 \\ 9 & 4 \end{bmatrix} = \begin{bmatrix} 15+9 & -3-6+4 \end{bmatrix}$

$= \begin{bmatrix} 24 & -5 \end{bmatrix}$

25. $\begin{bmatrix} 3 & 7 \end{bmatrix}\begin{bmatrix} 0 & -8 \\ 5 & 6 \end{bmatrix} = \begin{bmatrix} 0+35 & -24+42 \end{bmatrix}$

$= \begin{bmatrix} 35 & 18 \end{bmatrix}$

27. Not possible. The number of columns in the first matrix factor must be the same as the number of rows in the second matrix factor.

29. Not possible. The number of columns in the first matrix factor must be the same as the number of rows in the second matrix factor.

31. $BA + B = \begin{bmatrix} 8 & -5 \end{bmatrix}\begin{bmatrix} -3 & 1 \\ 2 & 3 \end{bmatrix} + \begin{bmatrix} 8 & -5 \end{bmatrix}$

$= \begin{bmatrix} -24-10 & 8-15 \end{bmatrix} + \begin{bmatrix} 8 & -5 \end{bmatrix}$

$= \begin{bmatrix} -34 & -7 \end{bmatrix} + \begin{bmatrix} 8 & -5 \end{bmatrix}$

$= \begin{bmatrix} -26 & -12 \end{bmatrix}$

33. Not possible. The number of columns in the first matrix factor C must be the same as the number of rows in the second matrix factor C.

35. $D^2 = \begin{bmatrix} -5 & 4 \\ -1 & -1 \end{bmatrix}\begin{bmatrix} -5 & 4 \\ -1 & -1 \end{bmatrix}$

$= \begin{bmatrix} 25-4 & -20-4 \\ 5+1 & -4+1 \end{bmatrix}$

$= \begin{bmatrix} 21 & -24 \\ 6 & -3 \end{bmatrix}$

37. $DA = \begin{bmatrix} -5 & 4 \\ -1 & -1 \end{bmatrix}\begin{bmatrix} -3 & 1 \\ 2 & 3 \end{bmatrix}$

$= \begin{bmatrix} 15+8 & -5+12 \\ 3-2 & -1-3 \end{bmatrix}$

$= \begin{bmatrix} 23 & 7 \\ 1 & -4 \end{bmatrix}$

39. DB is not possible. The number of columns of D is not equal to the number of rows of B.

41. Let a = number of customers in Store A.
b = number of customers in Store B.

$\begin{cases} 0.9a + 0.2b \\ 0.1a + 0.8b \end{cases}$ is the basis for the transition matrix

$\begin{bmatrix} 0.9 & 0.2 \\ 0.1 & 0.8 \end{bmatrix}\begin{bmatrix} 250 \\ 750 \end{bmatrix} = \begin{bmatrix} 375 \\ 625 \end{bmatrix}$ (February 1)

$\begin{bmatrix} 0.9 & 0.2 \\ 0.1 & 0.8 \end{bmatrix}\begin{bmatrix} 375 \\ 625 \end{bmatrix} = \begin{bmatrix} 463 \\ 537 \end{bmatrix}$ (March 1)*

*rounded

43. $$P\begin{bmatrix} x \\ y \end{bmatrix} = \begin{bmatrix} x \\ y \end{bmatrix}$$

$$\begin{bmatrix} 0.7 & 0.45 \\ 0.3 & 0.55 \end{bmatrix}\begin{bmatrix} x \\ y \end{bmatrix} = \begin{bmatrix} x \\ y \end{bmatrix}$$

$$\begin{cases} 0.7x + 0.45y = x \\ 0.3x + 0.55y = y \end{cases}$$

$$\begin{cases} -0.3x + 0.45y = 0 \\ 0.3x - 0.45y = 0 \end{cases} \text{ or}$$

$\begin{cases} -0.3x + 0.45y = 0 \\ -0.3x + 0.45y = 0 \end{cases}$; i.e., they are identical.

Testing whether

$$-0.3(600) + 0.45(400) \stackrel{?}{=} 0$$

Yes, $-180 + 180 = 0$

End of Section 10.4

Chapter Ten: Section 10.5
Solutions to Odd-numbered Exercises

1. $\begin{cases} 14x-5y=7 \\ x+9y=2 \end{cases}$

$$\begin{bmatrix} 14 & -5 \\ 1 & 9 \end{bmatrix}\begin{bmatrix} x \\ y \end{bmatrix}=\begin{bmatrix} 7 \\ 2 \end{bmatrix}$$

3. $\begin{cases} \dfrac{3x-8y}{5}=2 \\ y-2=0 \end{cases} = \begin{cases} \dfrac{3}{5}x-\dfrac{8}{5}y=2 \\ y=2 \end{cases}$

$$\begin{bmatrix} \frac{3}{5} & -\frac{8}{5} \\ 0 & 1 \end{bmatrix}\begin{bmatrix} x \\ y \end{bmatrix}=\begin{bmatrix} 2 \\ 2 \end{bmatrix}$$

5. $\begin{cases} 4x=3y-9 \\ 13-2x=-4y \end{cases} = \begin{cases} 4x-3y=-9 \\ -2x+4y=-13 \end{cases}$

$$\begin{bmatrix} 4 & -3 \\ -2 & 4 \end{bmatrix}\begin{bmatrix} x \\ y \end{bmatrix}=\begin{bmatrix} -9 \\ -13 \end{bmatrix}$$

7. $\begin{cases} -6-2y=x \\ 9x+14=3y \end{cases} = \begin{cases} x+2y=-6 \\ 9x-3y=-14 \end{cases}$

$$\begin{bmatrix} 1 & 2 \\ 9 & -3 \end{bmatrix}\begin{bmatrix} x \\ y \end{bmatrix}=\begin{bmatrix} -6 \\ -14 \end{bmatrix}$$

9. $\begin{cases} 3x_1-7x_2+x_3=-4 \\ x_1-x_2=2 \\ 8x_2+5x_3=-3 \end{cases}$

$$\begin{bmatrix} 3 & -7 & 1 \\ 1 & -1 & 0 \\ 0 & 8 & 5 \end{bmatrix}\begin{bmatrix} x_1 \\ x_2 \\ x_3 \end{bmatrix}=\begin{bmatrix} -4 \\ 2 \\ -3 \end{bmatrix}$$

11. $\begin{cases} 2x-y=-3z \\ y-x=17 \\ 2+z+4x=5y \end{cases} = \begin{cases} 2x-y+3z=0 \\ -x+y=17 \\ 4x-5y+z=-2 \end{cases}$

$$\begin{bmatrix} 2 & -1 & 3 \\ -1 & 1 & 0 \\ 4 & -5 & 1 \end{bmatrix}\begin{bmatrix} x \\ y \\ z \end{bmatrix}=\begin{bmatrix} 0 \\ 17 \\ -2 \end{bmatrix}$$

13. Given $A=\begin{bmatrix} 0 & 4 \\ -5 & -1 \end{bmatrix}$, first check if $|A|=0$. It does not. $|A|=0+20=20$.

Therefore, A^{-1} exists:

Using the "short-cut":

$$A^{-1}=\frac{1}{20}\begin{bmatrix} -1 & -4 \\ 5 & 0 \end{bmatrix}=\begin{bmatrix} -\frac{1}{20} & -\frac{1}{5} \\ \frac{1}{4} & 0 \end{bmatrix}$$

15. Given $A=\begin{bmatrix} 3 & 4 \\ -4 & -5 \end{bmatrix}$, first check if $|A|=0$. It does not. $|A|=-15+16=1$.

Therefore, A^{-1} exists:

Using the "short-cut":

$$A^{-1}=\frac{1}{1}\begin{bmatrix} -5 & -4 \\ 4 & 3 \end{bmatrix}=\begin{bmatrix} -5 & -4 \\ 4 & 3 \end{bmatrix}$$

17. Given $A=\begin{bmatrix} -\frac{1}{5} & 0 \\ \frac{1}{5} & \frac{1}{2} \end{bmatrix}$, first check if $|A|=0$. It does not. $|A|=-\frac{1}{10}$

Therefore, A^{-1} exists:

Using the "short-cut":

$$A^{-1}=\frac{1}{\frac{1}{-10}}\begin{bmatrix} \frac{1}{2} & 0 \\ -\frac{1}{5} & -\frac{1}{5} \end{bmatrix}$$

$$=-10\begin{bmatrix} \frac{1}{2} & 0 \\ -\frac{1}{5} & -\frac{1}{5} \end{bmatrix}$$

$$=\begin{bmatrix} -5 & 0 \\ 2 & 2 \end{bmatrix}$$

19. Given $A=\begin{bmatrix}-2 & -4 & -2\\ 1 & -4 & 1\\ 4 & -3 & 4\end{bmatrix}$.

Notice that 2 of the columns are identical. Therefore, A is not invertible. Moreover, you could also conclude this by showing that $|A|=0$.

21. Given $A=\begin{bmatrix}-\frac{5}{11} & -\frac{8}{11} & 1\\ \frac{13}{11} & \frac{12}{11} & -2\\ -\frac{2}{11} & -\frac{1}{11} & 0\end{bmatrix}$, find A^{-1}

through row operations:

$$\left[\begin{array}{ccc|ccc}-\frac{5}{11} & -\frac{8}{11} & 1 & 1 & 0 & 0\\ \frac{13}{11} & \frac{12}{11} & -2 & 0 & 1 & 0\\ -\frac{2}{11} & -\frac{1}{11} & 0 & 0 & 0 & 1\end{array}\right]\xrightarrow{\substack{11R_1\\ 11R_2\\ 11R_3}}$$

$$\left[\begin{array}{ccc|ccc}-5 & -8 & 11 & 11 & 0 & 0\\ 13 & 12 & -22 & 0 & 11 & 0\\ -2 & -1 & 0 & 0 & 0 & 11\end{array}\right]\xrightarrow{-3R_3+R_1}$$

$$\left[\begin{array}{ccc|ccc}1 & -5 & 11 & 11 & 0 & -33\\ 13 & 12 & -22 & 0 & 11 & 0\\ -2 & -1 & 0 & 0 & 0 & 11\end{array}\right]\xrightarrow{\substack{2R_1+R_3\\ -13R_1+R_2}}$$

$$\left[\begin{array}{ccc|ccc}1 & -5 & 11 & 11 & 0 & -33\\ 0 & 77 & -165 & -143 & 11 & 429\\ 0 & -11 & 22 & 22 & 0 & -55\end{array}\right]\xrightarrow{-\frac{1}{11}R_3}$$

$$\left[\begin{array}{ccc|ccc}1 & -5 & 11 & 11 & 0 & -33\\ 0 & 77 & -165 & -143 & 11 & 429\\ 0 & 1 & -2 & -2 & 0 & 5\end{array}\right]\xrightarrow{\substack{5R_3+R_1\\ \frac{1}{11}R_2}}$$

$$\left[\begin{array}{ccc|ccc}1 & 0 & 1 & 1 & 0 & -8\\ 0 & 7 & -15 & -13 & 1 & 39\\ 0 & 1 & -2 & -2 & 0 & 5\end{array}\right]\xrightarrow{R_2\leftrightarrow R_3}$$

$$\left[\begin{array}{ccc|ccc}1 & 0 & 1 & 1 & 0 & -8\\ 0 & 1 & -2 & -2 & 0 & 5\\ 0 & 7 & -15 & -13 & 1 & 39\end{array}\right]\xrightarrow{-7R_2+R_3}$$

$$\left[\begin{array}{ccc|ccc}1 & 0 & 1 & 1 & 0 & -8\\ 0 & 1 & -2 & -2 & 0 & 5\\ 0 & 0 & -1 & 1 & 1 & 4\end{array}\right]\xrightarrow{-R_3}$$

$$\left[\begin{array}{ccc|ccc}1 & 0 & 1 & 1 & 0 & -8\\ 0 & 1 & -2 & -2 & 0 & 5\\ 0 & 0 & 1 & -1 & -1 & -4\end{array}\right]\xrightarrow{\substack{-R_3+R_1\\ 2R_3+R_2}}$$

$$\left[\begin{array}{ccc|ccc}1 & 0 & 0 & 2 & 1 & -4\\ 0 & 1 & 0 & -4 & -2 & -3\\ 0 & 0 & 1 & -1 & -1 & -4\end{array}\right]$$

Then, $A^{-1}=\begin{bmatrix}2 & 1 & -4\\ -4 & -2 & -3\\ -1 & -1 & -4\end{bmatrix}$

23. Given $A=\begin{bmatrix}-1 & 2 & -1\\ 0 & 3 & -1\\ 0 & 4 & -1\end{bmatrix}$, find A^{-1}

through row operations:

$$\left[\begin{array}{ccc|ccc}-1 & 2 & -1 & 1 & 0 & 0\\ 0 & 3 & -1 & 0 & 1 & 0\\ 0 & 4 & -1 & 0 & 0 & 1\end{array}\right]\xrightarrow{\substack{-R_1\\ \frac{1}{3}R_2}}$$

$$\left[\begin{array}{ccc|ccc}1 & -2 & 1 & -1 & 0 & 0\\ 0 & 1 & -\frac{1}{3} & 0 & \frac{1}{3} & 0\\ 0 & 4 & -1 & 0 & 0 & 1\end{array}\right]\xrightarrow{-4R_2+R_3}$$

$$\left[\begin{array}{ccc|ccc}1 & -2 & 1 & -1 & 0 & 0\\ 0 & 1 & -\frac{1}{3} & 0 & \frac{1}{3} & 0\\ 0 & 0 & \frac{1}{3} & 0 & -\frac{4}{3} & 1\end{array}\right]\xrightarrow{2R_2+R_1}$$

$$\left[\begin{array}{ccc|ccc}1 & 0 & \frac{1}{3} & -1 & \frac{2}{3} & 0\\ 0 & 1 & -\frac{1}{3} & 0 & \frac{1}{3} & 0\\ 0 & 0 & \frac{1}{3} & 0 & -\frac{4}{3} & 1\end{array}\right]\xrightarrow{\substack{R_3+R_2\\ 3R_3\\ -R_3+R_1}}$$

$$\left[\begin{array}{ccc|ccc}1 & 0 & 0 & -1 & 2 & -1\\ 0 & 1 & 0 & 0 & -1 & 1\\ 0 & 0 & 1 & 0 & -4 & 3\end{array}\right]$$

$$A^{-1}=\begin{bmatrix}-1 & 2 & -1\\ 0 & -1 & 1\\ 0 & -4 & 3\end{bmatrix}$$

25. Given $A=\begin{bmatrix} -\frac{6}{5} & -\frac{2}{5} & -1 \\ \frac{3}{5} & \frac{1}{5} & 1 \\ 1 & 0 & 1 \end{bmatrix}$, find A^{-1} through row operations:

$$\left[\begin{array}{ccc|ccc} -\frac{6}{5} & -\frac{2}{5} & -1 & 1 & 0 & 0 \\ \frac{3}{5} & \frac{1}{5} & 1 & 0 & 1 & 0 \\ 1 & 0 & 1 & 0 & 0 & 1 \end{array}\right] \xrightarrow[\substack{R_1 \leftrightarrow R_3 \\ 5R_2}]{5R_1}$$

$$\left[\begin{array}{ccc|ccc} 1 & 0 & 1 & 0 & 0 & 1 \\ 3 & 1 & 5 & 0 & 5 & 0 \\ -6 & -2 & -5 & 5 & 0 & 0 \end{array}\right] \xrightarrow[6R_1+R_3]{-3R_1+R_2}$$

$$\left[\begin{array}{ccc|ccc} 1 & 0 & 1 & 0 & 0 & 1 \\ 0 & 1 & 2 & 0 & 5 & -3 \\ 0 & -2 & 1 & 5 & 0 & 6 \end{array}\right] \xrightarrow{2R_2+R_3}$$

$$\left[\begin{array}{ccc|ccc} 1 & 0 & 1 & 0 & 0 & 1 \\ 0 & 1 & 2 & 0 & 5 & -3 \\ 0 & 0 & 5 & 5 & 10 & 0 \end{array}\right] \xrightarrow{\frac{1}{5}R_3}$$

$$\left[\begin{array}{ccc|ccc} 1 & 0 & 1 & 0 & 0 & 1 \\ 0 & 1 & 2 & 0 & 5 & -3 \\ 0 & 0 & 1 & 1 & 2 & 0 \end{array}\right] \xrightarrow[-R_3+R_1]{-2R_3+R_2}$$

$$\left[\begin{array}{ccc|ccc} 1 & 0 & 0 & -1 & -2 & 1 \\ 0 & 1 & 0 & -2 & 1 & -3 \\ 0 & 0 & 1 & 1 & 2 & 0 \end{array}\right]$$

$$A^{-1}=\begin{bmatrix} -1 & -2 & 1 \\ -2 & 1 & -3 \\ 1 & 2 & 0 \end{bmatrix}$$

27. Given $A=\begin{bmatrix} 0 & 1 & 1 \\ 1 & 1 & 0 \\ 0 & 1 & 2 \end{bmatrix}$, find A^{-1} through row operations:

$$\left[\begin{array}{ccc|ccc} 0 & 1 & 1 & 1 & 0 & 0 \\ 1 & 1 & 0 & 0 & 1 & 0 \\ 0 & 1 & 2 & 0 & 0 & 1 \end{array}\right] \xrightarrow{R_2 \leftrightarrow R_1}$$

$$\left[\begin{array}{ccc|ccc} 1 & 1 & 0 & 0 & 1 & 0 \\ 0 & 1 & 1 & 1 & 0 & 0 \\ 0 & 1 & 2 & 0 & 0 & 1 \end{array}\right] \xrightarrow[-R_2+R_3]{-R_3+R_1}$$

$$\left[\begin{array}{ccc|ccc} 1 & 0 & -2 & 0 & 1 & -1 \\ 0 & 1 & 1 & 1 & 0 & 0 \\ 0 & 0 & 1 & -1 & 0 & 1 \end{array}\right] \xrightarrow[2R_3+R_1]{-R_3+R_2}$$

$$\left[\begin{array}{ccc|ccc} 1 & 0 & 0 & -2 & 1 & 1 \\ 0 & 1 & 0 & 2 & 0 & -1 \\ 0 & 0 & 1 & -1 & 0 & 1 \end{array}\right]$$

$$A^{-1}=\begin{bmatrix} -2 & 1 & 1 \\ 2 & 0 & -1 \\ -1 & 0 & 1 \end{bmatrix}$$

29. Given $A=\begin{bmatrix} \frac{2}{3} & \frac{8}{9} & \frac{1}{9} \\ -\frac{1}{3} & \frac{2}{9} & -\frac{2}{9} \\ -\frac{1}{3} & -\frac{7}{9} & -\frac{2}{9} \end{bmatrix}$ find A^{-1} through row operations:

$$\left[\begin{array}{ccc|ccc} \frac{2}{3} & \frac{8}{9} & \frac{1}{9} & 1 & 0 & 0 \\ -\frac{1}{3} & \frac{2}{9} & -\frac{2}{9} & 0 & 1 & 0 \\ -\frac{1}{3} & -\frac{7}{9} & -\frac{2}{9} & 0 & 0 & 1 \end{array}\right] \xrightarrow{9 \times \text{all Rows}}$$

$$\left[\begin{array}{ccc|ccc} 6 & 8 & 1 & 9 & 0 & 0 \\ -3 & 2 & -2 & 0 & 9 & 0 \\ -3 & -7 & -2 & 0 & 0 & 9 \end{array}\right] \xrightarrow[\frac{1}{2}R_1+R_2]{-R_2+R_3}$$

$$\left[\begin{array}{ccc|ccc} 6 & 8 & 1 & 9 & 0 & 0 \\ 0 & 6 & -\frac{3}{2} & \frac{9}{2} & 9 & 0 \\ 0 & -9 & 0 & 0 & -9 & 9 \end{array}\right] \xrightarrow[\frac{1}{9}R_3]{\frac{1}{6}R_1}$$

$$\left[\begin{array}{ccc|ccc} 1 & \frac{4}{3} & \frac{1}{6} & \frac{3}{2} & 0 & 0 \\ 0 & 6 & -\frac{3}{2} & \frac{9}{2} & 9 & 0 \\ 0 & -1 & 0 & 0 & -1 & 1 \end{array}\right] \xrightarrow[-R_3]{\frac{1}{6}R_2}$$

$$\left[\begin{array}{ccc|ccc} 1 & \frac{4}{3} & \frac{1}{6} & \frac{3}{2} & 0 & 0 \\ 0 & 1 & -\frac{1}{4} & \frac{3}{4} & \frac{3}{2} & 0 \\ 0 & 1 & 0 & 0 & 1 & -1 \end{array}\right] \xrightarrow{R_2 \leftrightarrow R_3}$$

$$\left[\begin{array}{ccc|ccc} 1 & \frac{4}{3} & \frac{1}{6} & \frac{3}{2} & 0 & 0 \\ 0 & 1 & 0 & 0 & 1 & -1 \\ 0 & 1 & -\frac{1}{4} & \frac{3}{4} & \frac{3}{2} & 0 \end{array}\right] \xrightarrow[-R_2+R_3]{-\frac{4}{3}R_2+R_1}$$

$$\left[\begin{array}{ccc|ccc} 1 & 0 & \frac{1}{6} & \frac{3}{2} & -\frac{4}{3} & \frac{4}{3} \\ 0 & 1 & 0 & 0 & 1 & -1 \\ 0 & 0 & -\frac{1}{4} & \frac{3}{4} & \frac{1}{2} & 1 \end{array}\right] \xrightarrow{-4R_3}$$

$$\left[\begin{array}{ccc|ccc} 1 & 0 & \frac{1}{6} & \frac{3}{2} & -\frac{4}{3} & \frac{4}{3} \\ 0 & 1 & 0 & 0 & 1 & -1 \\ 0 & 0 & 1 & -3 & -2 & -4 \end{array}\right] \xrightarrow{-\frac{1}{6}R_3+R_1}$$

$$\left[\begin{array}{ccc|ccc} 1 & 0 & 0 & 2 & -1 & 2 \\ 0 & 1 & 0 & 0 & 1 & -1 \\ 0 & 0 & 1 & -3 & -2 & -4 \end{array}\right]$$

$$A^{-1} = \begin{bmatrix} 2 & -1 & 2 \\ 0 & 1 & -1 \\ -3 & -2 & -4 \end{bmatrix}$$

31. Solve $\begin{cases} -2x - 2y = 9 \\ -x + 2y = -3 \end{cases}$

$$\begin{bmatrix} -2 & -2 \\ -1 & 2 \end{bmatrix}\begin{bmatrix} x \\ y \end{bmatrix} = \begin{bmatrix} 9 \\ -3 \end{bmatrix}$$

Use the short-cut for 2 x 2 matrices:

$$A^{-1} = \begin{bmatrix} -2 & -2 \\ -1 & 2 \end{bmatrix}^{-1} = \frac{1}{\begin{vmatrix} -2 & -2 \\ -1 & 2 \end{vmatrix}}\begin{bmatrix} 2 & 2 \\ 1 & -2 \end{bmatrix}$$

$$= \frac{1}{-4-2}\begin{bmatrix} 2 & 2 \\ 1 & -2 \end{bmatrix}$$

$$= \frac{1}{-6}\begin{bmatrix} 2 & 2 \\ 1 & -2 \end{bmatrix}$$

$$= \begin{bmatrix} -\frac{1}{3} & -\frac{1}{3} \\ -\frac{1}{6} & \frac{1}{3} \end{bmatrix}$$

$$\begin{bmatrix} x \\ y \end{bmatrix} = \begin{bmatrix} -\frac{1}{3} & -\frac{1}{3} \\ -\frac{1}{6} & \frac{1}{3} \end{bmatrix}\begin{bmatrix} 9 \\ -3 \end{bmatrix} = \begin{bmatrix} -3+1 \\ -\frac{3}{2}-1 \end{bmatrix} = \begin{bmatrix} -2 \\ -\frac{5}{2} \end{bmatrix}$$

$x = -2;\ y = -\frac{5}{2}$

33. Solve $\begin{cases} -2x + 3y = 1 \\ 4x - 6y = -2 \end{cases}$

$$\begin{bmatrix} -2 & 3 \\ 4 & -6 \end{bmatrix}\begin{bmatrix} x \\ y \end{bmatrix} = \begin{bmatrix} 1 \\ -2 \end{bmatrix}$$

The inverse matrix method does not work because $|A| = 0$. In fact, the system is dependent. Solve for one of the variables in terms of the other:

$$-2x + 3y = 1$$
$$-2x = -3y + 1$$
$$x = \frac{3y-1}{2}$$

Solution Set: $\left\{\left(\frac{3y-1}{2}, y\right) \mid y \in \mathbb{R}\right\}$

35. Solve $\begin{cases} -5x = 10 \\ 2x + 2y = -4 \end{cases}$

$$\begin{bmatrix} -5 & 0 \\ 2 & 2 \end{bmatrix}\begin{bmatrix} x \\ y \end{bmatrix} = \begin{bmatrix} 10 \\ -4 \end{bmatrix}$$

Use the short-cut for 2 x 2 matrices:

$$A^{-1} = \begin{bmatrix} -5 & 0 \\ 2 & 2 \end{bmatrix}^{-1} = \frac{1}{\begin{vmatrix} -5 & 0 \\ 2 & 2 \end{vmatrix}}\begin{bmatrix} 2 & 0 \\ -2 & -5 \end{bmatrix}$$

$$= \frac{1}{-10}\begin{bmatrix} 2 & 0 \\ -2 & -5 \end{bmatrix}$$

$$= \begin{bmatrix} -\frac{1}{5} & 0 \\ \frac{1}{5} & \frac{1}{2} \end{bmatrix}$$

$$\begin{bmatrix} -\frac{1}{5} & 0 \\ \frac{1}{5} & \frac{1}{2} \end{bmatrix}\begin{bmatrix} 10 \\ -4 \end{bmatrix} = \begin{bmatrix} -2 \\ 2-2 \end{bmatrix} = \begin{bmatrix} -2 \\ 0 \end{bmatrix}$$

$x = -2;\ y = 0$

37. Solve $\begin{cases} -x+2y-z=6 \\ 3y-z=-1 \\ 4y-z=5 \end{cases}$ by finding the inverse of the coefficient matrix:

$$\begin{bmatrix} -1 & 2 & -1 \\ 0 & 3 & -1 \\ 0 & 4 & -1 \end{bmatrix}\begin{bmatrix} x \\ y \\ z \end{bmatrix}=\begin{bmatrix} 6 \\ -1 \\ 5 \end{bmatrix}$$

$$\left[\begin{array}{ccc|ccc} -1 & 2 & -1 & 1 & 0 & 0 \\ 0 & 3 & -1 & 0 & 1 & 0 \\ 0 & 4 & -1 & 0 & 0 & 1 \end{array}\right] \xrightarrow{-R_1}$$

$$\left[\begin{array}{ccc|ccc} 1 & -2 & 1 & -1 & 0 & 0 \\ 0 & 3 & -1 & 0 & 1 & 0 \\ 0 & 4 & -1 & 0 & 0 & 1 \end{array}\right] \xrightarrow{-R_2+R_3 \to R_2}$$

$$\left[\begin{array}{ccc|ccc} 1 & -2 & 1 & -1 & 0 & 0 \\ 0 & 1 & 0 & 0 & -1 & 1 \\ 0 & 4 & -1 & 0 & 0 & 1 \end{array}\right] \xrightarrow{-4R_2+R_3}$$

$$\left[\begin{array}{ccc|ccc} 1 & -2 & 1 & -1 & 0 & 0 \\ 0 & 1 & 0 & 0 & -1 & 1 \\ 0 & 0 & -1 & 0 & 4 & -3 \end{array}\right] \xrightarrow[-R_3]{2R_2+R_1}$$

$$\left[\begin{array}{ccc|ccc} 1 & 0 & 1 & -1 & -2 & 2 \\ 0 & 1 & 0 & 0 & -1 & 1 \\ 0 & 0 & 1 & 0 & -4 & 3 \end{array}\right] \xrightarrow{-R_3+R_1}$$

$$\left[\begin{array}{ccc|ccc} 1 & 0 & 0 & -1 & 2 & -1 \\ 0 & 1 & 0 & 0 & -1 & 1 \\ 0 & 0 & 1 & 0 & -4 & 3 \end{array}\right]$$

$$A^{-1}=\begin{bmatrix} -1 & 2 & -1 \\ 0 & -1 & 1 \\ 0 & -4 & 3 \end{bmatrix}$$

$$\begin{bmatrix} x \\ y \\ z \end{bmatrix}=\begin{bmatrix} -1 & 2 & -1 \\ 0 & -1 & 1 \\ 0 & -4 & 3 \end{bmatrix}\begin{bmatrix} 6 \\ -1 \\ 5 \end{bmatrix}$$

$$\begin{bmatrix} x \\ y \\ z \end{bmatrix}=\begin{bmatrix} -6-2-5 \\ 1+5 \\ 4+15 \end{bmatrix}=\begin{bmatrix} -13 \\ 6 \\ 19 \end{bmatrix}$$

$x=-13,\ y=6,\ z=19$

39. Solve $\begin{cases} -2x-4y-2z=1 \\ x-4y+z=0 \\ 4x-3y+4z=-1 \end{cases}$

$$\begin{bmatrix} -2 & -4 & -2 \\ 1 & -4 & 1 \\ 4 & -3 & 4 \end{bmatrix}\begin{bmatrix} x \\ y \\ z \end{bmatrix}=\begin{bmatrix} 1 \\ 0 \\ -1 \end{bmatrix}$$

Since $A^{-1}=0$ (notice that two columns are identical), the inverse of the coefficient matrix does not exist. Using Cramer's Rule, you can readily determine that $D=0$ and that $D_x=-1$, i.e., $D_x \neq 0$. This means that there is no solution to the system.

41. Solve $\begin{cases} 2x-y-3z=-10 \\ 2y-z=11 \\ -x+4z=0 \end{cases}$ by finding the inverse of the coefficient matrix:

$$\left[\begin{array}{ccc|ccc} 2 & -1 & -3 & 1 & 0 & 0 \\ 0 & 2 & -1 & 0 & 1 & 0 \\ -1 & 0 & 4 & 0 & 0 & 1 \end{array}\right] \xrightarrow[\frac{1}{2}R_2]{R_1 \leftrightarrow -R_3}$$

$$\left[\begin{array}{ccc|ccc} 1 & 0 & -4 & 0 & 0 & -1 \\ 0 & 1 & -\frac{1}{2} & 0 & \frac{1}{2} & 0 \\ 2 & -1 & -3 & 1 & 0 & 0 \end{array}\right] \xrightarrow{-2R_1+R_3}$$

$$\left[\begin{array}{ccc|ccc} 1 & 0 & -4 & 0 & 0 & -1 \\ 0 & 1 & -\frac{1}{2} & 0 & \frac{1}{2} & 0 \\ 0 & -1 & 5 & 1 & 0 & 2 \end{array}\right] \xrightarrow{R_2+R_3}$$

$$\left[\begin{array}{ccc|ccc} 1 & 0 & -4 & 0 & 0 & -1 \\ 0 & 1 & -\frac{1}{2} & 0 & \frac{1}{2} & 0 \\ 0 & 0 & \frac{9}{2} & 1 & \frac{1}{2} & 2 \end{array}\right] \xrightarrow{\frac{2}{9}R_3}$$

$$\left[\begin{array}{ccc|ccc} 1 & 0 & -4 & 0 & 0 & -1 \\ 0 & 1 & -\frac{1}{2} & 0 & \frac{1}{2} & 0 \\ 0 & 0 & 1 & \frac{2}{9} & \frac{1}{9} & \frac{4}{9} \end{array}\right] \xrightarrow{\frac{1}{2}R_3+R_2}$$

$$\left[\begin{array}{ccc|ccc} 1 & 0 & -4 & 0 & 0 & -1 \\ 0 & 1 & 0 & \frac{1}{9} & \frac{5}{9} & \frac{2}{9} \\ 0 & 0 & 1 & \frac{2}{9} & \frac{1}{9} & \frac{4}{9} \end{array}\right] \xrightarrow{4R_3+R_1}$$

$$\left[\begin{array}{ccc|ccc} 1 & 0 & 0 & \frac{8}{9} & \frac{4}{9} & \frac{7}{9} \\ 0 & 1 & 0 & \frac{1}{9} & \frac{5}{9} & \frac{2}{9} \\ 0 & 0 & 1 & \frac{2}{9} & \frac{1}{9} & \frac{4}{9} \end{array}\right]$$

$$\begin{bmatrix} x \\ y \\ z \end{bmatrix} = \frac{1}{9}\begin{bmatrix} 8 & 4 & 7 \\ 1 & 5 & 2 \\ 2 & 1 & 4 \end{bmatrix}\begin{bmatrix} -10 \\ 11 \\ 0 \end{bmatrix} = \frac{1}{9}\begin{bmatrix} -80+44 \\ -10+55 \\ -20+11 \end{bmatrix}$$

$$= \frac{1}{9}\begin{bmatrix} -36 \\ 45 \\ -9 \end{bmatrix} = \begin{bmatrix} -4 \\ 5 \\ -1 \end{bmatrix}$$

$x = -4, y = 5, z = -1$

43. Solve the given sets of equations by using row operations on the coefficient matrix to find its inverse:

$$\left[\begin{array}{ccc|ccc} 1 & 2 & -1 & 1 & 0 & 0 \\ 3 & 3 & -1 & 0 & 1 & 0 \\ 4 & 4 & -1 & 0 & 0 & 1 \end{array}\right] \xrightarrow[-4R_1+R_3]{-3R_1+R_2}$$

$$\left[\begin{array}{ccc|ccc} 1 & 2 & -1 & 1 & 0 & 0 \\ 0 & -3 & 2 & -3 & 1 & 0 \\ 0 & -4 & 3 & -4 & 0 & 1 \end{array}\right] \xrightarrow{-R_3+R_2}$$

$$\left[\begin{array}{ccc|ccc} 1 & 2 & -1 & 1 & 0 & 0 \\ 0 & 1 & -1 & 1 & 1 & -1 \\ 0 & -4 & 3 & -4 & 0 & 1 \end{array}\right] \xrightarrow{4R_2+R_3}$$

$$\left[\begin{array}{ccc|ccc} 1 & 2 & -1 & 1 & 0 & 0 \\ 0 & 1 & -1 & 1 & 1 & -1 \\ 0 & 0 & -1 & 0 & 4 & -3 \end{array}\right] \xrightarrow{-R_3}$$

$$\left[\begin{array}{ccc|ccc} 1 & 2 & -1 & 1 & 0 & 0 \\ 0 & 1 & -1 & 1 & 1 & -1 \\ 0 & 0 & 1 & 0 & -4 & 3 \end{array}\right] \xrightarrow[R_3+R_2]{-2R_2+R_1}$$

$$\left[\begin{array}{ccc|ccc} 1 & 0 & 1 & -1 & -2 & 2 \\ 0 & 1 & 0 & 1 & -3 & 2 \\ 0 & 0 & 1 & 0 & -4 & 3 \end{array}\right] \xrightarrow{-R_3+R_1}$$

$$\left[\begin{array}{ccc|ccc} 1 & 0 & 0 & -1 & 2 & -1 \\ 0 & 1 & 0 & 1 & -3 & 2 \\ 0 & 0 & 1 & 0 & -4 & 3 \end{array}\right]$$

The inverse matrix is $\begin{bmatrix} -1 & 2 & -1 \\ 1 & -3 & 2 \\ 0 & -4 & 3 \end{bmatrix}$.

It is used to operate on the constant matrix to find the solutions:

Set 1:

$$\begin{bmatrix} x \\ y \\ z \end{bmatrix} = \begin{bmatrix} -1 & 2 & -1 \\ 1 & -3 & 2 \\ 0 & -4 & 3 \end{bmatrix}\begin{bmatrix} 2 \\ -5 \\ 1 \end{bmatrix} = \begin{bmatrix} -2-10-1 \\ 2+15+2 \\ 20+3 \end{bmatrix} = \begin{bmatrix} -13 \\ 19 \\ 23 \end{bmatrix}$$

Solution: $(-13, 19, 23)$

Set 2:

$$\begin{bmatrix} x \\ y \\ z \end{bmatrix} = \begin{bmatrix} -1 & 2 & -1 \\ 1 & -3 & 2 \\ 0 & -4 & 3 \end{bmatrix}\begin{bmatrix} 1 \\ 1 \\ 1 \end{bmatrix} = \begin{bmatrix} -1+2-1 \\ 1-3+2 \\ -4+3 \end{bmatrix} = \begin{bmatrix} 0 \\ 0 \\ -1 \end{bmatrix}$$

Solution: $(0, 0, -1)$

Set 3:

$$\begin{bmatrix} x \\ y \\ z \end{bmatrix} = \begin{bmatrix} -1 & 2 & -1 \\ 1 & -3 & 2 \\ 0 & -4 & 3 \end{bmatrix}\begin{bmatrix} 0 \\ 1 \\ 1 \end{bmatrix} = \begin{bmatrix} 2-1 \\ -3+2 \\ -4+3 \end{bmatrix} = \begin{bmatrix} 1 \\ -1 \\ -1 \end{bmatrix}$$

Solution: $(1, -1, -1)$

45. Solve the given sets of equations by using row operations on the coefficient matrix to find its inverse:

$$\left[\begin{array}{ccc|ccc} -1 & 0 & 1 & 1 & 0 & 0 \\ -1 & 3 & 2 & 0 & 1 & 0 \\ 2 & -4 & -3 & 0 & 0 & 1 \end{array}\right] \xrightarrow[-R_1]{2R_1+R_3}$$

$$\left[\begin{array}{ccc|ccc} 1 & 0 & -1 & -1 & 0 & 0 \\ -1 & 3 & 2 & 0 & 1 & 0 \\ 0 & -4 & -1 & 2 & 0 & 1 \end{array}\right] \xrightarrow{R_1+R_2}$$

$$\left[\begin{array}{ccc|ccc} 1 & 0 & -1 & -1 & 0 & 0 \\ 0 & 3 & 1 & -1 & 1 & 0 \\ 0 & -4 & -1 & 2 & 0 & 1 \end{array}\right] \xrightarrow{\frac{1}{3}R_2}$$

$$\left[\begin{array}{ccc|ccc} 1 & 0 & -1 & -1 & 0 & 0 \\ 0 & 1 & \frac{1}{3} & -\frac{1}{3} & \frac{1}{3} & 0 \\ 0 & -4 & -1 & 2 & 0 & 1 \end{array}\right] \xrightarrow{4R_2+R_3}$$

$$\left[\begin{array}{ccc|ccc} 1 & 0 & -1 & -1 & 0 & 0 \\ 0 & 1 & \frac{1}{3} & -\frac{1}{3} & \frac{1}{3} & 0 \\ 0 & 0 & \frac{1}{3} & \frac{2}{3} & \frac{4}{3} & 1 \end{array}\right] \xrightarrow[-R_3+R_2]{3R_3}$$

$$\left[\begin{array}{ccc|ccc} 1 & 0 & -1 & -1 & 0 & 0 \\ 0 & 1 & 0 & -1 & -1 & -1 \\ 0 & 0 & 1 & 2 & 4 & 3 \end{array}\right] \xrightarrow{R_3+R_1}$$

$$\left[\begin{array}{ccc|ccc} 1 & 0 & 0 & 1 & 4 & 3 \\ 0 & 1 & 0 & -1 & -1 & -1 \\ 0 & 0 & 1 & 2 & 4 & 3 \end{array}\right]$$

The inverse matrix is $\begin{bmatrix} 1 & 4 & 3 \\ -1 & -1 & -1 \\ 2 & 4 & 3 \end{bmatrix}$.

It is used to operate on the constant matrix to find the solutions:

Set 1:

$$\begin{bmatrix} x \\ y \\ z \end{bmatrix} = \begin{bmatrix} 1 & 4 & 3 \\ -1 & -1 & -1 \\ 2 & 4 & 3 \end{bmatrix}\begin{bmatrix} 6 \\ -11 \\ 13 \end{bmatrix} = \begin{bmatrix} 6-44+39 \\ -6+11-13 \\ 12-44+39 \end{bmatrix} = \begin{bmatrix} 1 \\ -8 \\ 7 \end{bmatrix}$$

Solution: $(1,-8,7)$

Set 2:

$$\begin{bmatrix} x \\ y \\ z \end{bmatrix} = \begin{bmatrix} 1 & 4 & 3 \\ -1 & -1 & -1 \\ 2 & 4 & 3 \end{bmatrix}\begin{bmatrix} -2 \\ 2 \\ -1 \end{bmatrix} = \begin{bmatrix} -2+8-3 \\ 2-2+1 \\ -4+8-3 \end{bmatrix} = \begin{bmatrix} 3 \\ 1 \\ 1 \end{bmatrix}$$

Solution: $(3,1,1)$

Set 3:

$$\begin{bmatrix} x \\ y \\ z \end{bmatrix} = \begin{bmatrix} 1 & 4 & 3 \\ -1 & -1 & -1 \\ 2 & 4 & 3 \end{bmatrix}\begin{bmatrix} -4 \\ 2 \\ 0 \end{bmatrix} = \begin{bmatrix} -4+8 \\ 4-2 \\ -8+8 \end{bmatrix} = \begin{bmatrix} 4 \\ 2 \\ 0 \end{bmatrix}$$

Solution: $(4, 2, 0)$

End of Section 10.5

Chapter Ten: Section 10.6
Solutions to Odd-numbered Exercises

1. Partial fraction decomposition of

$f(x)=\dfrac{p(x)}{x^2-x-6}$

Factor the denominator:

$x^2-x-6=(x-3)(x+2)$

Then,

$f(x)=\dfrac{A_1}{x-3}+\dfrac{A_2}{x+2}$

3. Partial fraction decomposition of

$f(x)=\dfrac{p(x)}{x^3+11x^2+40x+48}$

Factor the denominator: The possible factors of 48 are $\pm\begin{Bmatrix}1,2,3,4,6,\\8,12,16,24\end{Bmatrix}$

Use synthetic division to determine the factors of $x^3+11x^2+40x+48$:

$$\begin{array}{r|rrrr} -3 & 1 & 11 & 40 & 48 \\ & & -3 & -24 & -48 \\ \hline & 1 & 8 & 16 & 0 \end{array} \Rightarrow x+3 \text{ is a factor}$$

$$\begin{array}{r|rrr} -4 & 1 & 8 & 16 \\ & & -4 & -16 \\ \hline & 1 & 4 & 0 \end{array} \Rightarrow x+4 \text{ is a factor}$$

$$\begin{array}{r|rr} -4 & 1 & 4 \\ & & -4 \\ \hline & 1 & 0 \end{array} \Rightarrow x+4 \text{ is a factor (twice)}$$

Then,

$f(x)=\dfrac{A_1}{x+3}+\dfrac{A_2}{x+4}+\dfrac{A_3}{(x+4)^2}$

5. Partial fraction decomposition of

$f(x)=\dfrac{p(x)}{(x+3)\left(x^2-4\right)}$

Complete the factorization of x^2-4:

$x^2-4=(x-2)(x+2)$

Then,

$f(x)=\dfrac{A_1}{x+3}+\dfrac{A_2}{x+2}+\dfrac{A_3}{x-2}$

7. $\dfrac{2x-1}{(x+2)^3(x-2)}$ matches with the decomposition in d. [The denominator in d satisfies the guideline and the factors of the given rational expression.]

9. $\dfrac{2x-1}{x^3-4x^2+4x}$ matches with the decomposition in h. [The denominator in h satisfies the guideline and the factors of the given rational expression. $x\left(x^2-4x+4\right)=x(x-2)^2$]

11. $\dfrac{2x-1}{x^3(x-2)^2}$ matches with the decomposition in a. [The denominator in a satisfies the guideline and the factors of the given rational expression.]

13. $\dfrac{2x-1}{x^3(x-2)}$ matches with the decomposition in c. [The denominator in c satisfies the guideline and the factors of the given rational expression.]

15. Given: $f(x)=\dfrac{3x^2+4}{x^3-4x}$

Factor the denominator:

$x^3-4x=x(x-2)(x+2)$

By the guidelines for decomposition:

$\dfrac{3x^2+4}{x^3-4x}=\dfrac{A_1}{x}+\dfrac{A_2}{x-2}+\dfrac{A_3}{x+2}$

Solve for A_1, A_2 and A_3.

Start by multiplying both sides by the factors of the denominator:

$$3x^2+4=(x^2-4)A_1+(x^2+2x)A_2+(x^2-2x)A_3$$
$$=A_1x^2-4A_1+A_2x^2+2xA_2+A_3x^2-2xA_3$$
$$=(A_1+A_2+A_3)x^2+(2A_2-2A_3)x-4A_1$$

Then, we can write a system of equations and solve for the A's:

$$\begin{cases} A_1+A_2+A_3=3 \\ 2A_2-2A_3=0 \\ -4A_1=4 \end{cases}$$

Equation 3 $\Rightarrow A_1=-1$

Substitute $A_1=-1$ in equation 1, so with equation 2, we have a system of two equations in two variables:

$$\begin{cases} -1+A_2+A_3=3 \\ 2A_2-2A_3=0 \end{cases}$$

Rewrite.

$$\begin{cases} A_2+A_3=4 \\ A_2=A_3 \end{cases}$$

Then,

$2A_3=4 \Rightarrow A_3=2 \Rightarrow A_2=2$

Then, we can write the partial decomposition of the given expression:

$$f(x)=\frac{-1}{x}+\frac{2}{x-2}+\frac{2}{x+2}$$

17. Given: $f(x)=\dfrac{4x+2}{(x^3+8x)(x^2+2x-8)}$

Factor the denominator:

$$(x^3+8x)(x^2+2x-8)=x(x^2+8)(x+4)(x-2)$$

By the guidelines for decomposition:

$$\frac{4x+2}{(x^3+8x)(x^2+2x-8)}=\frac{A_1}{x}+\frac{A_2x+B}{x^2+8}+\frac{A_3}{x+4}+\frac{A_4}{x-2}$$

Solve for A_1, A_2, B, A_3 and A_4.

Start by multiplying both sides by the factors of the denominator:

$$4x+2=A_1(x^2+8)(x+4)(x-2)+(A_2x+B)(x)(x+4)(x-2)+A_3(x)(x^2+8)(x-2)+A_4(x)(x^2+8)(x+4)$$

Regrouping:

$$4x+2=(x^4+2x^3+16x-64)A_1+(x^4+2x^3-8x^2)A_2+(x^3+2x^2-8x)B+(x^4-2x^3+8x^2-16x)A_3+(x^4+4x^3+8x^2+32x)A_4$$

Regrouping (again):

$$4x+2=(A_1+A_2+A_3+A_4)x^4+(2A_1+2A_2+B-2A_3+4A_4)x^3+(-8A_2+2B+8A_3+8A_4)x^2+(-8B+16A_1-16A_3+32A_4)x-64A_1$$

Then, we can write a system of equations and solve for the A's:

$$\begin{cases} A_1+A_2+A_3+A_4=0 \\ 2A_1+2A_2+B-2A_3+4A_4=0 \\ 2B-8A_2+8A_3+8A_4=0 \\ -8B+16A_1-16A_3+32A_4=4 \\ -64A_1=2 \end{cases}$$

Equation 5 $\Rightarrow A_1=-\dfrac{1}{32}$

Back substitute $A_1=-\dfrac{1}{32}$ in the equations yields the following system of four equations:

$$\begin{cases} A_2+A_3+A_4=\frac{1}{32} \\ B+2A_2-2A_3+4A_4=\frac{1}{16} \\ B-4A_2+4A_3+4A_4=0 \\ B+2A_3-4A_4=-\frac{9}{16} \end{cases}$$

Put into augmented matrix form and solve the system:

$$\left[\begin{array}{cccc|c} 0 & 1 & 1 & 1 & \frac{1}{32} \\ 1 & 2 & -2 & 4 & \frac{1}{16} \\ 1 & -4 & 4 & 4 & 0 \\ 1 & 0 & 2 & -4 & -\frac{9}{16} \end{array}\right] \longrightarrow$$

Interchange rows to start:

$$\left[\begin{array}{cccc|c} 1 & 0 & 2 & -4 & -\frac{9}{16} \\ 0 & 1 & 1 & 1 & \frac{1}{32} \\ 1 & 2 & -2 & 4 & \frac{1}{16} \\ 1 & -4 & 4 & 4 & 0 \end{array}\right] \xrightarrow[-R_1+R_4]{-R_1+R_3}$$

$$\left[\begin{array}{cccc|c} 1 & 0 & 2 & -4 & -\frac{9}{16} \\ 0 & 1 & 1 & 1 & \frac{1}{32} \\ 0 & 2 & -4 & 8 & \frac{10}{16} \\ 0 & -4 & 2 & 8 & \frac{9}{16} \end{array}\right] \xrightarrow[4R_2+R_4]{-2R_2+R_3}$$

$$\left[\begin{array}{cccc|c} 1 & 0 & 2 & -4 & -\frac{9}{16} \\ 0 & 1 & 1 & 1 & \frac{1}{32} \\ 0 & 0 & -6 & 6 & \frac{18}{32} \\ 0 & 0 & 6 & 12 & \frac{11}{16} \end{array}\right] \xrightarrow[R_3+R_4]{-\frac{1}{6}R_3}$$

$$\left[\begin{array}{cccc|c} 1 & 0 & 2 & -4 & -\frac{9}{16} \\ 0 & 1 & 1 & 1 & \frac{1}{32} \\ 0 & 0 & 1 & -1 & -\frac{3}{32} \\ 0 & 0 & 0 & 18 & \frac{40}{32} \end{array}\right] \xrightarrow{\frac{1}{18}R_4}$$

$$\left[\begin{array}{cccc|c} 1 & 0 & 2 & -4 & -\frac{9}{16} \\ 0 & 1 & 1 & 1 & \frac{1}{32} \\ 0 & 0 & 1 & -1 & -\frac{3}{32} \\ 0 & 0 & 0 & 1 & \frac{5}{72} \end{array}\right]$$

Row 4 $\Rightarrow A_4 = \frac{5}{72}$.

Then, backsubstituting:

$$A_3 = \frac{5}{72} - \frac{3}{32} = -\frac{7}{288}$$

$$A_2 = \frac{7}{288} - \frac{5}{72} + \frac{3}{32} = -\frac{1}{72}$$

$$B = -2\left(-\frac{7}{288}\right) + 4\left(\frac{5}{72}\right) - \frac{9}{16} = -\frac{17}{72}$$

Then, we can write the partial decomposition of the given expression:

$$f(x) = \frac{-\frac{1}{32}}{x} + \frac{-\frac{1}{72}x - \frac{17}{72}}{x^2+8} + \frac{-\frac{7}{288}}{x+4} + \frac{\frac{5}{72}}{x-2}$$

$f(x) =$

$$-\frac{1}{32x} - \frac{x+17}{72(x^2+8)} - \frac{7}{288(x+4)} + \frac{5}{72(x-2)}$$

19. Given: $f(x) = \frac{5x}{x^2-6x+8}$

Factor the denominator:

$x^2 - 6x + 8 = (x-4)(x-2)$

By the guidelines for decomposition:

$$f(x) = \frac{5x}{x^2-6x+8} = \frac{A_1}{x-4} + \frac{A_2}{x-2}$$

Solve for A_1 and A_2.

Start by multiplying both sides by the factors of the denominator:

$$5x = (x-2)A_1 + (x-4)A_2$$
$$= (A_1 + A_2)x - 2A_1 - 4A_2$$

Then, $\begin{cases} A_1 + A_2 = 5 \\ -2A_1 - 4A_2 = 0 \end{cases}$

Equation 1 $\Rightarrow A_1 = 5 - A_2$

Substitute in equation 2:

$$-2(5 - A_2) - 4A_2 = 0$$
$$-10 + 2A_2 - 4A_2 = 0$$
$$-2A_2 = 10$$
$$A_2 = -5$$

Then, $A_1 = 10$

The decomposition is

$$f(x) = \frac{10}{x-4} - \frac{5}{x-2}$$

21. Given: $f(x)=\dfrac{6x}{x^3+8x^2+9x-18}$

Factor the denominator:

$x^3+8x^2+9x-18=(x+3)(x-1)(x+6)$

By the guidelines for decomposition:

$$\frac{6x}{x^3+8x^2+9x-18}=\frac{A_1}{x+3}+\frac{A_2}{x-1}+\frac{A_3}{x+6}$$

Solve for A_1, A_2 and A_3.

Start by multiplying both sides by the factors of the denominator:

$$6x=A_1(x+6)(x-1)+A_2(x+3)(x+6)+A_3(x-1)(x+3)$$

Regrouping:

$$6x=\left(x^2+5x-6\right)A_1+\left(x^2+9x+18\right)A_2+\left(x^2+2x-3\right)A_3$$

Regrouping (again):

$$6x=\left(A_1+A_2+A_3\right)x^2+\left(5A_1+9A_2+2A_3\right)x+\left(-6A_1+18A_2-3A_3\right)$$

Then, we can write a system of equations and solve for the A's:

$$\begin{cases} A_1+A_2+A_3=0 \\ 5A_1+9A_2+2A_3=6 \\ -6A_1+18A_2-3A_3=0 \end{cases}$$

First, divide equation 3 by -3 to simplify. Put into augmented matrix form and solve the system:

$$\left[\begin{array}{ccc|c} 1 & 1 & 1 & 0 \\ 5 & 9 & 2 & 6 \\ 2 & -6 & 1 & 0 \end{array}\right] \xrightarrow[-5R_1+R_2]{-2R_1+R_3} \left[\begin{array}{ccc|c} 1 & 1 & 1 & 0 \\ 0 & 4 & -3 & 6 \\ 0 & -8 & -1 & 0 \end{array}\right] \xrightarrow[\frac{1}{4}R_2]{2R_2+R_3} \left[\begin{array}{ccc|c} 1 & 1 & 1 & 0 \\ 0 & 1 & -\frac{3}{4} & \frac{3}{2} \\ 0 & 0 & -7 & 12 \end{array}\right] \Rightarrow A_3=-\frac{12}{7}$$

Back substitute:

$$A_2=\frac{3}{4}\left(-\frac{12}{7}\right)+\frac{3}{2}=-\frac{9}{7}+\frac{3}{2}=\frac{3}{14}$$

$$A_1=\frac{12}{7}-\frac{3}{14}=\frac{21}{14}=\frac{3}{2}$$

The decomposition is

$$f(x)=\frac{3}{2(x+3)}+\frac{3}{14(x-1)}-\frac{12}{7(x+6)}$$

23. Given: $f(x)=\dfrac{1}{x^2-1}$

Factor the denominator:

$x^2-1=(x+1)(x-1)$

By the guidelines for decomposition:

$$\frac{1}{x^2-1}=\frac{A_1}{x-1}+\frac{A_2}{x+1}$$

Solve for A_1 and A_2.

Start by multiplying both sides by the factors of the denominator:

$$1=(x+1)A_1+(x-1)A_2=\left(A_1+A_2\right)x+\left(A_1-A_2\right)$$

Then, we can write a system of two equations and solve for the A's:

$$\begin{cases} A_1+A_2=0 \\ A_1-A_2=1 \end{cases}$$

Equation $1 \Rightarrow A_1=-A_2$.

Substitute in equation 2:

$$-2A_2=1$$

$$A_2=-\frac{1}{2}$$

Then, $A_1=\dfrac{1}{2}$

The decomposition is

$$f(x)=\frac{1}{2(x-1)}-\frac{1}{2(x+1)}$$

25. Given: $f(x)=\dfrac{x^2-4}{(x^4-16)(x^2+2x-8)}$

Notice that $f(x)=\dfrac{1}{(x^2+4)(x+4)(x-2)}$

The following solution is based on the reduction of the rational expression to simplest form:

By the guidelines for decomposition:

$$\frac{1}{(x^2+4)(x+4)(x-2)}=$$
$$\frac{A_1x+B}{x^2+4}+\frac{A_2}{x+4}+\frac{A_3}{x-2}$$

Solve for A_1, B, A_2 and A_3.

Start by multiplying both sides by the factors of the denominator:

$$1=(A_1x+B)(x^2+2x-8)+$$
$$A_2(x^3-2x^2+4x-8)+$$
$$A_3(x^3+4x^2+4x+16)$$

Regrouping:

$$1=(A_1+A_2+A_3)x^3+$$
$$(2A_1+B-2A_2+4A_3)x^2+$$
$$(-8A_1+2B+4A_2+4A_3)x+$$
$$(-8B-8A_2+16A_3)$$

Then, we can write a system of equations and solve for the A's:

$$\begin{cases} A_1+A_2+A_3=0 \\ 2A_1+B-2A_2+4A_3=0 \\ -8A_1+2B+4A_2+4A_3=0 \\ -8B-8A_2+16A_3=1 \end{cases}$$

Put into augmented matrix form and solve the system:

$$\left[\begin{array}{cccc|c} 1 & 0 & 1 & 1 & 0 \\ 2 & 1 & -2 & 4 & 0 \\ -8 & 2 & 4 & 4 & 0 \\ 0 & -8 & -8 & 16 & 1 \end{array}\right] \xrightarrow[-2R_1+R_2]{\frac{1}{2}R_3,\,-\frac{1}{8}R_4}$$

$$\left[\begin{array}{cccc|c} 1 & 0 & 1 & 1 & 0 \\ 0 & 1 & -4 & 2 & 0 \\ -4 & 1 & 2 & 2 & 0 \\ 0 & 1 & 1 & -2 & -\frac{1}{8} \end{array}\right] \xrightarrow[-R_2+R_4]{4R_1+R_3}$$

$$\left[\begin{array}{cccc|c} 1 & 0 & 1 & 1 & 0 \\ 0 & 1 & -4 & 2 & 0 \\ 0 & 1 & 6 & 6 & 0 \\ 0 & 0 & 5 & -4 & -\frac{1}{8} \end{array}\right] \xrightarrow[\frac{1}{5}R_4]{-R_2+R_3}$$

$$\left[\begin{array}{cccc|c} 1 & 0 & 1 & 1 & 0 \\ 0 & 1 & -4 & 2 & 0 \\ 0 & 0 & 10 & 4 & 0 \\ 0 & 0 & 1 & -\frac{4}{5} & -\frac{1}{40} \end{array}\right] \xrightarrow[\frac{1}{10}R_3]{}$$

$$\left[\begin{array}{cccc|c} 1 & 0 & 1 & 1 & 0 \\ 0 & 1 & -4 & 2 & 0 \\ 0 & 0 & 1 & \frac{2}{5} & 0 \\ 0 & 0 & 1 & -\frac{4}{5} & -\frac{1}{40} \end{array}\right] \xrightarrow{-R_3+R_4}$$

$$\left[\begin{array}{cccc|c} 1 & 0 & 1 & 1 & 0 \\ 0 & 1 & -4 & 2 & 0 \\ 0 & 0 & 1 & \frac{2}{5} & 0 \\ 0 & 0 & 0 & -\frac{6}{5} & -\frac{1}{40} \end{array}\right] \xrightarrow{-\frac{5}{6}R_4}$$

$$\left[\begin{array}{cccc|c} 1 & 0 & 1 & 1 & 0 \\ 0 & 1 & -4 & 2 & 0 \\ 0 & 0 & 1 & \frac{2}{5} & 0 \\ 0 & 0 & 0 & 1 & \frac{1}{48} \end{array}\right]$$

Row 4 $\Rightarrow A_3=\dfrac{1}{48}$.

Then, backsubstituting:

$$A_2=-\frac{2}{5}\left(\frac{1}{48}\right)=-\frac{1}{120}$$

$$B=4\left(-\frac{1}{120}\right)-2\left(\frac{1}{48}\right)=-\frac{3}{40}$$

$$A_1=-\left(-\frac{1}{120}\right)-\left(\frac{1}{48}\right)=-\frac{1}{80}$$

Then, we can write the partial decomposition of the given expression:

$$f(x)=\frac{-\frac{1}{80}x-\frac{3}{40}}{x^2+4}-\frac{1}{120(x+4)}+\frac{1}{48(x-2)}$$

$$f(x)=\frac{-x-6}{80(x^2+4)}-\frac{1}{120(x+4)}+\frac{1}{48(x-2)}$$

27. Given: $f(x)=\frac{x+3}{x^2-4}$

Factor the denominator:

$x^2-4=(x-2)(x+2)$

By the guidelines for decomposition:

$$f(x)=\frac{x+3}{x^2-4}=\frac{A_1}{x-2}+\frac{A_2}{x+2}$$

Solve for A_1 and A_2.

Start by multiplying both sides by the factors of the denominator:

$$x+3=(x+2)A_1+(x-2)A_2$$
$$=(A_1+A_2)x+2A_1-2A_2$$

Then, $\begin{cases} A_1+A_2=1 \\ 2A_1-2A_2=3 \end{cases}$

Equation 1 $\Rightarrow A_1=1-A_2$

Substitute in equation 2:

$$2(1-A_2)-2A_2=3$$
$$2-2A_2-2A_2=3$$
$$-4A_2=1$$
$$A_2=-\frac{1}{4}$$

Then, $A_1=\frac{5}{4}$

The decomposition is

$$f(x)=\frac{5}{4(x-2)}-\frac{1}{4(x+2)}$$

29. Given: $f(x)=\frac{x}{x^4-16}$

Factor the denominator:

$x^4-16=(x^2+4)(x-2)(x+2)$

By the guidelines for decomposition:

$$\frac{x}{x^4-16}=\frac{A_1x+B}{x^2+4}+\frac{A_2}{x-2}+\frac{A_3}{x+2}$$

Solve for A_1, B, A_2 and A_3.

Start by multiplying both sides by the factors of the denominator:

$$x=(A_1x+B)(x^2-4)+A_2(x^3+2x^2+4x+8)+A_3(x^3-2x^2+4x-8)$$

Regrouping:

$$x=(A_1+A_2+A_3)x^3+(B+2A_2-2A_3)x^2+(-4A_1+4A_2+4A_3)x+(-4B+8A_2-8A_3)$$

Then, we can write a system of equations and solve for the A's and B:

$$\begin{cases} A_1+A_2+A_3=0 \\ B+2A_2-2A_3=0 \\ -4A_1+4A_2+4A_3=1 \\ -4B+8A_2-8A_3=0 \end{cases}$$

Put into augmented matrix form and solve the system:

First, divide the equations by common factors found in each term to simplify.

$$\left[\begin{array}{cccc|c} 1 & 0 & 1 & 1 & 0 \\ 0 & 1 & 2 & -2 & 0 \\ -1 & 0 & 1 & 1 & \frac{1}{4} \\ 0 & 1 & -2 & 2 & 0 \end{array}\right] \xrightarrow[-R_2+R_4]{R_1+R_3}$$

$$\left[\begin{array}{cccc|c} 1 & 0 & 1 & 1 & 0 \\ 0 & 1 & 2 & -2 & 0 \\ 0 & 0 & 2 & 2 & \frac{1}{4} \\ 0 & 0 & -4 & 4 & 0 \end{array}\right] \xrightarrow[-\frac{1}{4}R_4]{\frac{1}{4}R_4+R_3}$$

$$\left[\begin{array}{cccc|c} 1 & 0 & 1 & 1 & 0 \\ 0 & 1 & 2 & -2 & 0 \\ 0 & 0 & 1 & 3 & \frac{1}{4} \\ 0 & 0 & 1 & -1 & 0 \end{array}\right] \Rightarrow \text{After adding}$$

$-R_3 + R_4$, rows 3 and 4 give us a system of two equations to solve:

$$\left[\begin{array}{cc|c} 1 & 3 & \frac{1}{4} \\ 0 & -4 & -\frac{1}{4} \end{array}\right] \xrightarrow{-\frac{1}{4}R_2} \left[\begin{array}{cc|c} 1 & 3 & \frac{1}{4} \\ 0 & 1 & \frac{1}{16} \end{array}\right] \Rightarrow$$

$$A_3 = \frac{1}{16}$$

Backsubstitution:

$$A_2 = \frac{1}{4} - 3\left(\frac{1}{16}\right) = \frac{1}{16}$$

From row 2: $B = -2\left(\frac{1}{16}\right) + 2\left(\frac{1}{16}\right) = 0$

From row 1: $A = -\frac{1}{16} - \frac{1}{16} = -\frac{1}{8}$

Then, we can write the partial decomposition of the given expression:

$f(x) =$

$$-\frac{x}{8(x^2+4)} + \frac{1}{16(x-2)} + \frac{1}{16(x+2)}$$

31. Given: $f(x) = \dfrac{2x+3}{(x^2-9)(x^2+4x-12)}$

Factor the denominator:

$$(x^2-9)(x^2+4x-12) = (x-3)(x+3)(x+6)(x-2)$$

By the guidelines for decomposition:

$$\frac{2x+3}{(x^2-9)(x^2+4x-12)} = \frac{A_1}{x-3} + \frac{A_2}{x+3} + \frac{A_3}{x+6} + \frac{A_4}{x-2}$$

Solve for A_1, A_2, A_3 and A_4.

Start by multiplying both sides by the factors of the denominator:

$$\begin{aligned} 2x+3 = {} & A_1(x^3+7x^2-36) + \\ & A_2(x^3+x^2-24x+36) + \\ & A_3(x^3-2x^2-9x+18) + \\ & A_4(x^3+6x^2-9x-54) \end{aligned}$$

Regroup:

$$\begin{aligned} 2x+3 = {} & (A_1+A_2+A_3+A_4)x^3 + \\ & (7A_1+A_2-2A_3+6A_4)x^2 + \\ & (-24A_2-9A_3-9A_4)x + \\ & (-36A_1+36A_2+18A_3-54A_4) \end{aligned}$$

Then, we can write a system of equations and solve for the A's:

$$\begin{cases} A_1+A_2+A_3+A_4 = 0 \\ 7A_1+A_2-2A_3+6A_4 = 0 \\ -24A_2-9A_3-9A_4 = 2 \\ -36A_1+36A_2+18A_3-54A_4 = 3 \end{cases}$$

Put into augmented matrix form and solve the system:

$$\left[\begin{array}{cccc|c} 1 & 1 & 1 & 1 & 0 \\ 7 & 1 & -2 & 6 & 0 \\ 0 & -24 & -9 & -9 & 2 \\ -36 & 36 & 18 & -54 & 3 \end{array}\right] \xrightarrow{\frac{1}{3}R_4}$$

$$\left[\begin{array}{cccc|c} 1 & 1 & 1 & 1 & 0 \\ 7 & 1 & -2 & 6 & 0 \\ 0 & -24 & -9 & -9 & 2 \\ -12 & 12 & 6 & -18 & 1 \end{array}\right] \xrightarrow[-7R_1+R_2]{12R_1+R_4}$$

$$\left[\begin{array}{cccc|c} 1 & 1 & 1 & 1 & 0 \\ 0 & -6 & -9 & -1 & 0 \\ 0 & -24 & -9 & -9 & 2 \\ 0 & 24 & 18 & -6 & 1 \end{array}\right] \xrightarrow{R_3+R_4}$$

$$\left[\begin{array}{cccc|c} 1 & 1 & 1 & 1 & 0 \\ 0 & -6 & -9 & -1 & 0 \\ 0 & -24 & -9 & -9 & 2 \\ 0 & 0 & 9 & -15 & 3 \end{array}\right] \xrightarrow[\substack{-4R_2+R_3 \\ -\frac{1}{6}R_2}]{\frac{1}{3}R_4}$$

$$\left[\begin{array}{cccc|c} 1 & 1 & 1 & 1 & 0 \\ 0 & 1 & \frac{3}{2} & \frac{1}{6} & 0 \\ 0 & 0 & 27 & -5 & 2 \\ 0 & 0 & 3 & -5 & 1 \end{array}\right] \xrightarrow[R_4 \leftrightarrow R_3]{-9R_4+R_3}$$

$$\left[\begin{array}{cccc|c} 1 & 1 & 1 & 1 & 0 \\ 0 & 1 & \frac{3}{2} & \frac{1}{6} & 0 \\ 0 & 0 & 3 & -5 & 1 \\ 0 & 0 & 0 & 40 & -7 \end{array}\right]$$

Then, $A_4 = -\dfrac{7}{40}$

Backsubstituting:

$$3A_3 + \frac{35}{40} = 1 \Rightarrow A_3 = \frac{1}{24}$$

$$A_2 = -\frac{3}{2}\left(\frac{1}{24}\right) - \frac{1}{6}\left(-\frac{7}{40}\right) \Rightarrow A_2 = -\frac{1}{30}$$

$$A_1 = \frac{1}{30} - \frac{1}{24} + \frac{7}{40} \Rightarrow A_1 = \frac{1}{6}$$

Then, we can write the partial decomposition of the given expression:

$$\frac{2x+3}{(x^2-9)(x^2+4x-12)} = \frac{1}{6(x-3)} - \frac{1}{30(x+3)} + \frac{1}{24(x+6)} - \frac{7}{40(x-2)}$$

33. Given: $f(x) = \dfrac{x^2}{x^3+5x^2+3x-9}$

Factor the denominator:

$$x^3+5x^2+3x-9 = (x+3)^2(x-1)$$

By the guidelines for decomposition:

$$\frac{x^2}{x^3+5x^2+3x-9} = \frac{A_1}{x-1} + \frac{A_2}{x+3} + \frac{A_3}{(x+3)^2}$$

Solve for A_1, A_2 and A_3.

Start by multiplying both sides by the factors of the denominator:

$$x^2 = A_1(x+3)^2 + A_2(x+3)(x-1) + A_3(x-1)$$

Regrouping:

$$x^2 = (x^2+6x+9)A_1 + (x^2+2x-3)A_2 + A_3x - A_3$$

Regrouping (again):

$$x^2 = (A_1+A_2)x^2 + (6A_1+2A_2+A_3)x + (9A_1-3A_2-A_3)$$

Then, we can write a system of equations and solve for the A's:

$$\begin{cases} A_1 + A_2 = 1 \\ 6A_1 + 2A_2 + A_3 = 0 \\ 9A_1 - 3A_2 - A_3 = 0 \end{cases}$$

Put into augmented matrix form and solve the system:

$$\left[\begin{array}{ccc|c} 1 & 1 & 0 & 1 \\ 6 & 2 & 1 & 0 \\ 9 & -3 & -1 & 0 \end{array}\right] \xrightarrow[-6R_1+R_2]{-9R_1+R_3} \left[\begin{array}{ccc|c} 1 & 1 & 0 & 1 \\ 0 & -4 & 1 & -6 \\ 0 & -12 & -1 & -9 \end{array}\right] \xrightarrow[-\frac{1}{4}R_2]{-3R_2+R_3}$$

$$\left[\begin{array}{ccc|c} 1 & 1 & 0 & 1 \\ 0 & 1 & -\frac{1}{4} & \frac{3}{2} \\ 0 & 0 & -4 & 9 \end{array}\right] \xrightarrow[-\frac{1}{4}R_3]{} \left[\begin{array}{ccc|c} 1 & 1 & 0 & 1 \\ 0 & 1 & -\frac{1}{4} & \frac{3}{2} \\ 0 & 0 & 1 & -\frac{9}{4} \end{array}\right] \Rightarrow A_3 = -\frac{9}{4}$$

Back substitute:

$$A_2 = \frac{1}{4}\left(-\frac{9}{4}\right) + \frac{3}{2} = -\frac{9}{16} + \frac{24}{16} = \frac{15}{16}$$

$$A_1 = \frac{-15}{16} + 1 = \frac{1}{16}$$

The decomposition is

$$f(x) = \frac{1}{16(x-1)} + \frac{15}{16(x+3)} - \frac{9}{4(x+3)^2}$$

35. Given: $f(x)=\dfrac{2x}{x^2-9}$

Factor the denominator:

$x^2-9=(x-3)(x+3)$

By the guidelines for decomposition:

$$f(x)=\frac{2x}{x^2-9}=\frac{A_1}{x-3}+\frac{A_2}{x+3}$$

Solve for A_1 and A_2.

Start by multiplying both sides by the factors of the denominator:

$2x=(x+3)A_1+(x-3)A_2$

$2x=(A_1+A_2)x+3A_1-3A_2$

Then, $\begin{cases} A_1+A_2=2 \\ 3A_1-3A_2=0 \end{cases}$ or $\begin{cases} A_1+A_2=2 \\ A_1-A_2=0 \end{cases}$

Equation $2 \Rightarrow A_1=A_2$

Substitute in equation 1:

$A_2+A_2=2$

$A_2=1$

$A_1=1$

The decomposition is

$$f(x)=\frac{1}{x-3}+\frac{1}{x+3}$$

37. Given: $f(x)=\dfrac{x^2}{x^3+4x^2-12x}$

Factor the denominator:

$x^3+4x^2-12x=x(x-2)(x+6)$

By the guidelines for decomposition:

$$\frac{x^2}{x^3+4x^2-12x}=\frac{A_1}{x}+\frac{A_2}{x+6}+\frac{A_3}{x-2}$$

Solve for A_1, A_2 and A_3.

Start by multiplying both sides by the factors of the denominator:

$x^2=A_1(x^2+4x-12)+A_2(x^2-2x)+A_3(x^2+6x)$

Regrouping:

$x^2=(A_1+A_2+A_3)x^2+(4A_1-2A_2+6A_3)x-12A_1$

The last equation implies that $-12A_1=0 \Rightarrow A_1=0$.

Then, the system reduces to

$\begin{cases} A_2+A_3=1 \\ -A_2+3A_3=0 \end{cases}$

Add the two equations to eliminate A_2.

Then, $4A_3=1 \Rightarrow A_3=\dfrac{1}{4}$

From equation 1, $A_2=1-\dfrac{1}{4}=\dfrac{3}{4}$

The decomposition is

$$f(x)=\frac{0}{x}+\frac{3}{4(x+6)}+\frac{1}{4(x-2)}=\frac{3}{4(x+6)}+\frac{1}{4(x-2)}$$

39. Use a graphing calculator to graph the left and right hand sides to test the given decomposition. Use the $\langle y=\rangle$ and $\langle$Graph$\rangle$ utilities.

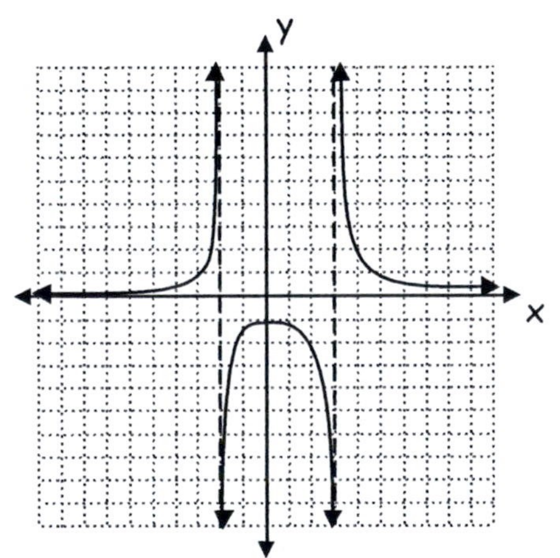

The decomposition is true.

41. Use a graphing calculator to graph the left and right hand sides to test the given decomposition. Use the $\langle y = \rangle$ and $\langle \text{Graph} \rangle$ utilities.

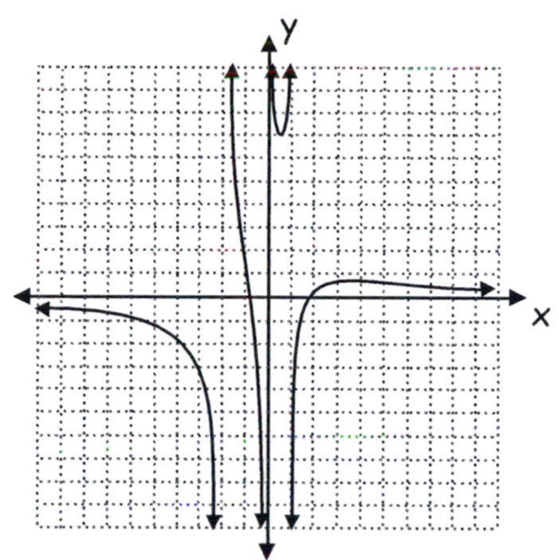

The decomposition is true.

43. Use a graphing calculator to graph the left and right hand sides to test the given decomposition. Use the $\langle y = \rangle$ and $\langle \text{Graph} \rangle$ utilities.

You should notice that the two graphs do not coincide exactly. Therefore, the two decompositions do not represent the same function. The decomposition is false.

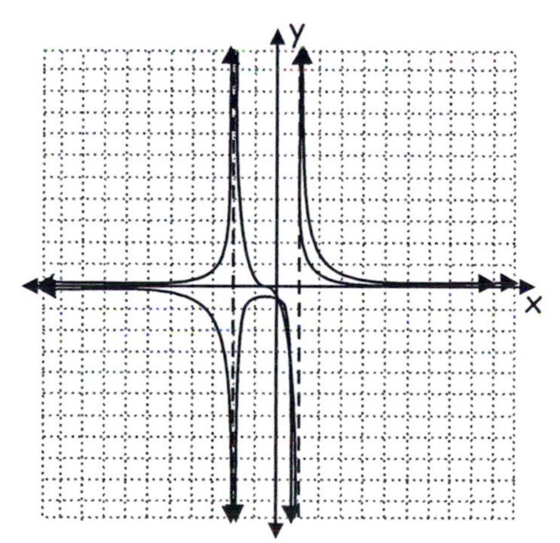

45. Given: $f(x) = \dfrac{1}{x(x+a)}$

By the guide for decomposition

$$\frac{1}{x(x+a)} = \frac{A_1}{x} + \frac{A_2}{x+a}$$

Multiply by $x(x+a)$

$$1 = A_1(x+a) + A_2(x) = (A_1 + A_2)x + A_1 a$$

$$\Rightarrow \text{the system } \begin{cases} A_1 + A_2 = 0 \\ A_1 a = 1 \end{cases} \Rightarrow A_1 = \frac{1}{a}$$

and $A_2 = -\dfrac{1}{a}$

These values yield the decomposition

$$f(x) = \frac{1}{ax} - \frac{1}{a(x+a)}$$

47. Given: $f(x) = \dfrac{1}{a^2 - x^2} = \dfrac{1}{(a-x)(a+x)}$

By the guide for decomposition

$$\frac{1}{a^2 - x^2} = \frac{A_1}{a-x} + \frac{A_2}{a+x}$$

Multiply by $(a-x)(a+x)$

$$1 = A_1(x+a) + A_2(a-x)$$

$$= (A_1 - A_2)x + A_1 a + A_2 a$$

$$\Rightarrow \text{the system } \begin{cases} A_1 - A_2 = 0 \\ A_1 a + A_2 a = 1 \end{cases}$$

$$\Rightarrow A_1 = A_2 \text{ and } 2A_1 = \frac{1}{a}$$

Then, $A_1 = \dfrac{1}{2a}$ and $A_2 = \dfrac{1}{2a}$

Therefore, $f(x) = \dfrac{1}{2a(a-x)} + \dfrac{1}{2a(a+x)}$

49. Given: $f(x)=\dfrac{1}{(x+a)(x+1)}$

By the guide for decomposition

$$\frac{1}{(x+a)(x+1)}=\frac{A_1}{x+a}+\frac{A_2}{x+1}$$

Multiply by $(x+a)(x+1)$:

$$1=A_1(x+1)+A_2(x+a)$$
$$=(A_1+A_2)x+A_1+A_2a$$

$$\Rightarrow \text{the system } \begin{cases} A_1+A_2=0 \\ A_1+A_2a=1 \end{cases}$$

Subtracting the first equation from the second:

$$A_2a-A_2=1$$
$$A_2(a-1)=1$$
$$A_2=\frac{1}{a-1}$$
$$A_1=-\frac{1}{a-1}$$

Then,

$$f(x)=-\frac{1}{(a-1)(x+a)}+\frac{1}{(a-1)(x+1)}$$

or, $f(x)=\dfrac{1}{a-1}\left[\dfrac{1}{(x+1)}-\dfrac{1}{(x+a)}\right]$

End of Section 10.6

Chapter Ten: Section 10.7
Solutions to Odd-numbered Exercises

1. Let $x =$ number of water containers.
Let $y =$ number of food containers.
Constraints:

$$\begin{cases} 60x+50y \le 50{,}000 \\ x+10y \le 6000 \\ x \ge 0 \\ y \ge 0 \end{cases}$$

Draw the equation part of each of the constraints (e.g., $60x+50y=50{,}000$ which is equivalent to $6x+5y=5000$). Mark the region that meets all constraints.

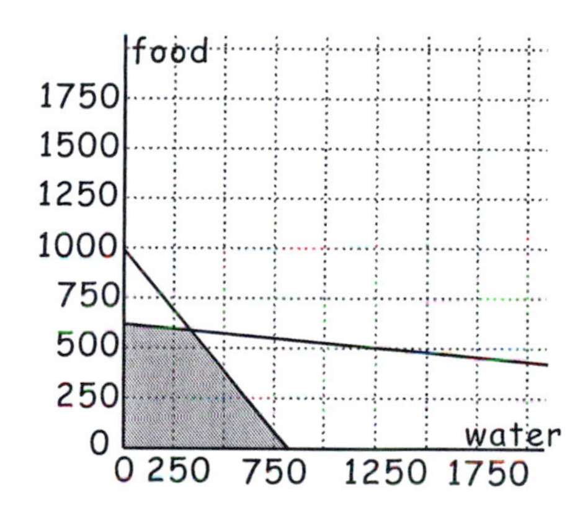

3. Let $x =$ number of shirts.
Let $y =$ number of pants.
Constraints:

$$\begin{cases} 12x+32y \le 80 \\ 2x+3y \le 10 \\ x \ge 0 \\ y \ge 0 \end{cases}$$

Draw the equation part of each of the constraints (e.g., $12x+32y=80$ which is equivalent to $3x+8y=20$). Mark the region that meets all constraints.

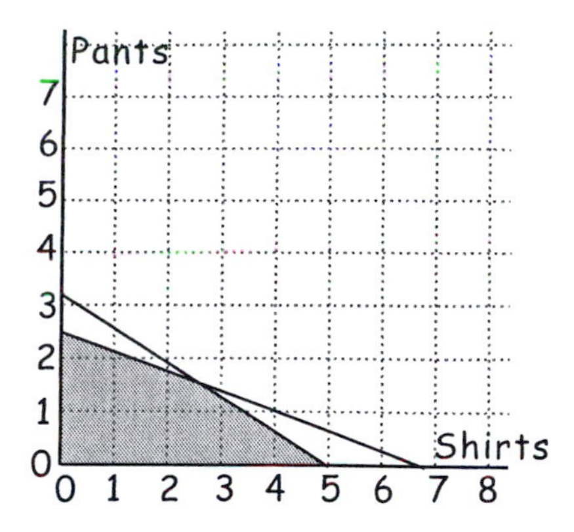

5. Find the minimum and maximum values of the given function $f(x,y)=2x+3y$
Constraints:

$$\begin{cases} x \ge 0 \\ y \ge 0 \\ x+y \le 7 \end{cases}$$

Sketch the region of constraint:

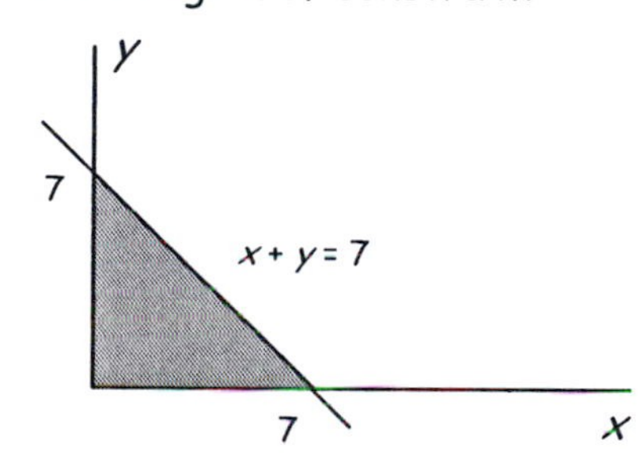

The three vertices give rise to 3 linear systems:

$$\begin{cases} x=0 \\ x+y=7 \end{cases} \quad \begin{cases} x=0 \\ y=0 \end{cases} \quad \begin{cases} y=0 \\ x+y=7 \end{cases}$$

Solving for the vertices, we get $(0,7)$, $(0,0)$, $(7,0)$

Substituting each of the ordered pairs (vertices) in the objective function, we can compare the results to find the maximum and minimum values:

$f(0,7)=2(0)+3(7)=21$

$f(0,0)=2(0)+3(0)=0$

$f(7,0)=2(7)+3(0)=14$

Min $=0$ at $(0,0)$

Max $=21$ at $(0,7)$

7. Find the minimum and maximum values of the given function $f(x, y) = 2x + 5y$

Constraints:

$$\begin{cases} x \geq 0 \\ y \geq 0 \\ 2x + y \leq 6 \end{cases}$$

Sketch the region of constraint:

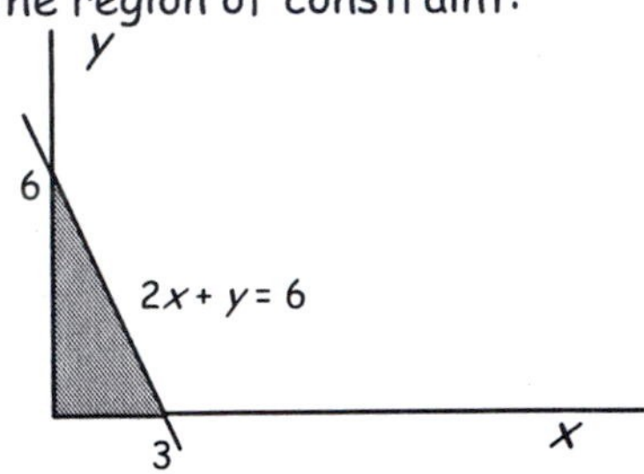

The three vertices give rise to 3 linear systems:

$$\begin{cases} x = 0 \\ 2x + y = 6 \end{cases} \quad \begin{cases} x = 0 \\ y = 0 \end{cases} \quad \begin{cases} y = 0 \\ 2x + y = 6 \end{cases}$$

Solving for the vertices, we get

$(0, 6)$, $(0, 0)$, $(3, 0)$

Substituting each of the ordered pairs (vertices) in the objective function, we can compare the results to find the maximum and minimum values:

$f(0, 6) = 2(0) + 5(6) = 30$

$f(0, 0) = 2(0) + 3(0) = 0$

$f(3, 0) = 2(3) + 5(0) = 6$

Min $= 0$ at $(0, 0)$

Max $= 30$ at $(0, 6)$

9. Find the minimum and maximum values of the given function $f(x, y) = 5x + 6y$

Constraints:

$$\begin{cases} 0 \leq x \leq 7 \\ 0 \leq y \leq 10 \\ 8x + 5y \leq 40 \end{cases}$$

Sketch the region of constraint:

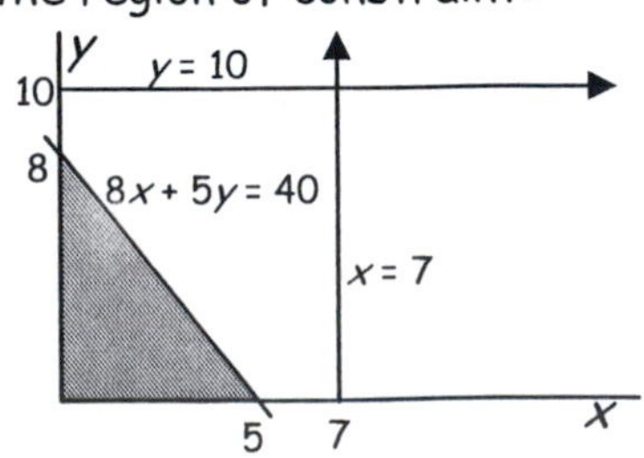

The three vertices give rise to 3 linear systems:

$$\begin{cases} x = 0 \\ 8x + 5y = 6 \end{cases} \quad \begin{cases} x = 0 \\ y = 0 \end{cases} \quad \begin{cases} y = 0 \\ 8x + 5y = 40 \end{cases}$$

Solving for the vertices, we get

$(0, 8)$, $(0, 0)$, $(5, 0)$

Substituting each of the ordered pairs (vertices) in the objective function, we can compare the results to find the maximum and minimum values:

$f(0, 8) = 5(0) + 6(8) = 48$

$f(0, 0) = 5(0) + 3(0) = 0$

$f(3, 0) = 2(3) + 5(0) = 6$

Min $= 0$ at $(0, 0)$

Max $= 48$ at $(0, 8)$

11. Find the minimum and maximum values of the given function $f(x,y) = 6x + 4y$

Constraints:

$$\begin{cases} 0 \leq x \leq 4 \\ 0 \leq y \leq 5 \\ 4x + 3y \leq 10 \end{cases}$$

Sketch the region of constraint:

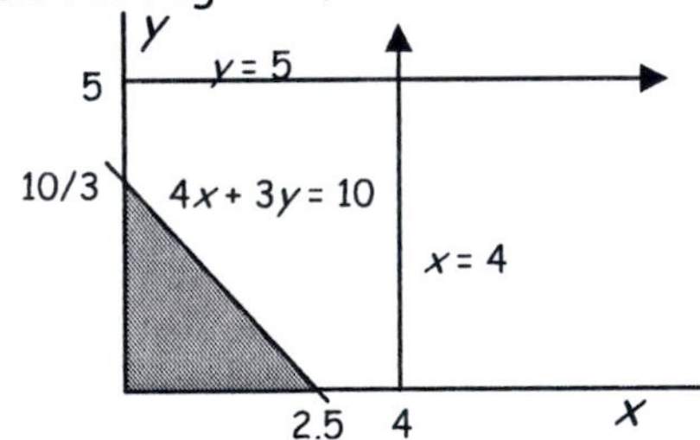

The three vertices give rise to 3 linear systems:

$$\begin{cases} x = 0 \\ 4x + 3y = 10 \end{cases} \quad \begin{cases} x = 0 \\ y = 0 \end{cases} \quad \begin{cases} y = 0 \\ 4x + 3y = 10 \end{cases}$$

Solving for the vertices, we get

$\left(0, \frac{10}{3}\right)$, $\left(\frac{5}{2}, 0\right)$, $(0, 0)$

Substituting each of the ordered pairs (vertices) in the objective function, we can compare the results to find the maximum and minimum values:

$f\left(0, \frac{10}{3}\right) = 6(0) + 4\left(\frac{10}{3}\right) = \frac{40}{3}$

$f\left(\frac{5}{2}, 0\right) = 6\left(\frac{5}{2}\right) + 4(0) = 15$

$f(0, 0) = 6(0) + 4(0) = 0$

Min $= 0$ at $(0, 0)$

Max $= 15$ at $\left(\frac{5}{2}, 0\right)$

13. Find the minimum and maximum values of the given function $f(x, y) = 6x + 8y$

Constraints:

$$\begin{cases} x \geq 0 \\ y \geq 0 \\ 4x + y \leq 16 \\ x + 3y \leq 15 \end{cases}$$

Sketch the region of constraint:

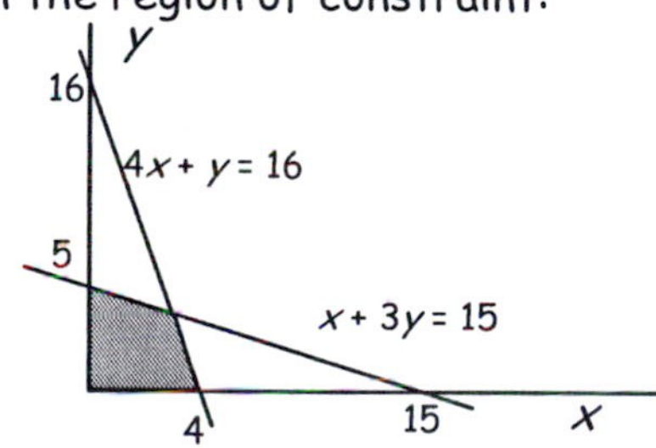

The four vertices give rise to 4 linear systems:

$$\begin{cases} x = 0 \\ x + 3y = 15 \end{cases} \quad \begin{cases} x = 0 \\ y = 0 \end{cases}$$

$$\begin{cases} 4x + y = 16 \\ x + 3y = 15 \end{cases} \quad \begin{cases} y = 0 \\ 4x + y = 16 \end{cases}$$

Solving for the vertices, we get $(0, 5)$, $(0, 0)$, $(3, 4)$, $(4, 0)$

Substituting each of the ordered pairs (vertices) in the objective function, we can compare the results to find the maximum and minimum values:

$f(0, 5) = 6(0) + 8(5) = 40$

$f(0, 0) = 6(0) + 8(0) = 0$

$f(3, 4) = 6(3) + 8(4) = 50$

$f(4, 0) = 6(4) + 8(0) = 24$

Min $= 0$ at $(0, 0)$

Max $= 50$ at $(3, 4)$

15. Find the minimum and maximum values of the given function $f(x, y) = 6x + y$

Constraints:

$$\begin{cases} x \geq 0; y \geq 0 \\ 3x + 4y \leq 24 \\ 3x + 4y \leq 48 \end{cases}$$

Sketch the region of constraint:

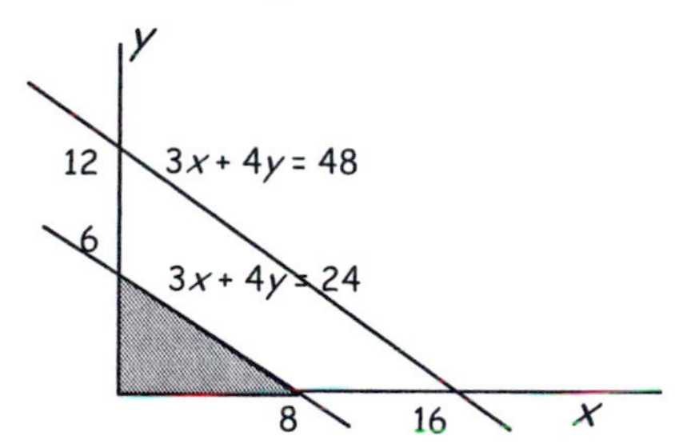

The three vertices give rise to 3 linear systems:

$$\begin{cases} x = 0 \\ 3x + 4y = 24 \end{cases} \quad \begin{cases} y = 0 \\ 3x + 4y = 24 \end{cases} \quad \begin{cases} x = 0 \\ y = 0 \end{cases}$$

Solving for the vertices, we get $(0, 6)$, $(8, 0)$, $(0, 0)$

Substituting each of the ordered pairs (vertices) in the objective function, we can compare the results to find the maximum and minimum values:

$f(0, 6) = 6(0) + 6 = 6$

$f(0, 0) = 6(0) + 0 = 0$

$f(8, 0) = 6(8) + 0 = 48$

Min $= 0$ at $(0, 0)$

Max $= 48$ at $(8, 0)$

17. Find the minimum and maximum values of the given function $f(x,y)=3x+10y$

Constraints:

$$\begin{cases} x \geq 0 \\ 2x+4y \geq 8 \\ 5x-y \leq 10 \\ x+3y \leq 40 \end{cases}$$

Sketch the region of constraint:

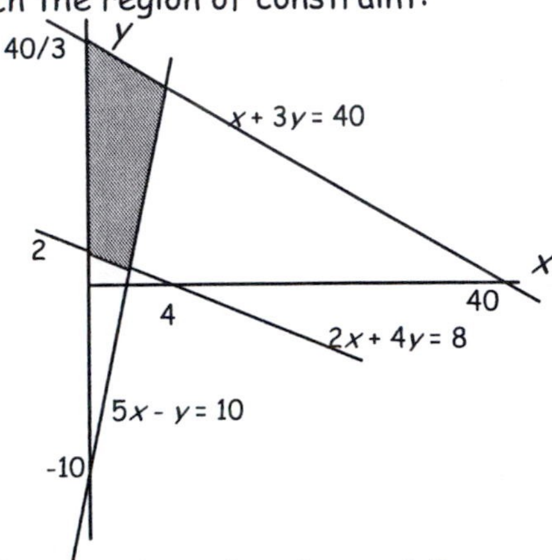

The four vertices give rise to 4 linear systems:

$$\begin{cases} x=0 \\ 2x+4y=8 \end{cases} \qquad \begin{cases} 5x-y=10 \\ 2x+4y=8 \end{cases}$$

$$\begin{cases} x+3y=40 \\ 5x-y=10 \end{cases} \qquad \begin{cases} x=0 \\ x+3y=40 \end{cases}$$

Solving for the vertices, we get

$$(0,2), \left(\frac{24}{11},\frac{10}{11}\right), \left(\frac{35}{8},\frac{95}{8}\right), \left(0,\frac{40}{3}\right)$$

Substituting each of the ordered pairs (vertices) in the objective function, we can compare the results to find the maximum and minimum values:

$$f(0,2)=3(0)+10(2)=20$$

$$f\left(\frac{24}{11},\frac{10}{11}\right)=3\left(\frac{24}{11}\right)+10\left(\frac{10}{11}\right)\approx 15.64$$

$$f\left(\frac{35}{8},\frac{95}{8}\right)=3\left(\frac{35}{8}\right)+10\left(\frac{95}{8}\right)\approx 131.9$$

$$f\left(0,\frac{40}{3}\right)=3(0)+10\left(\frac{40}{3}\right)\approx 133.3$$

$$\text{Min} \approx 15.64 \text{ at } \left(\frac{24}{11},\frac{10}{11}\right)$$

$$\text{Max} \approx 133.33 \text{ at } \left(0,\frac{40}{3}\right)$$

19. Let $x =$ number of Model X computers.
Let $y =$ number of Model Y computers.

Constraints:

$$\begin{cases} x \geq 0;\ y \geq 0 & (1) \\ 2.5x+3y \leq 4000 & (2) \\ 2x+y \leq 2500 & (3) \\ 0.75x+1.25y \leq 1500 & (4) \end{cases}$$

Profit function: $f(x,y)=50x+52y$

Sketch the graphs and find the vertices:

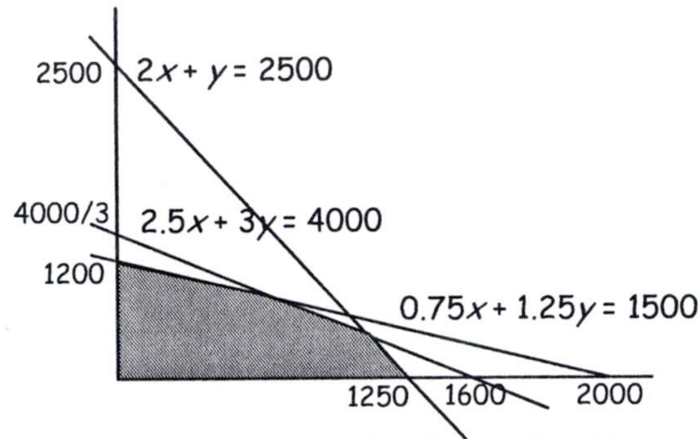

The vertices can be determined by solving each of the linear systems of two equations (i.e., finding the 3 intersections of the 3 equations and the xy-intercepts.) Here is the system for equations 2 and 3 from the list of constraints:

$$\left[\begin{array}{cc|c} \frac{5}{2} & 3 & 4000 \\ 2 & 1 & 2500 \end{array}\right] \xrightarrow[\frac{1}{2}R_2]{\text{interchange}}$$

$$\left[\begin{array}{cc|c} 1 & \frac{1}{2} & 1250 \\ \frac{5}{2} & 3 & 4000 \end{array}\right] \xrightarrow{-\frac{5}{2}R_1+R_2} \left[\begin{array}{cc|c} 1 & \frac{1}{2} & 1250 \\ 0 & \frac{7}{4} & 875 \end{array}\right]$$

The system yields the vertex (1000, 500).

The remaining four vertices can be determined similarly or by substitution of $x=0$ in equation 4, or of $y=0$ in equation 3. Note that the intersection of equations 3 and 4 (approx. (928.6, 642.9)does not satisfy inequality 2 of the constraints.

The vertices are
(1000, 500) for inequalities 2 and 3,
(571.4, 857.1) for inequalities 2 and 4,
(0, 1200) on the y-axis,
(1250, 0) on the x-axis, and
(0, 0) at the origin.
Compare the value of the profit function for each vertex:

$f(1000, 500) = 50(1000) + 52(500) = \$76,000$

$f(571.4, 857.1) = 50(591.4) + 52(857.1) = \$74,139.2$

$f(0, 1200) = \$62,400$

$f(1250, 0) = 50(1250) + 52(0) = \$62,500$

$f(0, 0) = 50(0) + 52(0) = 0$

$Max = \$76,000$ @ $(1000, 500)$
I.e., 1000 Model X and 500 Model Y.

21. Let $x =$ number of olive-based bundles.
Let $y =$ number of cranberry bundles.
Given (in table form):

Material	Olive (x)	Cranberry (y)
Solid	8	2
Striped	1	1
Flowered	2	7

Cost function: $f(x, y) = 10x + 20y$
Constraints:

$$\begin{cases} x \ge 0;\ y \ge 0 \\ 8x + 2y \ge 16 \\ x + y \ge 5 \\ 2x + 7y \ge 20 \end{cases}$$

Sketch the graphs of the constraints. Notice that there are four possible vertices to test for minimum value of the cost function:

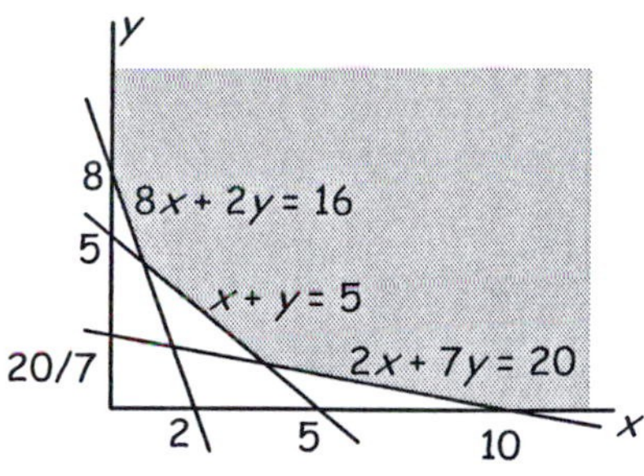

The vertices to test are
(0, 8), (1, 4), (3, 2), (10, 0)

$f(0, 8) = 10(0) + 20(8) = \$160$

$f(1, 4) = 10(1) + 20(4) = \$90$

$f(3, 2) = 10(3) + 20(2) = \$70$

$f(10, 0) = 10(10) + 20(0) = \$100$

Min $= \$70$ @ $(3, 2)$
She should buy 3 olive bundles and 2 cranberry bundles. The cost would be \$70.

23. Let $x =$ amount in municipal bonds.
Let $y =$ amount in treasury bills.
Income function:
$f(x, y) = 0.06x + 0.09y$
Constraints:

$$\begin{cases} x \ge 0;\ y \ge 0 \\ x + y = 25,000 \\ x \ge 15,000 \\ y \le 5000 \end{cases}$$

Sketch the graphs and find the vertices:

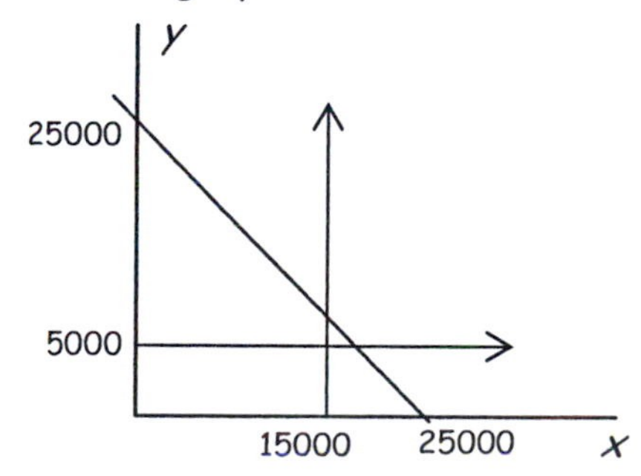

Notice that the solution is constrained by the equation $x + y = 25{,}000$. This means that only two vertices are possible:
(25000, 0) and (20000, 5000)
Test the vertices for Max and Min:

$$f(25000, 0) = 0.06(25000) + 0.09(0)$$
$$= \$1500$$
$$f(20000, 5000) = 0.06(20000) + 0.09(5000)$$
$$= \$1650$$

Conclusion:
Invest $20,000 at 6% and $5000 at 9%.

End of Section 10.7

Chapter Ten: Section 10.8
Solutions to Odd-numbered Exercises

1. Solve $\begin{cases} 3x-2y=6 \\ \dfrac{x^2}{4}+\dfrac{y^2}{9}=1 \end{cases}$

Graph: The graph of E_1 is a line and that of E_2 is an ellipse.

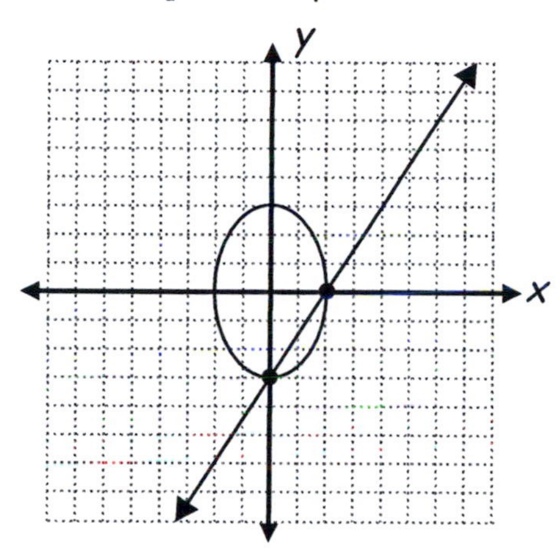

Algebraically:

Solve for x in E_1 and substitute in E_2 :

Substitute $x=\dfrac{6+2y}{3}$ in E_2

$$\frac{\left(\frac{6+2y}{3}\right)^2}{4}+\frac{y^2}{9}=1$$

$$\frac{36+24y+4y^2}{36}+\frac{y^2}{9}=1$$

$$8y^2+24y+36=36$$

$$y(8y+24)=0$$

$$y=0;\ y=-3$$

Substituting in E_1 and solving for x :

$$3x-2(0)=6$$

$$3x=6$$

$$x=2$$

$$3x-2(-3)=6$$

$$3x+6=6$$

$$x=0$$

Solution Set: $\{(0,-3),(2,0)\}$

3. Solve $\begin{cases} x^2+4y^2=5 \text{ or } \dfrac{x^2}{5}+\dfrac{y^2}{\frac{5}{4}}=1 \\ x^2+y^2=2 \end{cases}$

Graph: The graph of E_1 is an ellipse and that of E_2 is a circle.

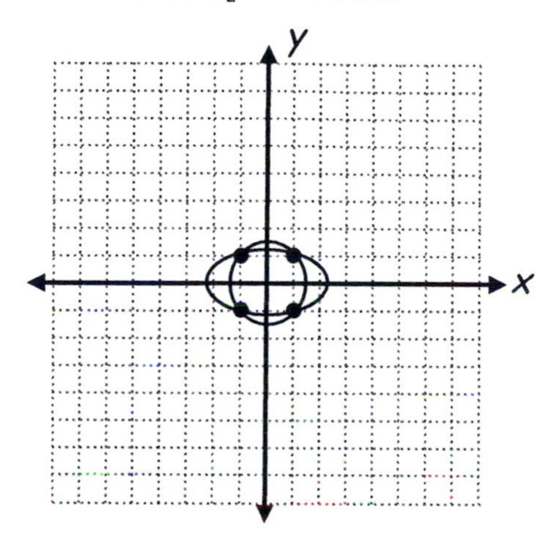

Algebraically, $E_2 : x^2=2-y^2$

Substituting in E_1 :

$$2-y^2+4y^2=5$$

$$3y^2=3$$

$$y^2=1 \Rightarrow y=\pm 1$$

For $y=1$, substitute in $x^2+4y^2=5$:

$x^2=1 \Rightarrow x=\pm 1$.

This result yields the solutions $(1,1)$ and $(-1,1)$.

For $y=-1$, substitute in $x^2+4y^2=5$:

$x^2=1 \Rightarrow x=\pm 1$.

This result yields the solutions $(1,-1)$ and $(-1,-1)$.

5. Solve $\begin{cases} y = x^2 \\ 2 - y = x^2 \text{ or } y = 2 - x^2 \end{cases}$

Graph: The graphs of E_1 and E_2 are parabolas.

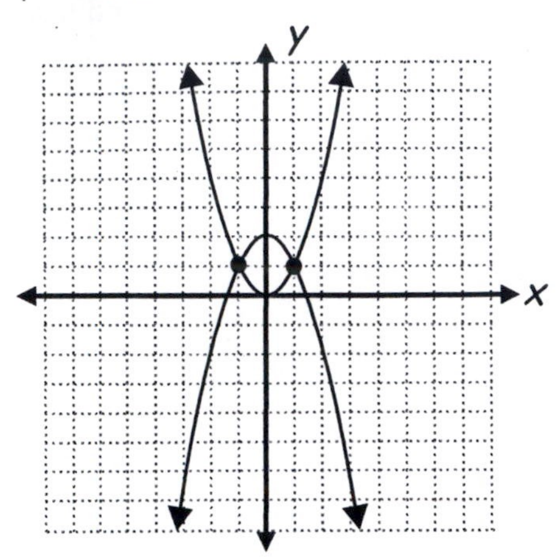

Algebraically, substituting E_1 in E_2:

$2 - x^2 = x^2$

$2x^2 = 2 \Rightarrow x = \pm 1$

For $x = 1$, $y = 1$.

For $x = -1$, $y = 1$.

Solution Set: $\{(1,1), (-1,1)\}$.

7. Solve $\begin{cases} y - x^2 = 1 \quad \text{or} \quad y = x^2 + 1 \\ y + 2 = 4x^2 \quad \text{or} \quad y = 4x^2 - 2 \end{cases}$

Graph: The graphs of E_1 and E_2 are parabolas.

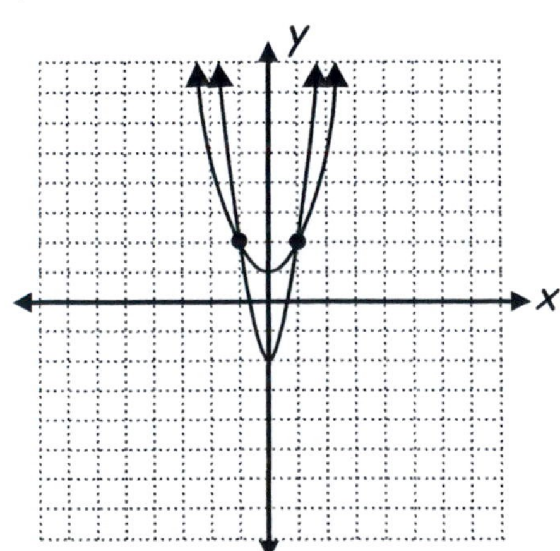

Algebraically, substituting E_1 in E_2:

$x^2 + 1 = 4x^2 - 2$

$3x^2 = 3 \Rightarrow x^2 = 1 \Rightarrow x \pm 1$

For $x = 1$, $y = 2$.

For $x = -1$, $y = 2$.

Solution Set: $\{(1, 2), (-1, 2)\}$.

9. Solve $\begin{cases} x = y^2 - 3 \\ x^2 + 4y^2 = 4 \text{ or } \dfrac{x^2}{4} + \dfrac{y^2}{1} = 1 \end{cases}$

Graph: The graph of E_1 is a parabola. The graph of E_2 is an ellipse. The graphs suggest that there is no solution to the system.

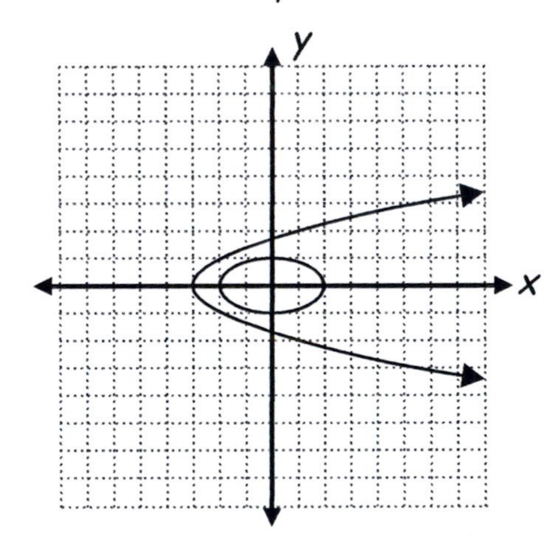

Algebraically, substituting E_1 in E_2:

$(y^2 - 3)^2 + 4y^2 = 4$

$y^4 - 6y^2 + 9 + 4y^2 = 4$

$y^4 - 2y^2 + 5 = 0$

Let $u = y^2$.

$u^2 - 2u + 5 = 0$

$u = \dfrac{2 \pm \sqrt{4 - 20}}{2} \Rightarrow u = 1 \pm 2i \Rightarrow$ no real solution.

11. Solve $\begin{cases} (x+1)^2+(y-1)^2=4 \\ (x+1)^2+4(y-1)^2=4 \end{cases}$

E_2 can be re-written as

$$\frac{(x+1)^2}{4}+\frac{(y-1)^2}{1}=1$$

Graph: The graph of E_1 is a circle and that of E_2 is an ellipse.

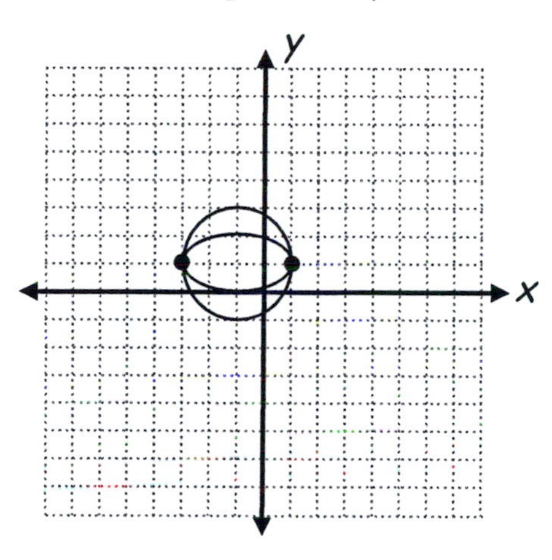

Algebraically, substitute

$(x+1)^2=4-(y-1)^2$ from E_1 into E_2:

$$4-(y-1)^2+4(y-1)^2=4$$
$$4+3(y-1)^2=4$$
$$3y^2-6y+3=0$$
$$y^2-2y+1=0$$

Solving for y:

$y-1=0 \Rightarrow y=1$

Substituting in E_1:

$$(x+1)^2=4$$
$$x+1=\pm 2$$
$$x=-1\pm 2$$

Solution Set: $\{(1,1),(-3,1)\}$

13. Given: $\begin{cases} (x-2)^2+y^2=4 \\ (x+2)^2+y^2=4 \end{cases}$

The first equation represents a circle with center $(2,0)$ and radius 2. The second represents a circle with center $(-2,0)$ and radius 2.

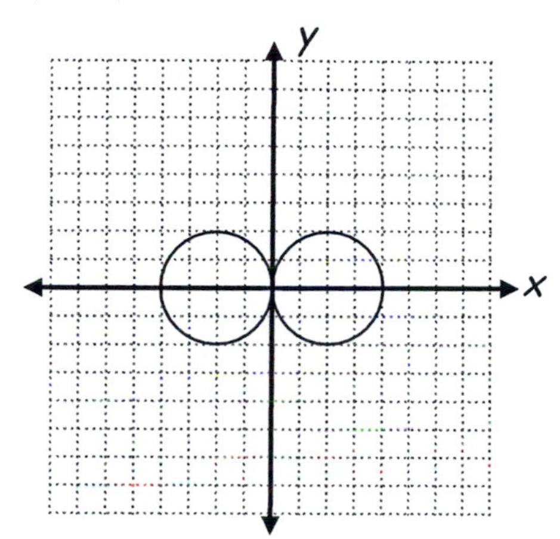

The graphs intersect at $(0,0)$.

Verification

Solve for y^2 in equation 1:

$y^2=4-(x-2)^2$.

Then, substitute in equation 2:

$$(x+2)^2+4-(x-2)^2=4$$
$$x^2+4x+4-x^2+4x-4=0$$
$$8x=0$$
$$x=0$$
$$y=0$$

15. Given: $\begin{cases} x^2+y^2=9 \\ \dfrac{x^2}{9}+\dfrac{y^2}{25}=1 \end{cases}$

The first equation represents a circle with center $(0,0)$ and radius 3. The second represents an ellipse with center $(0,0)$ and major axis along the y-axis. The minor axis along the x-axis with $b=3$ indicates intersections at $(-3,0)$ and $(3,0)$.

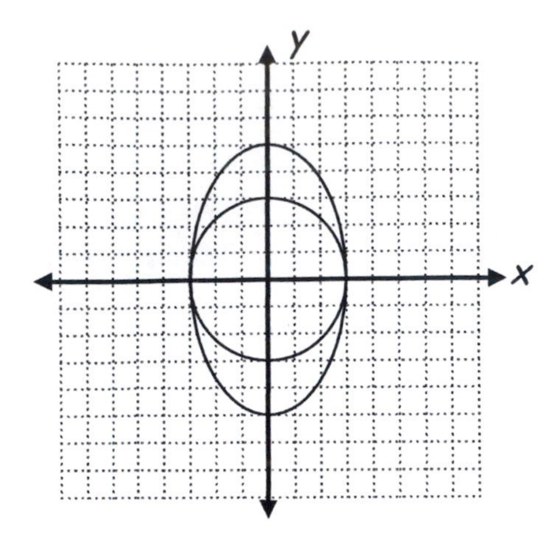

Verification

Solve for y^2 in equation 1:

$y^2 = 9 - x^2$.

Then, substitute in equation 2:

$$\frac{x^2}{9} + \frac{9 - x^2}{25} = 1$$

$$25x^2 + 81 - 9x^2 = 225$$
$$16x^2 = 144$$
$$x^2 = 9$$
$$x = \pm 3$$
$$y^2 = 9 - 9 = 0$$

The points of intersection are $(3, 0)$ and $(-3, 0)$.

17. Given: $\begin{cases} x^2 + \dfrac{y^2}{4} = 1 \\ y = 0 \end{cases}$

The first equation represents an elllipse with center $(0, 0)$ and major axis along the y-axis. The second is the x-axis. The ellipse intersects the x-axis along the minor axis at $(-1, 0)$ and $(1, 0)$.

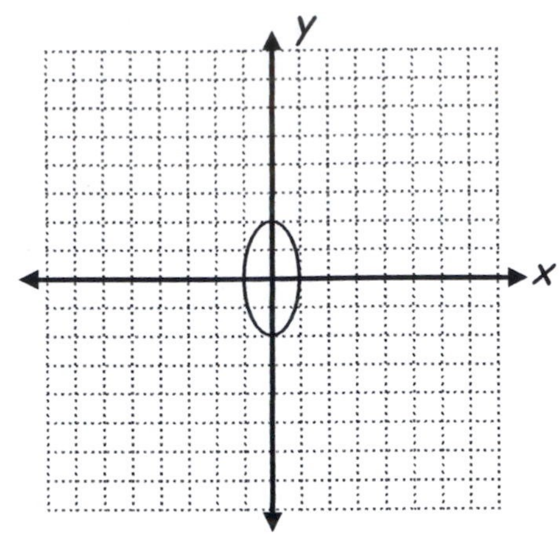

Verification

Substitute $y = 0$ in equation 1:

$x^2 = 1 \Rightarrow x = \pm 1$.

19. Given: $\begin{cases} 2y^2 - 3x^2 = 6 \\ 2y^2 + x^2 = 22 \end{cases}$

The first equation represents a hyperbola with center $(0, 0)$ and axis of symmetry the x-axis. The second represents an ellipse with center $(0, 0)$ and major axis along the x-axis.

Solve for x^2 in equation 2 and substitute in equation 1:

$$x^2 = 22 - 2y^2$$
$$2y^2 - 3(22 - 2y^2) = 6$$
$$8y^2 - 66 = 6$$
$$8y^2 = 72$$
$$y^2 = 9$$
$$y = \pm 3$$
$$x^2 = 22 - 2(3)^2$$
$$x^2 = 4$$
$$x = \pm 2$$

The solution set:

$\{(-2, -3), (-2, 3), (2, -3), (2, 3)\}$

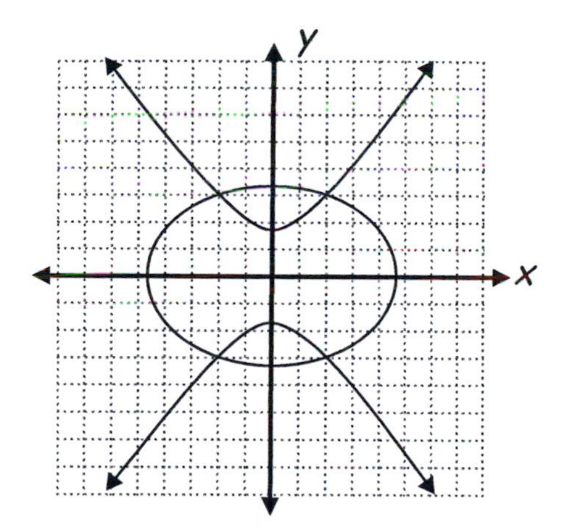

21. Given: $\begin{cases} x = y^2 - 4 \\ x + 13 = 6y \end{cases}$

The first equation represents a parabola that opens to the right with vertex $(-4, 0)$. The second equation represents a line with y-intercept $\left(0, \frac{13}{6}\right)$ and x-intercept $(-13, 0)$.

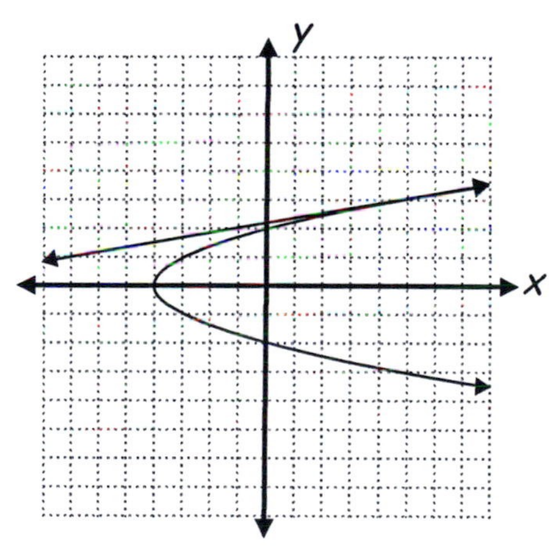

Substituting the first equation in the second:

$$y^2 - 4 + 13 = 6y$$
$$y^2 - 6y + 9 = 0$$
$$(y-3)^2 = 0$$
$$y = 3$$
$$x = (3)^2 - 4 = 5$$

Solution: $(5, 3)$

23. Given: $\begin{cases} 2y = x^2 - 4 \\ x^2 + y^2 = 4 \end{cases}$

The first equation represents a parabola that opens upward with vertex $(0, -2)$ and axis of symmetry the y-axis. The second represents a circle with center $(0, 0)$ and radius 2.

Solve for x^2 in equation 1 and substitute in equation 2:

$$x^2 = 2y + 4$$
$$2y + 4 + y^2 = 4$$
$$y^2 + 2y = 0$$
$$y(y+2) = 0 \Rightarrow y = 0;\ y = -2$$

For $y = 0$:

$$x^2 = 2(0) + 4$$
$$x = \pm 2$$

For $y = -2$:

$$x^2 = 2(-2) + 4$$
$$x = 0$$

The solution set:

$\{(2, 0), (-2, 0), (0, -2)\}$

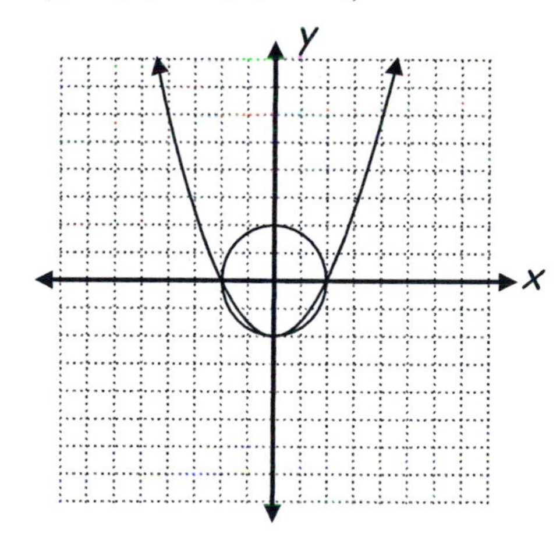

25. Solve $\begin{cases} x^2+y^2=30 \\ x^2=y \end{cases}$

Substitute E_2 in E_1:

$$y+y^2=30$$
$$(y+6)(y-5)=0$$
$$y=-6 \text{ or } y=5$$

For $y=-6$, $x^2=-6 \Rightarrow x=\pm i\sqrt{6}$

For $y=5$, $x^2=5 \Rightarrow x=\pm\sqrt{5}$

Solution Set:

$\{(-i\sqrt{6},-6),(i\sqrt{6},-6),(\sqrt{5},5),(-\sqrt{5},5)\}$

27. Solve $\begin{cases} x^2-1=y \\ 4x+y=-5 \end{cases}$

Solve E_2 for y and substitute in E_1:

$y=-4x-5$

Solve for x:

$x^2-1=-4x-5$

$x^2+4x+4=0$

$(x+2)^2=0 \Rightarrow x=-2$

Substitute in E_2:

$$4(-2)+y=-5$$
$$y=3$$

Solution Set: $\{(-2,3)\}$

29. Solve $\begin{cases} y=\dfrac{4}{x} \\ 2x^2+y^2=18 \end{cases}$

Substitute E_1 into E_2:

$$2x^2+\left(\frac{4}{x}\right)^2=18$$
$$2x^2+\frac{16}{x^2}=18$$
$$2x^4+16=18x^2$$

Let $u=x^2$ and re-write the equation:

$2u^2-18u+16=0$

$u^2-9u+8=0$

$(u-8)(u-1)=0$

$u-8=0 \Rightarrow x^2-8=0 \Rightarrow x=\pm 2\sqrt{2}$

$u-1=0 \Rightarrow x^2-1=0 \Rightarrow x=\pm 1$

For $x=1, y=4$

For $x=-1, y=-4$

For $x=2\sqrt{2}$, $y=\dfrac{4}{2\sqrt{2}}=\sqrt{2}$

For $x=-2\sqrt{2}$, $y=\dfrac{4}{-2\sqrt{2}}=-\sqrt{2}$

Solution Set:

$\{(1,4),(-1,-4),(-2\sqrt{2},-\sqrt{2}),(2\sqrt{2},\sqrt{2})\}$

31. Solve $\begin{cases} y-x^2=4 \\ x^2+y^2=16 \end{cases}$

Solve E_1 for x^2 and substitute in E_2.

Solve for y:

$$x^2=y-4$$
$$y-4+y^2=16$$
$$y^2+y-20=0$$
$$(y+5)(y-4)=0$$

$y=-5$ or $y=4$

For $y=-5$, $x^2=16-(-5)^2 \Rightarrow x=\pm 3i$

For $y=4$, $x^2=16-(-4)^2 \Rightarrow x=0$

Solution Set:

$\{(-3i,-5),(3i,-5),(0,4)\}$

33. Solve $\begin{cases} 2x^2+3y^2=6 \\ x^2+3y^2=3 \end{cases}$

Solve E_2 for x^2 and substitute in E_1.

Solve for y:

$$x^2=3-3y^2$$
$$2(3-3y^2)+3y^2=6$$
$$-3y^2=0 \Rightarrow y=0$$

For $y=0$, $x^2=3 \Rightarrow x=\pm\sqrt{3}$

Solution Set:

$\{(-\sqrt{3},0),(\sqrt{3},0)\}$

35. Solve $\begin{cases} 3x^2 - y = 3 \\ 9x^2 + y^2 = 27 \end{cases}$

Solve for y in E_1 and substitute in E_2:

$$y = 3x^2 - 3$$

$$9x^2 + (3x^2 - 3)^2 = 27$$

$$9x^4 - 9x^2 + 9 = 27$$

$$x^4 - x^2 - 2 = 0$$

$$(x^2 - 2)(x^2 + 1) = 0$$

$$x^2 - 2 = 0 \Rightarrow x = \pm\sqrt{2}$$

$$x^2 + 1 = 0 \Rightarrow x = \pm i$$

For $x = \sqrt{2}$, $y = 3(\sqrt{2})^2 - 3 = 3$.

For $x = -\sqrt{2}$, $y = 3(-\sqrt{2})^2 - 3 = 3$.

For $x = i$, $y = 3(i)^2 - 3 = -6$.

For $x = -i$, $y = 3(-i)^2 - 3 = -6$.

Solution Set:

$$\{(-\sqrt{2}, 3), (\sqrt{2}, 3), (i, -6), (-i, -6)\}$$

37. Solve $\begin{cases} x + y^2 = 2 \\ 2x^2 - y^2 = 1 \end{cases}$

Solve y^2 in E_1 and substitute in E_2:

$$y^2 = 2 - x$$

$$2x^2 - (2 - x) = 1$$

$$2x^2 + x - 3 = 0$$

$$x = \frac{-1 \pm \sqrt{1^2 - 4(2)(-3)}}{4}$$

$$x = \frac{-1 \pm \sqrt{25}}{4}$$

$$x = \frac{-1 \pm 5}{4}$$

For $x = 1$, $y^2 = 1 \Rightarrow y = \pm 1$.

For $x = -\frac{3}{2}$, $y^2 = \frac{7}{2} \Rightarrow y = \pm\frac{\sqrt{14}}{2}$.

Solution Set:

$$\left\{(1, 1), (1, -1), \left(-\frac{3}{2}, \frac{\sqrt{14}}{2}\right), \left(-\frac{3}{2}, -\frac{\sqrt{14}}{2}\right)\right\}$$

39. Solve $\begin{cases} y - 2 = (x - 2)^2 \\ y + 2 = (x - 1)^2 \end{cases}$

Solve for y in E_1 and substitute in E_2:

$$y = (x - 2)^2 + 2$$

$$(x - 2)^2 + 2 + 2 = (x - 1)^2$$

$$x^2 - 4x + 4 + 4 = x^2 - 2x + 1$$

$$2x = 7$$

$$x = \frac{7}{2}$$

$$y = \left(\frac{7}{2} - 2\right)^2 + 2 = \frac{9}{4} + 2 = \frac{17}{4}$$

Solution Set: $\left\{\left(\frac{7}{2}, \frac{17}{4}\right)\right\}$

41. Solve $\begin{cases} (x + 1)^2 + y^2 = 10 \\ \frac{(x - 2)^2}{4} + y^2 = 1 \end{cases}$

Solve for y^2 in E_2 and substitute in E_1:

$$y^2 = 1 - \frac{(x - 2)^2}{4}$$

$$(x + 1)^2 + 1 - \frac{(x - 2)^2}{4} = 10$$

$$4x^2 + 8x + 4 + 4 - x^2 + 4x - 4 = 40$$

$$3x^2 + 12x - 36 = 0$$

$$x^2 + 4x - 12 = 0$$

$$(x + 6)(x - 2) = 0$$

$x = -6$ or $x = 2$

Substitute in E_1:

For $x = -6$, $25 + y^2 = 10 \Rightarrow y = \pm i\sqrt{15}$

For $x = 2$, $y^2 = 1 \Rightarrow \pm 1$

Solution Set:

$$\{(2, 1), (2, -1), (-6, i\sqrt{15}), (-6, -i\sqrt{15})\}$$

43. Solve $\begin{cases} 2x = y - 1 \\ \dfrac{x^2}{25} + y^2 = 1 \end{cases}$

Solve for y in E_1 and substitute in E_2:

$y = 2x + 1$

$$\frac{x^2}{25} + (2x+1)^2 = 1$$

$$x^2 + 100x^2 + 100x + 25 = 25$$

$$101x^2 + 100x = 0$$

$$x(101x + 100) = 0$$

$x = 0$ or $x = -\dfrac{100}{101}$

For $x = 0$, $y = 1$

For $x = -\dfrac{100}{101}$, $y = 2\left(-\dfrac{100}{101}\right) + 1 = -\dfrac{99}{101}$

Solution Set: $\left\{(0,1), \left(-\dfrac{100}{101}, -\dfrac{99}{101}\right)\right\}$

45. Solve $\begin{cases} x^2 + 7y^2 = 14 \\ x^2 + y^2 = 3 \end{cases}$

Solve for x^2 in E_2 and substitute in E_1:

$x^2 = 3 - y^2$

$3 - y^2 + 7y^2 = 14$

$3 + 6y^2 = 14$

$$y^2 = \frac{11}{6} \Rightarrow y = \pm\frac{\sqrt{66}}{6}$$

Substituting in E_2:

For $y = \dfrac{\sqrt{66}}{6}$, $x^2 = 3 - \dfrac{11}{6} = \dfrac{7}{6} \Rightarrow$

$x = \pm\dfrac{\sqrt{42}}{6}$.

Also, for $y = -\dfrac{\sqrt{66}}{6}$, $x^2 = 3 - \dfrac{11}{6} = \dfrac{7}{6} \Rightarrow$

$x = \pm\dfrac{\sqrt{42}}{6}$.

Solution Set:

$$\left\{\left(\frac{\sqrt{42}}{6}, \frac{\sqrt{66}}{6}\right), \left(-\frac{\sqrt{42}}{6}, -\frac{\sqrt{66}}{6}\right), \left(-\frac{\sqrt{42}}{6}, \frac{\sqrt{66}}{6}\right), \left(\frac{\sqrt{42}}{6}, -\frac{\sqrt{66}}{6}\right)\right\}$$

47. Given: $\begin{cases} y = \dfrac{8}{x^2} \\ x^2 + y^2 = 8 \end{cases}$

From equation 2, $x^2 = 8 - y^2$.

Substituting in equation 1:

$$y = \frac{8}{8 - y^2}$$

$-y^3 + 8y - 8 = 0$

Test the factor $y - 2$:

2	−1	0	8	−8
		−2	−4	8
	−1	−2	4	0

Then, $y - 2$ is a factor and $y = 2$ yields a solution. Substitute in equation 2 and solve for x:

$x^2 + 4 = 8 \Rightarrow x = \pm 2$

From the last line in the synthetic division, solve for y:

$-y^2 - 2y + 4 = 0$, or $y^2 + 2y - 4 = 0$

Use the quadratic formula:

$$y = \frac{-2 \pm \sqrt{4 - 4(1)(-4)}}{2(1)}$$

$$y = \frac{-2 \pm \sqrt{20}}{2}$$

$y = -1 \pm \sqrt{5} \approx 1.236$ or -3.236

Substitute in equation 2 and solve:

For $y = 1.236$

$x^2 = 8 - (1.236)^2 = 6.472 \Rightarrow$

$x \approx \pm 2.544$

For $y = -3.236$

$x^2 = 8 - (-3.236)^2 = -2.472 \Rightarrow$

$x \approx \pm 1.572i$

Solution Set:

$\{(\pm 2, 2), (\pm 2.544, 1.236), (\pm 1.572i, -3.236)\}$

49. Given: $\begin{cases} xy = 6 \\ (x-2)^2 + (y-2)^2 = 1 \end{cases}$

Solve for y in equation 1 and substitute in equation 2:

$$y = \frac{6}{x}$$

$$x^2 - 4x + 4 + \left(\frac{6}{x} - 2\right)^2 = 1$$

$$x^2 - 4x + 4 + \frac{36}{x^2} - \frac{24}{x} + 4 = 1$$

Multiply both sides by x^2 :

$$x^4 - 4x^3 + 4x^2 + 36 - 24x + 4x^2 = x^2$$

$$x^4 - 4x^3 + 7x^2 - 24x + 36 = 0$$

Use synthetic division to test factors of 36:

3⌋	1	–4	7	–24	36
		3	–3	12	–36 ⇒
	1	–1	4	–12	0

$x - 3$ is a factor

2⌋	1	–1	4	–12
		2	2	12 ⇒
	1	1	6	0

$x - 2$ is a factor

The last result also yields $x^2 + x + 6$ as a factor. Set the factor equal to zero and use the quadratic formula to solve:

$$x = \frac{-1 \pm \sqrt{1 - 4(1)(6)}}{2} = \frac{-1 \pm \sqrt{23}i}{2}$$

Substitute each of the values in the original (first equation) and solve for y $\left(\text{i.e., } y = \frac{6}{x}\right)$:

For $x = 2$, $y = 3$.

For $x = 3$, $y = 2$.

For $x = \frac{-1 + i\sqrt{23}}{2}$, $y = \frac{12}{-1 + i\sqrt{23}}$

For $x = \frac{-1 - i\sqrt{23}}{2}$, $y = \frac{12}{-1 - i\sqrt{23}}$

The solution set:

$$\left\{(2, 3), (3, 2), \left(\frac{-1 + i\sqrt{23}}{2}, \frac{12}{-1 + i\sqrt{23}}\right), \left(\frac{-1 - i\sqrt{23}}{2}, \frac{12}{-1 - i\sqrt{23}}\right)\right\}$$

51. Given: $\begin{cases} y = x^3 - 1 \\ 3y = 2x - 3 \end{cases}$

Substitute the solution for y from equation 1 in equation 2, and solve for x:

$$3(x^3 - 1) = 2x - 3$$

$$3x^3 - 3 = 2x - 3$$

$$3x^3 - 2x = 0$$

$$x(3x^2 - 2) = 0$$

Set each factor equal to zero:

$x = 0 \Rightarrow y = -1$

$3x^2 - 2 = 0 \Rightarrow x^2 = \frac{2}{3} \Rightarrow x = \pm\frac{\sqrt{6}}{3}$

Substitute in equation 1 and solve for y:

For $x = \frac{\sqrt{6}}{3}$, $y = \left(\frac{\sqrt{6}}{3}\right)^3 - 1 = \frac{2\sqrt{6}}{9} - 1$

For $x = -\frac{\sqrt{6}}{3}$, $y = \left(-\frac{\sqrt{6}}{3}\right)^3 - 1 = -\frac{2\sqrt{6}}{9} - 1$

Solution Set:

$$\left\{(0, -1), \left(\frac{\sqrt{6}}{3}, \frac{2\sqrt{6}}{9} - 1\right), \left(-\frac{\sqrt{6}}{3}, \frac{-2\sqrt{6}}{9} - 1\right)\right\}$$

53. Given: $\begin{cases} xy - y = 4, \text{ or } y = \frac{4}{x-1} \\ (x-1)^2 + y^2 = 10 \end{cases}$

Substitute the solution for y from equation 1 in equation 2:

$$(x-1)^2 + \left(\frac{4}{x-1}\right)^2 = 10$$

$$(x-1)^2 + \frac{16}{(x-1)^2} = 10$$

$$(x-1)^4 + 16 = 10(x-1)^2$$

Let $u = x - 1$.
Then, $u^4 - 10u^2 + 16 = 0$
Let $v = u^2$
Then, $v^2 - 10v + 16 = 0$
Solve for v by using the quadratic formula:

$$v = \frac{10 \pm \sqrt{100 - 4(1)(16)}}{2}$$

$$v = \frac{10 \pm 6}{2} \Rightarrow v = 8;\ 2$$

Then, $u = \pm\sqrt{8} = \pm 2\sqrt{2}$, or $u = \pm\sqrt{2}$
Backsubstituting in $x = u + 1$:
$x = 1 \pm 2\sqrt{2}$, or $x = 1 \pm \sqrt{2}$
Substituting in $y = \dfrac{4}{x-1}$,
For $x = 1 + 2\sqrt{2}$, $y = \dfrac{4}{2\sqrt{2}} = \sqrt{2}$
For $x = 1 - 2\sqrt{2}$, $y = \dfrac{4}{-2\sqrt{2}} = -\sqrt{2}$
For $x = 1 + \sqrt{2}$, $y = \dfrac{4}{\sqrt{2}} = 2\sqrt{2}$
For $x = 1 - \sqrt{2}$, $y = \dfrac{4}{-\sqrt{2}} = -2\sqrt{2}$
Solution Set:
$\{(1+2\sqrt{2}, \sqrt{2}), (1-\sqrt{2}, -2\sqrt{2}), (1+\sqrt{2}, 2\sqrt{2}), (1-2\sqrt{2}, -\sqrt{2})\}$

55. Given: $\begin{cases} y = \sqrt{x-4} + 1 \\ (x-3)^2 + (y-1)^2 = 1 \end{cases}$

Substitute the solution for y from equation 1 in equation 2:

$$(x-3)^2 + \left(\sqrt{x-4} + 1 - 1\right)^2 = 1$$
$$x^2 - 6x + 9 + x - 4 = 1$$
$$(x-4)(x-1) = 0$$

Then, $x = 4$ or $x = 1$.
For $x = 4$, $y = \sqrt{4-4} + 1 = 1$
For $x = 1$, $y = \sqrt{-3} + 1 = \sqrt{3}i + 1$
Solution Set:
$\{(4, 1), (1, i\sqrt{3} + 1)\}$

57. Given: $\begin{cases} y^2 - y - 12 = x - x^2 \\ y - 1 + \dfrac{2x - 12}{y} = 0 \end{cases}$

The system can be rewritten as
$\begin{cases} y^2 - y - 12 = x - x^2 \\ y^2 - y - 12 = -2x \end{cases}$
Then,
$x - x^2 = -2x$, or $x^2 - 3x = 0 \Rightarrow$
$x = 0$ or $x = 3$
For $x = 0$, $y^2 - y - 12 = 0 \Rightarrow$
$(y-4)(y+3) = 0 \Rightarrow y = 4$ or $y = -3$
For $x = 3$,
$y^2 - y - 12 = 3 - (3)^2 \Rightarrow$
$y^2 - y - 12 = -6 \Rightarrow (y-3)(y+2) = 0 \Rightarrow$
$y = 3$ or $y = -2$
Solution Set:
$\{(0, 4), (0, -3), (3, 3), (3, -2)\}$

59. Given: $\begin{cases} \dfrac{(y+2)^2}{x+y} = 1 \Rightarrow x = y^2 + 3y + 4 \\ x = y^2 + 5y + 4 \end{cases}$

Set the solution for x in equation 1 equal to equation 2:

$$y^2 + 3y + 4 = y^2 + 5y + 4$$
$$2y = 0$$
$$y = 0$$

Substitute in equation 1:
$x = 0^2 + 3(0) + 4 = 4$
Solution: $(4, 0)$

61. Given: $\begin{cases} x = \sqrt{6y + 1} \\ y = \sqrt{\dfrac{x^2 + 7}{2}} \end{cases}$

Substitute the expression for x in the first equation into the second:

$$y = \sqrt{\frac{\left(\sqrt{6y+1}\right)^2 + 7}{2}}$$
$$y = \sqrt{\frac{6y + 1 + 7}{2}}$$

$y = \sqrt{3y + 4}$

$y^2 = 3y + 4$

Rearrange and solve for y:

$$y^2 - 3y - 4 = 0$$

$$(y - 4)(y + 1) = 0 \Rightarrow y = 4;\ y = -1$$

Substitute in equation 1:

For $y = 4$, $x = \sqrt{6(4) + 1} = 5$

Solution: $(5, 4)$

[Note: $y = -1$ does not lead to a solution for the second equation. It was introduced extraneously in the squaring step.]

63. Given: $\begin{cases} x^2 + 3x - 2y^2 = 5 \\ -4x^2 + 6y^2 = 3 \end{cases}$

Solve for y^2 in equation 1 and substitute in equation 2:

$$2y^2 = x^2 + 3x - 5$$

$$y^2 = \frac{x^2 + 3x - 5}{2}$$

$$-4x^2 + 6\left(\frac{x^2 + 3x - 5}{2}\right) = 3$$

$$-4x^2 + 3x^2 + 9x - 15 = 3$$

$$-x^2 + 9x - 18 = 0$$

$$x^2 - 9x + 18 = 0$$

$$(x - 3)(x - 6) = 0$$

Then, $x = 3$; $x = 6$

For $x = 3$,

$$y^2 = \frac{(3)^2 + 3(3) - 5}{2}$$

$$y^2 = \frac{13}{2} \Rightarrow y = \pm\frac{\sqrt{26}}{2}$$

For $x = 6$,

$$y^2 = \frac{(6)^2 + 3(6) - 5}{2}$$

$$y^2 = \frac{49}{2} \Rightarrow y = \pm\frac{7\sqrt{2}}{2}$$

Solution Set:

$$\left\{\left(3, +\frac{\sqrt{26}}{2}\right), \left(3, -\frac{\sqrt{26}}{2}\right), \left(6, -\frac{7\sqrt{2}}{2}\right), \left(6, +\frac{7\sqrt{2}}{2}\right)\right\}$$

65. Graph: $\begin{cases} y \le 2x + 1 \\ y < 4 \\ y > x \end{cases}$

First, graph the lines (boundaries of the system).

$\begin{cases} y = 2x + 1 \\ y = 4 \\ y = x \end{cases}$ (the boundary lines)

Then, use $(0, 0)$ to test the region and to "mark" each region that fits the direction of the inequalities. $(0, 0)$ is on the line $y = x$, so $(1, 2)$ could be used for its region.

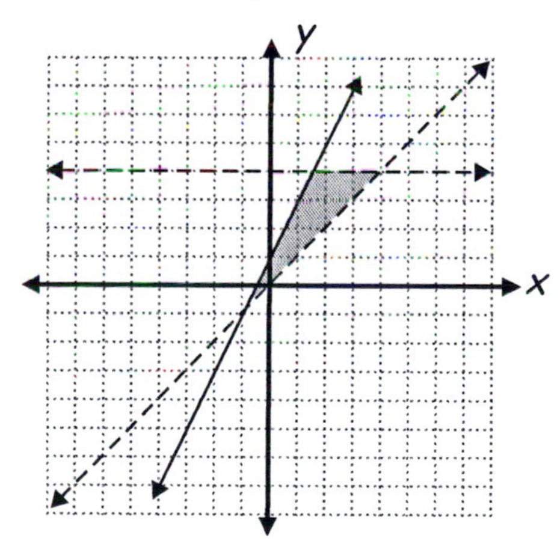

a. Yes. $(1, 2)$ satisfies all 3 conditions.

$2 \le (2)(1) + 1 = 3$ True

$2 < 4$ True

$2 > 1$ True

b. No. $(3, 4)$ does not satisfy all 3 conditions.

$4 \le (2)(3)+1 = 7$ True

$4 < 4$ False

$4 > 3$ True

c. No. $(-1, -1)$ does not satisfy $y > x$.

d. No. $(3, 3)$ does not satisfy $y > x$.

67. Given: $\begin{cases} y \ge x^2 - 2 \\ y \le (x-2)^2 \\ 3y > 2x + 12 \end{cases}$

First, graph the lines (boundaries of the system).

$$\begin{cases} y = x^2 - 2 \\ y = (x-2)^2 \\ 3y = 2x + 12 \end{cases}$$

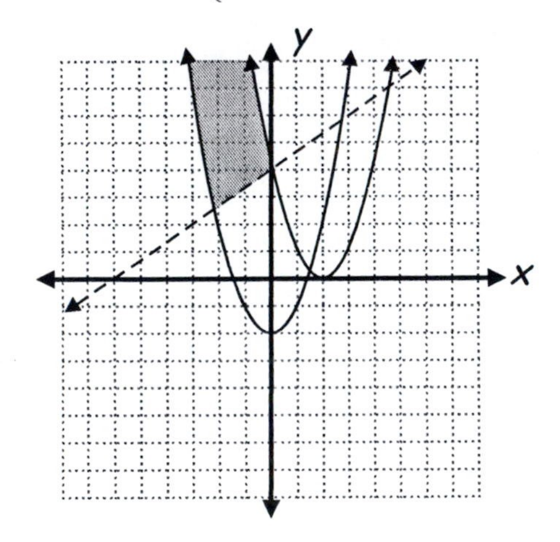

a. No. $(2, 5)$ does not satisfy 2 of the conditions.

$5 \ge 2^2 - 2 = 2$ True

$5 \le (2-2)^2$ False

$3(5) > 2(2) + 12$ False

b. No. $(7, 8)$ does not satisfy the first condition.

$8 \ge 7^2 - 2$ False

c. No. $(5, 0)$ does not satisfy the first condition.

$0 \ge 5^2 - 2$ False

d. No. $(3, 4)$ does not satisfy the first condition.

$4 \ge 3^2 - 2$ False

69. Given: $\begin{cases} y < 2x \\ y > x^2 \end{cases}$

First, graph the curves (boundaries of the system).

$$\begin{cases} y = 2x \\ y = x^2 \end{cases}$$

Then, use test points to shade the proper side of each graph. The solution is the intersection of the two regions.

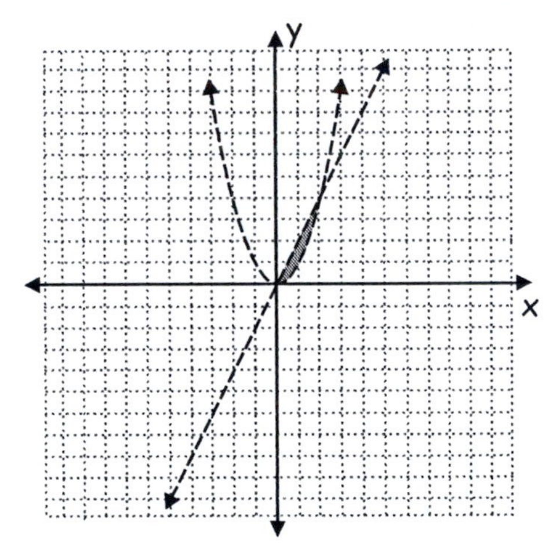

71. Given: $\begin{cases} y > x^2 \\ -3y \le x - 9 \end{cases}$

First, graph the curves (boundaries of the system).

$$\begin{cases} y = x^2 \\ -3y = x - 9 \end{cases}$$

Then, use test points to shade the proper side of each graph. The solution is the intersection of the two regions.

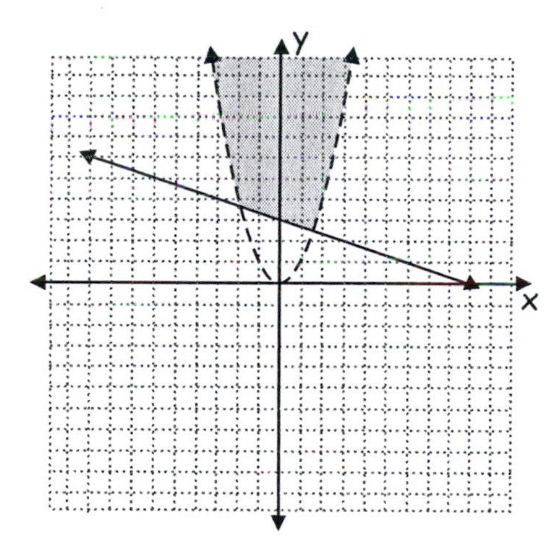

73. Given: $\begin{cases} y \le \sqrt{x} \\ 2y > (x-1)^2 - 4 \end{cases}$

First, graph the curves (boundaries of the system).

$$\begin{cases} y = \sqrt{x} \\ 2y = (x-1)^2 - 4 \end{cases}$$

Then, use test points to shade the proper side of each graph. The solution is the intersection of the two regions.

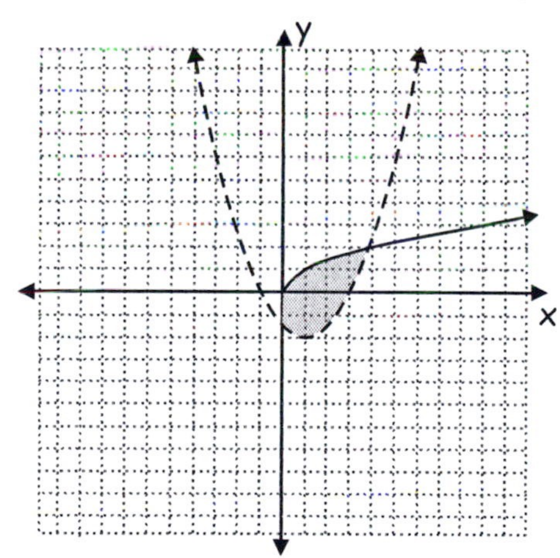

75. Given: $\begin{cases} y \ge x^3 \\ y \ge -x^3 \\ y < 2(x+1) \end{cases}$

First, graph the curves (boundaries of the system).

$$\begin{cases} y = x^3 \\ y = -x^3 \\ y = 2(x+1) \end{cases}$$

Then, use test points to shade the proper side of each graph. The solution is the intersection of the two regions.

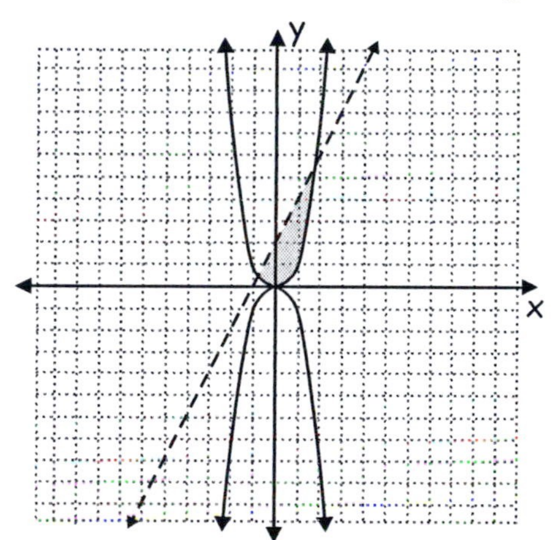

77. Given: $\begin{cases} y \le \sin x \\ y \ge 0 \end{cases}$

First, graph the curves (boundaries of the system).

$$\begin{cases} y = \sin x \\ y = 0 \end{cases}$$

Then, use test points to shade the proper side of each graph. The solution is the intersection of the two regions.

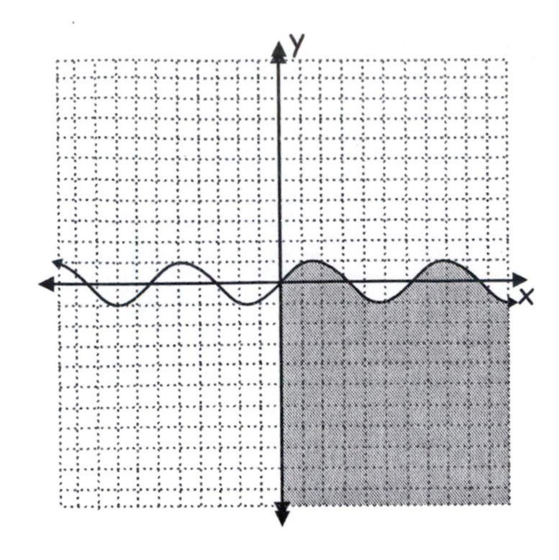

79. Let x and y be the two integers.

Solve $\begin{cases} xy = 88 \\ x + y = 19 \end{cases}$

Solve for x in E_2 and substitute in E_1 :

$x = 19 - y$

$(19 - y)y = 88$

$-y^2 + 19y = 88$

$y^2 - 19y + 88 = 0$

$(y - 8)(y - 11) = 0$

$y = 8$ or $y = 11$

Substitute in E_2 :

For $y = 8$, $x = 11$.

The two integers are 8 and 11.

81. Let $x =$ width and $y =$ length.

Solve $\begin{cases} x + 2x + 2y = 48.5 \\ xy = 95 \end{cases}$

Solve for x in E_1 and substitute in E_2 :

$3x + 2y = 48.5$

$x = \dfrac{48.5 - 2y}{3}$

$\left(\dfrac{48.5 - 2y}{3}\right)y = 95$

$-2y^2 + 48.5y = 285$

$2y^2 - 48.5y + 285 = 0$

$y = \dfrac{48.5 \pm \sqrt{(48.5)^2 - 4(2)(285)}}{4}$

$y = \dfrac{48.5 \pm \sqrt{72.25}}{4} = \dfrac{48.5 \pm 8.5}{4}$

$y = 10$ or $14\frac{1}{4}$

Substitute in E_2:

For $y = 10$ m, $x = 9\frac{1}{2}$ m

For $y = 14\frac{1}{4}$ m, $x = 6\frac{2}{3}$ m

83. Let $R =$ Maria's rate of speed and $r =$ Paul's rate. Use $d = rt$ and the conversion of 20 min $= \frac{1}{3}$ hour and 10-min. $= \frac{1}{6}$ hour.

E_1: Maria's Distance after 20 min. = Paul's distance after 20 minutes plus 4 miles.

E_2: Paul's time is 10 minutes more than Maria's time traveling 24 miles

$\begin{cases} \left(\frac{1}{3}\right)R = \left(\frac{1}{3}\right)r + 4 \\ \frac{24}{R} + \frac{1}{6} = \frac{24}{r} \end{cases}$

$\begin{cases} R = r + 12 \\ 144r + Rr = 144R \end{cases}$

Substitute R from E_1 and solve for r:

$144r + (r + 12)r = 144(r + 12)$

$144r + r^2 + 12r = 144r + 1728$

$r^2 + 12r - 1728 = 0$

$(r - 36)(r + 48) = 0$

$r = 36$ mph and $R = 48$ mph

End of Section 10.8

Chapter Ten Test
Solutions to Odd-Numbered Exercises

1. Solve $\begin{cases} 3x - y + z = 2 \\ -x + y - 2z = -4 \\ -6x + 2y - 2z = -7 \end{cases}$

Using Cramer's Rule:

Adding $2E_1 + E_3$ we have:

$$\begin{array}{r} 6x - 2y + 2z = 4 \\ -6x + 2y - 2z = -7 \\ \hline 0 = -3 \end{array}$$

The result is a false statement. Thus the two equations have no points in common and the system has no solution.

Solution set: $\varnothing$

3. Solve $\begin{cases} x + y - z = 1 \\ 3x - 4y - 5z = -1 \\ 6x - 3y + z = 20 \end{cases}$

Adding $E_1 + E_2$, we have:

$$\begin{array}{r} x + y - z = 1 \\ 6x - 3y + z = 20 \\ \hline 7x - 2y = 21 \end{array}$$

Adding $E_2 + 5E_3$, we have:

$$\begin{array}{r} 3x - 4y - 5z = -1 \\ 30x - 15y + 5z = 100 \\ \hline 33x - 19y = 99 \end{array}$$

Forming a system with the two resulting equations, we have:

$$\begin{cases} 7x - 2y = 21 \\ 33x - 19y = 99 \end{cases}$$

Adding $19E_1 - 2E_2$ in the new system, we have:

$$\begin{array}{r} 91x - 38y = 273 \\ -66x + 38y = -198 \\ \hline 25x = 75 \end{array}$$

Solving for x, $x = 3$.

Back substituting, we have:

$$7(3) - 2y = 21 \Rightarrow y = 0$$

and

$$3 + 0 - z = 1 \Rightarrow z = 2.$$

Solution: $(3, 0, 2)$

5. Re-write and simplify each equation.

$$\begin{cases} 4x + 5y - z = 0 \\ x + 3y + 2z = 3 \\ 10x - y - 6z = 0 \end{cases} \longrightarrow \left[\begin{array}{ccc|c} 4 & 5 & -1 & 0 \\ 1 & 3 & 2 & 3 \\ 10 & -1 & -6 & 0 \end{array}\right]$$

7. $\left[\begin{array}{cc|c} 2 & 3 & 5 \\ -4 & -1 & 2 \end{array}\right] \xrightarrow{2R_1 + R_2} \left[\begin{array}{cc|c} 2 & 3 & 5 \\ 0 & 5 & 12 \end{array}\right]$

9. $\begin{vmatrix} -1 & 3 & 1 \\ 1 & -4 & 0 \\ 0 & 2 & 3 \end{vmatrix} = -1\begin{vmatrix} -4 & 0 \\ 2 & 3 \end{vmatrix} - 1\begin{vmatrix} 3 & 1 \\ 2 & 3 \end{vmatrix}$

$$= -1(-12) - (9 - 2) = 12 - 7 = 5$$

11. $\begin{vmatrix} x^4 & x & x & 2x \\ 0 & x & x^3 & x \\ 0 & 0 & x & x \\ 0 & 0 & 0 & x^2 \end{vmatrix}$

$$= x^4 \begin{vmatrix} x & x^3 & x \\ 0 & x & x \\ 0 & 0 & x^2 \end{vmatrix}$$

$$= x^4 \left[x \begin{vmatrix} x & x \\ 0 & x^2 \end{vmatrix} \right] = x^4 \left[x\left(x^3\right) \right] = x^8$$

13. $\begin{bmatrix} 4x & 2y^2 & z \end{bmatrix} = \begin{bmatrix} 12 \\ 18 \\ -2 \end{bmatrix}$. This is impossible to apply. The matrices must be of the same order.

15. $\begin{bmatrix} 4 \\ 5 \\ 6 \end{bmatrix} \begin{bmatrix} -3 & 2 & 3 \end{bmatrix} = \begin{bmatrix} -12 & 8 & 12 \\ -15 & 10 & 15 \\ -18 & 12 & 18 \end{bmatrix}$

17. $\begin{cases} x_1 - x_2 + 2x_3 = -4 \\ 2x_1 - 3x_2 - x_3 = 1 \\ -3x_1 + 6x_3 = 5 \end{cases} \longrightarrow$

$$\begin{bmatrix} 1 & -1 & 2 \\ 2 & -3 & -1 \\ -3 & 0 & 6 \end{bmatrix}\begin{bmatrix} x_1 \\ x_2 \\ x_3 \end{bmatrix} = \begin{bmatrix} -4 \\ 1 \\ 5 \end{bmatrix}$$

19. $\left[\begin{array}{cc|cc} 2 & 2 & 1 & 0 \\ \frac{1}{2} & 1 & 0 & 1 \end{array}\right] \xrightarrow[\frac{1}{2}R_1]{-\frac{1}{4}R_1+R_2}$

$\left[\begin{array}{cc|cc} 1 & 1 & \frac{1}{2} & 0 \\ 0 & \frac{1}{2} & -\frac{1}{4} & 1 \end{array}\right] \xrightarrow{2R_2}$

$\left[\begin{array}{cc|cc} 1 & 1 & \frac{1}{2} & 0 \\ 0 & 1 & -\frac{1}{2} & 2 \end{array}\right] \xrightarrow{-R_2+R_1}$

$\left[\begin{array}{cc|cc} 1 & 0 & 1 & -2 \\ 0 & 1 & -\frac{1}{2} & 2 \end{array}\right]$

$$\begin{bmatrix} 2 & 2 \\ \frac{1}{2} & 1 \end{bmatrix}^{-1} = \begin{bmatrix} 1 & -2 \\ -\frac{1}{2} & 2 \end{bmatrix}$$

21. Partial decomposition of

$$f(x) = \frac{p(x)}{9x^4 - 6x^3 + x^2}$$

First, factor the denominator:

$$9x^4 - 6x^3 + x^2 = x^2(3x-1)^2$$

Then,

$$f(x) = \frac{A_1}{x} + \frac{A_2}{x^2} + \frac{A_3}{3x-1} + \frac{A_4}{(3x-1)^2}$$

23. Partial decomposition of

$$f(x) = \frac{p(x)}{x^2 + 3x - 4}$$

First, factor the denominator:

$$x^2 + 3x - 4 = (x+4)(x-1)$$

Then,

$$f(x) = \frac{A_1}{x+4} + \frac{A_2}{x-1}$$

25. The partial decomposition of

$$f(x) = \frac{x-4}{(2x-5)^2}$$

Write

$$\frac{x-4}{(2x-5)^2} = \frac{A_1}{2x-5} + \frac{A_2}{(2x-5)^2}$$

Multiply by the denominator:

$$x - 4 = A_1(2x-5) + A_2$$
$$x - 4 = 2A_1x + A_2 - 5A_1$$

Then, set up a system of equations:

$$\begin{cases} 2A_1 = 1 \Rightarrow A_1 = \frac{1}{2} \\ -5A_1 + A_2 = -4 \end{cases}$$

Substitute $A_1 = \frac{1}{2}$ in equation 2:

$$-5\left(\frac{1}{2}\right) + A_2 = -4 \Rightarrow A_2 = -\frac{3}{2}$$

The decomposition is

$$f(x) = \frac{1}{2(2x-5)} - \frac{3}{2(2x-5)^2}$$

27. The partial decomposition of

$$f(x) = \frac{x+1}{x^3 + x}$$

Write

$$\frac{x+1}{x^3+x} = \frac{A_1x + B}{x^2+1} + \frac{A_2}{x}$$

Multiply by the denominator:

$$x + 1 = (A_1x + B)x + A_2(x^2 + 1)$$
$$x + 1 = (A_1 + A_2)x^2 + Bx + A_2$$

Then, set up a system of equations:

$$\begin{cases} A_1 + A_2 = 0 \\ B = 1 \\ A_2 = 1 \Rightarrow A_1 = -1 \end{cases}$$

The decomposition is

$$f(x) = \frac{-x+1}{x^2+1} + \frac{1}{x}, \text{ or}$$

$$f(x) = \frac{1-x}{x^2+1} + \frac{1}{x}$$

29. The partial decomposition of

$$f(x)=\frac{x^2+12x+12}{x^3-4x}$$

Write

$$\frac{x^2+12x+12}{x^3-4x}=\frac{A_1}{x}+\frac{A_2}{x-2}+\frac{A_3}{x+2}$$

Multiply by the denominator:

$$x^2+12x+12=A_1\left(x^2-4\right)+A_2\left(x^2+2x\right)+A_3\left(x^2-2x\right)$$

$$x^2+12x+12=\left(A_1+A_2+A_3\right)x^2+\left(2A_2-2A_3\right)x-4A_1$$

Then, set up a system of equations:

$$\begin{cases} A_1+A_2+A_3=1 \\ 2A_2-2A_3=12 \Rightarrow A_2-A_3=6 \\ -4A_1=12 \Rightarrow A_1=-3 \end{cases}$$

From the first two equations:

$$\begin{cases} -3+A_2+A_3=1 \\ A_2=A_3+6 \end{cases}$$

By substitution, we have:

$$-3+A_3+6+A_3=1$$
$$2A_3=-2$$
$$A_3=-1$$
$$A_2=5$$

The decomposition is

$$f(x)=\frac{-3}{x}+\frac{5}{x-2}-\frac{1}{x+2}$$

31. Let $x=$ number of hours for statistics

$y=$ number of hours for biology

Contraints:

$$\begin{cases} x\ge 6 \\ y\ge 6 \\ y\le 8 \\ x\le 10 \\ x+y\ge 15 \end{cases}$$

Graph the equations (boundaries):

$$\begin{cases} x=6 \\ y=6 \\ y=8 \\ x=10 \\ x+y=15 \end{cases}$$

Mark the region of constraint:

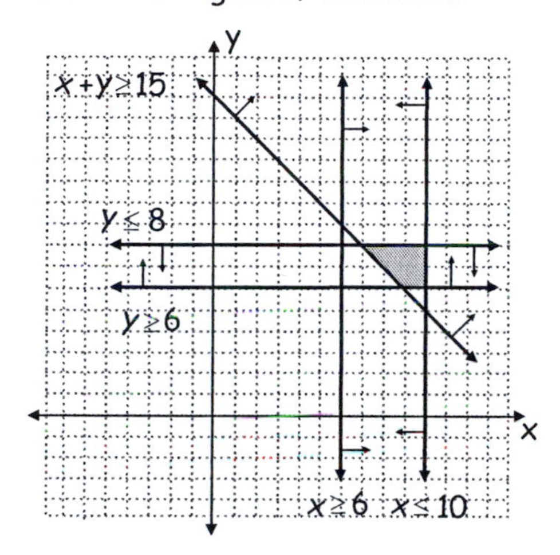

33. Find the minimum and maximum values of the given objective function on the region of constraints:

$f(x,y)=2x+y$

Constraints:

$$\begin{cases} x\ge 0 \\ x+4y\le 16 \\ 2x+3y\ge 6 \\ 3x-y\le 9 \end{cases}$$

Sketch the region of constraints:

First, graph the equations (boundaries).

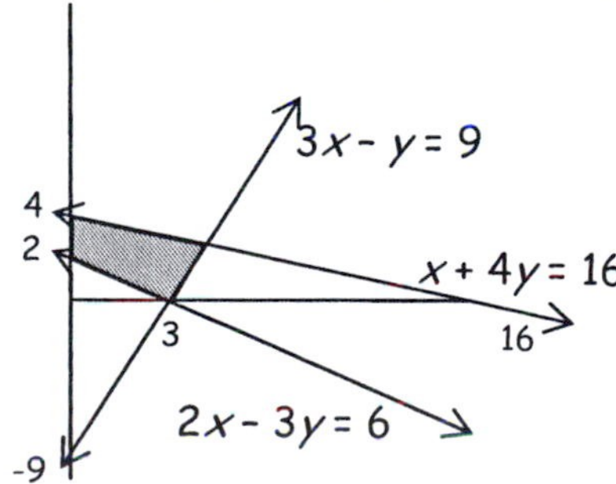

Check the value of $f(x, y)$ at each vertex: $(0,2),(0,4),(3,0)$ and $(4,3)$

$f(4,3)=2(4)+3=11$

$f(0,2)=2(0)+2=2$

$f(0,4)=2(0)+4=4$

$f(3,0)=2(3)+0=6$

$\text{Max}=11 \text{ @ } (4,3)$

$\text{Min}=2 \text{ @ } (0,2)$

35. Find the minimum and maximum values of the given objective function on the region of constraints:

$f(x, y)=5x+4y$

Constraints:

$$\begin{cases} x \geq 0; y \geq 0 \\ 2x+3y \leq 12 \\ 3x+y \leq 12 \\ x+y \geq 2 \end{cases}$$

Sketch the region of constraints:
First, graph the equations (boundaries).

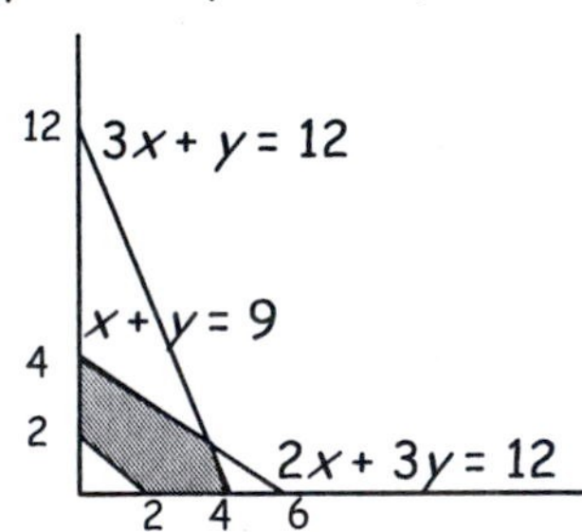

Check the value of $f(x,y)$ at each vertex: $\left(\frac{24}{7},\frac{12}{7}\right)$, $(0,4)$, $(0,2)$, $(4,0)$ and $(2,0)$.

$f\left(\frac{24}{7},\frac{12}{7}\right)=5\left(\frac{24}{7}\right)+4\left(\frac{12}{7}\right)=24$

$f(4,0)=5(4)+4(0)=20$

$f(0,4)=5(0)+4(4)=16$

$f(0,2)=5(0)+4(2)=8$

$f(2,0)=5(2)+4(0)=10$

$\text{Max}=24 \text{ @ } \left(\frac{24}{7},\frac{12}{7}\right)$

$\text{Min}=8 \text{ @ } (0,2)$

37. Find the minimum and maximum values of the given objective function on the region of constraints:

$f(x, y)=2x+4y$

Constraints:

$$\begin{cases} x \geq 0; y \geq 0 \\ 2x+2y \leq 21 \\ x+4y \leq 20 \\ x+y \leq 18 \end{cases}$$

Sketch the region of constraints:
First, graph the equations (boundaries).

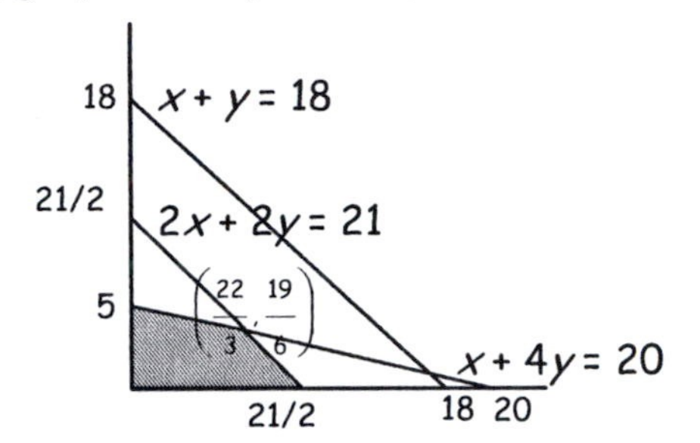

Check the value of $f(x, y)$ at each vertex: $(0,0)$, $\left(\frac{22}{3},\frac{19}{6}\right)$, $\left(\frac{21}{2},0\right)$, and $(0,5)$.

$f(0,0)=2(0)+4(0)=0$

$f\left(\frac{22}{3},\frac{19}{6}\right)=2\left(\frac{22}{3}\right)+4\left(\frac{19}{6}\right)=27.\overline{3}$

$f\left(\frac{21}{2},0\right)=2\left(\frac{21}{2}\right)+4(0)=21$

$f(0,5)=2(0)+4(5)=20$

$\text{Max}=27.\overline{3} \text{ @ } \left(\frac{22}{3},\frac{19}{6}\right)$

$\text{Min}=0 \text{ @ } (0,0)$

39. Cost function $f(x, y)=450x+550y$ for $x=$ number of bionic arms and $y=$ number of bionic legs.

$$\begin{cases} 20 \leq x \leq 60 \\ 15 \leq y \leq 40 \\ x+y \geq 50 \end{cases}$$

Sketch the region of constraints:
First, graph the equations (boundaries).

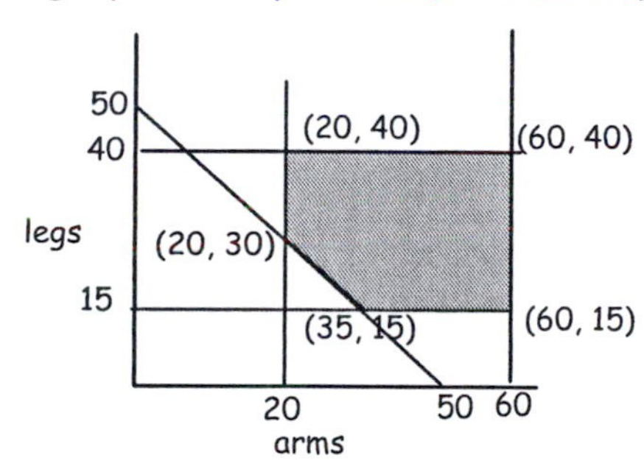

Check the value of $f(x, y)$ at each vertex: $(20, 30)$, $(20, 40)$, $(35, 15)$, $(60, 15)$ and $(60, 40)$.

$f(20, 30) = 450(20) + 550(30) = 25500$

$f(20, 40) = 450(20) + 550(40) = 31000$

$f(35, 15) = 450(35) + 550(15) = 24000$

$f(60, 15) = 450(60) + 550(15) = 35250$

$f(60, 40) = 450(60) + 550(40) = 49000$

Min = \$24,000 @ $(35, 15)$

35 bionic arms and 15 bionic legs

41. Solve $\begin{cases} x^2 + y^2 = 8 \\ 2x^2 + y^2 = 12 \end{cases}$

Substitute the expression for y^2 from equation 1 into equation 2:

$$2x^2 + 8 - x^2 = 12$$
$$x^2 = 4$$
$$x = \pm 2$$

For $x = 2$, $y^2 = 8 - 2^2 = 4 \Rightarrow y = \pm 2$

For $x = -2$, $y^2 = 8 - (-2)^2 = 4 \Rightarrow y = \pm 2$

Solution Set: $\{(-2, \pm 2), (2, \pm 2)\}$

43. Let x and y be the two integers.

$\begin{cases} \text{The product is 144} \\ \text{The sum is 25} \end{cases}$ translates to

$\begin{cases} xy = 144 \\ x + y = 25 \end{cases}$

Solve E_2 for y and substitute in E_1:

$y = 25 - x$

$$x(25 - x) = 144$$
$$-x^2 + 25x = 144$$
$$x^2 - 25x + 144 = 0$$
$$(x - 16)(x - 9) = 0$$

$x = 16$ and $y = 9$ are the two integers.

45. Graph the system of inequalities. First, graph the associated boundary equation:

$\begin{cases} y^2 \le 9 - x^2, \text{ or } x^2 + y^2 \le 9 \\ y < |x| \\ y > -|x| \end{cases}$

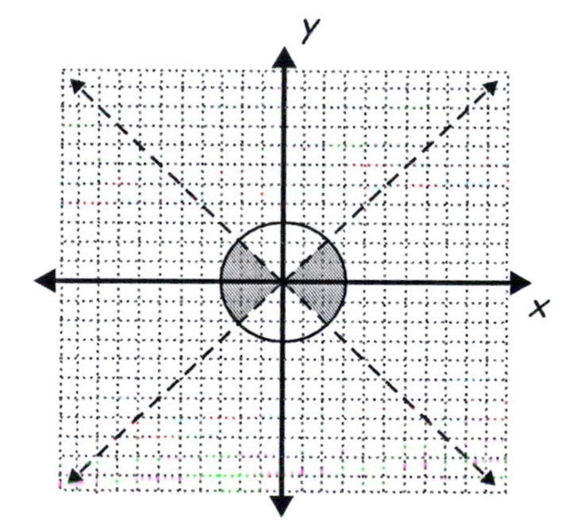

a. $(2, 5)$ No. The point is outside the intersection of the regions.

b. $(7, 8)$ No. The point is outside the intersection of the regions.

c. $(5, 0)$ No. The point is outside the intersection of the regions.

d. $(3, 4)$ No. The point is outside the intersection of the regions.

47. Graph $\begin{cases} y \le \sqrt{x+1} \\ y > x^2 - 1 \end{cases}$

First, graph the boundary equations

Graph $\begin{cases} y = \sqrt{x+1} \\ y = x^2 - 1 \end{cases}$

Mark each region from the boundary and observe the intersection of the regions:

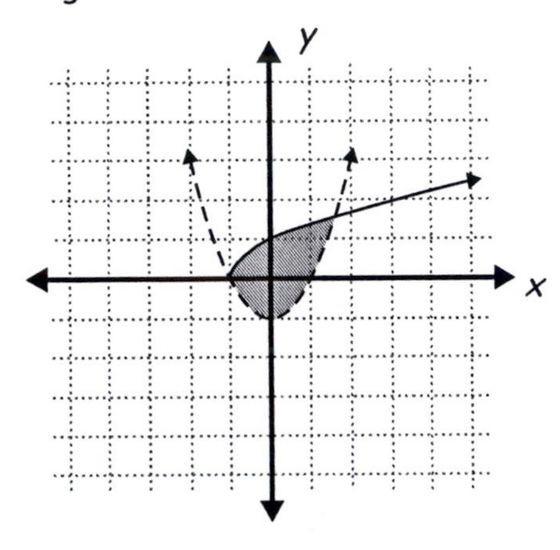

End of Chapter 10 Test

Chapter Eleven: Section 11.1
Solutions to Odd-numbered Exercises

1. $a_n = 7n - 3$

$a_1 = 7(1) - 3 = 4$

$a_2 = 7(2) - 3 = 11$

$a_3 = 7(3) - 3 = 18$

$a_4 = 7(4) - 3 = 25$

$a_5 = 7(5) - 3 = 32$

3. $a_n = (-2)^n$

$a_1 = (-2)^1 = -2$

$a_2 = (-2)^2 = 4$

$a_3 = (-2)^3 = -8$

$a_4 = (-2)^4 = 16$

$a_5 = (-2)^5 = -32$

5. $a_n = \frac{(-1)^n}{n^2}$

$a_1 = \frac{(-1)^1}{1^2} = -1$

$a_2 = \frac{(-1)^2}{2^2} = \frac{1}{4}$

$a_3 = \frac{(-1)^3}{3^2} = -\frac{1}{9}$

$a_4 = \frac{(-1)^4}{4^2} = \frac{1}{16}$

$a_5 = \frac{(-1)^5}{5^2} = -\frac{1}{25}$

7. $a_n = \left(-\frac{1}{3}\right)^{n-1}$

$a_1 = \left(-\frac{1}{3}\right)^{1-1} = 1$

$a_2 = \left(-\frac{1}{3}\right)^{2-1} = -\frac{1}{3}$

$a_3 = \left(-\frac{1}{3}\right)^{3-1} = \frac{1}{9}$

$a_4 = \left(-\frac{1}{3}\right)^{4-1} = -\frac{1}{27}$

$a_5 = \left(-\frac{1}{3}\right)^{5-1} = \frac{1}{81}$

9. $a_n = \frac{(n-1)^2}{(n+1)^2}$

$a_1 = \frac{(1-1)^2}{(1+1)^2} = 0$

$a_2 = \frac{(2-1)^2}{(2+1)^2} = \frac{1}{9}$

$a_3 = \frac{(3-1)^2}{(3+1)^2} = \frac{1}{4}$

$a_4 = \frac{(4-1)^2}{(4+1)^2} = \frac{9}{25}$

$a_5 = \frac{(5-1)^2}{(5+1)^2} = \frac{4}{9}$

11. $a_n = (-n+4)^3 - 1$

$a_1 = (-1+4)^3 - 1 = 26$

$a_2 = (-2+4)^3 - 1 = 7$

$a_3 = (-3+4)^3 - 1 = 0$

$a_4 = (-4+4)^3 - 1 = -1$

$a_5 = (-5+4)^3 - 1 = -2$

13. $a_n = (-1)^n \sqrt{n}$

$a_1 = (-1)^1 \sqrt{1} = -1$

$a_2 = (-1)^2 \sqrt{2} = \sqrt{2}$

$a_3 = (-1)^3 \sqrt{3} = -\sqrt{3}$

$a_4 = (-1)^4 \sqrt{4} = 2$

$a_5 = (-1)^5 \sqrt{5} = -\sqrt{5}$

15. $a_n = 4n - 3$

$a_1 = 4(1) - 3 = 1$

$a_2 = 4(2) - 3 = 5$

$a_3 = 4(3) - 3 = 9$

$a_4 = 4(4) - 3 = 13$

$a_5 = 4(5) - 3 = 17$

17. $a_n = 2^{n-2}$

$a_1 = 2^{1-2} = \frac{1}{2}$

$a_2 = 2^{2-2} = 1$

$a_3 = 2^{3-2} = 2$

$a_4 = 2^{4-2} = 4$

$a_5 = 2^{5-2} = 8$

19. $a_n = (3n)^n$

$a_1 = (3(1))^1 = 3$

$a_2 = (3(2))^2 = 36$

$a_3 = (3(3))^3 = 729$

$a_4 = (3(4))^4 = 20{,}736$

$a_5 = (3(5))^5 = 759{,}375$

21. $a_n = \frac{5n}{n+3}$

$a_1 = \frac{5(1)}{1+3} = \frac{5}{4}$

$a_2 = \frac{5(2)}{2+3} = 2$

$a_3 = \frac{5(3)}{3+3} = \frac{5}{2}$

$a_4 = \frac{5(4)}{4+3} = \frac{20}{7}$

$a_5 = \frac{5(5)}{5+3} = \frac{25}{8}$

23. $a_n = \frac{n^2 + n}{2}$

$a_1 = \frac{1^2 + 1}{2} = 1$

$a_2 = \frac{2^2 + 2}{2} = 3$

$a_3 = \frac{3^2 + 3}{2} = 6$

$a_4 = \frac{4^2 + 4}{2} = 10$

$a_5 = \frac{5^2 + 5}{2} = 15$

25. $a_n = \frac{(n+1)^2}{(n-1)^2}$

$a_1 = \frac{(1+1)^2}{(1-1)^2}$ is undefined

$a_2 = \frac{(2+1)^2}{(2-1)^2} = 9$

$a_3 = \frac{(3+1)^2}{(3-1)^2} = 4$

$a_4 = \frac{(4+1)^2}{(4-1)^2} = \frac{25}{9}$

$a_5 = \frac{(5+1)^2}{(5-1)^2} = \frac{9}{4}$

27. $a_n = \frac{2n-1}{3n}$

$a_1 = \frac{2(1)-1}{3(1)} = \frac{1}{3}$

$a_2 = \frac{2(2)-1}{3(2)} = \frac{1}{2}$

$a_3 = \frac{2(3)-1}{3(3)} = \frac{5}{9}$

$a_4 = \frac{2(4)-1}{3(4)} = \frac{7}{12}$

$a_5 = \frac{2(5)-1}{3(5)} = \frac{3}{5}$

29. $a_n = -(n-1)^2$

$a_1 = -(1-1)^2 = 0$

$a_2 = -(2-1)^2 = -1$

$a_3 = -(3-1)^2 = -4$

$a_4 = -(4-1)^2 = -9$

$a_5 = -(5-1)^2 = -16$

31. Given: $a_1 = 2,\ a_n = (a_{n-1})^2$ for $n \geq 2$

$a_1 = 2$

$a_2 = (2)^2 = 4$

$a_3 = (4)^2 = 16$

$a_4 = (16)^2 = 256$

$a_5 = (256)^2 = 65{,}536$

33. Given: $a_1 = 1,\ a_n = na_{n-1}$ for $n \geq 2$

$a_1 = 1$

$a_2 = 2(1) = 2$

$a_3 = 3(2) = 6$

$a_4 = 4(6) = 24$

$a_5 = 5(24) = 120$

35. Given: $a_1 = 2,\ a_n = \sqrt{(a_{n-1})^2 + 1}$ for $n \geq 2$

$a_1 = 2$

$a_2 = \sqrt{(2)^2 + 1} = \sqrt{5}$

$a_3 = \sqrt{(\sqrt{5})^2 + 1} = \sqrt{6}$

$a_4 = \sqrt{(\sqrt{6})^2 + 1} = \sqrt{7}$

$a_5 = \sqrt{(\sqrt{7})^2 + 1} = 2\sqrt{2}$

37. Given: 5, 12, 19, 26, 33, ...
Notice the increase of 7 successively from the first term.
Then, $a_n = 5 + (n-1)7$, or $a_n = 7n - 2$

39. Given: $-1, 2, -6, 24, -120, \ldots$.
Notice the alternating signs and the increase by 1 of the multiplier from one term to the next.
Then,

$a_1 = -1$

$a_2 = -2(a_1) = 2$

$a_3 = -3(a_2) = -6$

$a_4 = -4(a_3) = 24$

$a_5 = -5(a_4) = -120$

$\Rightarrow a_n = -na_{n-1}$

41. Given: $\frac{1}{4}, \frac{1}{7}, \frac{1}{12}, \frac{1}{19}, \frac{1}{28}, \ldots$

$a_1 = \frac{1}{1^2 + 3} = \frac{1}{4}$

$a_2 = \frac{1}{2^2 + 3} = \frac{1}{7}$

$a_3 = \frac{1}{3^2 + 3} = \frac{1}{12}$

... with the pattern established:

$a_n = \frac{1}{n^2 + 3}$

43. Given: $-34, -25, -16, -7, 2, \ldots$.
Notice the increase of 9 successively from the first term. Then,
$a_n = -34 + (n-1)9$, or $a_n = 9n - 43$

45. Given: $\frac{1}{4}, \frac{1}{2}, 1, 2, 4, \ldots$.
Notice that the terms suggest the powers of 2:

$a_1 = \frac{1}{4} = \left(\frac{1}{2}\right)^2 = (2)^{-2} = 2^{1-3}$

$a_2 = \frac{1}{2} = \left(\frac{1}{2}\right)^1 = (2)^{-1} = 2^{2-3}$

$a_3 = 1 = (2)^0 = 2^{3-3}$

Then,

$a_n = 2^{n-3}$

47. Given: $\frac{1}{2}, \frac{1}{2}, \frac{3}{8}, \frac{1}{4}, \frac{5}{32}, \ldots$

Notice that the terms suggest the powers of 2:

$$a_1 = \frac{1}{2} = \frac{1}{2^1}$$
$$a_2 = \frac{1}{2} = \frac{2}{2^2}$$
$$a_3 = \frac{3}{8} = \frac{3}{2^3}$$
$$a_4 = \frac{1}{4} = \frac{4}{2^4}$$

Then,

$$a_n = \frac{n}{2^n}$$

49. Given: 3, 5, 7, 9, 11, ...

Notice the successive odd whole numbers from 3.

$$a_1 = 3 = 2(1) + 1$$
$$a_2 = 5 = 2(2) + 1$$
$$a_3 = 7 = 2(3) + 1$$
$$a_4 = 9 = 2(4) + 1$$
$$a_5 = 11 = 2(5) + 1$$

Then, $a_n = 2n + 1$

51. Given: $1, \frac{1}{4}, \frac{1}{9}, \frac{1}{16}, \frac{1}{25}, \ldots$

Notice the successive squared terms in the denominators:

$$a_1 = 1 = \frac{1}{1^2}$$
$$a_2 = \frac{1}{4} = \frac{1}{2^2}$$
$$a_3 = \frac{1}{9} = \frac{1}{3^2}$$
$$a_4 = \frac{1}{16} = \frac{1}{4^2}$$
$$a_5 = \frac{1}{25} = \frac{1}{5^2}$$

Then, the pattern is $a_n = \frac{1}{n^2}$, or

$a_n = n^{-2}$

53. Given: $\frac{1}{9}, \frac{1}{3}, 1, 3, 9, ..$

Notice the powers of 3:

$$a_1 = \frac{1}{9} = 3^{-2} = 3^{1-3}$$
$$a_2 = \frac{1}{3} = 3^{-1} = 3^{2-3}$$
$$a_3 = 1 = 3^0 = 3^{3-3}$$
$$a_4 = 3 = 3^1 = 3^{4-3}$$
$$a_5 = 9 = 3^2 = 3^{5-3}$$

Then, the pattern is $a_n = 3^{n-3}$

55. $\sum_{i=1}^{7}(3i - 5) = -2 + 1 + 4 + 7 + 10 + 13 + 16$

$= 49$

57. $1 + 8 + 27 + \cdots + 216 = \sum_{i=1}^{6} i^3$

$= \frac{6^2(6+1)^2}{4}$

$= 441$

59. $\sum_{i=3}^{10} 5i^2 = 5(3)^2 + 5(4)^2 + 5(5)^2 + 5(6)^2 +$

$5(7)^2 + 5(8)^2 + 5(9)^2 + 5(10)^2$

$= 45 + 80 + 125 + 180 +$

$245 + 320 + 405 + 500$

$= 1900$

61. $\sum_{i=1}^{6}(-3)(2)^i = (-3)(2)^1 + (-3)(2)^2 + (-3)(2)^3 +$

$(-3)(2)^4 + (-3)(2)^5 + (-3)(2)^6$

$= -6 - 12 - 24 - 48 - 96 - 192$

$= -378$

63. Notice that 19,683 is divisible by 3 and with some experimentation you can determine that, like the first three terms of the series, it is a power of 3 $(3^9 = 19{,}683)$.
Why do we care? Because this means that 9 is the upper index in the summation that begins with a lower index of 2:

$$9+27+81+\cdots+19{,}683=\sum_{i=2}^{9}3^i=29{,}520$$

65.
$$\sum_{i=1}^{5}(-3)^i=(-3)^1+(-3)^2+(-3)^3+(-3)^4+(-3)^5$$
$$=-3+9-27+81-243$$
$$=-183$$

67.
$$\sum_{i=3}^{8}i^2-3=3^2-3+4^2-3+5^2-3+6^2-3+7^2-3+8^2-3$$
$$=6+13+22+33+46+61$$
$$=181$$

69.
$$\sum_{i=100}^{103}\frac{2i}{25}=\frac{2(100)}{25}+\frac{2(101)}{25}+\frac{2(102)}{25}+\frac{2(103)}{25}$$
$$=\frac{200}{25}+\frac{202}{25}+\frac{204}{25}+\frac{206}{25}$$
$$=\frac{812}{25}$$
$$=32.48$$

71.
$$3+\frac{33}{10}+\frac{36}{10}+\cdots+30=\sum_{i=1}^{91}\frac{3i+27}{10}$$
$$=\frac{1}{10}\sum_{i=1}^{91}(3i+27)$$
$$=\frac{3}{10}\sum_{i=1}^{91}(i+9)$$
$$=\frac{3}{10}\left[\sum_{i=1}^{91}i+\sum_{i=1}^{91}9\right]$$
$$=\frac{3}{10}\left[\frac{91(92)}{2}+9(91)\right]$$
$$=1501.5$$

73.
$$\sum_{i=1}^{100}\left(\frac{1}{i+3}-\frac{1}{i+4}\right)$$
$$S_1=\left(\frac{1}{1+3}-\frac{1}{1+4}\right)=\frac{1}{4}-\frac{1}{5}=\frac{1}{4\cdot5}$$
$$S_2=\left(\frac{1}{1+3}-\frac{1}{1+4}\right)+\left(\frac{1}{2+3}-\frac{1}{2+4}\right)$$
$$=\left(\frac{1}{4}-\frac{1}{5}\right)+\left(\frac{1}{5}-\frac{1}{6}\right)$$
$$=\frac{1}{4}-\frac{1}{6}=\frac{2}{4\cdot6}$$
$$=\frac{2}{4(2+4)}$$
$$S_3=\left(\frac{1}{1+3}-\frac{1}{1+4}\right)+\left(\frac{1}{2+3}-\frac{1}{2+4}\right)+\left(\frac{1}{3+3}-\frac{1}{3+4}\right)$$
$$=\left(\frac{1}{4}-\frac{1}{5}\right)+\left(\frac{1}{5}-\frac{1}{6}\right)+\left(\frac{1}{6}-\frac{1}{7}\right)$$
$$=\frac{1}{4}-\frac{1}{7}=\frac{3}{4\cdot(3+4)}$$

Notice the pattern.

$$S_n=\frac{n}{4(n+4)}$$
$$\sum_{i=1}^{100}\left(\frac{1}{i+3}-\frac{1}{i+4}\right)=S_{100}=\frac{100}{4(100+4)}$$
$$=\frac{25}{104}$$

75.
$$\sum_{i=1}^{\infty}\left(2^i-2^{i-1}\right)=\sum_{i=1}^{\infty}2^{i-1}(2-1)$$
$$=\sum_{i=1}^{\infty}2^{i-1}$$
$$S_1=2^0=1$$
$$S_2=1+2=3=2^2-1$$
$$S_3=1+2+4=7=2^3-1$$
$$S_4=1+2+4+8=15=2^4-1$$
$$\ldots$$
$S_n=2^n-1$, which does not converge as $n\to\infty$.

77. $\sum_{i=1}^{49}\left(\frac{1}{2i}-\frac{1}{2i+2}\right)=\sum_{i=1}^{49}\left(\frac{2i+2-2i}{2i(2i+2)}\right)$

$=\sum_{i=1}^{49}\left(\frac{2}{4i^2+4i}\right)$

$=\sum_{i=1}^{49}\frac{1}{2i(i+1)}$

$S_1=\frac{1}{2(1+1)}$

$S_2=\frac{1}{2(1+1)}+\frac{1}{4(2+1)}=\frac{2}{8}+\frac{2}{24}=\frac{1}{3}=\frac{2}{2(2+1)}$

$S_3=\frac{1}{3}+\frac{1}{24}=\frac{3}{8}=\frac{3}{2(3+1)}$

$S_4=\frac{3}{8}+\frac{1}{40}=\frac{2}{5}=\frac{4}{2(4+1)}$

...

$S_n=\frac{n}{2(n+1)}$

$S_{49}=\frac{49}{2(49+1)}=\frac{49}{100}$

79. According to the division property of logarithms,

$\sum_{i=1}^{100}\ln\left(\frac{i}{i+1}\right)=\sum_{i=1}^{100}\left[\ln i-\ln(i+1)\right]$

$S_1=\ln 1-\ln 2=-\ln 2$

$S_2=\ln 1-\ln 2+\ln 2-\ln 3=\ln 1-\ln 3$

$=-\ln 3$

$S_3=\ln 1-\ln 4=-\ln 4$

...

$S_n=-\ln(n+1)$

$S_{100}=-\ln(101)$

81. $a_1=4$

$a_2=7$

$a_3=4+7=11$

$a_4=7+11=18$

$a_5=11+18=29$

83. $a_1=10$

$a_2=20$

$a_3=10+20=30$

$a_4=20+30=50$

$a_5=30+50=80$

85. $a_1=13$

$a_2=-17$

$a_3=13+(-17)=-4$

$a_4=-17+(-4)=-21$

$a_5=-4+(-21)=-25$

87. Given $a_n=a_{n-1}a_{n-2}$

$a_1=1$

$a_2=-3$

$a_3=(1)(-3)=-3$

$a_4=-3(-3)=9$

$a_5=-3(9)=-27$

End of Section 11.1

Chapter Eleven: Section 11.2
Solutions to Odd-numbered Exercises

1. $a_1 = -2, d = 3$

$a_n = -2 + 3(n-1) = -5 + 3n$

3. $a_1 = 7, d = -2$

$a_n = 7 + (-2)(n-1) = 9 - 2n$

5. $a_1 = 5, a_5 = 41$

$41 = 5 + d(5-1)$

$d = \frac{41-5}{4} = 9$

$a_n = 5 + 9(n-1) = 9n - 4$

7. $a_3 = -9, d = -6$

$-9 = a_1 + (-6)(2) \Rightarrow a_1 = 3$

$a_n = 3 + (-6)(n-1) = 9 - 6n$

9. $a_5 = 100, d = 19$

Find a_1:

$100 = a_1 + 19(4)$

$a_1 = 100 - 76 = 24$

Find a_n:

$a_n = 24 + 19(n-1) = 19n + 5$

11. $a_1 = \frac{7}{2}, d = 1$

Find a_n:

$a_n = \frac{7}{2} + (n-1) = n + \frac{5}{2}$

13. $a_1 = 12, a_3 = -7$

Find d:

$-7 = 12 + d(2)$

$d = -\frac{19}{2}$

Find a_n:

$a_n = 12 + \left(-\frac{19}{2}\right)(n-1)$

$a_n = 12 - \frac{19}{2}n + \frac{19}{2} = \frac{43}{2} - \frac{19}{2}n$

15. $a_1 = -1, a_6 = -11$

Find d:

$-11 = -1 + d(5)$

$d = -\frac{10}{5} = -2$

Find a_n:

$a_n = -1 - 2(n-1) = 1 - 2n$

17. $a_4 = 17, d = -4$

Find a_1:

$17 = a_1 + (-4)(3)$

$a_1 = 29$

Find $a_n = 29 + (-4)(n-1) = 33 - 4n$

19. Given: $a_1 = 1, d = 2$

$a_n = a_1 + (n-1)d$

$a_7 = 1 + (7-1)(2) = 13$

21. Given: $a_1 = 0, d = \frac{1}{3}$

$a_n = a_1 + (n-1)d$

$a_7 = 0 + (7-1)\left(\frac{1}{3}\right) = 2$

23. Given: $a_1 = 8, d = -1$

$a_n = a_1 + (n-1)d$

$a_7 = 8 + (7-1)(-1) = 2$

25. Given: $\{5n-3\}$, Use $d = a_n - a_{n-1}$.

$d = 5n - 3 - (5(n-1) - 3) = -3 + 8$

$d = 5$

27. Given: $\{n+6\}$; Use $d = a_n - a_{n-1}$.

$d = n + 6 - ((n-1) + 6) = 6 - 5$

$d = 1$

29. Given: $\{\sqrt{2}-2n\}$, Use $d = a_n - a_{n-1}$.

$$d = \sqrt{2}-2n-\left(\sqrt{2}-2(n-1)\right) = -(2)$$
$$d = -2$$

31. $a_1 = -3,\ a_5 = 5$

Find d:

$$5 = -3 + d(4)$$
$$d = 2$$

Find a_{100}:

$$a_{100} = -3 + 2(99) = 195$$

33. $a_1 = 1,\ d = \frac{1}{3},\ a_n = 25$

Find n:

$$25 = 1 + \frac{1}{3}(n-1)$$
$$75 = 3 + n - 1$$
$$n = 73$$
$$a_{73} = 25$$

35. $a_1 = -16,\ d = 7$

Find a_{20}:

$$a_{20} = -16 + 7(19)$$
$$a_{20} = 117$$

37. Given: 2, 5, 8, 11, $\cdots$

$$a_n = 2 + 3(n-1) = 3n - 1$$
$$a_9 = 3(9) - 1 = 26$$

39. Given: 16, 12, 8, 4, $\cdots$

$$a_n = a_1 + (n-1)d$$
$$a_n = 16 + (n-1)(-4)$$
$$a_7 = 16 + (7-1)(-4) = -8$$

41. Given: -2, 1, 4, 7, $\cdots$

$$a_n = a_1 + (n-1)d$$
$$a_n = -2 + (n-1)(3)$$
$$a_6 = -2 + (6-1)(3) = 13$$

43. Given: 5, 10, 15, 20, $\cdots$

$$a_n = a_1 + (n-1)d$$
$$a_n = 5 + (n-1)(5)$$
$$a_{11} = 5 + (11-1)(5) = 55$$

45. $\sum_{i=1}^{100}(3i-8) = S_{100}$

$$a_1 = -5,\ a_2 = -2,\ d = 3$$
$$a_{100} = -5 + 3(99) = 292$$
$$S_{100} = \left(\frac{100}{2}\right)(-5+292) = 14{,}350$$

47. $\sum_{i=5}^{90}(4i+9) = S_{90} - S_4$

$$a_1 = 13,\ a_2 = 17,\ d = 4$$
$$a_{90} = 13 + 4(89) = 369$$
$$S_{90} = \frac{90}{2}(13+369) = 17{,}190$$
$$a_4 = 13 + 4(3) = 25$$
$$S_4 = \frac{4}{2}(13+25) = 76$$
$$S_{90} - S_4 = 17{,}190 - 76 = 17{,}114$$

49. $a_1 = 25,\ d = -7$

Find n for $a_n = -143$:

$$-143 = 25 + (-7)(n-1)$$
$$7n = 175$$
$$n = 25$$
$$a_{25} = -143$$
$$25 + 18 + \ldots + (-143) = \frac{25}{2}(25 + (-143))$$
$$= \frac{25}{2}(-118) = -1475$$

51. $\sum_{i=1}^{37}\left(-\frac{3}{5}i-6\right)=S_{37}$

$a_1=-\frac{33}{5},\ a_2=-\frac{36}{5}\Rightarrow d=-\frac{3}{5}$

Find a_{37}:

$a_{37}=-\frac{33}{5}+\left(-\frac{3}{5}\right)(36)=-\frac{141}{5}$

$S_{37}=\frac{37}{2}\left(-\frac{33}{5}-\frac{141}{5}\right)=-\frac{3219}{5}$

53. $\sum_{i=2}^{42}(2i-22)=S_{42}-a_1$

$a_1=-20,\ a_2=-18\Rightarrow d=2$

Find a_{42}:

$a_{42}=-20+2(41)=62$

$S_{42}=\frac{42}{2}(-20+62)=882$

$\sum_{i=2}^{42}(2i-22)=882-(-20)=902$

55. $a_1=7,\ d=-4$

Find n for $a_n=-101$

$-101=7+(-4)(n-1)$

$n=\frac{112}{4}=28$

$a_{28}=-101$

$7+3+\ldots+(-101)=\frac{28}{2}(7+(-101))$

$=-1316$

57. $a_1=1000,\ d=1000$

Use $S_n=na_1+d\left(\frac{(n-1)n}{2}\right)$ and solve for n:

$21{,}000=n(1000)+1000\left(\frac{n^2-n}{2}\right)$

$21=n+\frac{n^2-n}{2}$

$n^2+n-42=0$

$(n+7)(n-6)=0$

$n=6$ years

59. $a_1=100,\ d=-2,\ n=20$

Find S_{20}: (First, find a_{20})

$a_{20}=100+(-2)(19)=62$

$S_{20}=\frac{20}{2}(100+62)=1620$ lbs

61. Let $n=$ number of rows (terms)

$a_1=5,\ d=1$

$S_n=290=na_1+d\left(\frac{(n-1)n}{2}\right)$

$290=5n+\frac{n^2-n}{2}$

Solve the quadratic (note that n must be positive):

$n^2+9n-580=0$

$(n+29)(n-20)=0$

Set $n-20=0$.

Then, $n=20$

The number of rows is 20.

End of Section 11.2

Chapter Eleven: Section 11.3
Solutions to Odd-numbered Exercises

1. $a_1 = -3, r = 2$

Use $a_n = a_1 r^{n-1}$

$a_n = -3(2)^{n-1}$

3. $a_1 = 2, r = -\frac{1}{3}$

Use $a_n = a_1 r^{n-1}$

$a_n = 2\left(-\frac{1}{3}\right)^{n-1}$

5. $a_2 = -\frac{1}{4}, a_5 = \frac{1}{256}$

Find r:

$\left(-\frac{1}{4}\right)(r)^3 = \frac{1}{256}$

$r^3 = -\frac{4}{256} = -\frac{1}{64}$

$r = -\frac{1}{4}$

Find a_1:

$a_2 = a_1\left(-\frac{1}{4}\right)$

$-\frac{1}{4} = a_1\left(-\frac{1}{4}\right)$

$a_1 = 1$

$a_n = \left(-\frac{1}{4}\right)^{n-1}$

7. $a_2 = \frac{1}{7}, r = \frac{1}{7}$

Find a_1:

$a_2 = a_1\left(\frac{1}{7}\right)$

$\frac{1}{7} = a_1\left(\frac{1}{7}\right) \Rightarrow a_1 = 1$

$a_n = \left(\frac{1}{7}\right)^{n-1}$

9. $a_3 = 9, a_5 = 81, r < 0$

Find r:

$a_5 = a_3 r^2$

$81 = 9r^2 \Rightarrow r^2 = 9 \Rightarrow r = -3$

Find a_1:

$a_3 = a_1 r^2$

$9 = a_1(-3)^2$

$a_1 = 1$

$a_n = (-3)^{n-1}$

11. $a_1 = 3, r = \frac{2}{3}$

$a_n = 3\left(\frac{2}{3}\right)^{n-1}$

13. $a_3 = 28, a_6 = -224$

Find r:

$a_6 = a_3 r^3$

$-224 = 28r^3$

$r^3 = -8 \Rightarrow r = -2$

Find a_1:

$a_3 = a_1 r^2$

$28 = a_1(-2)^2 \Rightarrow a_1 = 7$

$a_n = 7(-2)^{n-1}$

15. $a_5 = 1, a_6 = 2$

Find r:

$a_6 = a_5 r$

$2 = 1(r)$

$r = 2$

Find a_1:

$a_5 = a_1 r^4$

$1 = a_1(2)^4 \Rightarrow a_1 = \frac{1}{16}$

$a_n = \frac{1}{16}(2)^{n-1}$

17. $a_2 = \frac{13}{17}, r = \frac{4}{3}$

Find a_1:

$$a_2 = a_1(r)$$

$$\frac{13}{17} = a_1\left(\frac{4}{3}\right)$$

$$a_1 = \frac{39}{68}$$

$$a_n = \frac{39}{68}\left(\frac{4}{3}\right)^{n-1}$$

19. $a_2 = -\frac{5}{2}, a_5 = \frac{5}{16}$

Find r:

$$a_5 = a_2 r^3$$

$$\frac{5}{16} = -\frac{5}{2}r^3$$

$$r^3 = -\frac{1}{8}$$

$$r = -\frac{1}{2}$$

Find a_1:

$$a_2 = a_1 r$$

$$-\frac{5}{2} = a_1\left(-\frac{1}{2}\right)$$

$$a_1 = 5$$

$$a_n = 5\left(-\frac{1}{2}\right)^{n-1}$$

$$a_{15} = 5\left(-\frac{1}{2}\right)^{14}$$

$$a_{15} = \frac{5}{16,384}$$

21. $a_3 = -2, a_4 = -16$

Find r:

$$a_4 = a_3 r$$

$$-16 = -2r \Rightarrow r = 8$$

Find a_1:

$$a_3 = a_1 r^2$$

$$-2 = a_1(8)^2 \Rightarrow a_1 = -\frac{1}{32}$$

$$a_{13} = -\frac{1}{32}(8)^{12} = -2,147,483,648$$

23. Given that $S_{10} = \sum_{i=1}^{10} 3\left(-\frac{1}{2}\right)^i$ is a partial sum of a geometric sequence. In expanded form we have that

$$S_{10} = 3\left(-\frac{1}{2}\right)^1 + 3\left(-\frac{1}{2}\right)^2 + 3\left(-\frac{1}{2}\right)^3 + \cdots$$
$$3\left(-\frac{1}{2}\right)^{10}$$

$$= -\frac{3}{2} + \left(-\frac{3}{2}\right)\left(-\frac{1}{2}\right) + \left(-\frac{3}{2}\right)\left(-\frac{1}{2}\right)^2 + \cdots$$
$$\left(-\frac{3}{2}\right)\left(-\frac{1}{2}\right)^9$$

$$a_1 = -\frac{3}{2}, r = -\frac{1}{2}$$

$$S_{10} = \frac{-\frac{3}{2}\left(1 - \left(-\frac{1}{2}\right)^9\right)}{1 - \left(-\frac{1}{2}\right)}$$

$$= \frac{-\frac{3}{2}\left(\frac{1024}{1024} - \left(\frac{1}{1024}\right)\right)}{\frac{3}{2}}$$

$$= -\frac{1023}{1024}$$

25. Find $\sum_{i=10}^{40} 2^i = \sum_{i=1}^{40} 2^i - \sum_{i=1}^{9} 2^i$

$$= S_{40} - S_9$$

Expand to find a_1 and r:

$$\sum_{i=1}^{40} 2^i = 2 + 2(2) + 2(2)^2 + 2(2)^3 + \cdots + 2(2)^{39}$$

$$a_1 = 2, r = 2$$

$$S_{40} = \frac{2(1-2^{40})}{1-2}$$

$$S_9 = \frac{2(1-2^9)}{1-2}$$

$$S_{40} - S_9 = -2(1 - 2^{40} + 2^9 - 1)$$

$$= -2(-2^9(2^{31} - 1))$$

$$= 2^{10}(2^{31} - 1)$$

$$= 2,199,023,254,528$$

27. Given $2+6+\cdots+39{,}366$

$a_1=2,\ r=3$

Find n:

$$a_n=a_1r^{n-1}$$
$$39{,}366=2(3)^{n-1}$$
$$3^{n-1}=19{,}683$$
$$3^9=19{,}683$$
$$n-1=9$$
$$n=10$$
$$S_{10}=\frac{2\left(1-3^{10}\right)}{1-3}=59{,}048$$

29. Given $1-3+\cdots+59{,}049$

$a_1=1,\ r=-3$

Find n for $a_n=59{,}049$:

$$a_n=a_1r^{n-1}$$
$$59{,}049=(1)(-3)^{n-1}$$
$$(-3)^{n-1}=59{,}049$$
$$(-3)^{10}=59{,}049\Rightarrow n-1=10\Rightarrow n=11$$
$$S_{11}=\frac{(1)\left(1-(-3)^{11}\right)}{1-(-3)}=\frac{177{,}148}{4}=44{,}287$$

31. Given $1+\frac{3}{5}+\cdots+\frac{243}{3125}$

$a_1=1,\ r=\frac{3}{5}$

Find n for $a_n=\frac{243}{3125}$:

$$a_n=a_1r^{n-1}$$
$$\frac{243}{3125}=\left(\frac{3}{5}\right)^{n-1}$$
$$\left(\frac{3}{5}\right)^5=\frac{243}{3125}$$
$$n-1=5\Rightarrow n=6$$
$$S_6=\frac{(1)\left(1-\left(\frac{3}{5}\right)^6\right)}{1-\frac{3}{5}}=\frac{7448}{3125}=2.38336$$

33. Given the series

$$\sum_{i=1}^{\infty}\left(\frac{4}{5}\right)^i=\frac{4}{5}+\frac{16}{25}+\cdots+\left(\frac{4}{5}\right)^n+\cdots$$

$a_1=\frac{4}{5},\ r=\frac{4}{5}$

Since $|r|<1$, the infinite geometric series converges:

$$\sum_{i=1}^{\infty}\left(\frac{4}{5}\right)^i=\frac{\frac{4}{5}}{1-\frac{4}{5}}=\frac{4}{5}\cdot 5=4$$

35. Given the series

$$\sum_{i=0}^{\infty}\left(-\frac{8}{9}\right)^i=1+(1)\left(-\frac{8}{9}\right)+(1)\left(-\frac{8}{9}\right)^2+\cdots$$

$a_1=1,\ r=-\frac{8}{9}$

Since $|r|<1$, the infinite geometric series converges:

$$\sum_{i=0}^{\infty}\left(-\frac{8}{9}\right)^i=\frac{1}{1-\left(-\frac{8}{9}\right)}=\frac{1}{\frac{17}{9}}=\frac{9}{17}$$

37. Given the series

$$\sum_{i=0}^{\infty}(-1)^i=1-1+1+\cdots$$

$a_1=1,\ r=-1$

Since $|r|\geq 1$, the infinite geometric series does not converge:

39. Given the series

$$\sum_{i=0}^{\infty}5\left(\frac{6}{11}\right)^i=5+5\left(\frac{6}{11}\right)+5\left(\frac{6}{11}\right)^2+\cdots$$

$a_1=5,\ r=\frac{6}{11}$

Since $|r|<1$, the infinite geometric series converges:

$$\sum_{i=0}^{\infty}\left(\frac{6}{11}\right)^i=\frac{5}{1-\frac{6}{11}}=5\left(\frac{11}{5}\right)=11$$

41. $1.\overline{65} = 1 + \frac{65}{100} + \frac{65}{10{,}000} + \frac{65}{100{,}000} + \cdots$

$$= 1 + 65\left(\frac{1}{100}\right) + 65\left(\frac{1}{100}\right)^2 + \cdots$$

$$= 1 + \sum_{i=1}^{\infty} 65\left(\frac{1}{100}\right)^i$$

$$a_1 = \frac{65}{100},\ r = \frac{1}{100}$$

Then, $1.\overline{65} = 1 + \dfrac{\frac{65}{100}}{1 - \frac{1}{100}} = 1\frac{65}{99}$

$$= \frac{164}{99}$$

43. $-0.\overline{5} = -\left(\frac{5}{10} + \frac{5}{100} + \frac{5}{1000} + \cdots\right)$

$$= -\left[5\left(\frac{1}{10}\right) + 5\left(\frac{1}{100}\right) + 5\left(\frac{1}{1000}\right) + \cdots\right]$$

$$= -\sum_{i=1}^{\infty} 5\left(\frac{1}{10}\right)^i$$

$$a_1 = \frac{5}{10},\ r = \frac{1}{10}$$

Then, $-0.\overline{5} = -\dfrac{\frac{5}{10}}{1 - \frac{1}{10}} = -\frac{5}{10} \cdot \frac{10}{9} = -\frac{5}{9}$

45. $0.\overline{029} = \frac{29}{1000} + \frac{29}{1{,}000{,}000} + \frac{29}{1{,}000{,}000{,}000} + \cdots$

$$= 29\left(\frac{1}{1000}\right) + 29\left(\frac{1}{1000000}\right) + \cdots$$

$$= \sum_{i=1}^{\infty} 29\left(\frac{1}{1000}\right)^i$$

$$a_1 = \frac{29}{1000},\ r = \frac{1}{1000}$$

$$0.\overline{029} = \frac{\frac{29}{1000}}{1 - \frac{1}{1000}} = \frac{29}{1000} \cdot \frac{1000}{999}$$

$$= \frac{29}{999}$$

47. Isolate the initial drop of 10 ft. in the infinite series of rebounds and drops from the subsequent rebounds and drops. Then, notice that the rebound and next "drop" are the same distance. Form the infinite series from this pairing:

$10 + 10(0.8)2 + 10(0.8)^2(2) + 10(0.8)^3(2) + 10(0.8)^4(2) + \cdots$

$$= 10 + 20\left[0.8 + 0.8^2 + 0.8^3 + \cdots\right]$$

$$= 10 + 20\left[0.8 + 0.8(0.8) + 0.8(0.8)^2 + \cdots\right]$$

$$= 10 + 20\sum_{i=1}^{\infty} 0.8(0.8)^{i-1}$$

$$a_1 = 0.8,\ r = 0.8$$

$$10 + 20\sum_{i=1}^{9} 0.8(0.8)^{i-1} =$$

$$10 + 20\left(\frac{0.8\left(1-(0.8)^9\right)}{1-0.8}\right) \approx$$

$$10 + 20\left(\frac{0.6926258}{0.2}\right) \approx$$

$$10 + 20(3.46313) \approx 79.26 \text{ ft.}$$

For the "theoretical" infinite bounce, we have that

$$10 + 20\sum_{i=1}^{\infty} 0.8(0.8)^{i-1} = 10 + 20\left(\frac{0.8}{1-0.8}\right)$$

$$= 10 + 20(4)$$

$$= 90 \text{ ft.}$$

49. a_1 = amount after 1 month.

Monthly interest rate $= \frac{0.04}{12} = \frac{1}{300}$.

The amount $a_1 = 10{,}000\left(1+\frac{1}{300}\right)$

$$= 10{,}000\left(\frac{301}{300}\right)$$

$$a_2 = 10{,}000\left(\frac{301}{300}\right)^2$$

$$\cdots$$

$$a_n = 10{,}000\left(\frac{301}{300}\right)^n$$

where n = the number of months

Ten years = 120 months

$$a_{120} = 10{,}000\left(\frac{301}{300}\right)^{120} = \$14{,}908.33$$

51. To describe the amount of shading, notice that, after the first step of shading, exactly $\frac{1}{4}$ of the remaining area is shaded at each subsequent step. The amount of shaded space is described by the infinite geometric series

$$S = \frac{1}{2} + \frac{1}{2}\left(\frac{1}{4}\right) + \frac{1}{2}\left(\frac{1}{4}\right)\left(\frac{1}{4}\right) + \cdots$$

$$= \sum_{i=1}^{\infty} \frac{1}{2}\left(\frac{1}{4}\right)^{i-1}$$

$$a_1 = \frac{1}{2},\ r = \frac{1}{4}$$

$$\sum_{i=1}^{\infty} \frac{1}{2}\left(\frac{1}{4}\right)^{i-1} = \frac{\frac{1}{2}}{1-\frac{1}{4}} = \frac{1}{2}\cdot\frac{4}{3}$$

$$= \frac{2}{3}$$

End of Section 11.3

Chapter Eleven: Section 11.4
Solutions to Odd-numbered Exercises

1. Given: $S_k = \dfrac{1}{3(k+2)}$. Find S_{k+1}:

$$S_{k+1} = \frac{1}{3((k+1)+2)} = \frac{1}{3(k+3)}$$

$$S_{k+1} = \frac{1}{3k+9}$$

3. Given: $S_k = \dfrac{k(k+1)(2k+1)}{4}$. Find S_{k+1}:

$$S_{k+1} = \frac{(k+1)(k+1+1)(2(k+1)+1)}{4}$$

$$S_{k+1} = \frac{(k+1)(k+2)(2k+3)}{4}$$

For Exercises 5 - 41, use the Principle of Mathematical Induction to prove the given statements.

5. Basis Step. Show $P(1)$ is true:

For $n=1$, $1=1$ and $\dfrac{1(1+1)}{2}=1$

Induction Step: Assume $P(k)$ is true.

Show that $P(k+1)$ is true.

Assume

$$1+2+3+\cdots+k = \frac{k(k+1)}{2}$$

Then,

$$1+2+3+\cdots+k+(k+1) = \frac{k(k+1)}{2}+(k+1)$$
$$= \frac{k^2+k+2k+2}{2}$$
$$= \frac{k^2+3k+2}{2}$$
$$= \frac{(k+1)(k+2)}{2}$$

Therefore, $P(k+1)$ is true.

7. Basis Step. Show $P(1)$ is true:

For $n=1$, $2(1)=2$ and $1(1+1)=2$

Induction Step: Assume $P(k)$ is true.

Show that $P(k+1)$ is true.

Assume

$$2+4+6+\cdots+2k = k(k+1)$$

Then,

$$2+4+6+\cdots+2k+2(k+1) =$$
$$k(k+1)+2(k+1) =$$
$$k^2+k+2k+2 =$$
$$k^2+3k+2 = (k+1)(k+2)$$

Therefore, $P(k+1)$ is true.

9. Basis Step. Show $P(1)$ is true:

For $n=1$, $4^{1-1}=1$ and $\dfrac{4^1-1}{3}=1$

Induction Step: Assume $P(k)$ is true.

Show that $P(k+1)$ is true.

Assume

$$4^0+4^1+4^2+\cdots+4^{k-1} = \frac{4^k-1}{3}$$

Then,

$$4^0+4^1+4^2+\cdots+4^{k-1}+4^{k+1-1} =$$
$$\frac{4^k-1}{3}+4^k =$$
$$\frac{4^k-1+3(4^k)}{3} =$$
$$\frac{4(4^k)-1}{3} = \frac{4^{k+1}-1}{3}$$

Therefore, $P(k+1)$ is true.

11. Basis Step. Show $P(1)$ is true:
For $n=1$,
$$\frac{1}{[3(1)-2][3(1)+1]}=\frac{1}{4} \text{ and } \frac{1}{3(1)+1}=\frac{1}{4}$$
Induction Step: Assume $P(k)$ is true.
Show that $P(k+1)$ is true.
Assume
$$\frac{1}{1\cdot 4}+\frac{1}{4\cdot 7}+\frac{1}{7\cdot 10}+\cdots+\frac{1}{(3k-2)(3k+1)}=$$
$\frac{k}{3k+1}$. Then,
$$\frac{1}{1\cdot 4}+\frac{1}{4\cdot 7}+\frac{1}{7\cdot 10}+\cdots+\frac{1}{(3k-2)(3k+1)}+$$
$$\frac{1}{(3(k+1)-2)(3(k+1)+1)}=$$
$$\left[\frac{1}{1\cdot 4}+\frac{1}{4\cdot 7}+\frac{1}{7\cdot 10}+\cdots+\frac{1}{(3k-2)(3k+1)}\right]+$$
$$\frac{1}{(3k+1)(3k+4)}$$
To simplify, substitute the $P(k)$ assumption in the bracketed expression. Add the resulting terms and factor the denominator. The final step is to re-write the denominator in the form needed to confirm $P(k+1)$:
$$\frac{k}{3k+1}+\frac{1}{(3k+1)(3k+4)}=$$
$$\frac{3k^2+4k+1}{(3k+1)(3k+4)}=\frac{(3k+1)(k+1)}{(3k+1)(3k+4)}=$$
$$\frac{k+1}{(3(k+1)+1)}$$
Therefore, $P(k+1)$ is true.

13. Basis Step. Show $P(1)$ is true:
For $n=1$, $5(1)=5$ and $\frac{5(1)(1+1)}{2}=5$
Induction Step: Assume $P(k)$ is true.
Show that $P(k+1)$ is true.
Assume
$$5+10+15+\cdots+5k=\frac{5k(k+1)}{2}.$$
Then,
$$5+10+15+\cdots+5k+5(k+1)=$$
$$(5+10+15+\cdots+5k)+5k+5=$$
$$\frac{5k(k+1)}{2}+5k+5=$$
$$\frac{5k^2+15k+10}{2}=$$
$$\frac{5(k+1)(k+2)}{2}=$$
$$\frac{5(k+1)[(k+1)+1]}{2}$$
Therefore, $P(k+1)$ is true.

15. Basis Step. Show $P(1)$ is true:
For $n=1$, $1+\frac{1}{1}=2$ and $1+1=2$
Induction Step: Assume $P(k)$ is true.
Show that $P(k+1)$ is true.
Assume
$$\left(1+\frac{1}{1}\right)\left(1+\frac{1}{2}\right)\left(1+\frac{1}{3}\right)\cdots\left(1+\frac{1}{k}\right)=k+1.$$
Then,
$$\left(1+\frac{1}{1}\right)\left(1+\frac{1}{2}\right)\left(1+\frac{1}{3}\right)\cdots\left(1+\frac{1}{k}\right)\left(1+\frac{1}{k+1}\right)=$$
$$(k+1)\left(1+\frac{1}{k+1}\right)=$$
$$k+1+\frac{k+1}{k+1}=$$
$$(k+1)+1$$
Therefore, $P(k+1)$ is true.

17. Basis Step. Show $P(1)$ is true:

For $n = 4$, $4! = 24$ and $2^4 = 16$.

So, $4! > 2^4$

Induction Step: Assume $P(k)$ is true.

Show that $P(k+1)$ is true.

Assume

$k! > 2^k$

Then,

$(k+1)! = k!(k+1) > 2^k(k+1)$

Then, since $k \ge 4$,

$2^k(k+1) \ge 2^k(4+1) > 2^k \cdot 2 = 2^{k+1}$

Therefore, $P(k+1)$ is true.

19. Basis Step. Show $P(1)$ is true:

For $n = 1$, $-(1+1) = -2$ and

$-\frac{1}{2}(1^2 + 3(1)) = -2$

Induction Step: Assume $P(k)$ is true.

Show that $P(k+1)$ is true.

Assume

$$-2 - 3 - 4 \cdots - (k+1) = -\frac{1}{2}(k^2 + 3k)$$

Then,

$$-2 - 3 - 4 \cdots - (k+1) - ((k+1)+1) =$$
$$-\frac{1}{2}(k^2 + 3k) - (k+2) =$$
$$-\frac{k^2 + 3k + 2k + 4}{2} =$$
$$-\frac{(k^2 + 2k + 1) + (3k + 3)}{2} =$$
$$-\frac{1}{2}\left((k+1)^2 + 3(k+1)\right)$$

Therefore, $P(k+1)$ is true.

21. Basis Step. Show $P(1)$ is true:

For $n = 1$, since $2^1 = 2$, $2^1 > 1$.

Induction Step: Assume $P(k)$ is true.

Show that $P(k+1)$ is true.

Assume

$2^k > k$ Then,

$2^{k+1} = 2^1 \cdot 2^k > 2k = k + k \ge k + 1$

Therefore, $P(k+1)$ is true.

23. Basis Step. Show $P(1)$ is true:

For $n = 1$, $1(1+1) = 2$ and

$\frac{1(1+1)(1+2)}{3} = 2$

Induction Step: Assume $P(k)$ is true.

Show that $P(k+1)$ is true.

Assume

$1 \cdot 2 + 2 \cdot 3 + 3 \cdot 4 + \cdots + k(k+1) =$

$\frac{k(k+1)(k+2)}{3}$. Then,

$$[1 \cdot 2 + 2 \cdot 3 + 3 \cdot 4 + \cdots + k(k+1)] +$$
$$(k+1)((k+1)+1) =$$
$$\frac{k(k+1)(k+2)}{3} + (k+1)(k+2) =$$
$$\frac{k(k+1)(k+2) + 3(k+1)(k+2)}{3} =$$
$$\frac{(k+1)(k+2)(k+3)}{3} =$$
$$\frac{(k+1)((k+1)+1)((k+1)+2)}{3}$$

Therefore, $P(k+1)$ is true.

25. Basis Step. Show $P(1)$ is true:

For $n = 5$, $2^5 = 32$ and $4 \cdot 5 = 20$.

So, $2^5 > 4 \cdot 5$.

Induction Step: Assume $P(k)$ is true.

Show that $P(k+1)$ is true.

Assume $2^k > 4k$.

Then,

$2^{k+1} = 2 \cdot 2^k = 2^k + 2^k > 4k + 4k$

Then, $4k + 4k > 4k + 4 = 4(k+1)$

Therefore, $P(k+1)$ is true.

27. Basis Step. Show $P(1)$ is true:

For $n=1$, $1^5=1$ and

$$\frac{1^2(1+1)^2\left(2(1)^2+2(1)-1\right)}{12}=1$$

Induction Step: Assume $P(k)$ is true.

Show that $P(k+1)$ is true.

Assume

$1^5+2^5+3^5+\cdots+k^5=$

$\frac{k^2(k+1)^2(2k^2+2k-1)}{12}$. Then,

$$1^5+2^5+3^5+\cdots+k^5+(k+1)^5=$$

$$\frac{k^2(k+1)^2(2k^2+2k-1)}{12}+(k+1)^5=$$

$$\frac{(k+1)^2\left[k^2(2k^2+2k-1)+12(k+1)^3\right]}{12}=$$

$$\frac{(k+1)^2\left[2k^4+14k^3+35k^2+36k+12\right]}{12}=$$

$$\frac{(k+1)^2(k+2)^2(2k^2+6k+3)}{12}=$$

$$\frac{(k+1)^2(k+2)^2\left(2(k+1)^2+2(k+1)-1\right)}{12}$$

Therefore, $P(k+1)$ is true.

29. Basis Step. Show $P(1)$ is true:

For $n=1$, $2(1)-1=1$ and $1^2=1$

Induction Step: Assume $P(k)$ is true.

Show that $P(k+1)$ is true.

Assume

$1+3+5+7+\cdots+2k-1=k^2$

Then,

$$(1+3+5+7+\cdots+2k-1)+(2(k+1)-1)=$$

$$k^2+(2k+1)=$$

$$k^2+2k+1=$$

$$(k+1)^2$$

Therefore, $P(k+1)$ is true.

31. Basis Step. Show $P(1)$ is true:

For $n=1$, $\left(a^m\right)^1=a^m$ and $a^{m\cdot 1}=a^m$

Induction Step: Assume $P(k)$ is true.

Show that $P(k+1)$ is true.

Assume

$\left(a^m\right)^k=a^{mk}$

Then,

$$\left(a^m\right)^{k+1}=\left(a^m\right)^k\left(a^m\right)^1$$

$$=a^{mk}\cdot a^m$$

$$=a^{mk+m}\cdot a^{m(k+1)}$$

Therefore, $P(k+1)$ is true.

33. Note that to say that 5 is a factor of N means that there is some integer p such that $N=5p$.

Basis Step. Show $P(1)$ is true:

For $n=1$, $\left(2^{2(1)-1}+3^{2(1)-1}\right)=5$ and 5 is a factor of 5.

Induction Step: Assume $P(k)$ is true.

Show that $P(k+1)$ is true.

Assume

$\left(2^{2(k)-1}+3^{2(k)-1}\right)=5p$, for some integer p.

Then,

$$\left(2^{2(k+1)-1}+3^{2(k+1)-1}\right)=\left(2^{2k+1}+3^{2k+1}\right)$$

$$=2^2\cdot 2^{2k-1}+3^2\cdot 3^{2k-1}$$

$$=4\cdot 2^{2k-1}+9\cdot 3^{2k-1}$$

$$=4\cdot 2^{2k-1}+(4+5)\cdot 3^{2k-1}$$

$$=4\cdot 2^{2k-1}+4\cdot 3^{2k-1}+5\cdot 3^{2k-1}$$

$$=4\left(2^{2k-1}+3^{2k-1}\right)+5\cdot 3^{2k-1}$$

$$=4(5p)+5\cdot 3^{2k-1}$$

$$=5\left(4p+3^{2k-1}\right)$$

Therefore, 5 is a factor and $P(k+1)$ is true.

35. Note that to say that 3 is a factor of n^3+3n^2+2n means that there is some integer p such that $n^3+3n^2+2n=3p$.
Basis Step. Show $P(1)$ is true:
For $n=1$, $\left(1^3+3(1)^2+2(1)\right)=6=3\cdot 2$
so 3 is a factor.
Induction Step: Assume $P(k)$ is true.
Show that $P(k+1)$ is true.
Assume
$\left(k^3+3k^2+2k\right)=3p$, for some integer p.
Then,

$$\begin{aligned}(k+1)^3+3(k+1)^2+2(k+1)&=\\ k^3+6k^2+11k+6&=\\ \left(k^3+3k^2+2k\right)+\left(3k^2+9k+6\right)&=\\ 3p+3\left(k^2+3k+2\right)&=\\ 3\left(p+k^2+3k+2\right)&\end{aligned}$$

Therefore, 3 is a factor and $P(k+1)$ is true.

37. Note that to say that 4 divides 5^n-1 means that there is some integer p such that $5^n-1=4p$.
Basis Step. Show $P(1)$ is true:
For $n=1$, $5^1-1=4=4\cdot 1$
so 4 divides 5^1-1.
Induction Step: Assume $P(k)$ is true.
Show that $P(k+1)$ is true.
Assume $5^k-1=4p$, for some integer p.
Then,

$$\begin{aligned}5^{k+1}-1&=5\cdot 5^k-5+4\\ &=5\left(5^k-1\right)+4\\ &=5(4p)+4\\ &=4(5p+1)\end{aligned}$$

Therefore, 4 divides $5^{k+1}-1$ and $P(k+1)$ is true.

39. The problem is to prove that $1+2+4+8+\cdots+2^{n-1}=2^n-1$, for all $n\geq 1$, by mathematical induction.
Basis Step. Show $P(1)$ is true:
For $n=1$, $2^{1-1}=1$ and $2^1-1=1$.
Induction Step: Assume $P(k)$ is true.
Show that $P(k+1)$ is true.
Assume $1+2+4+8+\cdots+2^{k-1}=2^k-1$
Then,

$$\begin{aligned}\left(1+2+4+8+\cdots+2^{k-1}\right)+2^{k+1-1}&=\\ \left(2^k-1\right)+2^k&=\\ 2\cdot 2^k-1&=2^{k+1}-1\end{aligned}$$

Therefore, $P(k+1)$ is true.

41. The hint suggests three cases to work through using mathematical induction. The three cases are the sets of values for t and n needed for k to be greater than or equal to 4:
Given that $k=2t+5n$:
Case 1: $\left[t\geq 1, n\geq 1, \text{ and } k=2t+5n\right]$

$$\begin{aligned}k+1&=2t+5n+1\\ &=2t+5n+(6-5)\\ &=(2t+6)+(5n-5)\\ &=2(t+3)+5(n-1)\end{aligned}$$

Case 2: $\left[t=0, n\geq 1, \text{ and } k=5n\right]$

$$\begin{aligned}k+1&=5n+1\\ &=5n+(6-5)\\ &=6+(5n-5)\\ &=2(3)+5(n-1)\end{aligned}$$

Case 3: $\left[t\geq 2, n=0, \text{ and } k=2t\right]$

$$\begin{aligned}k+1&=2t+1\\ &=2t+(5-4)\\ &=(2t-4)+5\\ &=2(t-2)+5(1)\end{aligned}$$

End of Section 11.4

Chapter Eleven: Section 11.5 Solutions to Odd-numbered Exercises

1. For each of the 3 missing digits, there are 10 possibilities:
$10 \cdot 10 \cdot 10 = 1000$ numbers

3. By eliminating the digit 9, there are $9^7 = 4,782,969$ such telephone numbers (This allows a zero in the first place).

5. $15! = 1.308 \times 10^{12}$

7. There is only one possibility for the first position, (k), three for the second (since k is already used), then two possibilities for the third, and one for the fourth:
$1 \cdot 3 \cdot 2 \cdot 1 = 3! = 6$

9. There are 5 ways to answer each of the ten items, or $5^{10} = 9,765,625$ ways in all.

11. $(26+10)^6 = 36^6 = 2,176,782,336$
(Use a calculator!)

13. $26 \cdot 25 \cdot 24 \cdot 10 \cdot 9 \cdot 8 = 11,232,000$

15. Number of permutations of 30 books, choosing 12 to arrange (permute) in a row: ${}_{30}P_{12} = 4.143 \times 10^{16}$.

17. There are 36 symbols to choose without repetition for 8 ordered positions: (Use a calculator!)
${}_{36}P_8 \approx 1.220 \times 10^{12}$

19. For 6 chairs in the room:
${}_7P_6 = 5,040$
For 7 chairs in the room:
${}_7P_7 = 5,040$
Having the 7th child stand is mathematically equivalent to having the 7th chair in the first case.

21. ${}_{26}P_3 = 15,600$ three-letter words

23. ${}_{15}P_2 = \dfrac{15!}{(15-2)!} = \dfrac{15!}{13!} = 210$

25. ${}_{19}P_{17} = \dfrac{19!}{(19-17)!} = \dfrac{19!}{2!}$
$\approx 60,822,550,204,416,000$

27. ${}_4C_2 = \dfrac{4!}{(4-2)!2!} = \dfrac{4!}{2!2!} = 6$

29. ${}_{21}C_{14} = \dfrac{21!}{(21-14)!14!} = \dfrac{21!}{7!14!} = 116,280$

31. This situation calls for a combination (different orderings does not apply):
${}_7C_3 = \dfrac{7!}{4!3!} = 7 \cdot 5 = 35$

33. Select 2 points from 9:
${}_9C_2 = \dfrac{9!}{7!2!} = 36$

35. ${}_{75}C_5 = 17,259,390$
(Use a calculator!)

37. There are 6 characters, but three of them are A's and two of them are N's, so we have,
$\dfrac{6!}{3!2!} = \dfrac{6 \cdot 5 \cdot 4}{2} = 60$

39. Number possible for each:

4 physics: ${}_{10}C_4 = \dfrac{10!}{6!4!} = 210$

4 computer science: ${}_{8}C_4 = \dfrac{8!}{4!4!} = 70$

4 mathematics: ${}_{13}C_4 = \dfrac{13!}{9!4!} = 715$

Total: Using the Counting Principle,
$210 \cdot 70 \cdot 715 = 10{,}510{,}500$

41. Use the multiplication principle of counting:
$4(28) = 112$ different single-scoop ice cream cones.

43. Use the multiplication principle of counting: $8(4)(3) = 96$ outfits.

45. Use the multiplication principle of counting:
$3(2)(4)(3)(4) = 288$ course schedules.

47. Use the multiplication principle of counting:
$5(4)(3)(2)(1) = 120$ 5-letter strings.

49. Use the multiplication principle of counting:
$25(24)(23)(22) = 303{,}600$ ways.

51. ${}_{12}C_4 = \dfrac{12!}{(12-4)!4!} = \dfrac{12!}{8!4!} = 495$ pizzas

53. ${}_{52}C_8 = \dfrac{52!}{(52-8)!8!} = \dfrac{52!}{44!8!}$
$= 752{,}538{,}150$ groups

55. $\dfrac{{}_7P_7}{2!1!1!3!} = \dfrac{7!}{2!3!} = 420$ ways

57. Expand $(x-2y)^7$:

$$(x-2y)^7 = \binom{7}{0}(x)^7(-2y)^0 + \binom{7}{1}(x)^6(-2y)^1 + \binom{7}{2}(x)^5(-2y)^2 + \binom{7}{3}(x)^4(-2y)^3 + \binom{7}{4}(x)^3(-2y)^4 + \binom{7}{5}(x)^2(-2y)^5 + \binom{7}{6}(x)^1(-2y)^6 + \binom{7}{7}(x)^0(-2y)^7$$

$$= x^7 - 14x^6y + 84x^5y^2 - 280x^4y^3 + 560x^3y^4 - 672x^2y^5 + 448xy^6 - 128y^7$$

59. Expand $(x^2-y^3)^4$:

$$(x^2-y^3)^4 = \binom{4}{0}(x^2)^4(-y^3)^0 + \binom{4}{1}(x^2)^3(-y^3)^1 + \binom{4}{2}(x^2)^2(-y^3)^2 + \binom{4}{3}(x^2)^1(-y^3)^3 + \binom{4}{4}(x^2)^0(-y^3)^4$$

$$= x^8 - 4x^6y^3 + 6x^4y^6 - 4x^2y^9 + y^{12}$$

61. Expand $(4x+5y^2)^6$:

$$(4x+5y^2)^6 = \binom{6}{0}(4x)^6(5y^2)^0 + \binom{6}{1}(4x)^5(5y^2)^1 + \binom{6}{2}(4x)^4(5y^2)^2 + \binom{6}{3}(4x)^3(5y^2)^3 + \binom{6}{4}(4x)^2(5y^2)^4 + \binom{6}{5}(4x)^1(5y^2)^5 + \binom{6}{6}(4x)^0(5y^2)^6 =$$

$$4096x^6 + 30{,}720x^5y^2 + 96{,}000x^4y^4 + 160{,}000x^3y^6 + 150{,}000x^2y^8 + 75{,}000xy^{10} + 15{,}625y^{12}$$

63. Expand $\left(x^3-y^2\right)^5$:

$$\left(x^3-y^2\right)^5=\binom{5}{0}\left(x^3\right)^5\left(-y^2\right)^0+\binom{5}{1}\left(x^3\right)^4\left(-y^2\right)^1+$$
$$\binom{5}{2}\left(x^3\right)^3\left(-y^2\right)^2+\binom{5}{3}\left(x^3\right)^2\left(-y^2\right)^3+$$
$$\binom{5}{4}\left(x^3\right)^1\left(-y^2\right)^4+\binom{5}{5}\left(x^3\right)^0\left(-y^2\right)^5$$
$$=x^{15}-5x^{12}y^2+10x^9y^4-$$
$$10x^6y^6+5x^3y^8-y^{10}$$

65. Expand $(a-2b+c)^3$. The terms of the expansion contain a^3, a^2b, ab^2, b^3, a^2c, abc, b^2c, ac^2, bc^2 and c^3:

$$(a-2b+c)^3=$$
$$\binom{3}{3\ 0\ 0}a^3+\binom{3}{2\ 1\ 0}a^2(-2b)+$$
$$\binom{3}{1\ 2\ 0}a(-2b)^2+\binom{3}{0\ 3\ 0}(-2b)^3+$$
$$\binom{3}{2\ 0\ 1}a^2c+\binom{3}{1\ 1\ 1}a(-2b)c+$$
$$\binom{3}{0\ 2\ 1}(-2b)^2c+\binom{3}{1\ 0\ 2}ac^2+$$
$$\binom{3}{0\ 1\ 2}(-2b)c^2+\binom{3}{0\ 0\ 3}c^3=$$
$$a^3-6a^2b+12ab^2-8b^3+3a^2c$$
$$-12abc+12b^2c+3ac^2-6bc^2+c^3$$

67. Expand $(2x+3y-z)^3$. The terms of the expansion contain x^3, x^2y, x^2z, xy^2, xz^2, xyz, y^3, y^2z, yz^2 and z^3:

$$(2x+3y-z)^3=$$
$$\binom{3}{3\ 0\ 0}(2x)^3+\binom{3}{2\ 1\ 0}(2x)^2(3y)^1+$$
$$\binom{3}{2\ 0\ 1}(2x)^2(-z)^1+\binom{3}{1\ 2\ 0}(2x)^1(3y)^2+$$
$$\binom{3}{1\ 1\ 1}(2x)^1(3y)^1(-z)^1+$$
$$\binom{3}{1\ 0\ 2}(2x)^1(-z)^2+$$
$$\binom{3}{0\ 3\ 0}(3y)^3+\binom{3}{0\ 2\ 1}(3y)^2(-z)^1+$$
$$\binom{3}{0\ 1\ 2}(3y)^1(-z)^2+\binom{3}{0\ 0\ 3}(-z)^3=$$
$$8x^3+36x^2y-12x^2z+54xy^2-36xyz+$$
$$6xz^2+27y^3-27y^2z+9yz^2-z^3$$

69. Coefficient of x^4y^3 in the expansion of $\left(x^2-2y\right)^5$: Base the combinatorial coefficient on $\left(x^2\right)^2=x^4$. Then,

$$\binom{5}{2}\left(x^2\right)^2(-2y)^3=\binom{5}{2}(-8)x^4y^3$$
$$=-80x^4y^3$$

The coefficient is -80.

71. The first three terms in the expansion of $(2x+3)^{13}$:

$$\binom{13}{13}(2x)^{13}(3)^0+\binom{13}{12}(2x)^{12}(3)^1+$$
$$\binom{13}{11}(2x)^{11}(3)^2+\cdots \text{ Simplifying ...}$$
$$\binom{13}{13}(2)^{13}x^{13}=8192x^{13}$$
$$\binom{13}{12}(2)^{12}(3)^1x^{12}=159{,}744x^{12}$$
$$\binom{13}{11}(2)^{11}(3)^2x^{11}=1{,}437{,}696x^{11}$$

73. The 11th term in the expansion of $(x+2)^{24}$ contains the factor x^{14} (You can verify this by counting down 11 terms from x^{24}.)

$$\binom{24}{14}x^{14}(2)^{10} = \frac{24!}{10!14!}(2^{10})x^{14}$$
$$= 1{,}961{,}256(1024)x^{14}$$
$$= 2{,}008{,}326{,}144x^{14}$$

75. The ninth term in the expansion of $(x-6y)^{12}$ contains the factor x^4.

$$\binom{12}{4}x^4(-6y)^8 = \frac{12!}{8!4!}x^4(-6)^8y^8$$
$$= 495(1{,}679{,}616)x^4y^8$$
$$= 831{,}409{,}920x^4y^8$$

77. Prove $\binom{n}{k} = \binom{n}{n-k}$:

$$\binom{n}{k} = \frac{n!}{(n-k)!k!} \quad \text{(by definition)}$$

$$\binom{n}{n-k} = \frac{n!}{(n-(n-k))!(n-k)!}$$
(by definition)

Simplify:

$$\binom{n}{n-k} = \frac{n!}{k!(n-k)!}$$
$$= \frac{n!}{(n-k)!k!}$$
$$= \binom{n}{k}$$

79. Prove $\binom{n}{0} + \binom{n}{1} + \cdots + \binom{n}{n} = 2^n$

The left side might suggest a binomial expansion without variables such as $(1+1)$. In fact,

$$(1+1)^n = \binom{n}{0}(1)^0(1)^n + \binom{n}{1}(1)^1(1)^{n-1} + \cdots + \binom{n}{n}(1)^n(1)^0$$
$$= \binom{n}{0} + \binom{n}{1} + \cdots + \binom{n}{n}$$

But, since $(1+1)^n = 2^n$,

$$\binom{n}{0} + \binom{n}{1} + \binom{n}{2} \cdots + \binom{n}{n} = 2^n$$

End of Section 11.5

Chapter Eleven: Section 11.6 Solutions to Odd-numbered Exercises

1. Given: $P(E) = \frac{2}{5}$. Then,

$P(E^c) = 1 - \frac{2}{5} = \frac{3}{5}$

3. Given: $P(E) = \frac{4}{13}$. Then,

$P(E^c) = 1 - \frac{4}{13} = \frac{9}{13}$

5. Given: $P(E) = \frac{2}{3}$. Then,

$P(E^c) = 1 - \frac{2}{3} = \frac{1}{3}$

7. Given: $S = 8, E = \{2, 5\}, F = \{3, 7, 9\}$

Note: $E \cap F = \phi, E \cup F = \{2, 3, 5, 7, 9\}$

(a) $P(E \cap F) = P(\phi) = 0$

(b) $P(E \cup F) = \frac{5}{8}$

9. Given: $S = 5, E = \{4, B\}, F = \{3\}$

Note: $E \cap F = \phi, E \cup F = \{3, 4, B\}$

(a) $P(E \cap F) = P(\phi) = 0$

(b) $P(E \cup F) = \frac{3}{5}$

11. Given: $S = 4, E = \{1, \beta\}, F = \{\alpha, 2\}$

Note: $E \cap F = \phi, E \cup F = \{1, 2, \alpha, \beta\}$

(a) $P(E \cap F) = P(\phi) = 0$

(b) $P(E \cup F) = P(\{1, 2, \alpha, \beta\}) = \frac{4}{4} = 1$

13. Given: $S = 16, E = \{1, 2, A, m, 13, Y, 8\}$, $F = \{1, 9, 11, m\}$

Note: $E \cap F = \{1, m\}$, $E \cup F = \{1, 2, A, m, Y, 8, 9, 11, 13\}$

(a) $P(E \cap F) = \frac{2}{16} = \frac{1}{8}$

(b) $P(E \cup F) = \frac{9}{16}$

15. Description of the sample space: The set of all 4-tuples consisting of H (Heads) and T (Tails). [Examples: (HTHT), (TTTH), (THTH)]. In all there are 16 such 4-tuples.

17. Description of the sample space: The set of all ordered pairs consisting of H (Heads) or T (Tails) in the first place and one of the 13 Heart cards in the second place. There are 26 such ordered pairs in the sample space.

19. In all there are $6 \cdot 6 \cdot 6 = 216$ ordered triples in this sample space. Each ordered triple consists of one of the six values in each place.

21. The sample space is the set of single values from 1 to 38.

23. **a.** There are 4 possibilities of 3 or higher on a die.

$P(n \geq 3) = \frac{4}{6} = \frac{2}{3}$.

b. There are two non-prime, even numbers in the set $\{1, 2, 3, 4, 5, 6\}$.

$P(4 \text{ or } 6) = \frac{2}{6} = \frac{1}{3}$

25. The sample space of equally likely outcomes has $2 \cdot 2 \cdot 2 = 8$ elements of ordered triples. Each ordered triple (i.e., outcome) consists of Hs and Ts.

a. There are three possible outcomes with exactly two heads: $\{(HHT), (HTH), (THH\}$

$P(\text{Exactly 2 Heads}) = \frac{3}{8}$

b. There is only one outcome that is $(HTH): P((HTH)) = \frac{1}{8}$

c. There are four outcomes that are two or more Heads: $\{(HHT), (THH), (HTH), (HHH)\}$

$P(\text{Two or More Heads}) = \frac{4}{8} = \frac{1}{2}$

27. $S = \{(w, x, y, z) \mid w, x, y, z \text{ are digits}\}$

$n(S) = 10 \cdot 10 \cdot 10 \cdot 10 = 10{,}000$

$E = \{(w, x, y, N) \mid w, x, y \text{ are digits; } N = 7, 8 \text{ or } 9\}$

$E = 10 \cdot 10 \cdot 10 \cdot 3 = 3000$

$P(E) = \frac{3000}{10{,}000} = \frac{3}{10}$

29. S = set of all 9-digit sequences.
E = set of all 9-digit sequences that have no 8 in the sequence.

$n(S) = 10^9;\ n(E) = 9^9$

$P(E) = \frac{9^9}{10^9} \approx 0.3874$

31. Let E = "rains Day 1 and not on Day 2"
Let F = "no rain Day 1 and rains Day 2"
E and F are mutually exclusive.

$P(E) = \frac{1}{4} \cdot \frac{3}{4} = \frac{3}{16}$ (assumes that the weather of Day 1 and Day 2 are independent).

Similarly, $P(F) = \frac{3}{4} \cdot \frac{1}{4} = \frac{3}{16}$.

Then, $P(E \cup F) = \frac{3}{16} + \frac{3}{16} = \frac{3}{8}$.

33. **a.** E = drawing a face card or a Diamond.

There are 13 Diamond cards--3 of which are face cards. There are 12 face cards in all--9 of which are not diamonds. Then,

$n(E) = 13 + 9 = 22.$

$P(E) = \frac{22}{52} = \frac{11}{26}$.

b. E = drawing a face card that is not a Diamond.

$n(E) = 12 - 3 = 9$

$P(E) = \frac{9}{52}$

c. There are six red face cards and two black Kings. Two Kings are also red.

E = drawing a red face card or King

$n(E) = 6 + 2 = 8$

$P(E) = \frac{8}{52} = \frac{2}{13}$

35. a. Experiment: Drawing two cards, one at a time, replacing the first card and drawing the second.
Let E = first draw is an Ace.
Let F = second draw is an Ace.
$$P(E \cup F) = \frac{4}{52} \cdot \frac{4}{52} = \frac{1}{169}$$

b. Experiment: Two cards are drawn at the same time.
E = both cards are Aces.
$n(E) = {}_2C_2$
$n(S) = {}_{52}C_2$
$$P(E) = \frac{{}_4C_2}{{}_{52}C_2} = \frac{6}{26 \cdot 51} = \frac{1}{221}$$

37. Let E = "it rains today"
Let F = "Bob forgets to put top up"
Let G = "Bob's car gets wet"
Then, $G = E \cap F$
$$P(G) = P(E) \cdot P(F) = \frac{3}{4} \cdot \frac{1}{4} = \frac{3}{16} = 0.1875$$
(This statement assumes that Bob's car gets wet by no other means and that E and F are independent.)

39. The sample space is defined by
$S = \{(x, y) \mid x, y \in \{1, 2, 3, 4, 5, 6\}\}$

a. Let $E = \{(x, y) \mid |x - y| = 0\}$
$n(E) = 6$
$n(S) = 6 \cdot 6 = 36$
$$P(E) = \frac{6}{36} = \frac{1}{6}$$

b. Let $E = \{(x, y) \mid |x - y| = 1\}$
$n(E) = 10$ (count them--e.g. (2, 3))
$n(S) = 6 \cdot 6 = 36$
$$P(E) = \frac{10}{36} = \frac{5}{18}$$

c. Let $E = \{(x, y) \mid |x - y| = 4\}$
$n(E) = 4$ (count them--e.g. (2, 6))
$n(S) = 6 \cdot 6 = 36$
$$P(E) = \frac{4}{36} = \frac{1}{9}$$

41. We focus on the number of 5-person groups (combinations), and the definition of probability. We must find the number of 5-person groups in which Jim could be a member. That is the numerator. The number of possible 5-person groups out of 100 persons is the denominator For Jim, put him on every possible group of 4 out of the other 99 employees: Number of 5-person groups with Jim as a member is ${}_{99}C_4$. The number of possible 5-person groups from the 100 employees is ${}_{100}C_5$.
$$P(\text{Jim is laid off}) = \frac{{}_{99}C_4}{{}_{100}C_5} = \frac{1}{20} = 0.05$$

43. $P(\text{Choosing peanut butter}) = \frac{2}{9}$

45. $P(\text{Picking a blue marble}) = \frac{3}{10}$

47. 20% of 135 = 0.2(135) = 27
i.e., 27 chances out of 135 is 20% probability.

End of Section 11.6

Chapter Eleven Test
Solutions to Odd-Numbered Exercises

1. Given $-7, -1, 5, 11, 17, \ldots$
The succession of terms is increasing by 6. Having $a_1 = -7$ and using $6n$ to generate successive terms requires that we have -13 added to $6n$ for each n: $a_n = 6n - 13$.

3. Given $0, 3, 8, 15, 24, 35, \ldots$
Notice that each number in the sequence is one less than a perfect square: $0 = 1 - 1$, $3 = 4 - 1$, $8 = 9 - 1$, $15 = 16 - 1$, or
$0 = 1^2 - 1$, $3 = 2^2 - 1$, $8 = 3^2 - 1$, $15 = 4^2 - 1$.
Then, in general for this sequence,
$a_n = n^2 - 1$

5. $$\sum_{i=2}^{7}(-2)^{i-1} = -2 + (-2)^2 + (-2)^3 + (-2)^4 + (-2)^5 + (-2)^6$$
$$= -2 + 4 - 8 + 16 - 32 + 64$$
$$= 42$$

7. $5, 2, -1, -4, -7, \ldots$
This arithmetic sequence decreases by 3 from term to term.
$d = -3$ and $a_1 = 5$
$a_n = 5 + (-3)(n-1)$
$a_n = -3n + 8$

9. $a_7 = -43$ and $d = -9$
Use $a_n = a_1 - 9(n-1)$ and solve for a_1:
$a_7 = a_1 - 9(7-1)$
$-43 = a_1 - 54$
$a_1 = 11$
$a_n = 11 - 9(n-1)$, or $a_n = 20 - 9n$

11. $$\sum_{i=1}^{60}(-4i+3) = S_{60} = \left[\frac{60}{2}\right](a_1 + a_{60})$$
$a_1 = -1$ and $a_{60} = -237$
$$S_{60} = \left[\frac{60}{2}\right](-1 + (-237)) = -7{,}140$$

13. $$5 + 10 + \ldots + 20{,}480 = \frac{a_1(1-r^n)}{1-r} = S_n$$
For $a_n = 20{,}480$, find n:
$a_1 = 5$ and $r = 2$
$20{,}480 = 5(2)^{n-1}$
$4096 = 2^{n-1}$
$4096 = 2^{12} \Rightarrow n - 1 = 12$ or $n = 13$
$$S_{13} = \frac{5(1-2^{13})}{1-2}$$
$S_{13} = 40{,}955$

15. $$\sum_{i=1}^{\infty}\left[-\frac{5}{4}\right]^i = \sum_{i=1}^{\infty}-\frac{5}{4}\left[-\frac{5}{4}\right]^{i-1}$$
$\left|-\frac{5}{4}\right| \geq 1 \Rightarrow$ series does not converge.

17. Basis Step. Show $P(1)$ is true:
For $n = 1$, $1^2 = 1$ and $\frac{1(1+1)(2(1)+1)}{6} = 1$
Induction Step: Assume $P(k)$ is true. Show that $P(k+1)$ is true.
Assume
$$1 + 4 + 9 + \cdots + k^2 = \frac{k(k+1)(2k+1)}{6}$$ Then,
$$1 + 4 + 9 + \cdots + k^2 + (k+1)^2 =$$
$$\frac{k(k+1)(2k+1)}{6} + (k+1)^2 =$$
$$\frac{(k+1)\left[2k^2 + k + 6(k+1)\right]}{6} =$$
$$\frac{(k+1)\left[2k^2 + 7k + 6\right]}{6} =$$
$$\frac{(k+1)(k+2)(2k+3)}{6} =$$
$$\frac{(k+1)((k+1)+1)(2(k+1)+1)}{6}$$
Therefore, $P(k+1)$ is true.

19. Basis Step. Show $P(1)$ is true:

For $n=1$, $(3(1)+2)=5$ and

$\frac{1(3(1)+7)}{2}=5$

Induction Step: Assume $P(k)$ is true.

Show that $P(k+1)$ is true.

Assume

$5+8+11+\cdots+(3k+2)=\frac{k(3k+7)}{2}$

Then,

$$5+8+11+\cdots+(3k+2)+(3(k+1)+2)=$$
$$\frac{k(3k+7)}{2}+(3k+5)=$$
$$\frac{3k^2+7k+6k+10}{2}=$$
$$\frac{(3k^2+13k+10)}{2}=$$
$$\frac{(k+1)(3k+10)}{2}=$$
$$\frac{(k+1)(3(k+1)+7)}{2}$$

Therefore, $P(k+1)$ is true.

21. Note that to say that 4 divides 11^n-7^n means that there is some integer p such that $11^n-7^n=4p$.

Basis Step. Show $P(1)$ is true:

For $n=1$, $11^1-7^1=4=4\cdot 1$

so 4 divides 11^1-7^1.

Induction Step: Assume $P(k)$ is true.

Show that $P(k+1)$ is true.

Assume $11^k-7^k=4p$, for some integer p.

This means the same as $\frac{11^k-7^k}{4}=p$.

Then,

$$11^{k+1}-7^{k+1}=11\cdot 11^k-7\cdot 7^k$$
$$=4\cdot 11^k+7\cdot 11^k-7\cdot 7^k$$
$$=4\cdot 11^k+7(11^k-7^k)$$
$$=4\cdot 11^k+7(4p)$$
$$=4(11^k+7p)$$

Therefore, 4 divides $11^{k+1}-7^{k+1}$ and $P(k+1)$ is true.

23. Use the multiplication principle for counting:

$8\cdot 10^3\cdot 26^3=140{,}608{,}000$

25. $\frac{8!}{3!2!}=\frac{8\cdot 7\cdot 6\cdot 5\cdot 4\cdot 3\cdot 2\cdot 1}{3\cdot 2\cdot 1\cdot 2\cdot 1}=3{,}360$

27. ${}_{21}P_5=2{,}441{,}880$

29. Expand $(x+2y+z)^3$, using the multinomial theorm:

$$\frac{3!}{3!0!0!}x^3+\frac{3!}{0!3!0!}(2y)^3+\frac{3!}{0!0!3!}z^3+$$
$$\frac{3!}{2!1!0!}x^2(2y)+\frac{3!}{2!0!1!}x^2z+\frac{3!}{1!2!0!}x(2y)^2$$
$$+\frac{3!}{0!2!1!}(2y)^2z+$$
$$\frac{3!}{1!0!2!}xz^2+\frac{3!}{0!1!2!}2yz^2+$$
$$\frac{3!}{1!1!1!}x(2y)z=$$
$$x^3+8y^3+z^3+6x^2y+3x^2z+12xy^2+$$
$$12y^2z+3xz^2+6yz^2+12xyz$$

31. Standard deck contains 12 face cards.

Let $E=$"draw 5-card hands--all face cards"

$n(E)={}_{12}C_5$

Let $S=$"draw 5-card hands"

$n(S)={}_{52}C_5$

$$P(E)=\frac{{}_{12}C_5}{{}_{52}C_5}=\frac{792}{2{,}598{,}960}=\frac{33}{108{,}290}$$

End of Chapter 11 Test